T0360570

## Demography and Evolutionary Ecology of Hadza Hunter-Gatherers

The Hadza, an ethnic group indigenous to northern Tanzania, are one of the few remaining hunter-gatherer populations. Archaeology shows 130,000 years of hunting and gathering in their land but Hadza are rapidly losing areas vital to their way of life. This books offers a unique opportunity to capture a disappearing lifestyle.

Blurton Jones interweaves data from ecology, demography, and evolutionary ecology to present a comprehensive analysis of the Hadza foragers. Discussion centers on expansion of the adaptationist perspective beyond topics customarily studied in human behavioral ecology, to interpret a wider range of anthropological concepts.

Analyzing behavioral aspects, with a specific focus on relationships and their wider impact on the population, this book reports the demographic consequences of different patterns of marriage and the availability of helpers such as husbands, children, and grandmothers. Essential for researchers and graduate students alike, this book will challenge pre-conceptions of human sociobiology. Supplementary material is available online at: www.cambridge.org/Hadza

**Nicholas Blurton Jones** is Professor Emeritus of Anthropology at the University of California Los Angeles. His research focuses on applying the methods and outlook of animal behavior research in direct studies of human behavior, specifically related to hunter-gatherer cultures. He has conducted numerous field visits to observe and study the last of the remaining hunter-gatherer tribes in Tanzania, and is the editor of *Ethological Studies of Child Behaviour*, published by Cambridge University Press in 1972.

## Cambridge Studies in Biological and Evolutionary Anthropology



Also available in the series

53. *Technique and Application in Dental Anthropology* Joel D. Irish and Greg C. Nelson (eds). 978 0 521 87061 0
54. *Western Diseases: An Evolutionary Perspective* Tessa M. Pollard 978 0 521 61737 6
55. *Spider Monkeys: The Biology, Behavior and Ecology of the Genus* Ateles Christina J. Campbell 978 0 521 86750 4
56. *Between Biology and Culture* Holger Schutkowski (ed). 978 0 521 85936 3
57. *Primate Parasite Ecology: The Dynamics and Study of Host–Parasite Relationships* Michael A. Huffman and Colin A. Chapman (eds.) 978 0 521 87246 1
58. *The Evolutionary Biology of Human Body Fatness: Thrift and Control* Jonathan C. K. Wells 978 0 521 88420 4
59. *Reproduction and Adaptation: Topics in Human Reproductive Ecology* C. G. Nicholas Mascie-Taylor and Lyliane Rosetta (eds.) 978 0 521 50963 3
60. *Monkeys on the Edge: Ecology and Management of Long-Tailed Macaques and their Interface with Humans* Michael D. Gumert, Agustín Fuentes, and Lisa Jones-Engel (eds.) 978 0 521 76433 9
61. *The Monkeys of Stormy Mountain: 60 Years of Primatological Research on the Japanese Macaques of Arashiyama* Jean-Baptiste Leca, Michael A. Huffman, and Paul L. Vasey (eds.) 978 0 521 76185 7
62. *African Genesis: Perspectives on Hominin Evolution* Sally C. Reynolds and Andrew Gallagher (eds.) 978 1 107 01995 9
63. *Consanguinity in Context* Alan H. Bittles 978 0 521 78186 2
64. *Evolving Human Nutrition: Implications for Public Health* Stanley Ulijaszek, Neil Mann, and Sarah Elton (eds.) 978 0 521 86916 4
65. *Evolutionary Biology and Conservation of Titis, Sakis and Uacaris* Liza M. Veiga, Adrian A. Barnett, Stephen F. Ferrari, and Marilyn A. Norconk (eds.) 978 0 521 88158 6
66. *Anthropological Perspectives on Tooth Morphology: Genetics, Evolution, Variation* G. Richard Scott and Joel D. Irish (eds.) 978 1 107 01145 8

67. *Bioarchaeological and Forensic Perspectives on Violence: How Violent Death is Interpreted from Skeletal Remains* Debra L. Martin and Cheryl P. Anderson (eds.) 978 1 107 04544 6
68. *The Foragers of Point Hope: The Biology and Archaeology of Humans on the Edge of the Alaskan Arctic* Charles E. Hilton, Benjamin M. Auerbach, and Libby W. Cowgill (eds.) 978 1 107 02250 8
69. *Bioarchaeology: Interpreting Behavior from the Human Skeleton,* 2nd edn. *Clark Spencer Larsen 978 0 521 83869 6 & 978 0 521 54748 2*
70. *Fossil Primates* Susan Cachel 978 1 107 00530 3
71. *Skeletal Biology of the Ancient Rapanui (Easter Islanders)* Vincent H. Stefan and George W. Gill (eds.) 978 1 107 02366 6

# Demography and Evolutionary Ecology of Hadza Hunter-Gatherers

NICHOLAS BLURTON JONES

Professor Emeritus

University of California Los Angeles

**CAMBRIDGE**
**UNIVERSITY PRESS**

University Printing House, Cambridge CB2 8BS, United Kingdom

One Liberty Plaza, 20th Floor, New York, NY 10006, USA

477 Williamstown Road, Port Melbourne, VIC 3207, Australia

314-321, 3rd Floor, Plot 3, Splendor Forum, Jasola District Centre, New Delhi - 110025, India

79 Anson Road, #06-04/06, Singapore 079906

Cambridge University Press is part of the University of Cambridge.

It furthers the University's mission by disseminating knowledge in the pursuit of education, learning and research at the highest international levels of excellence.

www.cambridge.org
Information on this title: www.cambridge.org/9781107069824



First published 2016

*A catalogue record for this publication is available from the British Library*

*Library of Congress Cataloging in Publication data*
Blurton Jones, N. G. (Nicholas G.)
Demography and evolutionary ecology of Hadza hunter-gatherers / Nicholas Blurton Jones (Professor Emeritus, University of California Los Angeles).
pages cm. – (Cambridge studies in biological and evolutionary anthropology ; 71)
Includes bibliographical references and index.
ISBN 978-1-107-06982-4 (Hardback)
1. Hatsa (African people)–Social life and customs. 2. Hatsa (African people)–Population. 3. Hunting and gathering societies–Tanzania–Eyasi, Lake, Region. 4. Demographic anthropology–Tanzania–Eyasi, Lake, Region. 5. Human ecology–Tanzania–Eyasi, Lake, Region. 6. Sociobiology–Tanzania–Eyasi, Lake, Region. 7. Human behavior–Environmental aspects–Tanzania–Eyasi, Lake, Region. 8. Eyasi, Lake, Region (Tanzania)–Social life and customs. 9. Eyasi, Lake, Region (Tanzania)–Environmental conditions. I. Title.
DT443.3.H37B55 2016
305.896–dc23 2015017917

ISBN 978-1-107-06982-4 Hardback
ISBN 978-1-107-65705-2 Paperback

Additional resources for this publication at www.cambridge.org/Hadza

# Contents

*Note*: The digital supplementary information (SI), references cited in each chapter, can be found at www.cambridge.org/Hadza

# Preface and acknowledgments

An effort of this size and duration, fieldwork spread over more than 15 years, data processing, analyzing, writing spread over another decade, depends on the help and support of many individuals and institutions. UCLA rescued me from the bleak economic prospects of the U.K. in the early 1980s, and although I came to realize that the efforts of many future colleagues were involved, Michael McGuire's leadership and guidance was instrumental.

As described in the introduction, Lars Smith, who began research among the Hadza while a graduate student at Harvard, was the key to my beginning to work among the Hadza. Numerous Hadza knew and liked Lars and were happy to receive some more friendly attention from someone he introduced to them. Tanzania in the 1980s was an almost impossible place to do anything. However, Lars had the skills and was willing to pass them on. He even had two prepared field assistants to offer. One of them, Gudo Mahiya, became vital to my demography project. His understanding of our project, and our wish for direct evidence grew rapidly. His quiet, unobtrusive manner was perfect for helping me interview women as well as men. Gudo also became an expert helper for Bonny Sands' language research, and encouraged by Jeanette Hanby and assisted by Jeanette and Bonny, he collected and recorded a number of traditional Hadza stories. These were assembled into a small booklet. By sending copies to Mangola (usually by hand with Frank Marlowe) for Gudo to sell to tourists, we were, for a while, able to provide Gudo with a useful pension after my fieldwork ended.

Permission for research in Tanzania was given by the Commission on Science and Technology. I am grateful to the commission and its officers for facilitating permissions and trouble-free passage through regional, district, and local levels of government. The fieldwork was supported by grants from National Science Foundation (BNS 8507859, BNS 8807436, DBS 9216088, SBR 9514214), and from UCLA Academic Senate Research Fund, and The Swan Fund. Another field visit was supported from a grant to Professor Peter Ladefoged to help him find some Hadza to collaborate in his phonetic studies. My 2000 visit was supported by funds raised by Annette Wagner to make her film of the Hadza.

Colleagues in UCLA Anthropology, Gradute School of Education, and Neuropsychiatric Institute were helpful, supportive, and informative in countless ways. I owe much to three outstanding collaborations at different stages in my career. Fae Hall, Mel Konner, and Kristen Hawkes have had profound and lasting influences on my thinking. I am grateful to Monique Borgerhoff Mulder for giving me the idea of a book, and to Nancy Howell and Kristen Hawkes for encouraging and helpful comments on a very early draft. Among my other good fortune has been a succession of great teachers and patrons, long gone but long remembered and to whom I am permanently grateful: Richard Schardt, Duncan Wood, David Vowles, Niko

Tinbergen, and Jim Tanner. Frank Marlowe, who, even if the dedication of his book implies a very inflated view of my ability as a teacher, was the only student who I instantly knew could handle fieldwork among the Hadza, although I failed at first to anticipate the great extent and success of his work in the years since 1995. Jim O'Connell's photographs from our 1985 to 1990 fieldwork have amazed and delighted me ever since his exhibition "Children of the Baobab". I wish we could include even more of them. At Cambridge University Press, I wish to thank Martin Griffiths, Katrina Halliday, Joanna Breeze, Victoria Parrin and Ilaria Tassistro for their encouragement and for guiding me through the intricacies of book production. I am grateful to Jeanette Mitchell of Traist Publishing Services for patient and thorough copy editing.

In Tanzania, David Bygott and Jeanette Hanby played several vital roles, especially after they settled in Mangola. Before each trip, they provided key information about conditions in Tanzania, ensured that our vehicle was up-and-running and available somewhere in Arusha. In Mangola, they allowed us to build our store shed near their house and readied a camp site and access to the clean water of the springs. All that in exchange for a few trinkets from the outside world, such as a bewildering array of teas that only the United States can offer. Residents of Karatu and Mangola provided friendship and frequent assistance. Thank you to Athumani, Paskali, Momwei Merus, Doctor Lyimo, Abdul Bariye, and Jovita Duko at Yaeda. They and several others helped us through a variety of adventures and misadventures. I still puzzle over how it was that everyone on the long chain of communication from the United States seemed to know that good news was on its way to me before I finally got the email with our oldest's crucial MLE results. Perhaps Jill had been trying to phone, normally a futile effort in those days. Officials at all levels of government were gracious and as helpful as the conditions under which they had to work allowed. Professor Chrys Kamuzora at University of Dar es Salaam Demography Unit helped immeasurably as our local contact, advising us through thick and thin. This project would not have happened without him.

Solo fieldwork is a peculiar exercise. Among the Hadza it was a mixture of sheer breathtaking magic, spells of total isolation from the outside world, and occasional exhaustion. In the Tanzania of last century, it began as a bewildering grind of toeing the line and serious tedium. We even had to go to Kenya for groceries. Gradually, everything began to work more smoothly in Tanzania, and by the end of my studies, I could reliably get through Dar es Salaam in exactly a week, and complete other preparations, buy supplies, and arrive in Mangola within another week. What kept me going in the hardest moments was the thought of my family back home. Half way through the project, Jill and I celebrated our 25th wedding anniversary. Except that, when we counted up my absences on fieldwork, we were two and a half years short. The lag is even greater now that we come within sight of our 50th. We seem to have been quite good at this "pair bonding" business and have had a wonderful life together. Younger readers need to constantly remind themselves that life goes by too fast. Looking back, it all seemed quite natural, even easy, although in Part Two of this book you will see how very poorly I understand how or why marriages work.

Jill managed to raise the children through and beyond their teenage years, and maintained her sanity despite all those seldom truly predictable absences (neither grants nor research permits were ever certain until the last minute). Furthermore, she has coped with the years of my retirement, during which I have worked at home well beyond normal commuter hours. Somehow we got away with raising three wonderful children, each successful in their chosen careers. With their wonderful partners, they have given us nine grandchildren. All arrived after my field days ended, and I have only once refused a visit on account of the book, even including visits to our itinerant daughter and her family in Finland and Czech Republic, two wonderful countries we would otherwise never have seen, and even to Moscow, home of the traffic jam and snow-bound playgrounds.

The grandchildren are the joy of our lives. If only the prospects of their Hadza age mates could be as good as theirs. Such contrasts are the great sting for the anthropologist. We exchange unsurpassable experiences, great colleagues, and absorbing puzzles about the bases of human nature for the first-hand knowledge that all is far from well with the world. Many anthropologists have established charities and attempted to work for the future of their study peoples. I have not. After a variety of efforts at the behest of various Hadza, I gave up. Even if one thinks one knows what the people want, for me the practical-political-personal obstacles proved overwhelming. The task is impossible for anyone not permanently resident in Tanzania. I can only wish good luck to those who try. In the case of the Hadza, their most enduring and effective support comes from the Dorobo Fund for Tanzania (dorobofund.squarespace.com/hunter-gatherers). In this book I attempt an in-depth analysis of its several topics, which leads to lengthy chapters. Some of the details are relegated to, some arguments expanded upon, and side issues explored in digital supplementary information, referenced in each chapter as, for example "SI 3.6" This material can be found at www.cambridge.org/Hadza.

The Hadza deserve a good outcome. Their ceaseless energy and good spirits, in a setting of hardship and increasing adversity, is endlessly impressive. They are a lesson to us all, at many, many levels. I can give no quick prescriptions for their future, no recipes for the decision makers. My work has not been aimed in that direction. Nonetheless, I do share the belief that more knowledge can lead to better decisions. The book reports academic westerners' questions about hunter-gatherers, and not just any academic but those of a particular school of thought, one that expects us humans to be highly opportunistic optimizers of our evolutionary fitness. It would be hard to think of a more "ivory tower" approach. However, here and there it does give insights that if we look humbly we can recognize in all of us humans.

# Part I

# Demography

# 1 Introduction

There are about a thousand people whose first language is Hadzane. At the time of my research between 1985 and 2000, most of them lived, hunted, and gathered plant foods in rocky hills in the eastern rift valley near Lake Eyasi in northern Tanzania. They call themselves Hadza, or Hadzabe (plural), or to a Kiswahili speaker, Wahadzabe, adding the Kiswahili plural animate noun prefix. They can be roughly divided into eastern and western sub-populations.

Hadza live in spectacular country. Many eastern Hadza camps are within sight of the outer wall of Ngorongoro crater, a UNESCO World Heritage Site, and many others are just across the rift valley from the equally well-known Serengeti National Park, Olduvai gorge, and Laetoli of fossil footprint fame (Photograph 1.1). Western Hadza lived adjacent to high-priced safari country around Maswa, south of Serengeti. James Woodburn, the first serious anthropologist to write about the Hadza in English (based initially on his intensive fieldwork in 1959–1961) (Woodburn, 1964) has reported on many aspects of Hadza life, and since 1988, our research group has added publications on behavioral ecology and life history. Now Frank Marlowe (2010) has collected his and other's recent research to give an excellent description of Hadza life. My aim here is more specialized. First, I want to set a detailed study of Hadza demography alongside the classic works on hunter-gatherer demography by Howell (1979) and Hill and Hurtado (1996) and other recent accounts such as those of Early and Headland (1998). Second, I want to use individual variation within the whole population to pursue some of our long-standing questions about how individuals, hunting and gathering in a sub-Saharan savanna environment, promote their reproductive success (RS). This should be useful to anyone interested in the evolution of our species.

From some hillside Hadza camps, one would be able to see, with a strongly wind-stabilized telescope, the tourist buses climbing the outer flank of Ngorongoro crater. Yet the majority of Hadza live by an economy as far removed from that of the tourists, and of most other Tanzanian citizens, as it is possible to get. Despite brief experiences with other lifestyles, most Hadza acquired the bulk of their food by hunting wild animals and gathering wild plants. When my colleagues Hawkes and O'Connell weighed and recorded all the food coming into Hadza camps in their 1985–1986 year in the field, they found that only 5% of the calories came from agricultural sources, traded from farmers. Ten years later in 1995–1996, when my student Frank Marlowe spent a year collecting similar data in a much wider array of Hadza camps, some very near non-Hadza villages and including camps routinely

**Photograph 1.1** View from Kidelu mountains across the lakeshore plain and Lake Eyasi to the western rift wall, beyond which lies Maswa game controlled area, Serengeti, and Laetoli. © James F. O'Connell, 2015. Reproduced with permission.

visited by tourists, he obtained a very similar result (6.6%; Marlowe, 2010, p. 126). The Hadza say they have always hunted and gathered, and often say that they wish to do so for ever. When the context arises, they explain to other Tanzanians that the bush is clean, peaceful, and safe, and that unlike farmers, they like to eat meat and the bush provides enough, even though their hunting is conducted only by traditional bow and arrow.

In 1911, Erich Obst, a German geographer, visited the Hadza and they told him that they had always lived there and always hunted and gathered. All were living this way during his time with them, but he wondered whether it was true that they always had. He discussed evidence such as whether Hadzane words for domestic animals were borrowed or not (they are borrowed but unlike his contemporary, the army doctor Dempwolff, Obst did not know enough of the languages of neighboring tribes to learn this). Obst also left one field site because there was "too much influence from Isanzu" and many "indistinct types." We must address similar issues of Hadza history and identity. Although we have yet to hear a Hadza express any doubt about his identity as a Hadza, we must ask ourselves who is a Hadza and who is not? Are they a sufficiently distinct and limited population to be the subject of a demographic study, or should we instead study the demography of an area such as an administrative or geographical district? Shall we identify people by their language, the identity that they and others attribute to them, their economy, or their location? Especially important for our demographic interest: does the population recruit and lose people only by birth and death, or also by migration and changes in ethnic identity? Are the Hadza a closed biological population that we could describe with stable population models, in which birth and death predominate massively over migration and change in ethnic identity, or could they be, as Cooper (1949) heard suggested, a floating population of dropouts and tax evaders from the surrounding populations? We also need to set the Hadza in their contemporary and historical contexts and describe more about the land they inhabit. Will our results reflect the demography of a forager lifestyle, or of some mixture of lifestyles, mixed both over time and across different locations?

Alongside their persistence as hunter-gatherers in the populous twentieth century, probably the most remarkable thing about the Hadza is their language. Some have related Hadzane to the southern African Khoisan languages, because it includes some of the click consonants found in those languages. However, the distinguished South African linguist Dorothea Bleek, after her several months with the Hadza in 1930, apparently could add little more than the existence of gender in the language to the clicks as links to the southern African languages. Recently, in a much more intensive study, Sands (Sands *et al.*, 1993; Sands, 1995) has shown that links to Khoisan are minimal. Even if links are eventually established, they will be remote. Three major language families are represented by the surrounding farming and herding neighbors of the Hadza: Iraqw (Cushitic, Afro-Asiatic), Isanzu and Sukuma (Bantu, Niger-Kordofanian), Datoga and Maasai (Nilotic, Nilo-Saharan). Hadzane belongs to none of these families and must be regarded as an isolate. It must have remained distinct from all the modern languages of Africa for many hundreds of years, an extraordinary feat of cultural survival. Even more remarkable are the results of genetic surveys by Sarah Tishkoff, Joanna Mountain, and colleagues (Scheinfeldt *et al.*, 2010). The Hadza are genetically extremely distant from southern African Khoisan speakers, and from others, even the nearby Sandawe. All three, the Hadza, Sandawe, and Khoisan, along with Pygmies, emerge as the most deep-rooted populations, and distinct from the Bantu, Nilotic, and Cushitic-speaking majority.

Mountain and colleagues (Knight *et al.*, 2003) argued that the most parsimonious claim is that the click consonants existed in humanity's first languages and were lost by all but Khoisan, Hadzane, and Sandawe. Alternative explanations were quickly offered, mainly emphasizing the ease of borrowings between languages, but the very great antiquity of the Hadza coalescence with the other persisting hunter-gatherer populations is undoubtedly another of their claims to distinction (Wells, 2006, p. 133).

While Hadza talk as if they had their land to themselves until quite recently, this may not be entirely true. Evidence such as the existence of stone irrigation channels at Endamagha at the north end of present-day Hadza country suggests the presence of farmers in the area around 1700 (Sutton, 1986), and farmers are there again today where irrigation is possible at the foot of the surrounding mountains. Herders have left archaeological traces in the area on and off during the last thousand or so years (Mabulla, 2007). Mabulla's review of the archaeology of the Eyasi basin indicates traces of hunters and gatherers in the Eyasi basin since 130,000 years ago. Much of the time, both prey and plant foods included species eaten by Hadza today.

The Hadza have been subjected to the curiosity of many researchers since Obst, yet they have not taken their place in the popular imagination nor in the anthropological literature in the same way as the !Kung and other Khoisan hunter-gatherers, or the Efe, Baka, or other "Pygmy" groups, or any of the Australian aboriginal cultures. The Hadza have remained little known outside the small band of anthropologists who specialize in hunters and gatherers, and for many years most Hadza have liked it that way. Today, the more worldly among them recognize that being better known by anthropologists and tourists may give them some protection from losing their land and their identity. This view coincides conveniently with the anthropologist's wish to learn as much as possible about their economy before it is lost.

To some extent, the Hadza remained little known because most of the early visitors wrote only in German, including Ludwig Kohl-Larsen who spent substantial amounts of time with the Hadza in the ill-fated 1930s and wrote several books about them. Recently his films were rediscovered, and stripped of the propaganda of their terrible era, formed the basis of a pair of TV films made for SWR by Annette Wagner. Showing Kohl-Larsen's films and photographs to contemporary Hadza, including children and grandchildren of individuals featured in the films, was both riotously enjoyable, and enormously helpful for our age estimations. Similarly important for age estimation were genealogies in Dorothea Bleek's notebooks and her photographs from 1930, kindly provided by the Cape Town University Archive. An excellent film was made by James Woodburn in 1970, and it is Woodburn who has brought the Hadza to the attention of English-speaking anthropologists, especially with his two very valuable papers in the "Man the Hunter" symposium (Lee and deVore, 1968), and a number of later landmark papers. Woodburn has continued to visit and write on the Hadza constantly since his original 1959–1961 fieldwork. Woodburn also shepherded a group of biological anthropologists led by Nigel Barnicot in a set of studies as part of the International Biological Program (IBP) in 1966–1967. Their papers, which include Dyson's demographic analysis, Hiernaux and Hartono on anthropometry, Bennett and others on diseases, injuries, age pattern of blood pressure and cholesterol levels, form an important basic knowledge of Hadza biology. Although I and my colleagues come from theoretical perspectives very different from Woodburn's, we find ourselves almost always in agreement with Woodburn's descriptions and insights about the Hadza.

In the early 1980s, I began to look around for a hunter-gatherer population where we might be able to study human biology and behavior from an adaptationist perspective. I was particularly interested in seeing whether it was possible to replicate the work I had done with Richard Sibly on birth spacing using Richard Lee's ecological data (Blurton Jones and Sibly, 1978), and Nancy Howell's demographic data on the !Kung (Blurton Jones, 1986, 1987). I tracked down Lars Smith, who had gone as a student of Irv DeVore to do fieldwork among the Hadza in the 1970s. After an excellent start and the acquisition of much knowledge about how to work among the Hadza, Lars had settled in East Africa. Lars agreed to help me come on a pilot visit in 1982 and assess the situation. The political and logistic situation could not have been worse, but the field situation was perfect. At that time, the Hadza were very much left to get on with their lives unimpeded. I began my East Africa fieldwork apprenticeship under Lars' expert guidance. Lars again guided me, Hawkes, and O'Connell in our 1984 pilot visit, and then in 1985 Lars and I completed the first of the series of censuses on which this book is based. All of the subsequent, Utah–UCLA and Marlowe group fieldwork owes its existence to Lars' skill at negotiating the Tanzania of the early 1980s.

At about this time, another initiative in the adaptationist perspective had been that of Kristen Hawkes on optimal foraging among the Ache with her then students Kim Hill, Hillard Kaplan, and Magdalena Hurtado, and by Hawkes and her colleague James O'Connell on hunting and use of space by Australian aboriginal people. It was natural that we should collaborate and after a pilot visit to the Hadza, Hawkes, O'Connell, and I quickly formed a team to collaborate on such investigations among the Hadza. Hawkes and O'Connell have published a number of papers on women's work, grandmothers as helpers, meat distribution, and the economic puzzle of big game hunting, spatial distribution of objects, bone transport, economics of scavenging, and so on. I collaborated in gathering some of the data for these projects. For example, large game are caught rather seldom, so anyone who was in the field collected the data on bone transport and meat distribution and we pooled these data. I have written on children's foraging, and development of foraging skills (with Frank Marlowe), marriage and divorce, residence patterns of grandmothers and grandfathers, as well as two papers on the early stages of our demographic research. The demographic research has continued slowly for the entire duration of our project. There are several reasons for the lack of speed. Most important is that demographic research among very mobile hunters and gatherers simply takes a long time. Nancy Howell recently commented that finally there is someone who knows why she took so long to publish her !Kung demography (Howell, 1979). I am consoled to think that, reciprocally, there is someone who may understand why it has taken me so long to publish this account of Hadza demography.

The first aim of the demographic study was to complete as thorough and accurate an account as possible of the demography of another hunter-gatherer population to set alongside the excellent studies of the !Kung by Howell (1979) and of the Ache by Hill and Hurtado (1996). This is a straightforward descriptive goal. Straightforward but not easy, for the Hadza are far from the ideal demographic subjects. They move every few weeks, change their names more than once in their lifetime, can be referred to differently by father and by mother. Although they were quick to appreciate that we wanted to learn everything we could about their ability to live in the bush, they were very unfamiliar with the idea of being interviewed out of harmless curiosity. Hence, my first demographic goal was merely to establish how many babies they had, at what ages they had them, at what ages people died – in other words, to estimate age-specific fertility and mortality. Along with this came other descriptive measures, the crude death rate and birth rate, age structure, and so on. In this kind of small, mobile, and non-literate population, these simple descriptive tasks are not so simple. One of the most difficult tasks is to find out how many Hadza there are, and it is especially important to try to make some estimates of the accuracy or error in our estimates of Hadza demographic parameters. I have addressed the issue of uncertainty and error as intensively as I can. Partly, I do this by generous use of resampling methods. I have also tried to fully exploit the interdependence of demographic measures collected by different, independent observations.

The demographic results bring back to mind some long-standing questions about hunter-gatherers. What, if anything, regulates their population? Because, as some of the contributors to the "Man the Hunter" symposium (Lee and DeVore, 1968) recognized, evolution has long pushed individuals to maximize their RS, it is unlikely that any "self-sacrificing" mechanism exists that closely regulates the population. However, it could be that the members of a population are all affected by density dependent factors, or instead by less predictable random fluctuations in the environment. In common with other modern hunter-gatherers, the Hadza have been increasing quite fast. The data suggest their population has been "stable" (fertility and mortality remaining constant) but far from "stationary" (total numbers neither increasing nor decreasing). At these rates of increase, hunter-gatherers could have filled the world many times over in as little as a few thousand years. Yet they clearly did not. Like Hassan (1975), Hill and Hurtado (1996) discussed this contradiction, pointing out that it would take mortality levels higher than any ever observed to keep forager populations stationary. We can think about and model some of the possibilities, past and present.

The second general aim is to use individual variation in a population-sized sample to illuminate some of our adaptationist questions. Thus, in Part II of this book I turn my attention to those selection pressures that probably always stood in the way of population "restraint." Besides my original interest in whether there is an optimal inter-birth interval and whether, as expected from the richer ecology, it is shorter than among the !Kung, we should be able to use our larger sample to test ideas about paternal investment and monogamy (amplifying and checking tests made for Blurton Jones *et al.*, 2000), the role of hunting in men's reproductive strategies, grandmothers as helpers, and so on. As Hill and Hurtado (1996) pointed out, helpers are an especially important topic for adaptationists who would use observed populations to test their ideas. Helpers could "dampen out" many important trade-offs.

Traditionally, anthropologists have taken the view that hunter-gatherers give us a special view on human evolution. Contemporary hunter-gatherers are widely accepted as one of four windows into the past: fossils and archaeology, contemporary primates, contemporary and recent hunter-gatherers, and quantitative molecular genetics, the newest and very important fourth window. As forager occupants of relatively rich sub-Saharan wooded savanna, the Hadza would seem to have as good a claim as any to the attention of anthropologists who are interested in the role of that ecology in the evolution of our species. However, we need to get more analytical about this belief. Often we work at our projects without much thought to exactly why we think hunter-gatherers are so informative, concentrating instead on the precision and detail of our studies; but there are pitfalls, criticisms, doubts, and questions about the contribution of hunter-gatherer studies, and we cannot avoid them indefinitely. It is my impression that researchers' intuitive feel of the way to work is usually good, and usually well ahead of their explicit rationale. Nonetheless, only careful analysis of what we are trying to say, and whether it is justified, can test this impression. Analytic guides include Foley (1996), Smith *et al.* (2001), and Bird and O'Connell (2006).

Reflecting my two basic aims, this book is organized in two parts. The first reports our methods and findings on basic demography. The second reports our more unusual and interesting questions about behavioral ecology. However, first I need to give more background information about the Hadza and their home. The photographs provide an introduction to both. Jim O'Connell's photographs excellently illustrate Hadza life and economy as we observed it between 1985 and 1990. I have selected pictures that introduce features of Hadza economy that we regard as the essential bases of their lives. A quick skim through the pictures and their legends may serve as an alternative orientation to the questions addressed in the book and the contexts in which they arise.

# 2 Geography and ecology in the Eyasi basin

Natural grasslands do occur in East Africa, but only in areas of unusual edaphic conditions.
A. Joy Belsky, 1990

Hadza live in spectacular and varied country. Rocky hills with wooded savanna are their most common residence; forest-capped mountains, the expanse of the lake, and the rift escarpments are their daily view. For the Hadza researcher, it is a relief to see the popular contrast of forest and plains corrected by Belsky among others. But let's begin at the beginning. The rift dominates the scenery, influences the climate and the soil, generates the water supply, and ultimately determines the population densities and economic activities of the region around Lake Eyasi. In this chapter, I aim to describe the climatic, floral, and faunal environment in which we observed the Hadza, suggest some implications, and align the current conditions with what we can determine about the historical and prehistoric conditions. The recent tome by Spinage (2012) has been a valuable backup to my own literature searches.

Most of the eastern Hadza live in a roughly rectangular area south and east of Lake Eyasi in the bottom of the rift valley, southwest of Ngorongoro crater. The area measures just 90 km × 40 km (3600 km$^2$) (55 miles × 25 miles), a little larger than the Los Angeles basin, or Long Island New York, and about the size of a middling English county. Figure 2.1 shows the general features of the area around Lake Eyasi. The map marks non-Hadza villages mentioned in the text, and shows the general areas inhabited by Hadzane speakers. Figure 2.2 shows the location of Hadza camps that we visited between 1985 and 2000. One of the most striking features is the proximity to Ngorongoro crater, Laetoli, Olduvai, and Serengeti. The main tourist road passes a mere 25 km from the northernmost Hadza camps, but the Hadza are separated from this road by 500–1000 m (1600–3300 ft) in altitude, and more important differences in habitat. The area where most eastern Hadza live had remained very sparsely populated and little visited until the final 20 years of the twentieth century. Three reasons stand out: (1) The main travel routes have followed the well-watered highlands and avoided the much drier lowlands next to Lake Eyasi. (2) The basin has very low rainfall; agriculture is only sustainable where irrigation is possible, using water from the highlands. (3) The tsetse fly was abundant, and a strong disincentive to herders.

Describing the ecology and climate is a natural preliminary to writing about any population, but there are some specific purposes to such a description. What sort of history of droughts, shortages, even famines, seems likely? Have game animals

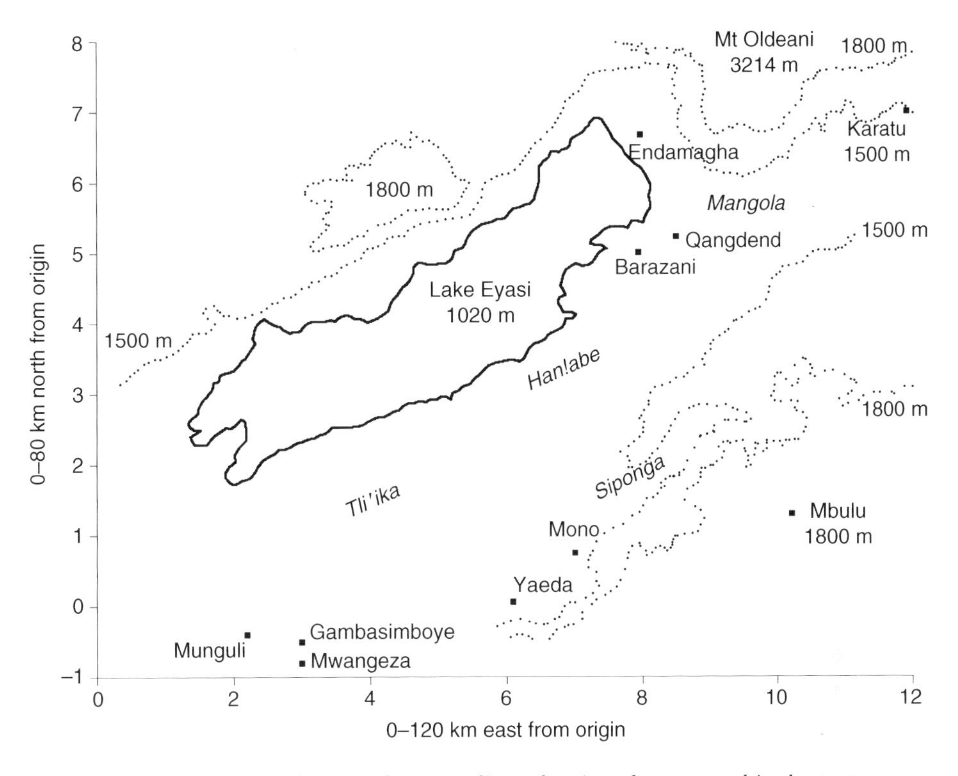


**Figure 2.1** Map of the Eyasi basin and surroundings showing places named in the text. Also labeled in their approximate location are the Hadza regions Tli'ika, Siponga, Mangola, Han!abe. The axes are based on Universal Transverse Mercator (UTM) coordinates, measured in kilometers from a point of origin close to Hadza country. 1500 m- and 1800 m-contour lines are marked as broken lines. Karatu and Mbulu are district capitals.

become more scarce as "bushmeat" trade has grown all over Africa? What other degradations of the environment may have occurred? What diseases are present and potentially affecting Hadza fertility and mortality? Hadza geography is also relevant to wider issues in human evolution, the ecology, and selection pressures on *Homo sapiens*. How does the Hadza environment compare to the environments of other African hunter-gatherers? What is there to eat in wooded savanna, and how do you obtain it? Have hunter-gatherer environments changed in ways that make a difference?

## 2.1 Altitude differences and landscape variety

The Eyasi basin, Lake Eyasi, and the country of the eastern Hadza lies at the bottom of a branch of the East African rift valley. The land east of the Eyasi fault sunk about a thousand meters (Foster *et al.*, 1997, fig. 3a; Le Gall *et al.*, 2008). The lake is 1020 m above sea level and most of eastern Hadza country is between 1020 m and 1500 m (3347 ft and 4921 ft) above sea level. On three sides, Hadza country is hedged in by highlands; to the north and northeast Oldeani mountain

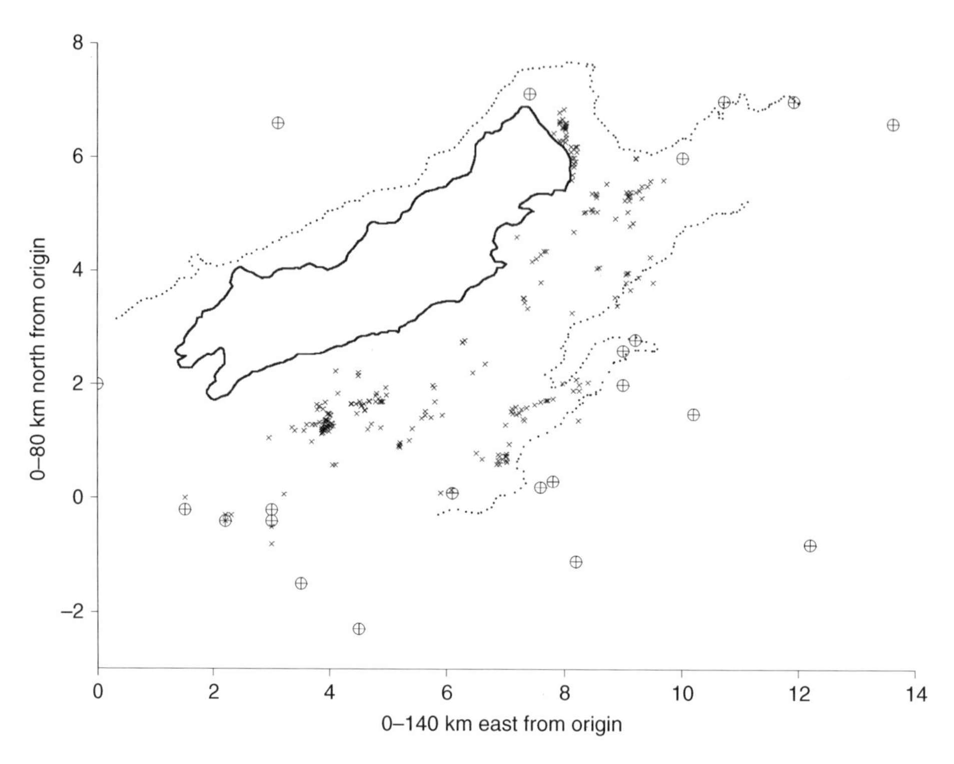


**Figure 2.2** Locations of Hadza camps visited during the study period (indicated by x). Also shown are the approximate locations of the few individual Hadza reported to be living outside the study area (shown as crossed circles).

(3214 m/10,545 ft) and Ngorongoro crater rim (2300 m/7546 ft), to the northwest the rift wall (1800 m/5900 ft) on the far side of the lake, to the southeast the eastern escarpment leads up to the Mbulu highlands at 1800–2200 m (5900–7220 ft) elevation. Seismic activity continues quite vigorously. An earthquake on May 6, 1964 was rated at 6.4 on the Richter Scale (UCLA Geography Department library). Many Hadza remember it vividly and are able to recount where they were camped, with whom, and even sometimes which infants they were then carrying. We were shown the immense boulders said to have fallen from the Kidelu mountains into the edge of the Yaeda plain during this quake. Marlowe, Bonny Sands, and Will Grundy felt a substantial tremor in the same mountains in August 1997. There are also hot springs beside Lake Eyasi.

Both escarpments gradually become less impressive as one moves southwest, less high, and less steep. At the southwest end of the Yaeda plain, the country changes gradually; the plain gives way to woodland, which gives way to farmland. One leaves Hadza country and approaches areas that have been the center of Isanzu occupation for as long as there is written history of the area. On the way, one passes areas around Munguli, Mwangeza, Domenik, which used to be more woodland, more bush, and more often used by the Hadza than today. In 1930, Dorothea Bleek stayed with the Hadza camped some "12 miles north of Mkalama" (Bleek 1931a, in her notebooks

**Photograph 2.1** Baobab trees seem to like rocky ground, and there is plenty of rocky ground in Hadza country. © James F. O'Connell, 2015. Reproduced with permission.

she says four hours walk northeast of Mkalama). The people with whom she stayed had kinship links with both the eastern and western Hadza, and Isanzu. Today, the Hadza moving between east and west walk through the same area but now it is occupied by Isanzu and Iramba, no longer by Hadza.

Eyasi is a landlocked lake, rivers flow into it but no river flows out, a feature that repeatedly fascinated the early German explorers. Because there is no outflow, Eyasi is a salt lake. The level of water in the lake varies markedly from year to year. Gudo Mahiya told me that in 1992 he and Joachim Wagner (a language researcher from the University of Hamburg) drove all the way across the lake; Gudo was impressed to see many animal tracks also making the crossing. Eastern Hadza country from time to time may be replenished with immigrations of fresh animals from Maswa and Serengeti. Obst (1915) also mapped his walk to a point about three quarters the way across the lake, reflecting the aridity of the first decade of the twentieth century.

The Eyasi basin includes several "landscapes": lakeshore plain, Kidelu mountains, Yaeda plain, Siponga and Lesaso mountains, Udahaya river, Balai river valley, Mangola villages and onion fields, and the northern lakeshore plain (web-based supplementary information, SI 2.1, www.cambridge.org/Hadza). Each of these "landscapes" includes or is within reach of a similar variety of basic plant resources: berry flats, extensive stands of tuber-bearing bushes on the hillsides, and baobab groves, which are most abundant in the mountains (Photograph 2.1). Dry, long-lasting *Grewia* fruit are found in the mountains and the wooded part of the Yaeda valley, and parts of the lakeshore plain. During my study period, large animals were most abundant in the Kidelu mountains and the savanna slopes of Oldeani mountain. None of the landscapes matches the stereotyped dichotomy between "open plains" and "dense rainforest" that plagues the consumer of much paleoanthropology literature. There is some open grassland in the Yaeda plain but Hadza make very little use of it and never camp there. Open grass plains are rather unusual in Africa (Belsky, 1990).

## 2.2 Water sources

Europeans, from the home of rain, may be prone to underestimate the importance of water for human habitation in Africa. Water is an issue for foragers, farmers, and herders alike; but foragers merely need enough to drink, and to cook with. Their plant foods rely on rain or ground water. Not all of their prey species need to drink, but most do. Herders need enough for their livestock as well as themselves and they maintain deep and effective wells. As long as their herds can reach water once every one to two days, it does not matter where the water lies. Cattle can travel to the wells, where herders can lift the water to drinking troughs. Farmers need either sufficient rain to fall on their fields, or the opportunity to irrigate. However, even for foragers, long daily walks to a water source can be an exhausting and time-consuming chore that usually falls to women and girls. In addition, because it makes sense to camp within easy reach of water, the distribution of water sources may shape the distance between camp and the best foraging patches.

Hadza country is blessed with numerous small water sources, which allows frequent camp moves, and avoids significant depletion of resources near camp. The Hadza usually camp quite near water (the mean of 10 cases with Universal Transverse Mercator Coordinate System (UTM) locations for water and camp was 0.5 km) but not too close, for they are eager to avoid mosquitos. They also do not want to disturb the game that uses the water, nor expose children to the predators that follow the game. Digging in a sand river sometimes produces water. Crevices and drainage areas from rocky outcrops and hills provide other small water holes, as do the hollow trunks of some baobab trees. In the rainy season, pools stay on the rocky outcrops for some weeks and are readily used. The two large rivers, Balai and Udahaya (SI 2.2), are of course dry most of the time but the Hadza and Datoga between them maintain quite deep water holes in the sandy river beds. The Hadza do not go to Lake Eyasi for water, or for salt, even though there are specialist salt extractors at the southwestern end of the Lake who have been there for at least a century (Senior, 1938), and salt is a gift frequently requested by the Hadza.

During the 1990s, some of the best water holes in Tliika were modified by Datoga herders. The Hadza complained that the modifications made it difficult for wild animals to reach the water and thus threatened the abundance of animals and their accessibility to Hadza hunters. In other areas, Datoga digging may make water easier for Hadza to access, although being fenced with thorns, the water holes may again be inaccessible to most wild animals. Cultural variation in the traditional technology of water access might repay closer study (Fosbrooke, 1950, gives an unusual example). Farmers and herders are well organized at digging and maintaining wells, Hadza and !Kung apparently less so. Small water sources raise issues of access and technology, perhaps more limiting for children and for our primate relatives. The distribution of foragers may once have been more restricted by their lack of technology or organization for accessing deeper water supplies.

The evergreen forests on Oldeani and the Ngorongoro crater rim are likely to be essential for the continued flow of the line of springs between Mangola and Barazani (Yanda and Madulu, 2005). These springs have been used by the Hadza since at least the time of Kohl-Larsen's fieldwork during the 1930s, but have also attracted outsiders, Tanzanian and European, since about that time. Large forests (Nou and Hanang) survive in the highlands east and south of Mbulu. The highland forests, although visible on the mountains from several places in Hadza country, seem not to have been used by the Hadza. Hadza may be afraid of surprise encounters with elephants (which they greatly fear) or buffalo, also dangerous. Even some Dorobo populations, generally known as montane forest dwellers, occupy land that includes lower forest margins (Kratz, 1994, fig. 6, p. 72; p. 71), do much of their hunting and gathering in savanna (Ichikawa, 1980; Cronk, 2004), using the forest as a place to build a house, as a superior honey source, while they acquire major plant foods in savanna. Montane forests may be of relatively little use to hunters and gatherers unless they are also specialist honey traders with a good supply of customers. Others have debated the foraging opportunities in dense lowland tropical forests (Headland, 1987; Bailey *et al.*, 1989).

## 2.3 Rainfall: seasonal, annual, and historical variation

The altitude differences of the Eyasi area shape the climate. In the basin it is hotter, it rains less, and strong diurnal mountain winds blow. Blown salt "whiteout" is an occasional feature of the lake margins. Temperature varies little around the year, averaging 35°C (95°F) in daytime and 14°C at night (Marlowe, 2010). Because rainfall shapes the flora and wild plant resources, and thus the fauna, its history and distribution is crucial to understanding the Hadza ecology and population.

### 2.3.1 Season

The annual mean 500 mm of rain falls in eastern Hadza country in a single season between November and April (Meindertsma and Kessler, 1997, fig. 1.7), much as Prins and Loth (1988) report for their Manyara area and the highlands between Manyara and the Eyasi basin. The rainy season is followed by a long dry season (roughly May to October). Lars Smith's air counts of camps in 1977–1978 showed a tendency for Hadza to be more scattered in the wet season and concentrated in the dry season. This may reflect greater availability of water sources for people in the wet season, and the concentration of game near fewer water sources in the dry season. Elsewhere. this pattern is more in evidence, as among the !Kung. According to Akie informants of Smith and myself in 1982, they had a similar pattern of wet season expansion, hunting across the Maasai steppe, and a dry season concentration near Kijungu. Savanna foragers may generally have followed such a pattern, leading to a greater chance of encounters with neighbors during the wet season, and comparative isolation during the dry season.

### 2.3.2 Altitude and rainfall

Rainfall in East Africa is strongly influenced by altitude (Brown and Cocheme, 1969; Fleer, 1981; Prins and Loth, 1988). The lowlands of the Eyasi basin, at 1000–1500 m, receive 400–600 mm in the average year (Figure 1.7 in Meindertsma and Kessler, 1997: mean 556 mm/yr at Yaeda, and 310 mm/yr at Mangola). The surrounding highlands at 1600–2300 m (5250–7550 ft) receive much higher rainfall, 700–1200 mm/yr, and support mixed farming, and dense montane forest where protected. Karatu receives a mean of 916 mm/yr, and Mbulu gets 827 mm/yr, while rainfall is even higher on the Ngorongoro crater rim (908–1200 mm). In the Eyasi basin, farming is only possible where irrigation is feasible, as at Mangola, and along the lakeshore between Barazani and Endamagha. Irrigation has declined at Yaeda; the people there say less water descends from the highlands these days.

### 2.3.3 Contemporary annual variation in rainfall

In Serengeti (Norton-Griffiths *et al.*, 1975; Pennycuik and Norton-Griffiths, 1976) and the Manyara area (Prins and Loth, 1988), both adjoining Hadza country, year-to-year variation in rainfall is extreme, and greater at lower mean rainfall, with a coefficient of variation about 30% at a 500 mm/yr mean. Thus, more than once in a lifetime, the Hadza experience a year with as little rainfall as 217 mm or as much as 723 mm. Hence, they commonly experience years with rainfall as low as that at Dobe (mean 405 mm, 1963–1969; range 239–597 mm, Lee, 1979, fig. 4.6) or the central Kalahari (392 mm for 1961–1971, Tanaka, 1980), and sometimes years with rainfall as high as that at Serengeti (mean, 803 mm, Norton-Griffiths *et al.*, 1975, table 1).

### 2.3.4 Variation in rainfall during the nineteenth and twentieth centuries

We have little direct evidence about variation in rainfall in the Eyasi basin. Being shallow, the area of Lake Eyasi should increase rapidly with high rainfall and increased runoff from the surrounding highlands. Obst's map appears to delineate the area of the lake on a basis other than water. He marks his walk across the lake in 1911 and Kohlschuttler's in 1900, but his outline of the lake must be based on the area of bare salt or the line of beaches. Baumann's map appears to include no survey data on the extent of the lake, just altitude measured at its northernmost end. The Mbulu annual rainfall series between 1935 and 1993 shown in Meinderstma and Kessler (1997) may be our best guide to recent rainfall history, although the Eyasi basin could be regarded as being in the rain shadow of the Mbulu highlands.

We can probably roughly extrapolate from the rapidly expanding literature on East African climate history (much of it summarized by Spinage, 2012), although spatial variation, even on a quite small scale, seems important. A concensus might be: dry in the early nineteenth century; wet in the 1870s, and severely drier from 1880 to 1960 (some report the dry period ending around 1920). The 1961 rise in Lake

Victoria coincided with higher than average rainfall at Mbulu and Loliondo, and a high level of Lake Emakat (Ryner *et al.*, 2008; Westerberg *et al.*, 2010). Spinage (2012) argued for large-scale (but quite complex) associations between observed rainfall and, for example, changes in the level of Lake Victoria, and rates of flow in the Congo or Nile Rivers (for which records date back to 5050 BP). Spinage includes extensive data and discussion of the severe decrease in glaciation all over East Africa as an indication of decreased precipitation. The dry period in the late nineteenth and early twentieth centuries may have been overlooked in discussions of the East African population in the early colonial years.

### 2.3.5 Long-term variation in rainfall: the last 2000 years

The lake core studies give quantitative data to great time depth, sometimes facilitate an inspiring multidisciplinary study such as that of Westerberg *et al.* (2010), but are not always easily translated into actual rainfall. Westerberg *et al.* (2010, p. 306) say of rainfall at Engaruka "Historically, the regional climate of eastern Africa has varied at amplitudes similar to those of the present day." At Engaruka, mean rainfall is 460 mm/yr with variability of 200–1000 mm/yr. The highland forest on Ngorongoro crater rim and the crater highlands has rainfall of about 1000 mm/yr, double the rainfall in the Eyasi basin (Anderson and Herlocker, 1973; Westerberg *et al.*, 2010), and at Kainam on the Mbulu plateau it reaches 1200 mm/yr. Given that rainfall in the highlands is about double that in the Eyasi basin, changes in rainfall sufficient to convert the basin to highland forest would have to be quite large, and to convert it to open near desert, less so. The central Kalahari had rainfall that averaged 392 mm/yr between 1961 and 1971 (Tanaka, 1980).

Given that baobab trees have been found by radiocarbon dating to live up to about 1000 years (Patrut *et al.*, 2007), we can suspect that the Hadza have not been without these and other of their staple food plants at any time in the last one thousand years. If the flora has remained roughly constant, then the wildlife species composition will have as well. Furthermore, variations in density of wildlife can be expected along with the fluctuations in rainfall (see following and Figure 2.3, from Coe *et al.*, 1976). Thus, the Hadza may have experienced fluctuations in wildlife density that surpassed variation during our studies, from much less abundant to somewhat more abundant.

### 2.3.6 Long-term variation in rainfall: tens of thousands of years

During the past 100,000 years, rainfall has often been very much lower than recent averages, seldom higher. Stager *et al.* (2011) gives us a clue about one major change in lake levels that we can translate to vegetation: "... the desiccation of Lake Victoria ... which today is the world's largest tropical lake. With rainfall over the watershed possibly reduced to less than a quarter of its present amount ... the lake dried out twice between 15 and 18 ka ...". At this time, Hadzaland being close to the Lake Victoria watershed, we might expect the Eyasi basin to have

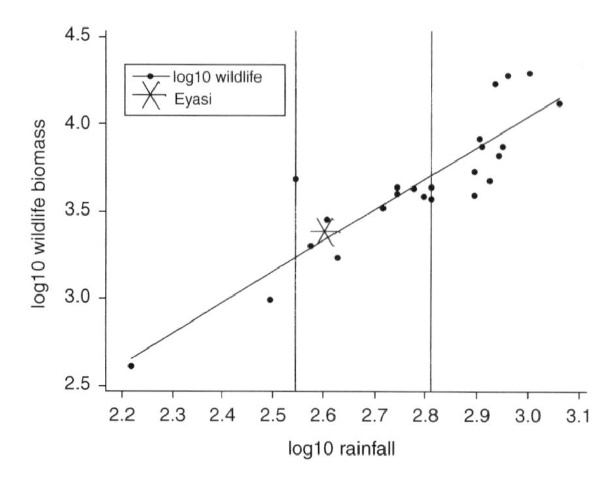


**Figure 2.3** Plot of Coe's data (Coe *et al.*, 1976) on wildlife biomass and rainfall in African game parks. Game biomass for the Tliika region in 1977 and rainfall in Eyasi basin are marked as a single star. Vertical reference lines mark Eyasi area rainfall ±30%.

become near desert (with rainfall perhaps as low as 125 mm/yr) except where water seeps down from the highlands. Because the springs at Mangola are fed from the highlands, they would likely endure quite extreme reductions in rainfall. Some Hadza may have concentrated around this area just as some appeared to do in the exceptional 1997 dry season.

However, given the statistical relationship of rainfall to altitude, we can assume that the Mbulu highlands would have come to approximate the flora and fauna of today's Eyasi basin. Hadza would have to move uphill a few miles. Thus, Hadza may have been able to maintain their economy through a quite extreme range of climate changes. The most notable effect might have been that in the mountains they came into contact with the Sandawe doing the same thing from the southeast. The coincidence with genetic findings is too close to ignore. Mitochondrial data suggest divergence between the Sandawe and Hadza about 21 ka, and Y-chromosome data suggest 15 ka (Tishkoff *et al.*, 2007, p. 2190) but Tishkoff *et al.* write: "our estimates of common ancestry of the Hadza at 15–20 kya may reflect longer term isolation of these 2 populations with high levels of interaction (gene flow) around 15–20 kya, and little subsequent gene flow" (p. 2191). The dry period may have brought the Hadza and Sandawe together by degrading their preferred habitat and removing the montane forest that had held them apart, while elsewhere in Africa the dry period may have separated southern and eastern African click-speakers as Tishkoff *et al.* suggest in a subsequent paragraph. Altitude differences in several parts of East Africa could have provided similar dry period refuges for other populations.

### 2.3.7 Droughts and the unusual weather of 1997 and 1998

Year-to-year predictability of rainfall is low (Norton-Griffiths *et al.*, 1975; Pennycuik and Norton-Griffiths, 1976; Prins and Loth, 1988), thus consecutive periods of dry years are uncommon but not unknown. A number of authors have claimed some regularity in annual variation, for example linking it to sunspots (Stager *et al.*, 2011). Earlier literature mentioned severe famine and drought in 1949 (Baker, 1974). Brooke (1967) lists this as severe for Mbulu and Singida. Ryner *et al.* (2008, fig. 2) show six periods of five years in which the five-year running average was below average rainfall in Mbulu, in 1930, 1945, 1953, 1973, 1981, and 1993. Three of these (1953, 1981, 1993) preceded low levels of Lake Emakat. A period of higher than usual running average rainfall was recorded from 1960 to 1970, and the years around 1990 seem to have had above average rainfall at Mbulu, but rainfall then declined into the early 2000s. Spinage (2012, table 2.2) lists droughts and famines recorded in East Africa since 1637. We can select seven droughts since 1917 as likely to have included the Eyasi area (1917–1918, 1920, 1929, 1933–1934, 1938, 1949, 1953). Later I will report on an attempt to find demographic correlates of these events.

Hadza informants have made little mention of droughts or famines. It seems likely that they manage during times that could be very harsh for farmers (Bagshawe, 1924, 1925 for 1917). The unusual weather of 1997–1998 gave a chance to look at how Hadza react to extremes of rainfall. Galvin *et al.* (2001) reported effects of the 1997 drought and 1998 floods on the Maasai in northern Tanzania. The rains of early 1997 almost completely failed, but in October 1997, exceptionally heavy rainfall was experienced all over East Africa, continuing throughout the early 1998 rainy season. In our June–September 1997 field season, we had noticed unusual movements of Hadza and indications of stress among them, the usually cheerful Hadza seemed grumpy and worried. The irrigation farmers at Mangola fared no better. The village clinic attendant told us that he was being called on to treat wounds, apparently from fights, something he had never witnessed there before. The fights were apparently over the irrigation ditches. While we could see no effects on the Hadza (or the Datoga) as dire as Galvin reports for the Maasai, I will later show some significant demographic differences in the years 1997, 1998, and other years.

While primary productivity increases as rainfall increases, so does the abundance of malaria (Githeko and Ndegewa, 2001; Craig *et al.*, 2004; Zhou *et al.*, 2004, but not Lindsay *et al.*, 2000) and rift valley fever (Patz *et al.*, 2005). Lemnge *et al.* (2001) confirm the seasonality of malaria in northern Tanzania but imply that the year-to-year relationship between rainfall and malaria in Tanzania can be complicated. Extremes of rainfall and flooding can wash away and drown the larvae. Fosbrooke (1972, p. 112) describes a similar relationship between plagues of *Stomoxys* flies and rainfall.

## 2.4 Flora: plant foods, distance from water, foraging return rates

Schultz (1971) described the plant communities and varied soil types of the area. The vegetation is classified as belonging to the Zambesian phytochorion, which includes most of the savanna country from Ethiopia to Namibia (Linder *et al.*, 2005). Within this phytochorion, a subdivision distinguishes wooded savanna from Miombo woodland (rainfall, 700–1500 mm/yr). By rainfall, at around 500 mm/yr, by herbivore species, and by tree species, the Eyasi basin clearly belongs in the wooded savanna category.

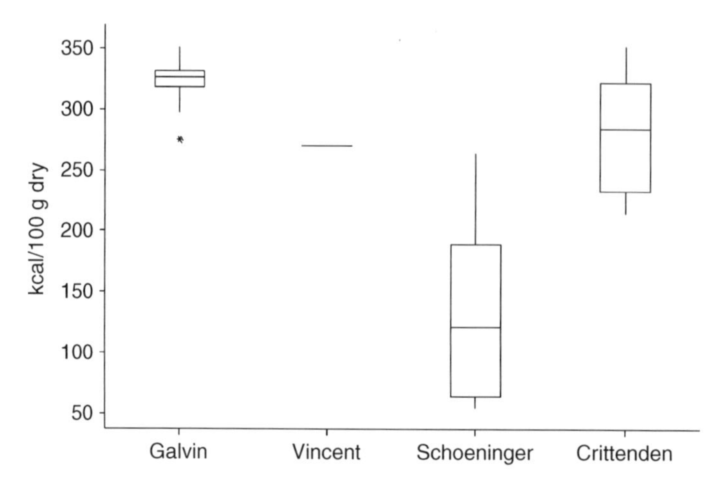


**Figure 2.4** Calorie values of four different samples of //ekwa tubers. Galvin *et al.*, nd, SI (28 specimens), Crittenden, 2009 (8 specimens), Schoeninger *et al.*, 2001 (6 specimens); Vincent, 1985a (3 specimens).

Being no botanist, I can only give an unashamedly Hadza-centered view of the plant resources in the Eyasi area, but a forager-centered view may be what we need for thinking about their adaptations and demography. The most used plant resources are found in all parts of eastern Hadza country but most are distributed in patches. Marlowe (2010, table 5.1) lists 30 species of plant food used by the Hadza, and Crittenden (2009, table 3) lists 22 taken by children. The use of plant and other resources varies by season, locality, and year. Marlowe (2010, p. 108) outlines the Hadza seasons, and McDowell (1981b) (SI) shows data on plant, honey, and meat use by season in the Mangola area in 1979–1980 (SI 2.3). SI 2.4 discusses a difference between our foraging measures and those of Marlowe. SI 2.5 gives calorie values of some of the most important plant foods and compares calorie values obtained by various authors. There is general agreement between these estimates for all but one resource, and with other authors listed by Vincent (1985a). The striking exception is given by Schoeninger *et al.* (2001) who reports great variation in the food value of six specimens of //ekwa (*Vigna frutescens*) tubers (Figure 2.4). She found some individual tubers of *Vigna frutescens* with less than one third of the value of the 28 cases analyzed by Galvin *et al.* (nd) (SI), and much lower than the estimates by all other authors. Schoeninger's figures would leave us wondering why Hadza women spend so much time on one of the least rewarding foods available to them. A key issue is the measurement of non-digestible fiber. Schoeninger used a "stomacher," which incubates the diced tuber in amylin, discarding the undissolved portion as equivalent to the "quid" of chewed fiber that Hadza consumers discard. Other food analysts continue to avoid this method, while identifying indigestible fiber by traditional methods. The recent re-analyses by Crittenden (2009) support the majority of assessments of the food value of //ekwa. Her subsequent analyses suggest somewhat lower food values than those of Galvin *et al.*, and she is currently investigating the extent to which Hadza may be able to digest more nutrition from tubers than otherwise expected (Crittenden, personal communication and Schnorr *et al.*, 2014).

### 2.4.1 Baobab

The baobab (*Adansonia digitata*) forests are the most striking feature of many parts of Hadzaland, these impressive trees being specially abundant in Tliika but also plentiful in Han!abe, parts of Siponga, the Endanawish area, and the southern slopes of the Balai valley. Baobab trees are frequent locations of honey bee (*Apis mellifera*) nests, and men climb them to get honey and sometimes knock down fruit. Most often the fallen fruit are collected off the ground by women and when seeds and pulp are ground together they form highly nutritious flour (Photograph 2.2; table SI 2.5). In most years, baobab fruit are available on the ground for a very large part of the year, at least eight months. In 2000, Gudo Mahiya and I were distressed to see near Yaeda that "Swahilis" were coming with sacks and donkeys to collect and take away baobab fruit, presumably to transport for sale in towns and villages. The developing "commercialization" of baobab (Kamatou *et al.*, 2011) may become the next way in which the Hadza will be deprived of their heritage.

Baobab fruit are abundant and can be collected within 5–10 min journey of many camps. From other camps we have seen Hadza make the occasional lengthy excursion to collect baobab, for instance the ~5 km from high in the center of Han!abe to Endanawish. If the easily broken shells are emptied in the grove and the pith and seeds carried home, several days worth of food can be collected in a single day trip

**Photograph 2.2** Women sitting together on a rock while pounding baobab pith and seeds, an abundant resource, easy to collect, but requiring lengthy processing. © James F. O'Connell, 2015. Reproduced with permission.

**Photograph 2.3** A child sneaks some baobab flour, despite frequent warnings from mother about the risk to his fingers. Behind them is a basket, and a gourd bowl. Some Hadza women are skilled basket makers. © James F. O'Connell, 2015. Reproduced with permission.

(for example, 10 kg at 3.81 kcal/g brings home 38,100 kcal to be processed during the following few days). Exploitation of baobab fruit by Hadza is probably limited mainly by the time it takes to process them (Photograph 2.3). Baobab seeds are slightly "springy"; they give way to pressure but do not crack easily. Apparently, neither elephants nor baboons can crack or digest baobab seeds; Hadza sometimes scavenge the seeds from feces of these animals. Even children can break the seeds by placing them on a rock and striking them repeatedly with hammer stones (fist-sized pebbles collected from a stream bed, or left on a rocky outcrop where women have regularly been processing baobab). Children use a "wet process," by pouring a little water onto the hammered pith and seeds, hammering some more, taking a handful, sucking out whatever is edible, spitting out the hard pieces of shell with seed adhering, hammering some more, sucking another handful, and so on, until nothing is left but inedible pieces of seed husk. Women use a more sophisticated "dry process." Pith and seeds are hammered and, from time to time, winnowed on a small piece of leather. The fibrous tissue holding the pod contents together is winnowed away, as eventually are pieces of seed casing. The remaining flour, a mixture of the pith and the seed kernels, is mixed with water and eaten fresh or cooked. Blurton Jones *et al.* (1989) calculated that children aged 5–10 acquire 629 kcal/hr by collecting and processing baobab while children aged 10–15 produced about 1014 kcal/hr. Children under age five occasionally work at baobab processing but seem to get very little. The smallest children cannot even break the pod with a stick or rock. The older teenagers and two women whose processing was analyzed obtained 1740–2243 kcal/hr of acquisition and processing.

### 2.4.2 Tubers

Rocky hillsides seem to be the most usual habitat for digging tubers. Judging by the behavior of the women, tubers tend to be patchily distributed. Vincent (1985a) published data on the distribution, abundance, accessibility, and nutrient values of tubers commonly used by the Hadza in the Mangola area. Her thesis (1985b) lists tuber-bearing plants elsewhere in Africa. Tubers were available and roughly equally nutritious in all seasons (SI 2.6). //ekwa gave the highest rate of return at 3240 kcal/hr digging and searching while working in the food patch (in patch). Travel time to tuber patches varied (between nine and 65 minutes from one camp) with the longest trip being 8 km and mean of 53 min, much further than, and twice as long as we experienced in Tliika or Lesaso (26.4 min, range 5–55 min). Vincent's women walked at 4.5 km/hr (range 2.7–7.0 km/hr, 14 trips). McDowell (1981b) and Tomita (1966) added more information on the Hadza resources in the Mangola area. Vincent (1985b, fig. 2) shows a transect of the Balai valley in which the locations of the most abundant tuber-bearing plants are marked. The amounts of tuber available are enormous. Vincent's table 2 shows means of 5.2 tons per ha of //ekwa gadabi, 1–17 tons of //ekwa hasa, and 63 tons of shumuko. While shumuko (*Vatovaea pseudolablab*) is used and is very abundant in the Mangola area, we did not see it used in Tliika. Its role is taken there by makalita (*Eminia antennulifera*); both have shallow roots and can be dug by children as well as adults. In the Mangola area the largest tubers (//ekwa) also grow in the sandier soil of the valley, and I have seen it dug in a river bottom near Lesaso between Mangola and Siponga. The tubers known as //ekwa in Tliika grow on rocky slopes and their extraction is a quite lengthy undergound mapping puzzle (Photograph 2.4). Hadza digging sticks are long and stout (Photograph 2.5), probably because they often have to be used to lever away boulders. In the late dry season of 1985 (September–October), at 1661 kcal/hr for women of childbearing age and 1786 kcal/hr in patch (1966 g/hr) for older women, the return rate for trips to dig //ekwa in Tliika was lower than in Vincent's Balai data, despite the shorter walking times. The late dry season of 1986 gave very similar figures to those of 1985. The late dry season of 1989 gave 1859 g/hr in patch (Hawkes *et al.*, 1995a). Women often took a lunch break in the bush to cook and eat some tubers. Groups of kin tended to have their own fire but "token" tubers were exchanged between groups. Sixty-one percent of the cooked weight of tubers was taken back to camp. Return rates for //ekwa in Tliika are closely matched by those of baobab, and in 1986 and 1989 women seemed to forage equally often on each. Readers should note that while Hawkes, O'Connell, and I have most often used "return rate" for acquisition and processing rates, Marlowe (2010) and colleagues have used "return rate" for the amount of food brought to camp per hour out of camp (SI 2.4).

### 2.4.3 Grewia

To those who have eaten them, *Grewia* are another very noticeable and less patchily distributed food. *Grewia* bushes grow to a little more than head height throughout the dry bush savanna, all over sub-Saharan Africa. However, their dry, brittle texture

**Photograph 2.4** Tubers of several kinds are abundant but the largest require strength and skill to excavate. A baby on the back is unlikely to make the task any easier. © James F. O'Connell, 2015. Reproduced with permission.

does not suggest edibility to the newcomer. They are available and edible long into the Hadza dry season. Lee (1979, p. 484) describes the extended seasonal rotation of the several species used by the Dobe !Kung. Some of these are the same species as used by the Hadza, and by at least some groups of East African "Dorobo" (Ichikawa, 1980; Marshall, 2001; Cronk, 2004). They are an important source of hard, dry berries, which, when Hadza mix them with water, form a sweet soup. They can also be eaten straight from the bush, and the fiber and seeds spat out or swallowed. The seeds are not digested and appear to be hard to crack (SI 2.7). In Hadza country, *Grewia* are found almost everywhere except the open grassland parts of the Yaeda plain. Hawkes *et al.* (1989) reports childbearing age women taking home the gains from an average 0.754 kg/hr in patch, and older women taking 0.819 kg/hr. At 2.34–3.32 kcal/g, these women

**Photograph 2.5** Digging tubers is a lengthy underground mapping puzzle. © James F. O'Connell, 2015. Reproduced with permission.

were bringing home 1764–1916 and 1916–2719 kcal, respectively, for every hour of picking in the berry patch (Photograph 2.6). These rates compare favorably with digging //ekwa and pounding baobab. Although grass grows among them, full grown *Grewia* seem very resistant to grazing, in heavily grazed areas grass can be seen surviving among the *Grewia* stems. However, regeneration of *Grewia* may be severely impaired by grazing (Prins and van der Jeugd, 1993).

### 2.4.4 Soft fruit

Important soft fruit (*Cordia gharaf* and *Salvadora persica*) is available only for a short season, beginning a week or two before the rains in October, November, or December (Photograph 2.7). Camps often move to the fruiting trees, but in 1988 Hawkes found Hadza walking for about 90 minutes from camp to fruit trees. She found that if one calculates the "team returns," the rate at which food is delivered to a woman's children, the walks by women and children to the fruit trees yielded

**Photograph 2.6** Grewia berries are abundant, widespread, and have a long season, but their seeds are not digested. Cordia (shown here) are tastier but have a short season. © James F. O'Connell, 2015. Reproduced with permission.

**Photograph 2.7** Anyone – men, women, and children – can collect seasonal soft fruit; here *Salvadora*, fruiting just before the rains begin. © James F. O'Connell, 2015. Reproduced with permission.

more food per hour than foraging on tubers near camp. "Team returns" also seemed important for understanding the foraging strategy of !Kung women and children in the dry season (Blurton Jones *et al.*, 1994). We think of *Cordia* and *Salvadora* as growing in "the berry flats" because they are most abundant in low flat bush country such as the lakeshore plain, the Balai valley, and the "mouth" of the Udahaya river. We have been especially familiar with *Cordia* at Miseu and Udahaya, and *Salvadora* lower on the lakeshore plain. Each is found, in patches, along the entire length of the lakeshore plain. They are also locally abundant in the Balai valley, although here they are being cut by the non-Hadza, not only for making fields but also to make charcoal. Vincent (1985a, table app10.1) gives mean travel times to fruit in the Mangola region as 38 min (6–103 m).

Hawkes *et al.* (1995a) gives the return rate for *Salvadora* at 1383 kcal/hr in patch, and at 2771 gm/hr in patch, and at 1.45 kcal/g, *Cordia* give 4018 kcal/hr in patch. No wonder the women I was camped with in November 1986 were so eager to move down to the berry flats as soon as they discovered they were ripening. No wonder three camps in November 1999 were placed in the cordia bushes, where even I could fill myself in a few minutes close to my tent. Perhaps only a little surprising was the movement of two camps of Siponga people far out of their region to the berries on the lakeshore plain not far from Barazani.

These small trees are also common nesting sites of weaver birds, sometimes in very extensive colonies, exploited for two to three weeks a year by Hadza during the rainy season (seen by me in February 1989, and Marlowe, 2010, p. 139; Crittenden, 2009). In the camp where I stayed in February 1989, the 11 children accompanied men and women on the very short walk to the nesting trees, in periods of 124–672 min, collecting 140–3600 g of nestlings and eggs, for a presumably protein rich, mean return rate of 369.5 kcal/hr.

In Hadzaland, water courses are sometimes the site of tamarind trees (*Tamarindus indicus*), now said to be indigenous most probably to Africa, not India as formerly believed (Diallo *et al.*, 2007). Tamarind fruit are used by the Hadza. As the pods age, they reveal crystalline matter that looks like sugar. Several pleasant concoctions can be made from tamarind. There is a substantial grove of tamarinds at the site of David Bygott and Jeanette Hanby's old house at Mangola. There is also the occasional tamarind in many parts of Hadza country. The larger riverbeds, Balai and Udahaya, bear (or have borne) gallery forest. The gallery forest along the Balai river has been largely removed for charcoal. The forest along the Udahaya river is being cleared at a great rate by Iraqw settlers moving down from the highlands. The acacia trees appear to bear little significance to the Hadza except for the baboons. Baboons frequent the gallery forest, using it for sleeping refuge, and Hadza sometimes hunt and eat baboons. Elsewhere in Hadza country, baboons and vervets are not very numerous, although we have seen the occasional lone vervet in Tliika. Both are quite abundant at Mangola near the springs. In addition, seeds of *Acacia nilotica* were recorded by Crittenden as being collected by Hadza children, and boiled, giving 256–267 kcal/100 g dry weight. *Acacia albida* was used by the Gwembe Tonga (Scudder, 1962) after extensive processing including three separate periods of boiling, one for an entire day. Palms (unidentified species) are found in a couple of locations. A particularly large grove is near Kisimangeda and one Hadza family often camped in this grove despite the intense wind noise. The large and very hard seeds are sometimes chewed on by Hadza.

Two very significant plants for the Hadza are the arrow poison trees panjube (*Adenium* sp.) and shanjo (*Strophanthus eminii*, illustrated in Woodburn (1970). Both are scarce and localized. One is more abundant in the southwest, the other more used in the northeast. Both kinds of poison are used quite widely in Africa. The tiny size of the San arrow points is associated by some with the use of poisons, and with the possible great antiquity of arrow poison use (discussed in Robbins *et al.*, 2012). Hadza poisoned arrows are relatively hefty, and Lars Smith has seen a buffalo felled in a single shot.

Savanna is sometimes described as dynamic and highly variable (Spinage, 2012). Most of Spinage's examples are from places previously settled and farmed or heavily herded. There is little indication of vegetation change in the Eyasi basin between Obst's 1911 fieldwork and ours (SI 2.8). His map marks vegetation patches in much the same places as we notice today, and his photographs appear completely familiar. The only noticeable changes are where immigrant farmers have cleared land, and charcoal traders have felled riverside gallery forest. It may be worth noting that all the Hadza plant foods discussed earlier derive from woody plants. For the Hadza, all good things grow on trees, or hide behind them (SI 2.9).

## 2.5 Fauna: prey and predators, abundance, decline, and ancient variation of animal populations

During our study, Hadza country was home to many of the well-known animals of the sub-Saharan savanna (table SI 2.10). We saw zebra (in the wooded savanna of Tliika), impala, buffalo, kudu (lesser and greater), giraffe, warthog, and all were sometimes taken by Hadza. At night, one regularly heard lions, hyenas, and leopards. Less familiar species were also present, or said to be. Among them we encountered klippspringer, bushbuck, gerenuk, eland, and traces of bush pig and elephant. Warthog and aardvark holes were a constant hazard of driving. Scorpions, cobra, black and green mambas, and puff-adders are common but seldom encountered.

All Hadza men regard themselves as hunters, although they vary greatly in success and the opportunity provided by their neighborhood. I have noticed that our readership sometimes finds it difficult to appreciate the rarity with which a Hadza hunter acquires an animal of any size (larger than, for example, an impala or warthog, 40 kg or larger). Most men go out hunting for several hours most days (a few hours in early morning, then again for a couple of hours in early evening), and come back empty handed (average 3.2 kg/man–day; Hawkes *et al.*, 1991). In late dry season they stay overnight in a hunting blind near a trail or water hole, the most successful form of hunting (7.5 kg/man–day). On average over all seasons, in 1985–1989, each hunter achieved his mean of 4.9 kg/day by succeeding on only about one day in every 30 (Hawkes *et al.*, 1991, table 2). Focal follows (following men on their hunts) showed men getting only 0.062 kg/hunter–day from small game. During our observations, men only occasionally brought home a small animal or bird, usually surreptitiously summoning us from inside their house to come and weigh it. They may have more often killed and eaten a small animal while out in the bush (as also described by Woodburn, 1968a). We came to regard Hadza as big game specialists (but see Wood and Marlowe, 2013, and Hawkes *et al.*, 2014), and were interested by Obst's remark that Hadza did not trap (did not use snares or pits), considering it "verpunt" (despised, forbidden, deVries, 1946) and "unwaidmannisch" (apparently an older spelling of *unweidmannisch*, perhaps best translated as "unsportsmanlike," denoting "not what a hunter does," according to de Vries and various online dictionaries). We have observed boys trapping birds with a noose. Hawkes points out that the !Kung and Ache attain their more reliable returns by including more small animals. From our results on hunting large and small animals, Hawkes derived the idea that big game hunting might benefit men by routes other than or in addition to providing for their offspring. Men noted for providing "public goods" from their rare catches of very large animals might be treated differently as neighbors or competitors. I will examine this idea at several points in Part II of the book, especially in Chapters 14 and 22.

Marlowe (1999, table 1) shows data on calories brought home by Hadza men in a variety of locations during his 1995 year in the field. He comments (p. 61), "These figures, it should be noted, are much lower than those reported in Hawkes *et al.* (1991) using data from 1985–1986, which supports Hadza claims that wildlife numbers have decreased due to encroachment by herders and farmers." Marlowe (2010, p. 36) writes, "The Hadza say that there is less game than in the past. As noted, Obst (1912) reported seeing large herds of big game in the Yaeda valley in 1911, whereas these days it is mostly Datoga cows and some gazelles." However, Fosbrooke (1972, pp. 88–89) shows the early travelers' reports of numbers of game in Ngorongoro were sometimes extreme exaggerations, the flora could only have supported a fraction of the numbers claimed. Based on his fieldwork between 1995 and 2010, Marlowe (2010, table 5.1) reports the same species taken by the Hadza as I listed previously. What can we determine about changes in abundance of the large prey species? Did the game density in which we observed Hadza hunting give us a misleading impression about the role of hunting in earlier times? Did the decline in wildlife affect Hadza demography in any way? There are several lines of evidence to help us determine whether large animals were decreasing or not.

### 2.5.1 Rainfall and wildlife biomass

O'Connell *et al.* (1988a, fig. 1) showed that the large animal biomass ($kg/km^2$) in the study area (from air counts in 1977) fell on the regression line of biomass on rainfall estimated by Coe *et al.* (1976) for 15 wildlife areas (SI 2.11). The abundance of large animals in Hadza country was thus about as expected for protected habitat at the observed rainfall. As rainfall varied, the abundance of large animals had probably also varied (Figure 2.3).

### 2.5.2 Aerial surveys

In 1977, 1978, and 1980, Lars Smith and Mike Norton-Griffiths had conducted aerial surveys of the area using the methods of Mike's company Ecosystems Ltd. (Norton-Griffiths, 1978). David Bygott and Jeanette Hanby were among the observers for these surveys. Lars made his results available to us early in our fieldwork. In 1989, O'Connell commissioned a similar survey by Marcus Borner and observers from Serengeti Research Institute (SRI), and they conducted another survey in 1992 (SI 2.12). The total wildlife densities for the areas surveyed were 4.45 individuals/km$^2$ in 1977, 2.09 in 1989, and 0.87 in 1992). There were fewer game animals in the area in 1989–1992 than in 1977–1980. Stoner *et al.* (2006) showed that according to air counts, most large herbivores decreased in eight census areas in Tanzania between 1988 and 2001.

### 2.5.3 Hunting success

Seasonal, local, and annual variation, and especially methodological issues, make comparisons difficult. My computations (SI 2.13) of the data shown in O'Connell *et al.* (1988a, 1988b) and Hawkes *et al.* (1991), gathered in 1985–1986, and by McDowell (1981b) in 1979–1980 suggest similar rates of success hunting large animals. Hawkes *et al.* observed a large animal catch once every 29 days (71 animals in 2072 hunter-days and an average of 4.89 kg/ man-day). McDowell reported one large animal in every 35 days, and 3.04 kg/day from large animals in Mangola region in 1979–1980. Smith (personal communication) from Tliika in late-dry 1977 suggests higher rates. Marlowe (2010, tables 5.4, 5.5) implies that hunting success may have declined between 1986 and 1995. Wood and Marlowe (2013) in their table 1 and table S 1, allow us to see that east of Lake Eyasi between 2005 and 2009, Hadza obtained large prey once every 103 hunter-days (19 animals in 1952 man-days) for an average of 1.083 kg/man-day. As Wood and Marlowe note, if we include their data from the west (where the Hadza lived adjacent to a vigorously protected tourist hunting area), the rate is similar to that of Hawkes *et al.* (1991). Wood and Marlowe (2013) report 0.15kg/ hunter-day from small game hunting in the east, which may differ from Hawkes *et al.*'s observed 0.062 kg/hunter-day, but both are lower than the 0.408 kg/day of men in Hawkes *et al.*'s experimental focal follows (follows of men paid to pursue only small game). Comments in Marlowe (2010) imply that since 1995, the Hadza may have been bringing to camp more small prey, honey, and plant foods than we observed during the 1980s.

### 2.5.4 Reasons for change

Thus, it seems likely that the amount of large game in eastern Hadza country declined between 1977 and 2009. It is not difficult to suggest some causes. After the 1986 economic reforms, fuel was no longer restricted to government use, game patrols were reduced, and country people began to have cash with which to buy wild meat. During the 1990s, signs of poachers with vehicles increased (motor vehicles heard at night, distant headlights seen, tracks running wildly around in open, hitherto undriven areas), as did local talk of their activities. Perhaps guns became more available by smuggling from Uganda or Rwanda. Bushmeat trade has been increasing all over Africa, with meat being transported to towns, and even to Europe (Zoological Society of London *ScienceDaily*, 2008). Its effect on large animals in western and northern Serengeti has been documented by Sinclair and Arcese (1995), Campbell and Borner (1995), Loibooki *et al.* (2002), and others.

There may be other reasons for change. Buffalo and impala tend to have strong population oscillations (Prins and Douglas-Hamilton, 1990). Removing a few 450-kg buffalo from computations of hunting yield makes a very large difference. During our pilot visit in 1984, we saw several buffalo taken by Hadza in a few days. They were still present in 1985–1986 but seemed to disappear after that. The Hadza in Tliika attributed the decline in large animals to Datoga fencing the water holes, making the area unattractive to wildlife. Apparently this has recently been reversed and access for wild animals is added to the watering holes. One decline in animal numbers apparently reversed itself. We often heard lions in the 1980s but they vanished in the early to mid-1990s, only to reappear or recover by the late 1990s. Hadza agreed with our impressions. In Serengeti, there was widespread distemper among lions at the time they disappeared from Hadzaland (Morell, 1994; Harder *et al.*, 1995). Their return implies that lions were apparently finding enough to eat in Hadzaland and we would likely have heard from Datoga acquaintances if they were taking more livestock. While eland are seldom seen and seldom caught by Hadza, one young man during the late 1990s described to us a location he enjoyed because it was a "kiwanja [Kiswahili: stadium/airfield/playing field] of eland." Eland and buffalo featured in Wood and Marlowe's 2013 food sharing data set, which includes observations west of Lake Eyasi.

*Apis mellifera*, the honey bee, should not be overlooked, nor confused with the more aggressive "Africanized" hybrid bees now found in the United States. The Hadza certainly do not overlook them; they are constantly alert to the opportunity to find honey of this and other species, watching for a cluster of bees and listening for the calls of the honey guide (Indicatoridae). Nests are found in trees, especially baobab, in rock crevices, and even underground. A tuft of burning grass is pushed into the hole and the hole enlarged with an axe (Marlowe has seen this done with a rock). Honey combs are taken out by hand, and much eaten on the spot. Honey is sometimes carried in a leather bag, sometimes in a baobab pod, or when nuclear families go honey hunting (as they say they sometimes do for most of June, very evident everywhere during our census in June 1985), in the cooking pot brought along for the purpose. We have quite often seen men bring a baobab pod full of honey back to their house where they quickly hide it in the grass wall. In Hawkes *et al.*'s small game hunting follows, men often stopped for honey and obtained 21.89 kg in 33 collecting events (0.663 kg/event). The calorie value of honey is commonly set at just over 3 kcal/gm, so these men obtained 65,670 kcal in their 33 attempts (1990 kcal/attempt). The average attempt took 41.4 min (0.69 hr), so the return rate on encounter was 2884 kcal/hr. The men Hawkes *et al.* followed acquired 0.78 kg/man-day (average day, 411 min), for 2340 kcal/day from honey, which is rather less than the 3000 kcal obtained by a 10-year-old boy in a three-hour trip observed by Crittenden (2009). The boys aged 13–18 followed by Blurton Jones in the 1986 dry season averaged much less in pursuit of kanoa ("sweat bee" honey, taxon not determined). Sweat bee nests are difficult to spot. The skill entails being able to see a one-inch translucent tube that descends from the trunk or low thick branch of a commiphora tree. Ducking under the branches, trying to dodge the thorns and keep up with Hadza boys searching for sweat bee honey is serious exercise for the anthropologist. Each nest provides 45 g (N=27) of honey, at an average of over two finds per hour, 130 g/hr, and 304 kcal/100 g, boys acquire about 339 kcal/hr (1424 J/hr), which makes its pursuit by any small boy who can get the use of an axe quite rewarding (Blurton Jones *et al.*, 1989).

### 2.5.5 Tsetse flies: guardians of Hadza cultural conservatism?

One piece of fauna made its presence felt out of all proportion to its size. The tsetse fly (*Glossina* sp.), carrying a commonly deadly cattle disease, could have been as important to Hadza cultural survival as the German and British pacification. The fly is an inconspicuous gray insect a little bigger than a housefly, but its bite can be quite painful. Although very attracted to Land Rovers (hot, bulky, and slow), the flies are also slow and can easily be killed if you see them settle on someone. Luckily, in the dry season they are confined to a very specialized habitat, although it is a habitat that is often convenient to drive through. They seemed to become less numerous during the period of our study, perhaps due to the disappearance of buffalo, which Spinage (2012) argues is an important target for tsetse. However, in previous times in Tanzania as a whole, the tsetse fly is said to have had a very large effect on human land use, keeping livestock owners and farmers away from the areas of highest tsetse density. Colonial governments instigated bush clearance schemes to reduce tsetse and facilitate ranching. Some authors indicate that experts underestimated the ability of traditional peoples and their livestock to accommodate to natural populations of tsetse (for example Giblin 1990; Mwangi *et al.*, 1998).

The International Biological Program (IBP) researchers in 1966–1967 trapped specimens of two tsetse fly species (*Glossina swynnertoni* and *G. pallidipes*) in eastern Hadza country, and Bennett *et al.* (1970, p. 876) comment that between 1963 and 1965, 207 non-Hadza cases of sleeping sickness were recorded at Mbulu hospital. Tsetse are abundant in Hadza country and every single Hadza must have been bitten many, many times by tsetse flies. Yet Bennett *et al.* (1970) found no trypanosomes (the sleeping sickness protozoan delivered by the tsetse fly) in blood films from 334 Hadza and only one person showing antibodies to trypanosomes. However, Ward (1999) reports an epidemic of sleeping sickness (confirmed by Haidom Lutheran hospital) at Munguli and other settlements some time after 1965, and presumably before he left Munguli in 1967. Sleeping sickness does not require the presence of livestock, wildlife are held to be a reservoir of the responsible trypanosomes, but the abundance of livestock may influence the rates of infection and sickness of people. Woodburn (1962, 1968a) mentions a tsetse bush clearance scheme encroaching on Hadza country in the years up to 1963. With a decline in wildlife, and availability of medicines for livestock, the proportion of tsetse flies carrying the trypanosomes may have declined. Sleeping sickness could also have been more abundant among the Hadza in the past. Boys might have been especially at risk, spending hours seeking sweat bee honey in the commiphora trees, which form a key part of tsetse habitat (Burtt, 1935) and on which tsetse rest (Jackson, 1945). In 1985, Lars Smith and I drove one elderly Hadza man to Haidom hospital where his sleeping sickness was confirmed and from which he later died. We have neither seen nor heard of other cases since then. Perhaps Hadza have some constitutional resistance to sleeping sickness reflected in Bennett's results.

## 2.6 Diseases ancient and modern: abundance in forest, savanna, desert

Bennett *et al.* (1970, 1973) reported the medical studies among the Hadza by the IBP group in 1966 and 1967, not long after the major efforts to settle the Hadza at Yaeda and Munguli (see Chapter 3). In the IBP Hadza file archived at University College London (UCL) anthropology department, there is also a manuscript report by Dr Aylett on the health of the Hadza at Yaeda in 1965, and Jelliffe (1962) examined 62 Hadza children before the settlement attempts. We made no such investigations, although an up-to-date repetition of Bennett *et al.*'s survey might have been highly informative. Bennett *et al.* could only test for diseases well known at the time of his research. Knowledge of tropical diseases has progressed, especially on zoonoses (Wolfe *et al.*, 2007).

Bennett *et al.* made serological tests for antibodies to 24 disease organisms (SI 2.14). These included well-known "crowd epidemic diseases," dangerous during the nineteenth century in Europe and North America, killers of thousands still today in the developing countries, diptheria, measles, mumps, whooping cough (pertussis), rubella, smallpox, plague, typhus, cholera, influenza. A significant percentage of the Hadza samples showed antibodies to these diseases (with the exception of diphtheria). This shows that many Hadza had been exposed to these diseases and the population was by no means isolated from them. Black *et al.* (1974) had a similar result among "remote" South American populations. Comparison of the Hadza in the settlements and in the bush gave no significant or consistent difference. Bennett *et al.* comment on the abundance of antibody to influenza A and B viruses that had been currently and recently circulating in the U.K., indicating very little isolation of the Hadza from the spread of such viruses.

Several of the crowd epidemic diseases are regarded as having evolved since the origin of agriculture. Are we to believe that the Hadza (and indigenous Australians and Americans) suffered lower mortality before the arrival of these diseases than they do now, some 100 or more years after these diseases took their initial toll? Hadza informants have reported no major outbreaks of the crowd epidemic diseases in the past but we witnessed one (measles in 1986, associated with a brief settlement attempt), and we cannot exclude an effect of initial exposure to them, which would most likely have occurred early in the nineteenth century. Anderson and May (1991, pp. 653–654) wrote, "Bands of hunter-gatherers probably ranged in size from around 20 to at most 100 individuals (Hassan, 1981) and these populations may have been sufficient to maintain many macroparasites as well as microparasites with long periods of infectiousness or (as is the case with hepatitis, herpes simplex, gonorrhea, and other infections) with asymptomatic carriers ... survive successfully in the soil or other inanimate environments (such as tetanus, gas gangrene, and other clostridiae) ... But the directly transmitted microparasites responsible for much mortality in historical times – smallpox, measles, cholera, and the like – have very high host threshold densities, and could not have been present in the pre-agricultural era."

We may have misunderstood the opportunities for transmission and maintenance of infectious disease in a world of hunter-gathers among hunter-gatherers. We too often forget the archaeological evidence of extensive trade networks from 130 kya. We tend to think only of "Bands of hunter-gatherers probably ranged in size from around 20 to at most 100 individuals" (Anderson and May 1991, pp. 653–654), and forget the likelihood of expanded wet season ranges bringing otherwise separated peoples into contact. There may have been unknown diseases, perhaps extinct, perhaps still present. Bennett *et al.* could only test for diseases known at the time and for which tests were available (SI 2.15).

More likely is that the greatest portion of Hadza mortality (which falls on the very young) is not a result of the infamous crowd diseases but of more "routine," elsewhere less significant, infections leading to diarrhea, coughs, and fevers. In Chapter 8, I report what we know about causes of death among the Hadza, which may help us assess how much the modern crowd diseases have changed the demographic picture.

Dunn (1968) described the variation in abundance of parasites with habitat. Twenty to 22 parasite species are found in Pygmies and Malayan Semang in their tropical forests, whereas only one to nine species are reported from studies of San and Australians in their drier environments. Bennett *et al.* (1970), in their paper on helminth and protozoal parasites, suggest that the Hadza, with 14–18 species (four commensals), are intermediate. Dunn links these findings to complexity or biodiversity of the ecosystem, and reports that they also apply to non-human primates. Thus, despite Dunn's recognition that "The epidemiological record for infection in hunter-gatherers is of course considerably 'contaminated' by contact with outsiders," it seems safe to assume considerable antiquity to this dimension of variation in parasitic disease. Something as simple as the abundance of diseases could account for significant variation in child mortality, the life history consequences of which were discussed by Migliano *et al.* (2007). Migliano *et al.*'s Pygmy populations mostly live in damper, more tropical climates than the Hadza and !Kung, and presumably suffer the higher parasite load described by Dunn. However, in this context, we should also not ignore the well-known interactions between nutritional status and infection (Scrimshaw, 2003).

Bennett observed positive Heaf tests for tuberculosis (TB) in 36% of Hadza men and 13% of women. Recent genetic studies have indicated that TB may be a much more ancient disease than hitherto believed, long pre-dating agriculture (Gutierrez *et al.*, 2005). It conforms to one of Anderson and May's (1991, p. 654) conditions for a disease to persist among hunter-gatherers – long periods of infectiousness with asymptomatic carriers.

Hadza commonly claim to be suffering from malaria and they are certainly often correct. Bennett *et al.* (1970) reported parasites in 26% of blood samples in June 1967, "spleen rates" (by manual examination) as 32%, and about equally frequent infection with *Plasmodium malariae* and *P. falciparum*. All measures declined with age, being highest in children aged one to nine years, and low in the under one-year-olds. Bennett *et al.* comment on the apparently lower rate found by Jelliffe *et al.* (1962) when he examined 62 children before the major settlement attempts. Bennett *et al.* note the existence of malaria in this nomadic population and contrasts this with the view that malaria would be unlikely to be endemic among small mobile groups. Bennett *et al.* suggest we should keep in mind the "good deal of interchange between camps which would disseminate the disease throughout the population." The history of malaria has been much investigated in recent years.

Genetic mechanisms that give some protection against falciparum malaria (such as *G6PD* deficiency) are said to be relatively recent and probably post-agricultural. Campbell and Tishkoff (2008) summarize the genetic evidence that special protective adaptations are recent, "Overall, these data are consistent with other evidence suggesting that the malaria parasite has had a significant impact on humans only within the past 10,000 years . . . possibly corresponding with the development of agriculture and/or pastoralism in Africa." Tills *et al.* (1982) report a low frequency of *G6PD* deficiency in the IBP sample of the Hadza (three people from 219 tested; Tills *et al.*, 1982, table 2), which they remark is much lower than among other African samples. This protective mechanism has apparently not been selected in the Hadza population.

There are other indications that malaria was less of a force before agriculture (as suggested by Frank Livingstone, 1958). In a series of papers using molecular genetic evidence Rich, Ayala, Wolfe, and others largely substantiate Livingstone's idea. Rich *et al.* (2009) show that *Plasmodium falciparum* evolved from *P. reichenowi*, the plasmodium that infects chimpanzees. The transfer may have occurred "as early as 2–3 million years ago, or as recently as 10,000 years ago." But they show that *P. falciparum* extended its worldwide range much more recently. Conclusions may change, and uncertainties become clarified as this fast-moving field progresses. Rich *et al.* point out that *P. falciparum* must have endured somehow, between its possibly ancient split from *P. reichenowi* and its massive expansion after the spread of agriculture. It is not clear how important malaria would have been to "hunters in a world of hunters."

Woodburn (Lee and DeVore, 1968, p. 91) discussed the case of a young man paralyzed by poliomyelitis. There have been many polio epidemics in East Africa and we may wonder about the exposure of the Hadza to this disease. Bennett *et al.* (1973) found 24/30 samples had antibody to "a poliovirus-type antigen," evidence of quite extensive exposure before 1966–1967. The most recent polio epidemics seem not to have reached the Hadza, although as far as we know, very few Hadza have been immunized. Tanzanian public health responses to polio have greatly expanded.

Sexually transmitted diseases (STDs) and disease of the reproductive tract are observed among the Hadza. The first case of AIDS was reported in Mangola (a non-Hadza patient) in 1991 and HIV was identified in a few Hadza in that area in 2010. The incidence among non-Hadza women in the Mbulu highlands and Singida was reported at 2% in 2003–2004 by Yahya-Malina *et al.* (2006). A few times, Hadza men have consulted us about genital lesions, including the inguinal bubes associated with chancroid (*Haemophilus ducreyi*) and *Lymphogranuloma venereum* (*Chlamydia trachomatis* strains). Bennett *et al.* (1973) say surprisingly little on this topic. They comment "The Hadza are now getting gonorrhea from neighboring tribes," mostly in the west, apparently assuming it could not be maintained in a small isolated population. Yet they cite "venereal disease" as seventh in a list of 13 most frequent reasons for Hadza seeking western medical help. There is no mention of chlamydia or haemophilus. Clearly, we will have to estimate primary and secondary infertility, common outcomes, and crude indicators of the frequency of STDs. Anderson and May (1991, p. 228, 654) include gonorrhea in their list of diseases that may have been maintained in hunter-gatherer-sized populations. "This suite of properties – virtual absence of a threshold density of hosts, long-lived carriers of infection, absence of lasting immunity – adds up to microparasites that can be well adapted to persisting in small, low-density aggregations of humans" (p. 228). Pennington (2001) pointed out the potential significance of STDs for hunter-gatherer demography, and I discuss this further in Chapters 11 and 17.

## 2.7 Summary and discussion: a rich savanna environment, ecology of children, vertical migrations

The Eyasi basin suffers very low rainfall but the Hadza have access to many small water sources all over the area. There is a strong difference between the single rainy season and the dry season, and hence abundant plant underground storage organs; a very brief season for soft fruits but a much longer season for the abundant and nutritious baobab fruit (the seeds are eaten) and *Grewia*. The abundance of water sources means that Hadza have many choices over where to stay and can easily move if foods become depleted. Distance between camp and food patches are small. Large game animals may have decreased since 1980, when their numbers were about as expected for the rainfall. The tsetse fly also appears to have decreased, and with the advent of cattle medicines, many herders have moved into the area.

Like most other modern hunter-gatherers, the Hadza have been exposed to modern crowd diseases. It is possible that epidemiologists underestimated the size of previous hunter-gatherer populations and the amount and frequency of contact between them. Tuberculosis and gonorrhea may have greater antiquity and have been present among hunter-gatherers for longer than we have often supposed.

### 2.7.1 A rich wooded savanna environment

The Hadza environment has been described as rich, which might lead us to expect demographic differences from the !Kung or central Kalahari San, where the environments seem very hard. Having worked among the !Kung in the Dobe and /ai/ai area, as well as among the Hadza (compared in the next paragraph), I concur with the description of Hadza habitat as rich, in plant foods as well as large animals. The most used plant foods are patchily distributed and women know where to go to get them. Patches of the most often used foods are available in each region. In Hadzaland there are no mongongos but baobab are vastly more widely distributed. The tubers used by the Hadza are more abundant and in denser patches. Both locations share an abundance of *Grewia* species. All these foods are the products of woody plants.

The data in Blurton Jones *et al.* (1994) allow us to compute that, in a good grove of mongongo trees, a woman can pick up enough nuts in an hour to give her, after the required 0.833 hours per kg. of processing, 1381 kcal/hr of work in patch and at home in camp. A Hadza woman gets 1500 kcal/hr (Hawkes *et al.*, 1995a) for comparably easy collecting and comparably lengthy processing of baobab. Both women have access to similarly valuable patches of food. However, there is a very large difference in the distance between food patches and dry season water, near which her camp must be placed for much of the year. In many parts of Hadza country, baobab trees are at most five or 10 minutes from camp, so the return rate for a baobab excursion is not much less than the 1500 kcal/hr collecting and processing rate. At Dobe, and at many other places in Ngamiland (because of the structure of the country and the habitat preference of mongongo trees), the nut groves are far from dry season water. The nearest and sometimes exhausted grove is 5.9 km from camp at Dobe, more than a three-hour round trip. If the grove is not depleted, the trip can at best yield 727 kcal/hr when travel and processing are included (Blurton Jones *et al.*, 1994, table 4). This is about half what the Hadza woman can expect.

This contrast may illustrate two important points. First, the labor and time involved in acquiring and processing food may be as limiting as the abundance of the food. Second, a key factor in African savanna forager women's work can be the distance between food and water. This distance will shape women's energy and time budgets in ways that may have an impact on the number of children they can bear and keep alive. Could the Hadza environment be any richer? If animals returned to their 1977 densities, it would be difficult to imagine a more productive environment for hunter-gatherers. The greater primary productivity of rainforest and moist savanna may not translate into more abundant and accessible foods. Laden and Wrangham (2005) contrast the 100 kg/km$^2$ in forest (Hladik *et al.*, 1984) with Vincent's (1985b, table 2) 40,000 kg/km$^2$ in Hadza savanna. As plants have more to gain from storage in severely seasonal rainfall, we might wonder about the intermediate rainfall of Miombo woodland (700–1500 mm/yr). Perhaps the scarcity of recorded hunter-gatherers in this habitat is not merely due to their replacement by farmers. I elaborate on African habitat variation in Chapter 22.

The number of large game animals in the Eyasi basin in 1977 was closely in line with expectations from rainfall (Coe *et al.*, 1976; O'Connell *et al.*, 1988a). It is thus unlikely that there were substantially more animals in previous periods. However, there are indications that the number of large ungulates decreased during the period of our studies. Greatly increased poaching was evident, and increased Datoga fencing and deepening of water sources constantly worried the Hadza for its likely effect on game animals. The abundance of game differed by region even in 1977, when the number counted in Mangola was much lower than elsewhere. During my study period, game were most often seen in Tliika and Han!abe, less often in Siponga, and seldom in Mangola. Vegetation in Mangola was the most depleted, by felling and clearing, and villagers have settled all around the main water sources. The higher parts of Siponga, occupied by Iraqw farmers, were also cleared.

If meat, as the hunting success rates reported previously suggest, has become less abundant since the 1980s, we might expect some demographic and social changes. Hadza may have broadened their diet, while still attempting to take any large game encountered. The broadened diet may have included catching or bringing home more small game that they may previously have ignored or eaten alone in the bush. Alternatively, the Hadza may have increased their effort, or suffered a reduction in the amount of meat that they eat. They may have increased their focus on honey, for consumption and trade. A naïve expectation might be that when there are more game, men will have more effect on their children's survival. An alternative expectation is that when there are few large game, men might turn their efforts from acquiring a rare, mountainous "public good" to pursuing small game that they keep for their wives and children. A change of this sort could account for some apparent contradictions between Marlowe's and my results on men, to be reported later in Part II.

### 2.7.2 Ecology of children, an opportunity for helpers

Hadza children are able to forage with some success (SI 2.16). From age five years, even in the dry season, they can dig shallow tubers like makalita, crush baobab seeds by pounding them on a rock, and snack on *Grewia* fruit (Blurton Jones *et al.*, 1989, 1994, 1997; Hawkes *et al.*, 1995a; Crittenden, 2009; Crittenden *et al.*, 2013). In the wet season, they can collect *Salvadora* and *Cordia* fruit, and join adults in collecting *Quelea* eggs and nestlings. Our data show that, at best, they can get about half their daily energy requirements by their own efforts. Only from about age 15 do they produce as much or more than they need. The remainder of their food has to come from other people. Children elsewhere also do some foraging, on shellfish in Merriam (Bird and Bird, 2002), and lizards in the western desert of Australia (Gould, 1969; Bird and Bliege Bird, 2005), tubers in Madagascar (Tucker and Young, 2005). However, they too seem to need more than they can acquire. !Kung children do relatively little foraging during the dry season, a few *Grewia* and tree gums being available near the dry season camp, and cracking nuts gathered by their mothers is a much more rewarding activity for the older children (Blurton Jones *et al.*, 1994). The more children can provide for themselves, the less the mother and others need to provide, and perhaps the more capacity the mother can divert to new babies, thus increasing her fertility. Perhaps the shellfishing expertise of children like those on Merriam contributed to the speed of the coastwise expansion to Australia proposed by palaeoanthropologists. Among the Hadza, we sometimes wondered whether depletion of "children's foods" near camps was felt as pressure on adult providers and prompted camp moves.

Kaplan (1997, fig. 10.4) and Hawkes *et al.* (1997) pointed out an important contrast between human diets and diets of other higher primates. While young non-human primates become self-sufficient soon after weaning, human weanlings do not. In some circumstances, they are entirely supported by adults, as perhaps in the !Kung dry season. These authors link this disparity to a difference between foods available in moist tropical forest and in dry savanna habitats. In savanna, you cannot live on soft fruit, shoots, and green leaves year round. Savanna has rich resources, abundant during long seasons, but they are difficult to get. Kaplan and colleagues have emphasized the abundant animal prey, only accessible by hunting or by scavenging from dangerous predators. Hawkes and colleagues emphasized plant resources, especially underground storage organs. As commented earlier, baobab seeds are difficult to process and are extremely strong, and like mongongo nuts, are eaten but undigested by elephants and baboons with the fruit digested but the nut unimpaired.

Kaplan and Hawkes, in their different ways, have linked the hard to get resources to the dependence of human offspring, giving an ecological cause to the radical differences between ape and human reproduction. They argue that because these resources are difficult to acquire and process, requiring both skill and strength, newly weaned offspring remain dependent on adults for food much longer than a newly weaned ape or other primate in forests. As a result, the "savanna" mother has to provide food for an overlapping series of offspring (among several authors who have made this point, a recent quantitative presentation is given by Gurven and Walker, 2006). Thus the resources and the economic dependence of the offspring create an opportunity for related helpers, whether fathers, siblings, aunts, or grandmothers. Hadza children give us an "exception that proves the rule," for despite their enthusiasm, they do not supply more than a portion of their requirements until the mid-teenage years. Their success at different ages with different foods (Blurton Jones *et al.*, 1989, 1997; Hawkes *et al.*, 1995a; Crittenden, 2009; and Crittenden *et al.*, 2013) may fill in important details in the Kaplan–Hawkes savanna resource hypothesis. Crittenden's observations of Hadza children competing with baboons for access to figs might suggest we think about forager children as suggestive if only partial models for Australopithecine foraging. They encompass the right size, and tend to use the simplest of tools.

### 2.7.3 Variation in resources

Severe variation in the African climate has been widely recognized and discussed (a good entry point to this extensive literature is Behrensmeyer, 2006). We see it in the short term through the many African populations that have suffered severely from droughts. Yet early in the 1900s, a time of severe droughts, the Hadza played host to starving Isanzu who fled to the bush to escape famine. By then, a dispossessed Datoga sect, the Kisamjeng (Bagshawe's spelling), had also moved into the area and lived by hunting and gathering until they rebuilt their herds. The environment was rich enough to support more people than the Hadza population of the time. The Hadza, like other hunter-gatherers, report no general experiences of famine. This may not mean they are entirely unaffected by drought. Serious droughts in northern Tanzania that were widely reported in the literature were listed earlier. I have not heard Hadza comment on them, but in the late dry season 1997, it was apparent that they were under some stress and I will report demographic effects in later chapters. The early part of my study period seems to have covered a fairly wet series of years. After 1997, exceptionally heavy rains fell in 1998, but in 1999, 2000, and subsequently, there was widespread and severe drought in East Africa. The highly variable rainfall should be considered in discussion of hunter-gatherer population regulation.

The contemporary 30% coefficient of variation of rainfall suggests that the Hadza have probably, several times in the last 2–500 years, experienced climates as dry as those of the central Kalahari San peoples. They could have experienced years with rainfall as high as in the Nairobi National Park or Serengeti. This implies only minor changes in flora. We can suspect that the Hadza have not been without their staple food plants at any time in the last thousand years, and probably much longer. From Dunn (1968) we can propose that their parasite loads could have varied between the low levels of desert dwellers to slightly higher than today. Distances between water and food patches could have been slightly greater from time to time.

### 2.7.4 Prehistory of the environment

Several much drier periods have been noted in distant African history. On the whole, however, their severity seems to be bounded roughly by the contemporary year-to-year 30% variation. Thus, the flora may have remained fairly constant, in which case the wildlife species composition will have changed little. Mabulla (2007) reports on Middle and Later Stone Age Eyasi fauna "the range of fauna represented in the assemblages is consistent with what is consumed today by Hadzabe hunter-foragers." But fluctuations in density of wildlife can be expected along with the fluctuations in rainfall (Figure 2.3). Thus, the Hadza may have experienced fluctuations in wildlife density that surpassed variation we have observed during our studies, from much less abundant to a little more abundant.

The rift valleys and volcanic peaks of East Africa offer much larger altitude differences than elsewhere on the continent. Because rainfall increases with altitude, flora follows rainfall, and fauna follows flora, wooded savanna habitats not unlike today's may have shifted up or downhill in accord with climate changes. The Hadza were probably not the only East African foragers who could respond to dry periods by moving to higher altitudes, or use more limited refuges where lowland springs were fed from the mountains.

# 3 History of the Hadza and the Eyasi basin

It was easy to believe, when you followed women and children digging tubers in the bush, that nothing has ever changed in Hadzaland. Digging sticks are undoubtedly an ancient tool, and tuber use may have been a key piece of our ecology since the time of *Homo erectus* or before (O'Connell *et al.*, 1999; Wrangham, 2009). It was easy to believe, sitting there in the bush watching their quiet industry, that the Hadza have been lost at the bottom of the rift valley for thousands of years. But the distinctive sound, 30,000 feet overhead, of the morning jet from Dar es Salaam to Mwanza jolts one's attention back to today, and the uncomfortable contrasts between the Hadza world and ours. Everything has changed. But here they were, day after day after day, sharpening their digging sticks, organizing the children, marching out to a tuber hillside, a berry patch, a favorite baobab tree.

One begins to look more carefully. Their digging sticks were sharpened with iron knives before we left camp (Photograph 3.1), and hardened on a fire that perhaps was lit with matches. The Hadza can make fire, and carefully keep glowing coals when needed but the matches were likely gifts from us, a small but popular present. Iron knives have been trade items for a long time, and iron arrow heads have been in use since well before Reche (1914) described Obst's material culture collection made 100 years ago. Iron has been used in East Africa since around 1000 BC, and in the Eyasi basin since AD 200 (Mabulla, 2007). There is a complicated truth somewhere between my tuber digging reverie and the morning jet to Mwanza.

Some of the "new" factors, like the jet to Mwanza, seem to be simply ignored. Someone did once ask Lars Smith who the people were that went in these "birds." White people and government, a good summary in the mid-1980s, was accepted and shrugged off. The airplane is irrelevant to daily life; but the "doctor who comes in a rocket," as it was described to me, was not. The short-lived attempt at a flying doctor service for the Hadza at Mangola could have saved some lives, could have persuaded a few people to move nearer to its landing strip. Outcomes might have shown up in our record of mortality and mobility. Few interventions, however, have lasted long in Hadzaland.

Neither the view of the Hadza as a "pristine," unspoiled, isolated society, nor the view that "if not pristine then spoiled," nor the view that surviving hunter-gatherers are a recent product of inequality and exploitation, will do justice to the situation. There is literature claiming that researchers are blind to, or even conceal, important factors in the contemporary circumstances of their study populations. An example, the "Kalahari Debate," is well known to all who study hunters and gatherers (Schrire, 1980; Wilmsen, 1989; Lee and Guenther, 1993). A more

**Photograph 3.1** A woman sharpening her digging stick with an iron knife. Archaeologists report iron from the Eyasi basin since AD 200. The tin can was eagerly salvaged from the observers. © James F. O'Connell, 2015. Reproduced with permission.

level-headed early approach to the impact of history is illustrated by Leacock (1982, and elsewhere). I intend to try to take my blinders off, especially with regard to two questions. What potential costs and benefits were Hadza exposed to while hunting and gathering in the twentieth century that they would not have met before? What has happened around the Hadza and been done to them that could have affected their population?

The Eyasi basin, where the Hadza live, is in one sense very wild and remote, and in another sense not far from anywhere. This complicates thinking and writing about history and the Hadza. Multidisciplinary research by historians, archaeologists, linguists, and geneticists has taught us quite a lot about the long and eventful history of Tanzania. Only a little of it has concerned the Eyasi basin directly, perhaps a sign of the basin's remoteness and "unimportance." Nonetheless, I will argue that the history shows us events that reached the Hadza, from the ivory and slave trades, from the Maasai expansion, and from the rapid recent growth of the Tanzanian population. We can see from the reports of a series of interested visitors to the Hadza, beginning with Obst in 1911, that most Hadza have acquired much of their food by hunting and gathering throughout the twentieth century, but they lived in a world surrounded by farmers. At first the neighboring farmers were some distance away in the higher rainfall country, then as their numbers increased and their farming land developed, they surrounded the Hadza more closely, encapsulating them (Woodburn, 1988). Finally, they have invaded traditional Hadzalands, although much more intensively around Mangola than elsewhere. Before I discuss these events in more detail in later sections, I will summarize other kinds of evidence that suggest an extraordinary antiquity to Hadza genetic and linguistic identity, and a near continuous ancient presence of hunters and gatherers in the Eyasi basin.

The topic of history in interior Tanzania is large, sprawling, and dangerously engrossing. Having read some of the source material, I can appreciate historians' difficulty in reaching consensus. One problem is the extreme local differentiation that seems to have marked the recent history of this nation of 200 languages. Lacking the historian's ability to craft a flowing and coherent narrative that covers all the required issues, I cannot do the topic justice. Instead, I placed in the SI several lengthy supplements: on the history of Tanzania (SI 3.1), on the chronology of events in and around Lake Eyasi, on the several externally imposed attempts to make the Hadza settle and farm, on material from previous accounts of the Hadza that indicate continued hunting and gathering and much constancy of behavior throughout the twentieth century, and on the neighboring societies and their histories.

Another difficulty encountered by historians has been the lack of information on climate change, remedied by the extensive recent lake deposit studies that are available to supplement the oral histories of droughts and famines. Multidisciplinary studies such as that by Westerberg *et al.* (2010) on Engaruka (at the foot of the Gregory rift wall north of Mto wa Mbu, just 100 km northeast over the volcano highlands from Mangola) and its interactions with climate and trade since the fourteenth to nineteenth centuries have set the pace. These offer the prospect of more accounts in which data overwhelm political predeliction. Warm wet and cold dry periods coincided with historical and archaeological events.

This chapter merely lists and summarizes some key points:

(i) The wealth of recently acquired knowledge about the ancient history of the Hadza and of the Eyasi basin. (ii) Evidence for Hadza hunting and gathering throughout the twentieth century; primary or secondary hunters and gatherers. (iii) Pots and pans: modern materials involved in Hadza daily life. (iv) Remoteness, contact, raids. Raids upon the Hadza reported by early informants may have reduced their numbers to a minumum around 1900. The cessation of these raids is associated with German colonial power. (v) The twentieth century progression of Hadza country from remote to encapsulated to invaded, in parallel with the late twentieth century increase in neighboring populations.

## 3.1 Prehistory: archaeology, linguistics, and genetics in the Eyasi basin

The written and oral history of the Hadza and the Eyasi basin is shallow but their genetic and linguistic history is remarkable, and the archaeology of the Eyasi basin where the Hadza live today is deep. Starting at 130,000 BP, it begins not very long after the origins of modern *Homo sapiens*. The first archaeological research in the Eyasi area was by Margit Kohl-Larsen (SI 3.2) at Mumba near Mangola in the 1930s. After a long gap, Mehlman conducted a major study at Mumba in the 1970s but this was difficult to access for some time (Mehlman, 1988). Mumba and other nearby sites were explored further by Dominguez-Rodrigo *et al.* (2007, 2008) and Audax Mabulla in the 2000s.

Mabulla (2007) has reviewed the Eyasi basin archaeological record. There is skeletal and archaeological evidence of *Homo sapiens* in the Eyasi basin at several time points from 130 kya through 91 kya, 45–60 kya, and 27–37 kya and on into the historical era. Thus, there is a record of hunting and gathering in the Eyasi basin on and off from 130 kya until today. The faunal remains associated with the earliest finds included species now extinct. The subsequent fauna resembled present-day fauna; "the range of fauna represented in the assemblages is consistent with what is consumed today by Hadzabe hunter-foragers" (Mabulla, 2007, p. 21). Finds indicating decorative arts include ochre, a palette with ochre color, inscribed ostrich eggshell fragments, and ostrich shell beads. There are also rock paintings of plants and some portray plants eaten by Hadza today. "The occurrence of a few obsidian pieces in Bed VIB dating to 130,000 BP suggest intensive curation of a high quality material through an exchange system between Naivasha (Kenya) and Eyasi (Tanzania) hunter foragers . . ." (Mabulla, 2007, p. 20).

Evidence of herders in the Eyasi basin begins around 2500 years ago (Mabulla, 2007, p. 23). Mabulla suggests close parallels between the archaeology of these herders and contemporary interactions between Hadza and herders and farmers. Iron first appears in the Eyasi basin around 1800 BP (about AD 200), and has been found associated with stone pipe fragments. Fosbrooke, (1956; Phillips, 1983) noted that stone pipes, like the ones Hadza still make for themselves, are quite widely found in Africa and pre-date the arrival of tobacco from the New World. Phillips (1983) suggests they were used for smoking hemp, which Burton (1859) found widely used in East Africa.

Linguistics and genetics, on their relevant time scales, suggest a very long history of Hadza identity. Linguists have come to regard Hadzane as an isolate. It was long classified as a Khoisan language, mainly because it contained click consonants. Bleek (1930, 1931a,b) claimed some other lines of evidence to link Hadzane to Khoisan languages but Sands (1998), based partly on a quantitative analysis, argued against such a link. The Hadza have been neighbored by Cushitic speakers since about 4000 BP, Nilotic speakers since about 2500 BP, and Bantu speakers for the last 1800 years. The survival of the Hadza language is both a remarkable feat of cultural survival, and some indication of their persisting independence from these other populations.

Geneticists have recently (SI 3.3) made the most striking contributions to our knowledge of history of the Hadza and the wider area (Tishkoff *et al.*, 2007). Their molecular genetics suggests that the Hadza are extremely distant not only from their Cushitic-, Nilotic-, and Bantu-speaking neighbors but also from San and Pygmy populations, and even from the nearby Sandawe (click-speakers, and foragers in the nineteenth century). Tishkoff and colleagues (Scheinfeldt *et al.*, 2010) summarize their studies of mitochondial DNA, Y-chromosome, and autosomal DNA as most likely indicating division between the Hadza and Sandawe around 20 kya, and between the Hadza and southern African Khoisan peoples around 56 kya. The possibilities include a low rate of intermarriage with neighboring Cushitic-, Nilotic-, and Bantu-speaking neighbors, and a very low rate between the Sandawe and Hadza in recent millennia, despite living only 150 km apart. It is also possible, according to Tishkoff *et al.*, that the Hadza and Sandawe diverged earlier but intermarried at a non-negligible rate for a while around 20 kya. In the last chapter, I discussed the opportunities for vertical migration in times of low rainfall. Perhaps during the extremes of drought when Lake Victoria dried out around 15–18 kya, Hadza and Sandawe met for a while in the mountain massif that separates them today.

Between 4000 years ago and AD 500, Cushitic, then Nilotic, then Bantu farmers and herders arrived in northern Tanzania. Bantu iron-using populations were settled around Lake Victoria between 500 BC and AD 200. Bantu populations spread east and south across Africa, reaching their southern extreme by about AD 500. We do not know when the current Bantu neighbors of Hadza (Sukuma, Iramba, Isanzu) arrived in their present locations. Iramba oral histories tell of migration from the region of Lake Victoria (Danielson, 1961). The Sukuma lived from west of Lake Eyasi north to the borders of Lake Victoria, and formed part of the Nyamwezi "empire" in the nineteenth century. We can suppose that iron reached Hadza by trade from the Lake region before these neighbors settled in their present locations. Herders were in place and their cattle tracks were heavily eroding the Ngorongoro highlands by AD 1300 (Fosbrooke, 1972, p. 151).

Thus, we know that people with some of the genetic distinctiveness of Hadza have lived in the broader region for tens of thousands of years. These people would have been hunters and gatherers whose neighbors were also hunter-gatherers until about 4000 years ago. We do not know what language they spoke (although it most likely contained click consonants, Knight *et al.*, 2003), nor what they called themselves. We do not know that they lived exactly where Hadza currently live although the archaeology shows that some hunter-gatherers lived there, and the genetics indicate that ancestors of today's Hadza most probably lived somewhere in this general area of East Africa. But it is probably not too imaginative to suspect the continuous presence of ancestors of Hadza hunting and gathering in and around the Eyasi basin for several tens of thousands of years. A counter-argument could be drawn from the rock paintings that can be seen in both eastern and western Hadza country. During my fieldwork, Hadza informants expressed ignorance about the artists but Mabulla (2003) reports that they claim these as made by their ancestors. Marlowe (2010, p. 180) reports that most Hadza assume the paintings were done by their ancestors but have no specific oral history about the paintings.

## 3.2 Continuity of Hadza hunting and gathering in the twentieth century

Although the researcher's job is to question everything, it is hard to doubt that the great majority of Hadza fed themselves almost entirely by hunting and gathering for almost all of the twentieth century. Since Obst (1912) and Dempwolff (1916), an increasing number of visitors have written accounts of Hadza (see SI 3.4 and Marlowe, 2010, table 2.1). These accounts allow a fairly detailed picture of events in Hadzaland during the twentieth century (SI 3.5). Hadza oral history covers much of the same period and I interviewed informants about some of the written history.

An important and surprising feature of the previous accounts is that they are independent reports of different field visits, some apparently made in ignorance of the accounts by previous visitors (for example, Bagshawe, 1924, 1925, and Cooper, 1949, both comment that they cannot read German). Thus, the reports are not an accumulation of repetitions of earlier reports, they are first-hand accounts of visits to Hadza; Obst and others include their own original photographs. All (sixteen accounts before our group) report Hadza as living by hunting and gathering, and occupying the parts of the Eyasi basin in which later researchers found them. None reports them as primarily farming or herding. There is just one hint in the other direction. Bleek (1931a, p. 274), on her visit to Hadza at the southern extremity of their range, commented that "The Hadzapi own no domestic animals and make no gardens." Some of the Hadza people she stayed with were married to Isanzus but she made no comment on their farming. However, she tellingly wrote "several families came from the east where they have no gardens." In Chapter 5, I will report the rates of intermarriage between the southernmost Hadza in Singida and the Isanzu.

Obst's descriptions of the area and Hadza are very familiar to modern eyes; little would appear to have changed except for the presence of modern clothing and absence of rhinoceros (SI 3.6). Marlowe (2002, 2010) has pointed out that, day to day, Hadza behavior, economy, and technology appear just as described by Obst in 1911. I agree (SI 3.7) and I think Marlowe is essentially correct in emphasizing the pressure of daily life. However, we need to think about some of the possible modern influences, even on daily life, if we are to report contemporary hunter-gatherers as representing or even merely illuminating anything about adaptation of past hunting and gathering economies in the variable environment of sub-Saharan Africa. Reaction to modern influences are themselves interesting and important data about behavior, adaptation, and demography, which as "natural experiments," can extend and test our models of human behavior.

Some earlier authors (Bagshawe, 1924; Cooper, 1949) discussed the claim of neighboring people's that Hadza were a floating population of runaways and tax evaders, perhaps changing their identity as they came and went. The genetic studies, and the genealogies gathered by Woodburn and reported by Stevens *et al.* (1977), as well as the genealogies developed by our group, show continuity of identity. They also allow us to see that Hadza, who we observed hunting and gathering in the final 20 years of the twentieth century, had parents who Woodburn described as hunting and gathering in the 1960s (Woodburn, 1968a, and PhD thesis), and grandparents who Kohl-Larsen and Bleek described as hunting and gathering in the 1930s. A few names from Bagshawe and Dempwolff suggest earlier links.

Some authors have made a distinction between "primary" and "secondary" hunters and gatherers, those who have "always" hunted and gathered with little to no farming, and those who were once farmers or herders but lost their fields or livestock and took to living by hunting and gathering. East African "Dorobo" have often been described as examples of secondary foragers (Chang, 1982). Wilmsen (1989) implied that we should consider !Kung and other San groups as peoples who had been deprived of their livestock by more powerful outsiders (despite minimal actual contact). The Eyasi basin has seen two recent examples of secondary hunters and gatherers: (1) Bagshawe (1925, p. 126) records, "During the 1918-1920 famine hundreds of natives belonging to other tribes took refuge in their country and lived on game which the Kangeju [Hadza] assisted them to kill" and on his page 127, "There are amongst the tribe a few foreign girls, mostly Anisanzu, who married Kangeju [Hadza] men during the famine." (2) Bagshawe (1924, part 1, p 31) says, "the Isimajig and Kisamjeng clans of the Tatoga, all of whom had been reduced to poverty by war, and were semi-bushmen" and (1925, part II, p. 117), Baghawe notes "Many years ago a large number of Kisamjeng Tatoga, under Saigile, took refuge from the Maasai on the borders of Eyasi." Recent informants, both Datoga and Hadza, have told us the same history. Kisamjeng rebuilt their herds and are now a widely recognized and successful Datoga subgroup. Obst (1912), surprised by his informants' reports of Maasai raids on people with no livestock, tried to contradict the Hadza view that they had always hunted and gathered. He suggested that the Hadza had their own names for livestock. However, Dempwolff (1916), familiar with more of the neighboring languages, showed the Hadza names for livestock to be clearly loan words.

Throughout the twentieth century, Hadza have traded with their neighbors, and while initially this was described as trade of meat and skins for iron, beads, and pots, during our observations the Hadza sometimes received a small amount of maize meal or sweet potatoes as part of their trade. Even late in my study, domesticates formed a very small part of the Hadza diet. In 1985-1986, Hawkes and O'Connell weighed food arriving at Hadza camps in Tliika; just 5% was from domestic plant foods. In 1995–1996, Marlowe (1998) did the same but in five camps all over eastern Hadza country; 6.6% of calories arriving in camp were from farm food. Marlowe (2010, p. 36) reports that in 14 camps up to 2005, 5.7% of food by weight was non-foraged food.

The persistence of hunting and gathering has been discussed by Woodburn (1988), Blurton Jones *et al.* (1996, 2015), and Marlowe (2002). Each of us notes the tsetse fly as an impediment to herding and herders, and the low rainfall as an obstacle for farming. Woodburn, after careful discussion of the small number of "delayed return" forager societies and the nature of relations between foragers and their neighbors, excludes the idea that encapsulation by powerful neighbors drives people back into an "immediate return" foraging economy. Blurton Jones (1996) discussed the idea that the richer environment of the Hadza favored the bush over farming in a poor agricultural environment, and suggested the difference in the availability of bush resources might account for the dissimilarity between !Kung and Hadza interactions with neighbors. In Blurton Jones (2015), I suggest that the Hadza had long had the opportunity to farm by irrigation near Mangola but have rejected farming, and rejected the settlement schemes because they conflict with the Hadza ethic of sharing and the mobility which may maintain it (also discussed in Chapter 22).

## 3.3 Pots and pans: modern materials and influences in Hadza daily life

The Hadza observed by anthropologists live in a world of hunters surrounded by farmers and herders. They have had small numbers of "Dorobo" neighbors in recent times

(Farler, 1882; Baumann, 1894b; Jaeger, 1911; Blackburn, 1982). It is not since about 4000 years ago when ancestors of the Cushitic-speaking peoples arrived in northern Tanzania that Hadza have lived in a world entirely of hunters among hunters. What might be some of the differences? Even in the most remote Hadza camps, there are obvious material features of life that cannot have great antiquity. We do not know how much difference they make but presumably people found it worthwhile to adopt them.

**Iron** For knives, axes, arrow heads, and the cold chisels used to make arrow heads, iron is probably the most important. Kusimba (2004) suggests that iron may have made it easier to kill large prey and accounted for an early change in the bone assemblages in the fortified rock shelters in Tsavo used between around 800 and 100 years ago. Speed and efficiency of butchering may also be affected. Stone projectile points may have needed more frequent repair or replacement and the manufacturing skills may have taken longer to acquire. Arrow poison may be an equally important factor, which Robbins *et al.* (2012) argue may be quite ancient.

**Cooking pots** During our study, Hadza used aluminum cooking pots ("soufrire"), often very old and battered, even patched (Photograph 3.2). During Obst's time, they used clay pots traded from the Isanzu. I have seen Hadza make dolls from clay but they did not fire them, nor did anyone shape a bowl at that time (plenty of soufrire were in camp). I know no indications about how the Hadza cooked before they had access to pots, other than cooking meat and tubers directly on the fire, which they still do routinely. Wrangham (2009, pp. 121–126) summarizes ethnographic data on cooking methods without pots.

**Photograph 3.2** A woman grinding baobab pith and seeds with a rock. Brush and skin winnowing pad are made by the woman herself from local materials. The highly valued large aluminum bowl behind her is obtained with difficulty by trade from neighbors or visitors. © James F. O'Connell, 2015. Reproduced with permission.

**Shoes** At the beginning of our study, a few people could be seen wearing leather sandals of their own manufacture, exactly as illustrated in Reche (1914). Some, especially children, went barefoot, and had thickly calloused soles of their feet to prove it. By the end of our study, most people wore "tyri," sandals made from disused motor tires, either presents from us, or traded at the traveling rural markets that became established after the economic reforms of 1986. The ground is rocky more often than sandy, and thorns are an abundant hazard. One 12-year-old boy in the camp where I stayed in 1986 spent several weeks confined to camp by a thorn in his foot that neither he, nor I, nor expert old ladies could dislodge. He recovered and became an active young man by the time my field visits ended.

**Clothing** Modern clothing presents the most obvious difference from early accounts. Early contacts and neighbors all seem to describe the Hadza as naked. Woodburn (1970) illustrates "traditional" men's clothing. Bleek describes women as wearing a three piece outfit identical to those she was accustomed to seeing among San in the Kalahari and which I also saw Hadza wearing in the 1980s, "a skin apron, round for a matron, in tassels for a girl, a skin back apron likewise hanging from the belt, and a kaross or skin cloak hanging from the shoulders and tied around the waist when a baby is carried in it. Infants wear no clothes, little boys only a belt and some ornament, little girls a tiny fringe apron to which a small kaross is added later" (Photograph 3.3). Modern woven clothing may make some parasites worse (see Kittler *et al.*, 2003 on clothing and the evolution of lice), or it may partially protect from

**Photograph 3.3** Ever cheerful and lively, Hadza children display a full range of clothing from the 1980s. Boys were commonly naked but were eager to eventually acquire "school shorts" (short pants). Three girls show a variety of worn-out western clothing or material. The girl third from the left is wearing a "traditional" girl's dress, the same as described by Bleek among the Hadza and among San peoples in the 1930s. © James F. O'Connell, 2015. Reproduced with permission.

biting insects and minor wounds that can become infected. The eagerness with which the Hadza seek modern clothing may initially have been promoted by the attitudes of their neighbors and their brief exposures to missionaries. Having clothes would make a significant difference in trading visits to neighbors outside Hadza country.

**Popular small presents** The Hadza readily accepted presents of matches, needles, cotton, and safety pins. They still make and use fire, transport hot coals, make fine thread from sinew, and rough string from *Sansevieria*. Beads are a more significant present but limited (because they were sometimes hard to buy, apparently because, for a while, they were regarded as "politically incorrect" to import) and have been traded as long as we have written records. They seem to have replaced ostrich egg shell necklaces. Some older women have extensive bands of beads hidden around their waists. The Hadza make and wear decorative headbands, and can trade beads with the Datoga who need beads for some of their ceremonial occasions. Around 1995, a Hadza woman invented a way to make beads out of discarded plastic containers by cutting them up and melting them on a wire spit over a fire. The art spread, probably because there were tourists coming who would buy the beads, and villagers who discarded enough plastic to give an easy supply of raw material.

**Modern medicine** Access to modern medicine was discussed in Blurton Jones *et al.* (2002), where we concluded that the impact was minimal during our study period. We have no information on the effectiveness of the limited traditional medicine used by the Hadza. If effective medical facilities have been provided, we may expect a noticeable reduction in mortality, at least among adults, depending on accessibility and efficiency. Orubuloye and Caldwell (1975) and others have shown how a hospital or modern health facility can reduce mortality, but that the effect is sometimes extremely localized (see also Armstrong Schellenberg *et al.*, 2008).

**Modern diseases** In Chapter 2, I discussed the incidence of diseases. The Hadza appear to have been exposed to modern crowd diseases for much of the twentieth century. In Chapter 8, I discuss causes of death; the impact of the modern diseases may only rarely be significant.

**Habitat destruction** Immigrants clear fields and trees. Wild plant resources are still abundant but there has been extensive clearing of *Salvadora* and *Cordia* bushes in the Mangola area. Gallery forest has been cleared for charcoal from the Balai valley, and farmers are cutting their way further down the Udahaya river. The processes outlined by McDowell (1981a, SI 3.8) and Tomita (1966) continue. Farmers clear fields, cut more trees for building and firewood, and charcoal sales. In some years, they get a harvest and others are encouraged to come into the area. In bad years, they survive with the charity of relatives in the rainy highlands. The country has no chance to recover. Hadzaland is in the grip of an environmental "ratchet" that tightens in rainy years, and does not loosen in dry years when recovery of natural vegetation is slow. The contrast with the excellent agricultural climate of the highlands (>800 mm rain/yr) probably accounted (along with the tsetse fly) for the lack of agricultural settlements in the area at the time of the earlier travelers. Goats are probably another impediment to regeneration of all significant plant species, and most immigrants, wisely for their short-term interests, wish to bring a few goats with them.

**Water hole modification** The Hadza in Tliika complained that the Datoga modify the water holes in ways that make it impossible for wild animals to drink. This was remedied sometime after 2000 (Marlowe personal communication).

**More trading partners** As neighbor populations increased and moved nearer and into Hadza country, opportunities for trade must have increased.

**Game laws** The Hadza hunt only with traditional weapons, a large bow, and a variety of arrows described in detail by Woodburn (1970) and Reche (1914). String snares are sometimes used by boys. The local game officers, and from time to time wider authorities, have recognized Hadza hunting as a special case, and not enforced the game laws for them. The difficulty and expense of traveling from the office into Hadza country also made enforcement impractical. President Nyerere granted the Hadza a "hunting license in perpetuity" but its recognition by local government has varied from time to time.

**Meat trade** We might expect relatively little meat trade between hunter-gatherer populations. Trade with farmers is more likely and widely observed on a very small scale. Meat, honey, bees wax, and skins are about the only things that the Hadza have to trade. With the recent upsurge in the "bushmeat" trade all over Africa, we might have expected more involvement by the Hadza. In my study period, there was very little (at an exchange rate very unfavorable to the Hadza), and in their recent paper on hunting between 2005 and 2009, Wood and Marlowe (2013) report only one instance of Hadza men attempting to dry meat in preparation for trade.

**Change in abundance of wild animals** I discussed this extensively in the last chapter. There does seem to have been a decrease in large game in eastern Hadza country between 1977 and the 2000s, probably mainly due to poaching by outsiders. However, the 1977 figures were as expected for the rainfall. Because rainfall varies across decades, centuries, and millennia, Hadza have experienced times of greater abundance and lesser abundance of game than during our studies. We might wonder whether, before Hadzaland was surrounded and infiltrated by farmers, the wild animal populations were better able to withstand variation in the pressure of Hadza hunting.

**Hypergamy** It has been widely observed that hunter-gatherer women tend to marry wealthy neighbors, while men can rarely take wives from the farmers or herders. The rates of intermarriage are examined in Chapter 5. As reported by Bagshawe (1924, 1925), Hadzaland saw a reversal of the usual pattern when, around 1917, the Isanzu came to the bush to avoid famine, and several Isanzu girls married Hadza men. Their descendants are still present in the Hadza population, as are descendants of Isanzu men who married Hadza women. The genetic data show that intermarriage of the Hadza and others was rare. Data from Australia and North America would be the only observational source of information on rates of intermarriage in a world of hunter-gatherers among hunter-gatherers. Wilkins and Marlowe (2006) have pointed out that the tendency for genetic data to show a predominance of philopatry among intermarrying couples may be restricted to post-agricultural circumstances.

**Room to expand** In the twentieth century, the Hadza had no room in which to expand if their population increased. The spatial variation in rainfall and habitat discussed in the last chapter may have set limits in antiquity both to expansion and to the likelihood and location of contact and competition with neighboring foragers.

**Political involvement and relative power** During our study period, the Hadza changed from being aware that the government was powerful, if capricious, to having a developing sense of its workings and the potential role of citizens (all Tanzanians are recognized as citizens) in its government. There was a small beginning in the early 1990s when, after the dispersal of the Mongo wa mono settlement, the Canadian Universities Service Overseas (CUSO) representatives energetically helped the Hadza gain official status for a large part of their land as an administrative "village." Sadly, the Hadza later became a minority in their own village. Only in the twenty-first century did they begin to gain some fragile value from these opportunities. Throughout our study, it was clear that Hadza were deferential and easily intimidated when encountering any government official even at the lowest levels. It was also evident that the Hadza believed these officers had little concern with Hadza daily life.

## 3.4 Remote but surrounded and raided: the Eyasi basin 1890–1940

The Hadza were first reported to the western world by Baumann (1894a) describing his 1892–1893 journey. There are fragments of information that may pre-date his visit. Oral history collected among several Iramba men by Daudi Kidamala (Danielson, 1961) mentions "The Wanisanzu moved to Myadu, Mwagala and to the land of the Wakindiga [Hadza] because they were defeated by the Anankali in warfare." No date is possible. Later we find "Long ago there was a Mmnyaturu, who became the slave of a Munisanzu. Every year they journeyed together to the Wakindiga. One time he discovered an elephant, which had died. He removed the tusks . . . used the ivory to trade for iron hoes from the Wanyamwezi. After . . . [he] . . . had grown up he was given his freedom and he returned home. After many years the Germans came to Kilimatinde." "Many years" plus how ever long it took him to become "grown up" may take us back well beyond Baumann hearing about the Hadza in 1892.

The Hadza (often known as "Tindiga") occupied an area that, even 30 or 40 years later, was described as remote and "uninhabited" (Barns, 1923; Johnson, 1923; Gilman, 1936). Although by the 1890s, Hadza and their neighbors were sufficiently acquainted for them to have had names to identify each other's ethnicity (table SI 3.9).

Nothing was written about the Eyasi area until Baumann's visit in 1892. However, Erhardt and Rebmann's 1855 "Slug Map" (Krapf, 1860), built from accounts by coastal and indigenous inland traders, appears to show two unlabeled lakes south of Oldonyo Lengai that, judged by their orientation, could represent lakes Manyara and Eyasi. Neither Krapf (1860) nor Wakefield (1870) report any comment by their informants about Lake Eyasi or the Hadza. Apparently, the major trade and tumult bypassed Lake Eyasi until the German exploration. However, this does not guarantee that the Eyasi basin and the Hadza were totally unaffected; local trade can far outreach the well-known trade routes.

The "remoteness" of the Eyasi basin can be illustrated by the reports of the early explorers. Coming from the north in 1892, Baumann (1894a) descended with difficulty, southward from the highlands near the peak of Oldeani mountain, to visit Lolpiro at the very north of Lake Eyasi in eastern Hadza country. Baumann also traveled through western Hadza country with a group of Sukuma elephant hunters in June and July 1892 but met no Hadza. He commented that the Hadza were so shy that even the elephant hunters seldom met them. On his map he marked "Tindiga" (Hadza) over a large expanse of recent western Hadza country.

Several German groups visited the Eyasi area between 1892 and Obst's 1911 visit. Baumann (1894a) was the first to report the existence of the Hadza but he, like Werther (1894, 1898), Jaeger (1905 visit, 1911), and Götzen (1895), each of whom traveled through Hadza country, met no Hadza. All commented on them. They heard about them from their guides, saw empty camps, and were shown the pegs that Hadza use to climb baobab trees. The Hadza were probably close by but hiding. These "close encounters" were scattered through the general area that we have referred to as "traditional Hadza country." These accounts show us that the guides from neighboring tribes knew about the Hadza. Obst was much more successful than these early visitors, spending eight weeks living with Hadza in their camps in 1911. Obst's informants were the first to show that relationships with neighbors had not all been peaceful. Raids by the Maasai, Isanzu, and Sukuma were described, including losses by death and by "taking away women and children." I discuss these more fully later.

The country surrounding the Hadza seemed quite sparsely populated. Obst, coming from the south, walked for four days to cover the 80 km from Mkalama to Baragu, a small hill just south of Yaeda. He describes passing through Isanzu villages during the first day, houses becoming sporadic and finally ceasing during the second day. During the third and fourth days he passed through apparently empty country "a real resort for lions and leopards that every night circled around the camp and once even attacked our lead herd of cattle in daylight" (Obst translation, p 2, SI 3.6). In the west, salt caravans visited Lake Eyasi from Sukumaland (Senior, 1938). As they passed through western Hadza country the "last 40 miles of the road was common to both caravans [routes] and led through uninhabited bush" Kjekshus (1977). Johnson (1923), discussing Kiniramba and the area where it is spoken writes, "The north is practically uninhabited for a considerable stretch excepting for a small bush tribe, the Kindiga [Hadza], who wander about the country round Lake Eyasi and in the Yaida valley." Barns (1923) marched from Lolpiro to Mbulu, stopping at the Matete (as the Balai river is often labeled) and makes no mention of the Hadza or any other inhabitants of the basin.

These early explorers do not state directly whether anyone other than the Hadza occupied the area. They describe it as "uninhabited," like much of the surrounding country. Neighboring populations, including the Iraqw, were quite localized, according to the early maps and published travel diaries. Nonetheless, we should wonder whether some of this "uninhabited" country (though heavily infested with tsetse flies) might have been used at least as seasonal grazing, and wonder to what extent the writers simply overlooked small settlements, isolated homesteads, and Datoga herders. I would expect their guides to have sought, rather than avoided, any such habitations, which would be linked by existing footpaths. Even if the surrounding country mapped as uninhabited really was truly uninhabited, we should note that several of the centers of population were in locations that became the administrative and commercial centers of today. Today, few Hadza camps are more than two full days' walk from one of these, with water available at several points en route. Thus, a hundred years ago, Hadza camps can have been no more than two to three days' walk from some center of population (Mbulu, Karatu, Mkalama, Dongobesh). Even if the intervening country filled up in the meantime, when first described, no Hadza lived much more than two days' walk from farmers, where iron, cotton, pots, beads, etc. were available. Baumann, Dempwolff, and Obst indicate that Hadza traded honey and skins for iron, beads, and metal decorations. I have seen no description of other than iron and wood arrow heads among the Hadza.

## Raids

Hadza have pointed out to many visitors a large rock beside which a Hadza family was once killed by the Maasai. Its location supports the claims that the Maasai attacked the Hadza from the northeast and Fosbrooke's claim that the Maasai inhabited the highlands between Karatu and Mto wa Mbu. Obst (1912, pp. 17–18 (translation p. 10) writes, "I attempted ... an investigation of the Wakindiga's history.... They told me that as far as they could think back they were living together with the Wahi in the area between the Mangola river and the Mumba mountains and continuously had to lead bloody feuds with their neighbors, especially with the Waissansu and the Wamburu. Their power was already weakened when the Maasai began intruding from the north-east." In some places when Obst writes "Wahi" he appears to mean the western Hadza (Wahi is the Sukuma name for Hadza), but here he may mean the Datoga, and his informant appears to be referring only to the Mangola area.

After discussing his surprise that the Maasai attacked people with no cattle, Obst continued (pp. 18, translation p. 10),

> Nevertheless, the Wakindiga still know about desperate fights with the Maasai. They attempted under their [the Hadza's] leaders Boiyoge and Wassaraguaiu to get rid of the troublesome intruders but finally had to retreat from the power of the Maasai. The poor remnants of the Wahi who were crushed by the fight escaped into the grass and bush savanna west of Lake Eyasi. Some of them [one part of the Wahi] united themselves with the Wakindiga who chose the rocky heights between Yaeda and Eyasi valleys as a refuge and live now in hordes of one to three families a poor existence as hunter and gatherers.

Here Obst seems to be writing about the Datoga, the Kisamjeng, who Bagshawe's informants and contemporary Hadza and Datoga informants describe as having lost their cattle and taken to a forager economy for a while.

> The Wakindiga still could not find peace and quiet. As long as there were herds of Elephants in the Eyasi and Yaeda valleys foreigners were coming and going constantly. The Wasukuma at least could come only in small troops, because of the longer distance from their home. They negotiated with the Wakindiga about a permit to hunt and to retain the meat in exchange

> for the delivery of iron hoes, knives and beads. Meanwhile, the closely located Waisansu used to arrive always ready for a war, raided the Wakindiga continuously, and took away any women and children they could find. Only after the elephants became more rare did the fights with the Waisansu stop. A quieter trade led to an exchange of the richnesses of nature and culture among the two peoples and initiated the Waisansu's peaceful invasion which we find at present everywhere.

Bagshawe reports the later peaceful Isanzu presence as a flight to the bush to avoid epidemics and famine. Hoes appear to have been a common way to trade iron.

The Hadza did not mention the slave trade, how could they know where the stolen women and children had gone? The export of ivory had its peak from 1840 to 1890, leading to a collapse in the trade and widespread scarcity of elephants, according to Spinage (1973). Not all contacts with the Isanzu had been unfriendly as indicated by the oral history quoted here. Relations between tribes may have included periods of friendly trade and occasional intermarriage interspersed with brief small wars that flared up over some incident. According to historians, much warfare in nineteenth century Tanzania was over control of the slave trade and foreign goods (by processes discussed by Ferguson and Whitehead, 1992, among others, and locally by Tippu Tib in Brode, 1907). We cannot extrapolate from the nineteenth century situation to the previous centuries.

Among later visitors, Bagshawe (1924, 1925) and Bleek (1931a,b, and notebooks) were told only of raids by the Maasai. Kohl-Larsen, however, was given a lengthy account, which included wars with the Datoga before the Maasai raids (these are the incidents listed by Fry and Soderberg, 2013). His informant mentioned the Isanzu only in the context of his having moved nearer to the Isanzu for safety.

Kohl-Larsen's (1958, based on his fieldwork of 1934–1936 and 1937–1939, translation p. 46) informant Schungwitsa told him, "In ancient times, when our grandfathers and grandmothers lived, they had enemies. The first enemies were the Mangati [Datoga]. It was during the night that the Mangati came. They surrounded us when we were celebrating the Epeme feast. The Mangati waited until we lay down to sleep. Then they broke out of the bush. Many of us were killed by their spears." A truce was made, and: "Us Hadzapi who were born last, have never heard again that a war or quarrel happened between the Mangati and us Hadzapi." Kohl-Larsen describes Schungwitsa as being in his 30s when interviewed, but he had several children (four of whom were still alive during our study, with the average year of birth being in 1933). The mean age at childbirth for a Hadza man is 35. This would make Schungwitsa 40 years old in 1938, and born in 1898. Let us also assume this average for his father and grandfather. Schungwitsa was born in 1898, his father in 1863, his grandfather in 1828. This is suggestively close to the time the Datoga (commonly called Mangati in northern Tanzania) were expelled from the crater by the Maasai (Borgerhoff Mulder *et al.*, 1989). It seems that around this time the Datoga passed through the Eyasi area before settling further south and in the highlands to the east (Danielson, 1961; Tomikawa, 1970). They may not have returned to the area until displaced from the highlands, beginning in the 1940s (Tomikawa, 1970).

Kohl-Larsen's informant Schungwitsa (from his children we came to know him as Sigwadzi), after describing the wars with the Maasai, ended his account thus, "–When we lived there the Germans arrived. When the Germans arrived, it became quiet in the land, and we returned to our old camps. We [had] lived there [near the Isanzu] a whole year. The fighting with the Maasai from this time on had died. We were very many people. We lived from the Usukuma mountains to the slopes of Oldeani." Hadza still do.

The numbers of deaths reported by the informants were very small but the fear of these intruders may have limited foraging and hunting. The loss of women and children would have been more demographically important. Slavers elsewhere captured women as well as men, and the slave trade continued internally until about 1910, mainly supplying plantations (cloves, sugar cane, sisal) on Zanzibar and the nearby coast.

Baumann's (1894b) map of his 1892 travels marks the Maasai as living near Karatu and in the Mbulu highlands (where not marked as "Urwald" – original forest), and mentions them living in Ngorongoro (along with Dorobo), and in Serengeti (where Dorobo were said to inhabit the western more wooded reaches). The Iramba oral history mentions Maasai raiding as far south as Kinyangiri (40 km due south of Mkalama), "But the sixth time the Maasai were defeated, as the Kinyangiri people used guns, provided by the Arabs" (probably at a time before the Germans arrived at Mkalama). Another noted Maasai raider was "Gussu and his Maasai lived near Lake Eyasi north of Mkalama. He conquered every time but finally discontinued the plunder of his own accord" (Danielson. 1961, pp. 74–75).

Maasai expansion was stopped by the Germans and the rinderpest epidemic of 1890, which was the most sudden and dramatic affliction of the Maasai (Homewood *et al.*, 2009). More than half their cattle died, as did the cattle of the Sukuma and other peoples (although the Iraqw seem to have escaped its devastation [Borjeson, 2004]). Baumann (1894a) describes the desperate plight of the Maasai in Ngorongoro in 1892. Sinclair (1979, pp. 3–6) summarizes the chain of devastation and points out that for a while cattle raiding increased as the Maasai and others attempted to rebuild their herds. This makes it difficult to date the periods during which the Maasai raided the Hadza.

The German administration in Mbulu was established in 1906. Its greatest effect would have been limiting raids upon the Iraqw. The boma at Mkalama would have had a similar benefit for the Isanzu and Iramba, but we can determine little about the cessation of Isanzu raids upon the Hadza, which may have been much more demographically effective. The raids are said to have declined with the population of elephants.

It is clear that by the beginning of the nineteenth century, the Hadza were surrounded by neighbors, people with different mother tongues, who knew themselves by other names, who had different economies and different remote histories (outlined in SI 3.10). The country of these neighbors is now completely unsuitable for a hunting and gathering economy. While this economy remains possible in much of eastern Hadza country, the Hadza have long lacked room to expand.

Farmer neighbors in the Eyasi basin were not unique to the nineteenth and twentieth centuries. Sutton (1986, p. 43, 48, footnote on p. 34; 1989; 1993, p. 54) described rock irrigation channels at Lolpiro and Endamagha, at the foot of Oldeani mountain, similar to those at Engaruka. He describes these as roughly contemporary with Engaruka and covering the period approximately AD 1500–1700 (see also Westerberg *et al.*, 2010). There is no evidence that Hadza farmed there. Their repeated rejection of farming in the twentieth century makes it seem unlikely. The Hadza's readiness to beg and trade makes it easy to believe that they found the irrigation farmers useful partners. However, as during much of the twentieth century, their involvement with these economies might have been quite limited. Exchange of meat, skins, and honey for farm food, pots, beads, and iron seems likely, as in recent times. Mabulla (2007) points out that the presence of this economy in the best-watered area does not imply its presence in the large, less well-watered remainder of Hadza country.

## 3.5 Encapsulated and invaded: the Eyasi basin 1940–2000

To understand the previous emptiness of the Eyasi basin and the modern encroachment on Hadzaland, we need to keep in mind two observations about Tanzania: the extreme heterogeneity of population density, and the recency of national population increase. Those of us who drive around in rural Tanzania are accustomed to finding that wherever we stop, however remote and desolate it appears, someone will have arrived at your side within minutes. He probably will really badly need a ride to just where you are going. Such hitchhikers always proved interesting and helpful. Only when one of them turned out to be the lookout for his entire village soccer team, reserves, and officials, did I have to refuse. Hence, it can be hard to believe Iliffe (1979, p. 8) when he describes Tanzania as "empty" around 1800. Others ask whether it was empty, or "emptied" by the slave trade and its consequences for farm labor, livestock theft, disease, and warfare (Kjekshus, 1977).

At all times during the twentieth century, the Tanzanian population has been very unevenly distributed. Well-watered areas, like the lower slopes of Kilimanjaro or the surroundings of Lake Victoria, have had high population density, the extensive dry regions have had very low population density. This shows in the German lexicon map, in Schnee (1920) [www.ub.bildarchiv-dkg.uni-frankfurt.de/Lexicon-Texte/karten], and in Gillman's (1936) map, and density during the period 1948–1967 was discussed by Hirst (1972). The modern national censuses continue to show the same pattern with, for example, the Kilimanjaro region densely and Arusha region thinly populated. Human dependence on water evidently has an enduring effect. Even with waxing and waning of well-digging technology (Fosbrooke 1950; Kashaigili, 2010, section 6.2), the pattern of population density is unlikely to have been very different in previous centuries. Population was clustered around the best-watered localities. It still is today. The Eyasi basin has been an area of very low rainfall.

While density remained uneven, in the second half of the twentieth century at around 3% per annum, the Tanzanian population grew as fast as any in the world

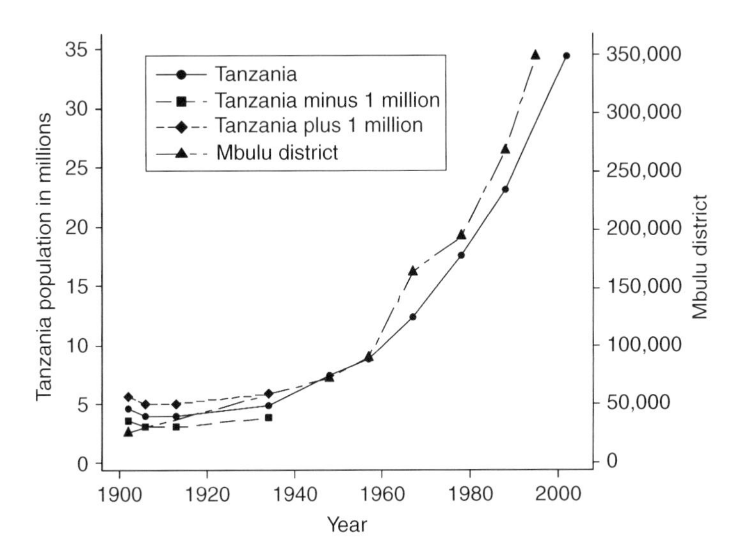


**Figure 3.1** Population of Mbulu district and of Tanzania, 1900–2002. The early estimates of the national population were based on methods that modern demographers regard as unsatisfactory. The broken lines between 1900 and 1935 show that even missing or overcounting a million people would make little difference to the overall picture. Mbulu district from Meindertsma and Kessler, 1997; Mbulu district, 1900, Walraqw, 1900, from Winter and Molyneaux, 1963. Districts have been divided from time to time. To the best of my knowledge, these numbers are for comparable boundaries.

(Figure 3.1). In 2002, the national population was 34.4 million, nearly five times its 1948 size (7.4 million). Earlier censuses were conducted by less than ideal methods but suggest a population of only about five million around 1910. The 1910 population took about 50 years to double in size. The doubling time at the end of the century was about 20 years. Many possible reasons have been debated. A clear increase in child survivorship that began in the 1940s has been persuasively linked to the arrival of effective antibacterial medicines such as sulphapyridine, M&B693 (see Walters, 2008, for a careful discussion and references).

The Hadza have been surrounded by non-forager neighbors for a long time. As discussed previously, at the turn of the nineteenth century, neighbors, while almost encircling the Hadza, were mostly quite far away, the Iraqw concentrated around their homeland at Iraq Da'aw, and the Isanzu some two days' walk or more (to judge from Obst's account). During the 1920s and 1930s, the Iraqw, encouraged by colonial powers, spread into good farming land to their north that had been vacated by the Maasai (Fosbrooke, 1972, pp. 22-24). In the 1940s, the Datoga began to move into the Mangola area and lakeshore plain from the Mbulu highlands, apparently under pressure of expanding farmland (Kohl-Larsen, 1958; Tomikawa, 1970).

During the twentieth century, the neighboring populations increased rapidly (Table 10.1, and SI 3.11), some localities faster than the national population, and non-Hadza have moved into parts of Hadzaland in increasing numbers late in the century. With fewer tsetse flies and better cattle medicine, it has become feasible country for herders. By 2000, the Datoga herded and settled almost everywhere in the Eyasi basin. Most of Hadza country is completely unsuitable for farming. Nonetheless, immigrants, mostly Isanzu and Iraqw, farm successfully where irrigation is possible, and perhaps because of the greatly increased density of people even in the moderately arable areas, seem prepared to clear a field even where there is little prospect of success. While a Hadza in worsening conditions might have to work longer hours, a farmer who cannot inherit enough land from his family can only go in search of new land, which he must rapidly clear. These fundamentally different economies may give rise to very different histories. Some of their modifications of the environment have been discussed in Chapter 2. While degrading the environment, these new neighbors offer a few other opportunities. Some Hadza who lived near farmers, especially at higher altitudes, spent parts of the growing season guarding maize fields, and eating the thieves (impala, warthog, and baboons). Others, between Mikocheni and Endamagha, have been paid in kind for digging sweet potatoes. Datoga often ask Hadza to supply honey.

In 1956, (Kohl-Larsen, 1958) people, both Iraqw and Datoga, were encouraged to move down to Mangola, and 1972 aerial photographs show several quite large fields very clearly. A German-owned farm was established by about this time. The population of Mangola ward increased dramatically between 1978 and 2002. The population of Mwangeza, to the south, in Isanzu country increased even faster (Table 10.1). The neighbors spread and moved closer, and the Hadza could truly be described as "encapsulated" by the mid-twentieth century. The expansions of the neighbors continued, especially in response to the major attempts to settle the Hadza in the mid-1960s. Non-Hadza quickly arrived to share the facilities. Eventually the Hadza left.

Earlier in the twentieth century, the pressure on Hadza country from their neighbors was clearly far lower than in the final 20 years or so. While it does not justify Iliffe's (1979, p. 8) description of central Tanzania in 1800 as "scarcely inhabited," the rate of increase suggests that the country was much "emptier" in the early decades of the twentieth century and before.

## 3.6 Repeated unsuccessful settlement attempts

An important aspect of twentieth century history has been a series of attempts to persuade the Hadza to settle and farm in order that schooling and medical services be made available (the mobile clinics that could be seen visiting the !Kung at Dobe in Botswana in 1988 were never tried). Most settlements were externally imposed. Typically, free food was offered, people were rounded up and taken to the settlement. After a year or two free food ceased, and people left. At no time were all the Hadza in a settlement. All the Hadza spent many years of their childhood and of their adult lives living in the bush as hunters and gatherers. No generation has grown up without experience of life in the bush. Nonetheless, there have been brief periods when a majority of the eastern Hadza were in a settlement, for example in the mid-1960s when Yaeda and Munguli settlements were at their peak. SI 3.12 describes the history of each settlement. Table 3.1 shows the numbers recorded at settlements.

We know of nothing happening to the Hadza in the twentieth century as profound as the settlement of the Ache, which brought about a sudden massive increase in deaths, followed by a rapid recovery (Hill and Hurtado, 1996). The arrival of larger numbers of Herero herders in the Dobe area in the 1950s had significant effects on !Kung fertility and mortality. Many !Kung settled with Herero, worked for them, married them, or begged from them. The Hadza have no such close relationship with the herders in their country (SI 3.10). We will see that there is little indication that Hadza fertility and mortality have undergone severe changes in recent years.

## 3.7 Tourism since 1995

At the end of the century, from about 1995, tourists began to come to the Eyasi basin, to see the bird life at the lakeshore, and to visit Hadza camps. Near Mangola and in the Balai valley the effects were quite severe, alcohol being readily bought with the cash left by the tourists. Elsewhere tourists were few, well managed by the tour companies, and effects were minimal but may have included a tendency toward larger camps at fewer and more accessible locations. In the twenty-first century the alcohol enabling effect of tourism worsened everywhere in eastern Hadza country (Marlowe personal communication). A more detailed sequential description of the course of events in the Eyasi basin in the twentieth century is given in SI 3.5.

## 3.8 Summary: lost at the bottom of the rift valley?

There is little doubt that the great majority of eastern Hadza lived by hunting and gathering throughout the twentieth century. They knew about their neighbors and their neighbors knew about them and, on a small scale, they traded meat, skins, and honey for tobacco, pots, iron, and cloth. In the final decades of the twentieth century, the populations of their neighbors increased as fast or faster than the national population.

We know from Hadza genes and language that their ancestors maintained themselves distinct from other populations for an extremely long time, and probably lived somewhere in the general area they inhabit today. We know from the archaeology that someone was hunting and gathering in the Eyasi basin on and off for the last 130,000 years. It seems reasonable to suppose that ancestors of the Hadza were living by hunting and gathering in the Eyasi basin for many tens of thousands of years.

We may be eager to determine whether our research subjects represent "secondary" hunter-gatherers, people who gave up farming or herding for a foraging life. A well-documented example from southeast Asia is given by Oota *et al.* (2005). In the Eyasi basin we have short-term examples from the Isanzu who joined the Hadza to escape a famine around 1917, and the Kisamjeng Datoga displaced and dispossessed by the Maasai, regaining their cattle and returning to a herder lifestyle later in the twentieth century. Although there are remains of irrigation channels at Endamagha

**Table 3.1** Approximate numbers of Hadza involved in settlement attempts. Few people have settled permanently. Many years with no information are omitted. *Sources*: Woodburn, 1968a, 1979; Bennett *et al.*, 1973; McDowell, 1981b; Lars Smith 1977 census, my censuses. Percent is total in settlements/total eastern Hadza population. Total population in the years 1985–2000 estimated as N of live people in the population register. Before 1985 by linear extrapolation. No settlement attempts since 2000. McDowell, and Bennett *et al.* report brief unsuccessful settlement attempts in 1924 and 1939 near Mangola

| Decade | Year | Yaeda | Munguli | Endamagha | Mikocheni | Mono | Domanga | % settled |
|---|---|---|---|---|---|---|---|---|
| 1960 | 1964 | | | | | | | 0.0 |
| | 1965 | 170 | 134 | | | | | 38.5 |
| | 1966 | 141 | 20 | | | | | 20.5 |
| | 1967 | 141 | 20 | | | | | 20.5 |
| 1970 | 1970 | | | | | | | |
| | 1971 | | | Started | | | | |
| | 1973 | | | c.155 (31 men × average family) | | | | |
| | 1974 | | | | | | | |
| | 1975 | | | "many had left" | | | | |
| | 1976 | | | | | | | |
| | 1977 | 35 | 56 | 11 | 4 | 0 | 0 | 16.6 |
| 1980 | 1980 | | | | Sought hoes and seeds | | | |
| | 1981 | | | | | | | |
| | 1982 | | | | c.25 | | | |
| | 1984 | | | | | | | |
| | 1985 | 13 | 105 | (24) | (5) | | | 12.4 |
| | 1986 | 150 | | | | | | 16.3 |
| | 1987 | | | | | | | |
| | 1988 | 0 | | | | | | |
| 1990 | 1990 | | 47 | (42) | 0 | 410 | 0 | 49.2 |
| | 1991 | | | | | 238 | 0 | 25.3 |
| | 1992 | | | | | 120 | 0 | 12.6 |
| | 1995 | | | | | 54 | 0 | 5.5 |
| | 1997 | | | | | 51 | 14 | 6.6 |
| | 1999 | | | | | 67 | 76 | 14.4 |
| 2000 | 2000 | | | | | 49 | 25 | 7.5 |

(Sutton, 1986 and other references cited previously), the archaeology suggests the farmers traded with local hunter-gatherers who continued their foraging lives during this period (Mabulla, 2007).

In this chapter I have summarized something of what happened to the Hadza during the twentieth century and what modern equipment they have (iron knives, axes, arrow heads, aluminum cooking pots, modern clothing). Despite increased encroachment by farmers and herders, and several quite forceful but unsuccessful attempts to make Hadza settle and farm, Marlowe (2002) and I (SI 3.7) summarize twentieth century accounts to show very little change in Hadza behavior. We both suggest that, day by day, Hadza have little to adapt to but their environment and each other. The chapters in Part I lead us to a picture of constancy of Hadza demographic parameters in the twentieth century. Although in the Hadza case we can argue that events in the impinging world did not induce much change in Hadza daily life, this cannot be generalized. Some of the most intriguing possibilities concern the nineteenth century, from which we have no information on the Hadza and very little on nearby events. In Blurton Jones (2015) I discussed possible reasons for Hadza cultural conservatism.

In his 1988 paper on "encapsulation," Woodburn asked whether "encapsulation" by numerically and economically more powerful populations could be responsible for Hadza egalitarianism, wide sharing, and rejection of ownership. Woodburn compared the "encapsulated" situation to the circumstances of some other hunters and gatherers, and closely examined its relationship to the "immediate returns system" that characterized such societies. He noted that among contemporary hunter-gatherers "immediate return societies" (characterized by sharing and egalitarianism) outnumber "delayed return societies." He suggests the latter probably adopted agriculture more easily and passed into the population of farming societies. A question about "immediate return societies" was whether they were prone to resist agriculture, or whether encapsulation had pushed them to become more oriented toward immediate returns. Woodburn (1988) thoroughly discussed a possible role of encapsulation in shaping the social systems of the Hadza, !Kung, and others toward an "immediate returns" system. It seems most likely that while "delayed return" foragers were able to adopt agriculture or herding quickly, "immediate return" foragers found their dominant ethic incompatible with these economies. The initial shaping of these economic systems was probably a result of the nature of the resources available (Suttles, 1968; Testart, 1982). Although undoubtedly once more numerous than today, delayed return forager societies may always have been scarce and local.

# 4 Research strategy, methods, and estimating ages

They must pass numerous empirical tests and only hope to gain acceptance according to the three inexorable phases of academic assessment: "This is rubbish"; "This is fascinating"; and "We knew it all the time."

Karl Niklas, 2001, *Trends in Ecology and Evolution* **16**, 468

## 4.1 The strategy of the demographic study

My aim in the first part of this book is to describe the basic demography of the eastern Hadza, and then debate some issues of hunter-gatherer population dynamics. Because hunter-gatherer populations such as the Hadza are small, mobile, difficult to study, have no written records or registered ages, and are the subjects of a number of uninformed and stubborn assumptions, I put a rather heavy emphasis on ways to check our results. This has shaped my central strategy: (1) estimate fertility, mortality, and migration; (2) look for evidence of stability in fertility and mortality; (3) if it is found, and migration is infrequent, use our fertility and mortality estimates as input to a stable population simulation that predicts age structure and structure-related measures; (4) match the predicted values to observed values. Ideally, these tests use data obtained by means that are different from those used to gather the data we used to estimate fertility and mortality in the first place.

If the checks fail badly, then we will know something is wrong with our methods or our estimates of fertility and mortality, or with the stable population assumption. If the checks match the predicted values, we may have more confidence in our results. I try to indicate the meaningfulness of the checks by using resampling methods, by comparing the observations to predictions from simulations that use higher or lower fertility or mortality as their input, and by comparison to other populations, especially the !Kung and Ache. I do not expect to confirm results to the great precision that one might expect in a large systematically recorded population such as an industrial nation. I do expect to be able to say whether we can confidently describe the Hadza as less or more fertile than the !Kung or Ache, or some other interesting population, and with higher or lower mortality than those populations.

One prerequisite for this strategy is that the Hadza are sufficiently near to a "closed" population. A closed population recruits only by birth, and loses individuals only by death. In any real population there is migration, and in populations of people there is intermarriage with people who hold other identities, and there is change in ethnic identity. Sometimes there are sub-populations, separate partially closed populations. Consequently, the demographic analysis begins with two short chapters that deal with migration, intermarriage, and Hadza regions.

Only after comparing the results to the predictions, and to other well-studied hunter-gatherers, particularly the !Kung (Howell, 1979, 2000) and Ache (Hill and Hurtado, 1996), do I compare our results to the Coale and Demeny (C&D) (1983) model populations. Thus, this study, in contrast to our earlier paper (Blurton Jones *et al.*, 1992), is not in any way dependent on the C&D models or any other published set of tables. Additionally, unlike Howell (1979), our method of estimating ages is completely independent of any such model. Other recent demographic studies of hunter-gatherers under a variety of conditions are important and will be referred to several times. Examples are Early and Headland (1998) on the Agta, Hill *et al.* (2007) on the Hiwi, Kramer and Greaves (2007) on the Pume. Important anthropological demographic studies of pastoralists include those of Pennington and Harpending (1993) on the Herero and of Borgerhoff Mulder (1992) on the Datoga neighbors of the Hadza. Migliano *et al.* (2007) have analyzed and extended previous studies of demography of several Pygmy groups in Africa and southeast Asia.

Concluding the first part, I reflect on the descriptive findings in a comparative perspective. I pay some attention to long-standing ideas about hunter-gatherers, and "primitive" people in general. It has been widely believed that they are disappearing, their populations decreasing rapidly; that their populations were exquisitely stationary, spared the explosions of the modern world; that they matured and began child-rearing very early; or that they bore few children, spaced widely. Most persistent of all is the belief that they died young. The last was extremely well tackled by Howell (1979), and by Hill and Hurtado (1996) on the Ache, but the idea persists or reappears interminably, usually without support or justification. Recently, a much more interesting claim of very short life expectancy among Pygmy populations has been offered by Migliano and colleagues (2007). Using the Weiss (1973) population models, they report life expectancies in the low 20s, a striking contrast with the low 30s of the !Kung, Ache, and Hadza. The Hadza resemble the !Kung and Ache in giving no support for early childbirth or early death. None offers any support for the view that, in a hunter-gatherer lifestyle, few if any adults lived past 45 years old.

We will see whether the Hadza data support the results from the Ache (and the comparative reviews of fertility by Campbell and Wood [1988]; Bentley *et al.* [1993a]), and question the ubiquity among hunter-gatherers of the low fertility, long inter-birth intervals, and almost zero-growth population of the !Kung. As others have pointed out (Hill and Hurtado, 1996; Keckler, 1997; Pennington, 2001), observed hunter-gatherer populations have fertility and mortality that generate a quite rapid population increase. Yet over the 200,000 or so years of modern *Homo sapiens*, the population cannot have increased at such rates; it showed close to zero growth long term (Hassan, 1975; Cox *et al.*, 2009). Even the growth rates after expanding out of Africa were, in the long term, much slower than we see among hunters and gatherers. This "hunter-gatherer population paradox" has received attention from anthropologists, geneticists, and archaeologists.

## 4.2 Demography fieldwork

The Hadza data come from a series of 10 field visits, each of two to four months, scattered between 1985 and 2000, and two pilot visits, in 1982 and 1984. Their timing did not adhere to a rigorous plan but was shaped by availability of funds and sabbatical leaves. Types and methods of data collection were also not uniform during the study period. I began reproductive history interviews only in 1992. Data were entered after each field season and a variety of printouts (by name, by age, by mother's name, and so on) were taken to the field each time to assist us in identifying individuals, running the "where are they now?" interviews, and so forth. I did not carry a laptop computer to the interviews; it was too intrusive, too difficult to use while sitting in the dust or cramped into a Hadza house or under a small shady bush. A laptop added little to a paper copy of the previous interview and the other printouts. I sometimes used a Psion hand-held computer (a little bigger than a cell/mobile phone) with an abbreviated version of the population register to check on identities of children, kin, and others.

Permits for research in Tanzania were issued by COSTECH (Commission on Science and Technology) in Dar es Salaam (SI 4.1). It is important to follow the requirements; there have been times when foreigners' failure to follow the procedures resulted in an embargo on research. Official procedures continued at regional and district levels.

Once the permits were all in order, supplies obtained and fuel stockpiled, work could begin. At the large village Mangola we had a store (a lockable green steel shed imported from Nairobi by O'Connell) and access to clean unsalty spring water and friendly company from David Bygott and Jeanette Hanby. They also stored our Land Rover (SI 4.2). The self-imposed exile of "the Hanbygotts" to "Mkwajuni" at Mangola in the late 1980s transformed the logistics of our project even more than Tanzania's economic change of direction in 1986. Toward the end of the project, we acquired e-mail access, installing a radio modem with the Hanbygott radio that linked to Arusha on a good day, if the baboons had not thrown David's antennae and solar panels out of the tamarind trees. Around Mangola I could always find my long-term Hadza field assistant Gudo Mahiya, who Jeanette Hanby had usually warned of my impending arrival, and from that base we drove out to stay wherever we needed to stay in the bush to the south and west.

There are several important, if obvious, practical differences between the fieldwork we have done on behavioral ecology and the fieldwork entailed in demography. To study behavior and ecology (mostly the economics of food acquisition), we stay in a camp for a long time. Except where required by study questions (as in Frank Marlowe's PhD study comparing behavior in large camps and small camps [Marlowe, 1998]), we usually chose to work in camps where little other than wild resources were available. For demography, I did not need to stay long in any one camp but I needed to try to go everywhere, to every camp, no matter where it was or what its economy was like. Each kind of fieldwork has its own joys and tribulations. Following foragers through the wild bush country is a magical experience. The climate is not too trying, even when you are following spirited little boys who are hard to keep up with. Getting to know a few people really well is rewarding and interesting, but the debts, or perceived debts and promises, can become very onerous, and one begins to wonder

**Photograph 4.1** We kept our camp simple but it still attracted interest. These two well-known men adopted it as a suitable place to work on arrows. The man on the right had injured his left wrist and hand in a dispute with a lion over a zebra. © James F. O'Connell, 2015. Reproduced with permission.

if other camps and other regions are the same or different from the present one. In the demographic data collection, staying a night or two in each camp reduces the debts and demands, you see all of the wonderful countryside, learn more bizarre off-road and automobile maintenance skills, and gather a lot of data rather quickly. Then it is time to break camp, pack the truck, and try to find the next camp, where, however bewildering, difficult, and puncture ridden the journey, you will be expected to instantly weigh and measure everybody and hand out their payment. You quickly become good at pitching your tent and cooking your supper in the dark, or doing without. In either kind of fieldwork, you lose a lot of weight, and sleep really well.

In both styles of fieldwork, we tried to minimize our intrusion, sleeping in small mountain tents, cooking on our own small fire, keeping all equipment and stores in the Land Rover. Minimizing our intrusion was not always easy (Photographs 4.1, 4.2). While in the field together, I always provided Gudo with a tent and food. I also helped him leave food with his wife to help during his absence. Providing his food in the field was a potential problem for him. During my brief census stays in any camp, there would always be visitors to my fire in the evening, and as a Hadza, Gudo was obliged to share what he had with the visitors. He began to lose weight and do too much return visiting! Although the visiting was an essential part of his job, catching up with the news, we evolved a system whereby I provided three plates full for supper, mine, Gudo's, and everybody else's. This seemed well accepted by all. Sometimes Gudo provided breakfast, raiding someone for their hard-earned baobab flour and mixing it with honey and water. Half a mug each of this paste gave a refreshing and sustaining start to the day.

**Photograph 4.2.** If you want to study children's time allocation, you must pretend that you do not have the field guide to the mammals of East Africa. Hadza children of all ages, and adults too sometimes, can spend many delighted hours poring over the pictures. © James F. O'Connell, 2015. Reproduced with permission.

Most of the camps in the Mangola region were visited on day trips from our base at Mkwajuni. Others were visited by a series of multi-day excursions. The general outline of each trip was planned in accord with feasible routes, and loads of water and diesel. The details depended on where people were living or reported to be living. The basic process was to visit a camp, and during our visit to ask or await information on where there were other camps. As this information accumulated, I filled in the details of the trip; sometimes this entailed a radical change of route. While in the 1985 census, we had several stretches of driving over completely trackless bush, by the mid-1990s, we usually knew some kind of track, or in the steeper areas simply walked, paying someone in the current camp to guard the vehicle and our tents.

## The population register

Individuals entered our population register by having their names and their parents' names listed in a census, presenting themselves to be weighed and measured, being listed in previous censuses to which we had access, or being described by their mother in her reproductive history interviews. At each census, we met a few people who we had not yet listed as far as we could tell. If they were clearly new people, they were given an ID (identity) number and added to the register. In each census there were a few people who we could not be sure that they did or did not match any entry in the population register. The file containing the register includes the age estimates for each individual.

## Anthropometry and household censuses

On our arrival in each camp, people expected us to weigh and measure everybody immediately, for which we paid with a mug of maize meal for each individual, including even the smallest babies. Details of the procedure are given in SI 16.2.

When the measuring was completed and things had settled down, Gudo would begin to visit around the camp inconspicuously and interview household members about the household composition. He (and everyone else) knows the names of adults and older children; the only new information concerns new children, changes in spouse, or visits by children, which are very common. Gudo recorded these and the age status and sex of each member, according to Hadza terminology (eight age grades for each sex), in a notebook of a type well familiar to Hadza from our previous censuses.

Gudo and I then reviewed this record and compared it with the anthropometry record, mainly to ensure that I knew the identity of the people recorded. Hadza use several names for themselves and their parents and children, and sometimes change their habitual name. Thus, ensuring who each person measured or listed was in the population register was an important task. Sometimes some remained who I could not trace in the population register, and usually Gudo knew or could find out something about them. For example, were they visitors only remotely related to anyone in the camp; were they a long-lost sibling and family visiting or moving back from life among "the Swahilis"; and so on? Probably the most common mystery category was young men who could not be fitted to the previously recorded children of their parents (who might be in another camp). Young men travel about, and often find an excellent new name on the way. These few people with no ID number were important for assessing migration (Chapter 5). The household list was corrected throughout the duration of our stay in each camp. Because Hadza move camp frequently, the camp locations shown on our map (Figure 2.2) are, of course, obsolete by many years, and the map is too small to allow anyone to find them.

## Reproductive history interviews

After the household list was as complete as we could make it, I prepared to interview women of reproductive age. This entailed finding the printouts of their previous interviews, searching the computer for any other information about each of them, and deciding which questions were most urgent. If I had interviewed them before, an update on the children was always the first priority; had they had any new births; were all the previously recorded children still alive? Details of the interview are given in Chapter 7, where the representativeness of the sample is also assessed. Interviews included information on parents and siblings, mother and children's ages relative to other familiar people, weaning status of each child, views on sharing different sized game, food availability in a new place or during husband's absence, and nominating men as experts in hunting, trading, etc. (Chapter 14).

## Group interviews: "where are they now?", relative age sessions, history sessions

From time to time, and not necessarily in every camp but at least once in every eastern Hadza region in every field season, I did a "where are they now?" interview. Sometimes the occasion for this would arise if the supper guests were people that I knew to be knowledgeable and communicative. Other times we would assemble three to five chosen people to come and work (in exchange for *ad lib* cups of heavily sweetened tea) at our camp or in some neutral piece of shade, or inside the vehicle later in the study when it inspired less awe. I worked through the population register asking about almost everyone except the smallest children, people we remembered that we had seen previously in the current field visit, or knew to be dead. Responses ranged from "don't know who that is" through "they say she lives with an Iraqw at Dirim," or "he died last rains," or "you already saw him at Sanola" to "she's here."

## History interviews

A few times on each field visit I interviewed older people about historical events or early foreign visitors. The main aim was information for our age estimates but we also acquired indirectly information about the historical events themselves, sometimes turning up new ones we had not heard about. Thus, "Melander, who came on a donkey" was later identified in Pastor Ward's book (Ward, 1999) as Lud Melander, a missionary who worked in the Isanzu area between 1945 and 1955. I also asked about people whose names and relationships were recorded by Bleek or by Kohl-Larsen in the 1930s.

## Relative age ranking of men

Toward the end of the series of visits, when it became evident that our information on men's ages had lagged behind that on women, I gathered two to three men at a time to rank everyone they knew by relative age, and talk about some of the less ancient historical markers.

## Moving to the next camp

When it was time to move to another camp, I packed as quietly as possible and we usually left early in the morning for our next destination, where the process began again. There were a few camps that we could not reach by Land Rover. We usually walked to them to write household lists and do interviews. Sometimes we carried the weighing machine with us, sometimes the stadiometer as well. About twice we asked people from the inaccessible camps to come with us to our current camp to be weighed. This worked quite well except when one young woman, to whom I gave a little more than her share (we had spent many days toiling around the bush together long ago when she was a child), announced tersely "We used to think Nicholas was a good person. Now we know he is not." This probably had nothing to do with the walk to our camp. It was not the only time that I noticed that giving someone more than others around them led to a noisy denouncement and lecture about how you really owed them much, much more. Everyone who has done fieldwork with hunter-gatherers has experienced the multiple difficulties of having so much in such an egalitarian society (for example, see the revelatory discussion in Lee, 1969). These Hadza tirades seemed especially hard to take but I think they are a way in which the individual who gets an extra gift can protect themselves from the demands, perhaps even ire, of the bystanders.

## 4.3 Methods of estimating ages

Very few Hadza know their age or birth date. Of course Hadza know their age status in Hadza terms (tseheya, clati, paanakwete, etc., see Marlowe, 2010, p. 55), but these have only loose relationships to chronological age. If asked to give a number, they are far from the mark. Hawkes and O'Connel used code names for individuals to avoid being overheard discussing them. One of their camp members gained the code "13" because this was the age this fully grown young woman had offered. Frank Marlowe mentions an elderly man who said his age was ten. Hadza know who is older than whom but relative age is not part of a formalized relationship as it is among the !Kung.

Age estimates are essential to any demographic study, and not always as perfect as we would wish, even in industrial society, so I devoted much effort to estimating people's ages (SI 4.3). Like Hill and Hurtado (1996), I especially wanted to do this independently of existing population models. I did not want to use Howell's method, which relied on linking age rankings to an age structure. Howell chose a C&D (1983) model to fit the numbers of known under five- and under 10-year-olds, and the mortality rates suggested by the data on the children of interviewed women. The model age structure shows the proportion of people in the oldest five-year age category and the oldest ranked people are fitted into that five-year age range, and so on down to the children of known age. I decided to use a mixture of relative age ranking, presence as infants in the series of censuses, and reported presence at various "historical events."

Relevant information was contained in:

1. Presence at historical marker dates. Absence signifies nothing, except where a woman was present and some of her children were reported, but not others (and if a woman of reproductive age were mentioned, I always asked about her children). Presence at a historical point was taken only as evidence that a person was born before the year in question, not as an indication that they were born during that year, unless informants clearly stated this (SI 4.4).

2. Census lists, including censuses from before our study period and to which we had access. Subjects listed in a census were coded as "seen by" the year of the census, if not listed they were coded as "not yet in" the census year (but could have simply been missed) (SI 4.5).
3. Relative ages. During the reproductive history interviews, women were asked about their siblings, and who was older or younger. They were also asked to name some women slightly older than themselves, and some slightly younger. Many also named people as "the same" in age. Women were then asked about each of their children and for names of some children slightly older or younger than each, and for the youngest, whether the birth was before or after the previous rains (most interviews were during the dry season). Small groups of men were asked to nominate a few men just a little older, and just a little younger than themselves, and about their age relative to a few others who I thought were close to their age (SI 4.6).
4. For infants, notes on the individual's developmental stage, and a note of whether the child was seen during my reproductive history interviews (SI 4.7).

Some individuals were given "known ages" (SI 4.8). These included 193 children among those born during the 15 years of my fieldwork (plus brief visits in 1982 and 1984). In addition, there were people who were clearly infants during a previous census or historical visit, or had had their birth recorded (one woman living in a village was able to show us her birth certificate). Four older individuals were allowed weaker criteria: two, securely recognized from Bleek's 1930 genealogies, had their childhood ages estimated by Bleek, and I used these as "known ages." Another was recognized by his older sister in a photograph by Kohl-Larsen and estimated by me to be aged four years in the photograph. Another was a baby named, with her mother, in Bleek's notebooks, photographs, and genealogy. These provide important, strong but not perfect, anchors for the older end of the age scale. The six men said to have joined the Kings African rifles during World War II were not treated as of known age but like others present at a historical event, treated only as born before that date.

These data were entered in a series of files, all indexed by the subject's ID number, that could be read by a visual basic program. The program can work through the individuals in the population register one by one, or it can present the information on any single individual specified by the user. I began by ensuring that data on "known age" individuals was entered. I then concentrated on babies and toddlers, then children. Next I attended to the interviewed women, and finally all others, in sequence of their ID numbers (SI 4.9).

The program presents on the monitor screen the following information about the individual.

1. Any existing age estimate for the subject and any others displayed.
2. Any "known age" estimate for the subject and others displayed.
3. "Seen by." The data of the first census in which the individual was recorded, and for a few individuals, the first history marker at which the individual was reported.
4. "Not yet in." The latest census in which the individual was not recorded, even if its mother or child were.

5. Any other history marker information that referred to the subject.
6. The subject's lists of people s/he nominated as older, equal, and younger than her/him (or if a child: the mother's lists for the child).
7. Those nominees, and others' rankings of the subject as older, equal, and younger.
8. Information about the nominees: any existing age estimate, seen by, not yet in, etc.
9. If the subject is a child, the program also displays notes on developmental status (sit, crawl, stand, walk).

Then the user (me) tries to use the following rules:

1. If an age estimate is already displayed, check it against "seen by" and "not yet in."
2. If no estimate is displayed, look at estimates for equals, olders, youngers.
3. If these can be reconciled with "seen by" and "not yet in," make a rough estimate, and note its quality on a scale of 1 to 4 (worst to best). Otherwise: skip to the next subject.

Move to next subject: repeat the process for another individual. Repeat the process for the individuals who were previously skipped because of inadequate information.

Thus, those with most, and most useful, information were given age estimates early in the process. Others were given estimates when more of their "olders" and "youngers" had received estimates. About 209 individuals out of the 1702 in the population register never received an estimate (12.3%); this includes people who died by 1985, who played little or no part in the data analyses presented in this book. Among those still alive in 1985, only 85 (5%) received no age estimate. Thirty-four of these individuals received an estimate of "roughyob" a different field in the file of estimated year of birth, that was only used when age mattered least and sample size mattered most (for example, in scoring the few apparent immigrants).

## Checks on the age estimates

I conducted a number of checks on the estimates, described in more detail in SI 4.10. The first set of checks led to re-estimation of ages for some individuals. These included checks on internal consistency.

Five individuals had inconsistent reports about their age relative to a historical marker or census visit. These five are from 2000 records (0.25%).

Seventeen children had an estimated year of birth later than when the child was first recorded in a census (1.9% of children out of 867 cases). Their estimate was revised accordingly. Thirty-two adults out of 1295 (2.5%) had a similar problem and their estimates were revised. Combining adults and children, 49/2162 cases (2.3%) were found with estimated birth years that were later than the year the individual was first recorded.

No records suggested that a child was born after its mother had died and only three were born after the father had died, two of these arose from errors in identification of the father (SI 4.11).

Informants contradicted themselves in 4.5% of repeat listings of olders and youngers (97/2136 repeat records).

Informants can also contradict each other. Out of 3318 pairs of individuals in which A had an estimated age older than B, there were 2441 relative age reports confirming A as older and 426 contradictory reports. When A was estimated to be younger than B, there were 2589/3923 reports of A as younger, and 736 contradictory reports. Informants had been asked to nominate people close to their own age but just a little older, and just a little younger.

The age gap between informant and nominee was compared with their relative age nominations. Pairs contradicted each other about their relative ages up to 10% of the time if the two people were estimated to be a year apart, but only about 5% if estimated to be three years apart in age, and around 2% when about five years apart.

I looked for implausible cases of mother's age at birth of the child (SI 4.10.6). This is problematic; we do not know at what age Hadza girls can first bear a child but we would be very surprised if it were less than age 10, and suspicious about cases where our estimates suggest an age of 15 or less. There were 57 cases where a child was estimated to have been born when its mother was estimated to be aged less than 16, 4.5% of mother–child pairs. Most of these cases concerned the earliest births of mothers of "children" who were already quite elderly. There was poor information for estimating the ages of their mothers. There were not enough of them, or any estimated securely enough, to make me wonder whether, say 70 years ago, Hadza women regularly bore children at a younger age than today.

We would be surprised if any woman over 50 years old bore a child. There were three cases in which a woman was estimated to be aged just over 50 when she gave birth. In Chapter 7, I discuss my unsuccessful efforts to change the age estimates for these women.

The age gap between mother and child was used by Howell (1979) as a test of the quality of her age estimates. If we had overestimated the ages of older women, the gap between their age and the age of their first child should be larger than among younger women. I looked to see if the age gap changed with the mother's year of birth. On the whole it did not, the regression line of age gap by mother's age was flat ($b=-0.0246$, $p=0.198$, adjusted $r^2=0.0021\%$). SI 4.10.7 gives results for particular parts of the age range. If we limit the sample to women estimated to have been born before 1946, the regression is significant but accounts for only 1.6% of the variance. Thus, although my estimates as a whole pass Howell's test, they begin to fail among the very oldest women. This could arise if some of the women were older than estimated but we might expect this to be cancelled by the tendency for their older children to have died unrecorded. The missing dead (people who were born but did not live long enough to enter our record) can be a problem for several analyses one might think of doing. The problem can be tackled with simulated populations.

## Final tests of the estimates

The previously discussed tests each led to re-estimation of the ages of 97 out of 1493 individuals (6.5%) who had received an age estimate. The next three tests

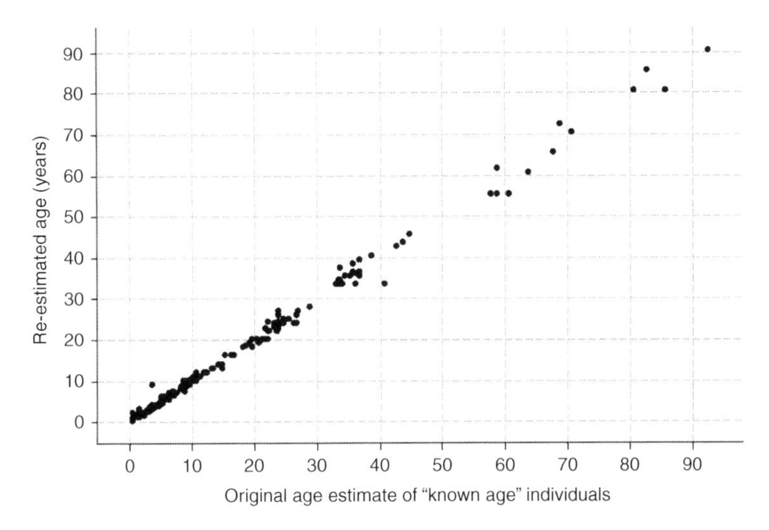


**Figure 4.1** Ages re-estimated without source of known age, plotted against original "known" age. Based on 205 individuals; $r^2$=99.5%; intercept, -0.2989; slope, 1.008.

did not. They represent the final tests of the final state of my age estimation. I re-estimated the ages of a randomly selected 200 individuals. Having selected an individual, the computer presented all the usual information except for the previous age estimate. I made another estimate and recorded it. Four individuals received no re-estimate because I knew them very well and could remember their ages. For the remaining 196, the regression of new estimate on original estimate gave an $r^2$ of 0.9944; the 95% confidence intervals (CI) covers about $\pm$0.25 years for babies and increase gradually to $\pm$1.25 years for 90-year-olds. Hence, the estimates are highly replicable. This does not mean the estimates are real, just that they are repeatable.

The next test was blindly to re-estimate the ages of the people of "known age." The computer randomly selected 100 of the children of known age and all the people over 15 years old who had a known age. It hid the previous age estimate, and in the case of children, hid the notes on developmental state made during the mother's interviews, leaving mainly the information on who else was older or younger than the subject. Figure 4.1 plots the "blind" re-estimates against the known ages. When the re-estimated ages are regressed on the known ages for all 205 cases, the adjusted $r^2$ is 0.9949 (p=1 $\times$ E-235) and intercept is-0.2989, p=0.0219, b is 1.008. Table 4.1 shows the fitted values and the 95% confidence intervals. This regression on the entire sample, which includes many babies, predicts that a re-estimated age of 90 would represent a known age of 90, and 95% of the cases would fall between 89.3 and 91.6 years.

Expecting accuracy to be best for children born during the study and worst for old people, I looked at regressions and confidence intervals for older sub-samples. The sample of 76 people aged 15–55 gave an adjusted $r^2$=0.9683 and intercept=0.3598 (p=0.5552, intercept not significantly different from zero) with

**Table 4.1** The ages of 200 people of known age were re-estimated with the known age hidden. The re-estimates were fitted by regression on known age. Linear regression gave the best fit; 95% CI are shown. Known age column indicates exact age, not an interval such as age 1–1.99. $r^2$=99.5%, intercept=–0.2989, slope=1.008, p=<0.00001

| Known age | Low 95 % CI | Fitted estimate | High 95% CI |
|---|---|---|---|
| 0 | -0.55 | -0.30 | -0.04 |
| 1 | 0.44 | 0.71 | 0.97 |
| 2 | 1.44 | 1.72 | 1.99 |
| 3 | 2.44 | 2.73 | 3.01 |
| 5 | 4.44 | 4.74 | 5.05 |
| 10 | 9.43 | 9.78 | 10.14 |
| 15 | 14.42 | 14.82 | 15.23 |
| 20 | 19.41 | 19.86 | 20.32 |
| 25 | 24.40 | 24.90 | 25.41 |
| 30 | 29.39 | 29.94 | 30.50 |
| 35 | 34.39 | 34.98 | 35.58 |
| 40 | 39.38 | 40.03 | 40.68 |
| 45 | 44.37 | 45.07 | 45.77 |
| 50 | 49.36 | 50.11 | 50.86 |
| 55 | 54.35 | 55.15 | 55.95 |
| 60 | 59.34 | 60.19 | 61.04 |
| 65 | 64.34 | 65.23 | 66.13 |
| 70 | 69.33 | 70.27 | 71.22 |
| 75 | 74.32 | 75.31 | 76.31 |
| 80 | 79.31 | 80.35 | 81.40 |
| 85 | 84.30 | 85.39 | 86.49 |
| 90 | 89.29 | 90.43 | 91.58 |

b=0.9852 (p=1.9 × E-57). The sample of 29 people aged 35–90 gave an adjusted $r^2$=0.9846, intercept=–1.84 (p=0.197), and b=1.03 (p=3.1 × E-26). These results are very encouraging. We can use the 95% CI (Tables 4.1 and SI 4.12 tables) to give us some indication of the possible range of error at different ages (Figure 4.2). Regression of the 29 individuals aged 35–90 suggests that a re-estimated age of 90 indicates a real age of 91, but that 95% of the cases would fall between 83.7 and 98.4 years. This seems much more realistic. Our estimates of really old people are not very accurate (there is some obvious "age-heaping" for the oldest people), but we should note that they are not consistent over- or underestimates. The analysis for 15–55-year-olds is important because it encloses the reproductive span of women (the analysis was done for men and women combined). This analysis suggested that people we estimate to be 15 years old are aged between 13.3 and 16.9, 35-year-olds are between 32.1 and 37.5, while 55-year-olds are between 51.1 and 57.9.

If we believe the "known ages," this test shows us that not only is the estimation procedure repeatable but that it adds no strong directional bias to the estimates. The procedure and data did not lead me to estimate people as mostly a bit older than

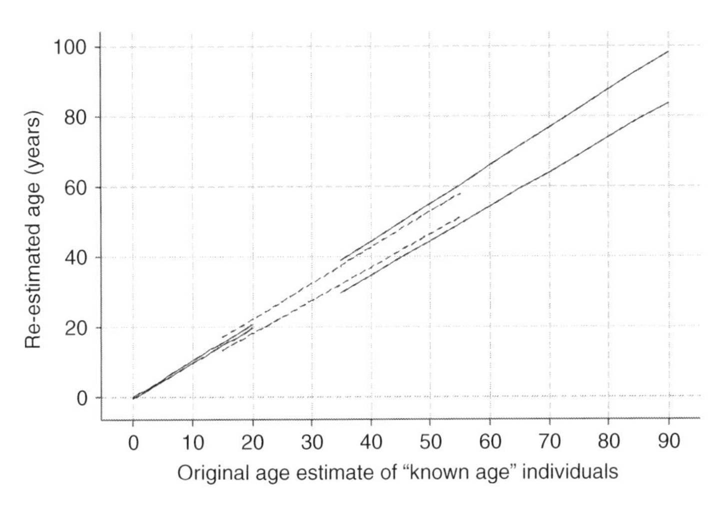


**Figure 4.2** 95% limits for re-estimated ages of sub-samples of different "known" ages: 0–20, 20–55, 35–90. Regressions of re-estimated age on "known age" suggest that children were re-estimated to be just under six months older than their "known age" (but small children had the most complete information, so many were in the "known age" group). Details of the regressions on each age group are shown in SI 4.12. Prediction intervals, confidence limits for an attempt to predict the age of the next child observed, for under 10-year-olds are ±1.6 years; for under 20-year-olds±1.7 years; for 15–55-year-olds±3.1 years, and for over 35s, ±6.0 years.

they were, or mostly a bit younger. A possible exception was discussed earlier; the ages of the oldest people and their mothers may have been "compressed" to give unrealistic age differences between mother and child. The age estimates from the 1985 age ranking were not used in the age estimation procedure described here. The 731 ages estimated by the two procedures agree closely (Pearson's correlation coefficient r=0.9854) but not perfectly (SI 4.13). The 1985 rankings place infants as a little too young and the oldest people as too old. Marlowe (2010) estimated ages by similar methods to mine before I gave him my estimates. He writes (p. 136) that the estimates agree closely (r=0.979, p=0.0005, N=482). I feel very confident about the age estimates. However, we should note that there are also estimates of their quality. We should keep errors in age estimation in mind when interpreting the results that depend upon them. It is important that for many analyses I restrict the sample to people we saw more often. These are also likely to be people with better age estimates than others.

## Great-grandmothers: a rough check of the age estimates for old people

Because so many people are convinced that in hunter-gatherer conditions no one lived into old age, often claimed for even very recent archaeological populations, I include another of our tests here that concentrated on the very old. Would we expect to see any great-grandmothers? And what about great-great-grandmothers?

If women had, on average, their first child around age 20, and we believe our estimates of age of young women for whom there is direct evidence, but most people had died by their 50s, we would not expect to see any great-grandparents. Calculating how many we would expect to see, based on our estimated ages and life table, is very complicated. We might, for instance, expect to see only the youngest great-grandparents still alive. However,we can quite easily use a population simulation to determine how many living maternal great-grandmothers we would expect to see if our estimates were correct. I used the simulation that keeps track of individuals and their kin, their year of birth, and their survival status in each year of the simulation. I scored how many children born during the final year of the simulation had a living maternal great-grandmother. This was converted to a percentage of the newborns. The simulation was run 100 times and the average percentage of newborns with a live great-grandmother, and the average age of these great-grandmothers, were reported. The simulation predicted that 15% of newborns should have a living maternal great-grandmother. The average age of living great-grandmothers was predicted to be 68.0.

If great-grandmothers existed, and were as old as I suggest, many of them would have lived too long ago to have been entered into my population register. To give a fair test we must restrict it to the most recently born infants, as long as we have a sufficiently large sample of infants. I chose to look at infants born during the final five years of my study, from 1995 onward. Among the 186 Hadza children born in or after 1995 (average year of birth, 1997.5), 24 had a living great-grandmother (12.9%, only 2.1% less than expected) and the average age of these great-grandmothers was 66.9, a negligible 1.1 years younger than predicted.

The procedure was repeated for great-great-grandmothers. Only about 0.31% of newborns were predicted to have a living great-great-grandmother and the expected age for any alive would be 75.8. The 186 Hadza children born in or after 1995 expected to have 0.58 ($186 \times 0.0031$) great-great-grandmothers, but had no living maternal great-great-grandmothers. Our oldest woman had an estimated age of 86 when she died, and there were only 11 who passed the age of 76 during our fieldwork.

If we change the mortality data used by the simulation so that chances of death increase rapidly from age 50 and the last surviving people die aged 60, we predict that only 2% of newborn children would have a living great-grandmother. This is substantially different from the prediction based on our Hadza life table and substantially different from what was observed in the real population. I conclude that I did not misclassify the ages of middle-aged people, 40- or 50-year-olds, as 60- and 70-year-olds.

## 4.4 Some definitions

### Eastern Hadza country

I use several phrases to denote particular areas. Frequently, I write of "Hadza country" or "eastern Hadza country," by which I mean the area inhabited by the majority of the eastern Hadza during the study. The eastern Hadza are very mobile within Hadza country but their range is not infinite. This limited area is shown by a plot of the Universal Transverse Mercator (UTM) coordinates of all the 243 camps that we visited between 1985 and 2000 (Figure 2.2); 1203 different known individuals were seen in this series of censuses.

Within this area there is a habitat separation. There is a tendency for Hadza to live in the rocky hills that divide the lake from the Yaeda plain, and the hills that border the Yaeda plain to the north and east. Datoga herders live for most of the year in the plains, both Yaeda and the lakeshore plain. Some Datoga families bring their cows into the highlands in the late dry season where water and grass are still plentiful. Empty spaces on the map mark the flat open country of the southern and central Yaeda plain. The easternmost end of the plain is covered in thick thorn scrub where Hadza quite often camp, and among which many Datoga also live. The lakeshore plain is much less used by Hadza. They forage there in berry seasons and *Quelea* nesting seasons, but even at these times, keep their camps mostly near the rocky, higher land.

Very few Hadza live outside this area. We have two ways to know about them. In our "where are they now?" interviews, we asked about anyone we had seen, or knew about from lists made by previous researchers. In our reproductive history interviews, we asked women to list all their siblings (children of their mother, by any husband/father); they report which are dead or alive and roughly where the living ones are living. We plotted the approximate locations for all except the five people who, at some time, had lived in Moshi or Arusha, on Figure 2.2.

### The core study area/the study area

"The study area" means the area covered by our repeated censuses. It corresponds to the area marked with camp locations in Figure 2.2, with the exception of "the Singida villages," Munguli, Mwangeza, Endasiku, which we seldom visited.

### Traditional Hadza country

By "traditional Hadza country" I refer to the slightly larger area described by Hadza informants as being formerly occupied by Hadza. Some localities so claimed can be confirmed by the observations of previous researchers and travelers or, less certainly, by the existence of Hadza names for particular places.

I examined the locations where earlier authors reported having met or heard about the Hadza (also known as the Tindiga) from Baumann (1894a and 1894b) onward. These include camps seen, places named and described by the early visitor, and places at which informants told us they were camped when the visitor came by (for example, pictures from Cooper, 1949). These locations mostly fall within the area where we found camps. One notable exception was the camp visited by Fosbrooke in 1950 (Fosbrooke, 1956). Two women who remembered his visit told my field assistant the location, which he later showed me, about 10 km from the road between

Karatu and Mbulu. These support informants' claims that they have long inhabited the land they occupy now. Outlying locations were reported by Dempwolff (in 1914), (Dempwolff, 1916–1917), Jaeger (1911), and Götzen (1895). These suggest that, despite the tendency to flee from strangers who came upon their bush camps, even 100 years ago, individual Hadza occasionally made excursions into the surrounding areas.

The early maps (Baumann 1894b; Jaeger 1911; Obst 1915) show Hadza ("Tindiga") occupying an area west of Lake Eyasi from somewhere southeast of the village Igulya, near the Semu river, to the escarpment, and on the east of Lake Eyasi, a roughly rectangular region appended to the eastern shore, that reaches to the eastern escarpment of the rift. Their southward extent is less clear. They have certainly lost ground in the south. Telling us how he lost his eye to a leopard, Mangashini described the location of his encounter, at a place now called Domenik, as wild bush country occupied only by Hadza. Today it is a thriving Isanzu village. Dorothea Bleek camped with some Hadza at a location recorded in her notebook as "4 hours walk north-east of Mkalama" and in her papers as "Some 12 miles north of Mkalama" (Bleek 1931a, p. 273). This would be somewhere near where "Endasiku Guard House" is marked on the 1:250,000 map (Series Y 503-Edition 1-TSO 1979 Lake Manyara), [UTM coordinates 694000 east, 9559000 north] not far from Matala and just northwest of Munguli. In Bleek's time, there were Isanzu living with the Hadza there and perhaps both were cultivating "gardens." This is only just outside the area Hadza inhabited during our study period. No Hadza live there today, although they would pass by on their way from eastern to western Hadza country. To the north, Hadza lived as far as the slopes of Oldeani in 1905 (Jaeger 1911) and the village of Lolpiro at the northernmost tip of Lake Eyasi, where Baumann (1894a) first encountered the Hadza, and where there has recently been at least one Hadza bush camp (Marlowe personal communication). The Hadza apparently comprised the majority of the population around Mangola during the 1930s, to judge by Kohl-Larsen's (1958) account. The Hadza have also lost the use of a large segment of the higher parts of the Siponga area, which has been cleared and planted by Mbulu farmers. We will see that, while the land available for foraging decreased in the final quarter of the twentieth century, the Hadza population has continued to increase quite rapidly.

Thus, "traditional Hadza country" is very similar to the area inhabited by the Hadza in recent times. There is no oral tradition of migration from some distant center of origin nor other evidence for any general migration of the Hadza during this century.

## "Swahilis," non-Hadza

Hadza often refer to non-Hadza as "WaSwahili." They do this despite having names for each neighboring "tribe" (mother-tongue group, if you prefer). The term denotes status as a villager, and usually farmer. It does not refer to the coastal people more generally known as the Swahili, who have a special place in Tanzanian history, language, and literature. I follow the Hadza usage in this book but try usually to write "non-Hadza." I realize that the topic of "tribe" is an issue in modern anthropology, and in Tanzanian politics, and that names that have been commonly used for peoples are now sometimes derogatory. However, I have generally followed common usage. For example, I refer to the !Kung, not Zhun/twa, because that is the name best known to anthropologists, their students, and their audience.

## 4.5 Special data sources

I have been able to use a number of valuable unpublished sources. Most notably Lars Smith's census lists from 1977–1978, conducted while he was a graduate student in anthropology at Harvard. Lars was instrumental in getting our research going, and thereby the entire corpus of recent work on the Hadza. He took me and Annie Vincent to the Hadza in 1982 when travel in East Africa was almost impossible, and again in 1984 for Hawkes, O'Connell, and my pilot visit. Lars' 1977–1978 census was very complete, included height and weight measurements, and names of every individual's parents, and whether the parents were dead or alive at the time. People's age category and reproductive status was recorded but there was no age ranking for the eastern Hadza (he did rank about 231 western Hadza, also covered by his census). After Lars' and my census in 1985, I joined him at his apartment in Djibouti to enter his 1977–1978 data into the computers in his office where he represented the company National Cash Registers (NCR) for Somalia and Djibouti. At that point, Lars' data had been on the verge of getting lost. Luckily he still had the printouts from an earlier computerization, as well as most of his original data sheets.

Lars also had paper copies of aerial surveys of the Eyasi basin that he conducted in 1977 with Mike Norton-Griffiths' company Ecosystems Ltd. These are not complete but seem to be the only surviving version. While there are maps showing approximate densities of livestock and houses of various kinds including Hadza camps in wet and dry seasons, the actual counts are missing. Only counts of wildlife are available. The results have been described in Chapter 2.

At one time, I believed I had permission to use some of the data from the 1966–1967 International Biological Program (IBP) Hadza archive at University College London (UCL). I used some of the information in my 1992 paper. Subsequently, a member of the original field team claimed that he had not agreed to give access to these records and I returned my copies to UCL and have made no more use of these raw data. The published accounts by Dyson (1977) and others in the IBP team are extensively cited throughout this book. As late as 2006, I heard that the archive remains intact at UCL.

William McDowell generously gave me permission to refer to his unpublished reports to the Tanzanian government from his fieldwork in 1980–1981, his list of Mangola people of that time, a list of newborns, and a dated photograph of a newborn baby with its mother and grandmother, which were in Lars Smith's possession. McDowell's reports include excellent information on foraging and hunting economics. His reports are included in the web-based supplementary information (SI).

The report on resources in Mbulu district edited by Meindertsma and Kessler (1997) is cited (with permission) as a publication in many chapters. It is hard to find and I was lucky enough to get both a preliminary copy and a final copy from David Bygott and Jeanette Hanby who produced it. They also made very many contributions to the substance of the report, which I think deserve more acknowledgment than they received.

For historical and age estimation, several other valuable sources were found. Upon my enquiry (at the suggestion of Carmel Schrire), the Cape Town University Library archivist was able to identify in their Dorothea Bleek collection the notebooks from Bleek's visit to the Hadza in 1930. They provided xeroxes of the notebooks and prints of her photographs. The partial genealogies in the notebooks were especially useful. The language notes were also of some interest to Bonny Sands during her fieldwork. Gudo Mahiya relished finding mistakes for her.

Annette Wagner, during her meticulous preparation for her filming visit to the Hadza, found numerous photographs by Kohl-Larsen from his visits in the 1930s. These were very valuable in age estimation and for eliciting genealogical information. Derek Jelliffe and Henry Fosbrooke also provided copies of photographs from their times with the Hadza.

The key works in German were translated for us. Obst was translated by Gabrielle Kopahl, a native German speaker with a bachelors degree in physical anthropology from Göttingen, and a graduate student at UCLA. She found Obst's language old fashioned, difficult, and unfamiliar (as too had my father-in-law, another native German speaker, aged four years when Obst was in the field and whose education in Germany spanned the years immediately after the time when Obst was writing). Gabi and I collaborated in the translation as follows. Gabi made a literal translation with comments and questions about the alternative meanings of the language. I read this preliminary translation and added further comments and questions from my fieldworker's perspective. Gabi went back to the original to try to ferret out solutions to our puzzles. Having eventually agreed on what we thought Obst was saying, we only then began to produce a readable English version. Gabi deserved more than her standard research assistant's wage. Frank Marlowe had Kohl-Larsen's 1958 book translated, and gave me a copy of the translation. Darlene Smucny, another graduate student who had studied in Germany, had translated several chapters of Kohl-Larsen's book for me (she was also defeated by Obst). I had made use in the field of the lists of names and sundry information available in Kohl-Larsen's book. It seems that Kohl-Larsen's 1950s' language posed none of the problems set by Obst. We still have not found anyone who can tell us what sickness Obst labeled "Darmkatarrhe." Nor does Obst's entangled German grammar enable us to be certain about one or two important historical claims.

There are several places where I refer to comments by James Woodburn published in the discussion sections of Lee and DeVore (1968). I generally cite them as Woodburn in Lee and DeVore, 1968, p. xx, where xx is a page number. Some might regard even published conference discussions as an unusual source but Woodburn's comments are especially valuable. I only cite them when I have found no comment on the topic in his later papers. They reflect his significant field experience at a time when no modern anthropologist had visited the Hadza and their lives had been much less disturbed than during most of my fieldwork. All of Woodburn's comments (in Lee and DeVore and his many subsequent publications) ring totally true of the Hadza in most localities during my study period. In some contexts, it is important that Woodburn was not polluted by any of the theoretical biases that rift contemporary human behavioral ecology. Our one disagreement would concern the amount of time women spend foraging, and the amount of time men spend gambling. Woodburn remarks that his comment was not based on systematic scheduled observation. Time allocation studies by McDowell, Hawkes and O'Connell, myself, and Marlowe (2010 , fig. 5.7) all record women foraging for much longer than a couple of hours a day, and in addition, spending much time processing baobab. I have only seen men gambling twice. The first time was on a day trip from Sanola to Engerao where we found men gambling. The other was at the Mongo wa mono settlement in 1990, where men were gambling with shoes!

## 4.6 Data processing and analysis

The population register is the most important data file. I described earlier how people appeared in the file. Important variables in the file are: the individual's names; sex; parents' names and tribal affiliations; and the individual's estimated year of birth, or month if known (entered as decimal years); whether dead or alive in the year 2000; and if dead, year of death, and year by which I learned that they were dead if the actual year of death were not known. Many people are in the file because they were recorded by Lars Smith in 1977 or 1978, and many of these had died by 1985. Some people were entered in the file twice. When this was recognized, their information was combined and one record marked as a duplicate. Each individual has an ID number (referred to several times as NID, Nick's ID number), and their parents' ID numbers are entered as well as their parents' names. Thus, it was easy to sort the file by mother, or by father, as well as by any of the other variables. This was useful in the field; for example, to ask a woman about some of her siblings.

Many other files are used. One contains all the household lists and individuals locations when measured (from the anthropometry records), enabling us to check if an individual was seen in a particular year. Another contains the cumbersome results of the "where are they now?" interviews, with a column for each year of a census containing a note about where the individual was reported. These two files allowed us to construct a third with a count of the number of times each individual was seen, and the number of times s/he was seen in the core study area.

Data extraction was usually done with programs written in Visual Basic 5 for each specific purpose. Important programs computed age-specific fertility, age-specific mortality, and resampling for both. An important program for Part I runs the population simulation that predicts age structure from observed (and resampled) fertility and mortality. This program extracts the predicted values of several measures from the simulated population, rate of increase, age structure, age at death, crude birth and death rates, and so on. Another important program scores these values from the real observed data; for instance, it reports the age structure of the sample of people who presented themselves to be measured. These anthropometry data are gathered almost completely independently from the data that gave the fertility and mortality estimates (reproductive interviews, "where are they now?" interviews). The sources of some of the other measures are less independent of the fertility and mortality data.

Crucial issues in programing were to try to use uniform methods of filtering data (SI 4.14). I excluded duplicate cases noted in the population register, and used the same checks each time for missing information, especially excluding individuals with no age estimate, and the few (13) cases for whom we did not know whether they had survived to 2000. It was important to use the same method of calculating age, from decimal year of birth to date or year of observation, and the same method for rounding these ages.

## 4.7 Notes on statistics

Almost all the statistical analyses are applied to quite specific questions that arise from theory in their respective fields. Demography has its questions, which direct Part I. Evolutionary Ecology and Behavioral Ecology, which direct Part II, are well recognized for their hypothesis-generating power. There is no "data trawling" here. There are about two places where I warn the reader that a question was provoked by "idle curiosity."

Compared to many quantitative anthropological field studies and even some other parts of the behavioral sciences, my sample is very large. By the standards of demography, my sample is minute. This is all the more reason for the close attention to checking my estimates of the demographic parameters and trying to generate some confidence intervals for the estimates.

Some of the statistical analyses were conducted at least twice by different methods. Sometimes this was a way to check on some of the statistical niceties (which seem not to be very important in practice). Sometimes it was because I saw two or more ways to answer the question and wanted to see if they gave the same answer. Often I report at least two of the methods. The results were always basically the same or very close. Any instances where different methods gave different results have been discarded as meaningless. For preliminary and, where relevant, final analyses, I used Minitab15 (Minitab Inc., State College, PA.). In reporting, I use the following abbreviations. Pearson correlation coefficient, r; beta coefficient in regression (slope), b; probability, p; adjusted r-squared to show percentage of variance accounted for by a model "adj rsqd" or "adjusted $r^2$". I indicate 95% CI, as for example (1.06–1.22) and odds ratio as "OR."

By far the most important statistical issue was the nested nature of many samples. Each woman has several children in the study, each child has some years of record. We especially want to know when a result is due to differences between mothers rather than differences between children. In part two we have more levels, each grandmother has one, two, or more daughters in the study, each of whom has one or more children. Which initial findings from traditional single-level regression are effects of events, which findings reflect characteristics of mothers, which reflect individual differences between grandmothers? Multilevel regression can separate out some of these effects. I chose the software MlWin 2.13 produced by Bristol University's Centre for Multilevel Modeling. I had two reasons for my choice. It is able to do multilevel logistic regression, needed for "yes/no" outcomes such as a child's birth or death, which some other software packages apparently did not at the time I made the choice. My other reason was largely personal. A leading role in the development of both the statistics and the software was taken by Harvey Goldstein. Early in my wandering career I briefly shared an office at London University's Institute of Child Health with Harvey who was embarking on his career – long development of multilevel methods, with educational research and longitudinal studies of child growth in mind. The choice was not merely a personal tribute, after we had graduated to shiny new offices of our own, with Harvey as built-in statistical guru for Jim Tanner's Department of Growth and Development, I learned to appreciate Harvey's insight into researchers' habits, dilemmas, and limitations. It also mattered that Mlwin comes with an excellent web course, and manuals with examples from the social sciences.

## 4.8 Population simulations

Because my mathematical education was limited and my grasp even more limited, but I greatly enjoy simple programing, I made generous use of simulations, mostly simple population simulations. Three population simulation programs are described in more detail in the chapters where they are first used.

1. Population simulation with resampling. The individuals that contribute the basic fertility and mortality data were randomly resampled. The program simulates 300 "years" of reproduction and death of the "seed population" and measures resulting rate of increase, age structure, age at death structure, and other measures such as gross birth and death rates during the final 50 "years" of the run. In Chapter 9 these predictions are compared with observation data on these predicted measures. Resampling was also used to give confidence intervals for the observations.
2. A simulation that keeps track of individuals and their kin. This was initially written to predict numbers of live siblings expected from different schedules of fertility and mortality. It has more uses in Part II to assess potential long-term effects of helpers.
3. A simplified simulation was mainly used to show effects on rate of increase from a variety of changes in mortality or fertility of various classes of people for a variety of durations and frequencies. It is principally used in Chapter 11.

Computer simulations have advantages and disadvantages. The main issue is how clearly the writer and the reader can see what has been done. Two comparisons should be kept in mind. One is between simulations and explicit mathematical models. The other is between personally written programs and "canned" simulation programs. It seems to me that canned simulation programs have all the disadvantages. The user probably does not know exactly what the program does or what it assumes, and the reader certainly does not. Personally written programs have the highest chance of the writer knowing what s/he did but are completely invisible to the reader. Indeed, a program that the writer has not used for a while can be extremely hard to decipher. My reaction to such an experience has been to write a new program, hopefully clearer and better documented. Because the failure to interpret an old program usually involves some question "how did it do this?", "did it score it this way, or that way?," in writing a new program one's attention to these details has been sharpened. As the essence of programing is breaking down a process into very small simple steps, writing these programs also helps one think about alternatives that were not evident beforehand.

For the writer, mathematical models have some of the same advantages, if you know how to write them! They have the advantage that for some readers they are probably equally clear. For others they are completely opaque and often leave the skeptical reader wondering exactly what went on and with little understanding of the argument, which leads to reservations about the outcome. In demography and life history, the Leslie matrix is a standard model successfully used by many. However, for my purposes, a simple iterative program that mimics it gives the same results whenever tested and has the advantage that special issues, such as age stratified migration, or helpers of particular ages who help targets of particular ages, can quite easily be incorporated in the program. It is essential, however, that the way the program does this is described in full, and well annotated in the program.

# 5 Migration and intermarriage: are the eastern Hadza a population?

Precise numbers are difficult to determine owing to the flexible ethnicity of peripheral persons.
Kaare and Woodburn, 1999

Are we justified in treating the eastern Hadza as a separate, self-contained population? The question is implicit in Kaare and Woodburn's remark. The issue is further emphasized by the change in anthropologists' attitudes to the term "tribe" (SI 5.1). Where once anthropologists wrote as if tribes were immutably identifiable, self-contained, and associated with particular locations, they have come to accept that tribes can be permeable and impermanent, that people sometimes change their location, and their identity, both in their own eyes and in the eyes of their neighbors. Few, if any, human populations are truly closed.

For demographic study, the important issue is how a population recruits new members and loses old ones. Stable population models in their simplest form assume recruitment is only by birth, and loss is only by death. These models are useful because of their internal consistency, the fact that several independently measurable variables (like age structure and rate of increase) follow inevitably from a specified fertility and mortality schedule. This allows many practical checks on the accuracy of field data and estimated population parameters.

Notwithstanding, every real population must include some migration, often of young adults adding to or subtracting from the number of people likely to produce children. Furthermore, human populations are uniquely prone to recruit or lose by change in ethnic identity. How many people "become Hadza"? How many people change their attribution or lifeway and "go to the Swahilis"? Marriages between members of one ethnic group and another raise additional questions. Hadza also migrate between east and west of Lake Eyasi. Can we study the eastern Hadza as a separate population? We need to measure the rate of migration and intermarriage to assess their effects.

The Hadza population seems to have intermarried with its neighbors at a very low rate for a very long time. In the twentieth century, genealogies collected by Woodburn between 1959 and 1967 showed a low level of intermarriage between Hadza and others, resulting in a surprisingly small inflow of genes from other tribes (Stevens *et al.*, 1977; Woodburn, 1988). Tishkoff *et al.* (2009) show that this has been the case for an extremely long time.

Among hunters and gatherers, most such marriages involve a forager woman marrying a man from a wealthier and more sedentary neighbor group (Bailey, 1988;

Headland and Reid, 1989). Among the Hadza, most of the modern mixed marriages follow this pattern. The woman may or may not leave the area of her kin. She may never return, or as among Hadza women, she may be quite likely to move back home. Her children may grow up as members of her husband's ethnic group, or her own. We need to try to estimate the frequency of these events in order to judge the coherence of our effort to describe the Hadza population. Bagshawe (1925) commented that, during the famine of 1917, some Isanzu girls married Hadza men. Around 1991, an Iraqw girl married a Hadza man and has lived in his camp ever since. These are interesting reversals of the common anthropological observation. Snyder (2005, p. 54) reports that Iraqw clans claim origins from men of several other ethnic groups, including the Hadza. Woodburn (1982, p. 194) tells how Hadza men were occasionally able to exploit an Iraqw tradition. A new Iraqw widow was regarded as "unclean" and unable to remarry until she had intercourse with a non-Iraqw man. Available Hadza solved the problem.

## 5.1 Migration, little but balanced

There was very little migration of Hadza in or out of eastern Hadza country, therefore I will merely summarize the data here and relegate the details to the web-based supplementary information (SI 5.2). Only two men appeared to have "become Hadza," marrying a Hadza woman, learning the language, and being accepted by other Hadza. Three more individuals have been seen during earlier phases of the process. In the core study area, around 10 people have been described by the Hadza as having "gone to the Swahilis," living among villagers and losing contact with the Hadzane-speaking majority.

The "where are they now?" interviews showed the location of people listed in the population register. Averaging over the eight censuses, at any time 15 known people were living in outlying areas, 2.1% of the total, 22 lived west of the lake (3.1%), and 622 lived in the study area (88.2% of the total) (Table 5.1, row 1). Asking 160 women about the location of all their 340 living siblings (Table 5.2, row 2) confirmed that these results were not merely a result of the failure of outlying people to have entered the population register. Figure 2.2 shows the locations of the outlying people.

Scoring the later location of people seen in Lars Smith's 1977 census and in our 1985 census gave a rate of emigration (Table 5.1, lines 3, 4). In SI 5.2, I calculate that, from the 1977 cohort of 126 women who would be of reproductive age between 1977 and 2000, about half a woman per year left. From the 1985 cohort of 187 women, about one woman left per year. Thus, just under 0.5% of the females of reproductive age left each year, moving outside Hadza country. Looking at new adults in subsequent censuses gave an indication of rates of immigration from the west, or return from outlying areas. The numbers migrating were few, and roughly balanced (Table 5.2). But the women who moved outside Hadza country tended to leave when young and return when older (mean age, 36.5). The loss of their reproductive capacity is reported later. Between 1990 and 1992, about 40 people left Munguli and moved to two locations in the study area.

**Table 5.1** Location of individuals at the time of the censuses. Information from "where are they now?" interviews, census household lists, and anthropometry. People were entered only if we, or our informants, knew their location in that year. The counts cannot be taken as summaries of the total study population because census effort varied (see Chapter 9). Each cell shows number and percentage of people "located in register": the person was identified and their location was reported. "West", number of individuals reported to be living in the country of the western Hadza, west of Lake Eyasi; "Far off", number of individuals reported to be living outside eastern Hadza country among non-Hadza, or permanently among non-Hadza inside Hadza country; "Study area", number of individuals seen in the core study area, the area visited during all censuses; "Singida villages", number of individuals reported living in or near villages in eastern Hadza country but outside the core study area, mostly in the Singida region; "Eastern Hadza country", number of individuals seen or reported to be living in the area recognized by Hadza as the country of the eastern Hadza: the study area plus the Singida villages

| Year | Located in register | West | Far off | Study area | Singida villages | Eastern Hadza country |
|---|---|---|---|---|---|---|
| Census mean | 706 | 22.04 | 14.9 | 622.5 | 46.75 | 669.2 |
| | 100% | 3.1% | 2.1% | 88.2% | 6.6% | 94.8% |
| Siblings | 340 | 22 | 7 | 294 | 17 | 311 |
| | 100% | 6.5% | 2.0% | 86.5% | 5.0% | 91.5% |
| 1977 | 412 | 17 | 23 | 363 | 5 | 368 |
| follow-up | 100% | 4.1% | 5.6% | 88.1% | 1.2% | 89.3% |
| 1985 | 539 | 12 | 25 | 476 | 13 | 489 |
| follow-up | 100% | 2.2% | 4.6% | 88.3% | 2.4% | 90.7% |

**Table 5.2** Combined emigrants and immigrants to eastern Hadza, between 1985 and 2000. An additional 11 people, not seen in 1985 but seen for the first time in a census after 1990, emigrated between 1990 and 2000. They have been added to the numbers of emigrants from the 1985 census

| 1985–2000 | To/from Singida villages | To/from the west | To/from "far off" |
|---|---|---|---|
| Emigrated | 3 male, 11 female | 3 male, 10 female | 16 male, 18 female |
| Immigrated | 22 male, 27 female | 12 male, 14 female | 12 male, 15 female |

These unmanageably small numbers of migrants imply that the use of stable population assumptions is probably valid, and that the eastern Hadza can be treated as a separate population. However, rates of intermarriage could change the picture.

## 5.2 Women married to non-Hadza men but living in Hadza country

In the remainder of this chapter, I report on the numbers of marriages to non-Hadza husbands and assess the demographic consequences. I also look at the history of intermarriage from the tribal identity noted in the population register for parents and grandparents of Hadza of different ages.

**Table 5.3** Married women in each census with a non-Hadza or Hadza husband. Some women are counted in more than one census. If we count each woman only once during the period 1985–2000, 33 of the 313 married Hadza women (11%) had a non-Hadza husband at some time during the period, three had a western Hadza husband, 277 had an eastern Hadza husband. Some women had more than one husband during the period. Mean of the nine censuses: 6.1% of married women had a non-Hadza husband

| Census year | Married Hadza women | Non-Hadza husband | Proportion |
|---|---|---|---|
| 1985 | 148 | 6 | 0.04 |
| 1990 | 119 | 7 | 0.058 |
| 1991 | 125 | 11 | 0.088 |
| 1992 | 95 | 7 | 0.073 |
| 1995 | 109 | 3 | 0.027 |
| 1997 | 100 | 8 | 0.08 |
| 1999 | 134 | 10 | 0.074 |
| 2000 | 118 | 6 | 0.05 |
| Totals | 948 | 58 | 0.061 |

The censuses comprise a good sample for estimating the number of Hadza married to non-Hadza in the study area (Table 5.3). On average, 6% of married Hadza women in the study area were married to non-Hadza men. There is a non-significant positive regression (r=0.158, p=0.708) between the proportion of married women with a non-Hadza husband and the year, indicating very little secular increase in the proportion of women marrying non-Hadza men during the 1985–2000 study period. There were very few non-Hadza wives of Hadza men, not more than three in the entire period 1985 and 2000. Women who appear in our censuses but are married to non-Hadza men appear to all be raising their children as Hadza. Thus, their intermarriage will have no effect on the population size.

Women who leave the study area to marry a non-Hadza man may be a different story. Eighty-three percent of them return later, at a mean age of 36.5. On average, they brought back 2.83 children, which is fewer than the 3.68 that a woman who stayed in the study area would have by age 36. Hence, the women who leave do subtract some of their reproductive potential from the Hadza population. Each woman that leaves, on average, removes 3.68 children from the Hadza population but she contributes $0.83 \times 2.83 = 2.35$ children, a net loss of 1.33 children over the years from maturity to age 35. With adult women leaving at the rate calculated, about one per year (0.5% of the reproductive aged women per year), or five women each 10 years, this subtracts $5 \times 1.33 = 6.65$ children every 10 years, a very small number. These levels of reproductive loss are assessed in the population simulation in Chapter 9 and found to be negligible.

While the current rate of intermarriage probably has a trivial effect on Hadza population growth within our study area, the inflow of non-Hadza genes is much greater, especially in recent decades. While the women who leave, and return later with their children, have only a small effect on population growth rate, they add substantially to the number of children growing up as Hadza but with a non-Hadza father. We can count the proportion of Hadza children with non-Hadza fathers in the censuses or population register.

## 5.3 Proportion of non-Hadza parents of today's Hadza

Nearly 16% (15.8%) of children in the population register and born between 1985 and 2000 were attributed by their mothers to a non-Hadza father. Many of them were living with a Hadza step-father. The figure varied starkly by locality, especially between the core study area and the Singida villages. When the sample is restricted to children or mothers seen in the core study area (defined in Chapter 4) on 70% or more of their records, the percentage is 10.4% (47 children out of 452 with an identified father). Those seen in the core area on less than 70% of their records, mainly children recorded at the Singida villages, were 48 with a non-Hadza father out of 111 children, 43.2%. Listing these families we see that many are from women who grew up in the Singida villages, married a non-Hadza man from that area and lived there most of the study period. We know that their children have been seen relatively less often in the core study area but we do not know what the future ethnic or economic identity of these children will be. Some may become foragers, some may become farmers. Some may grow up to marry Hadza, some to marry non-Hadza. These people probably represent the most common process (other than death) by which people of Hadza descent leave Hadza society. They represent the people whom Kaare and Woodburn (1999, p. 200) described as "peripheral persons" of "flexible ethnicity."

Previously, I reported that, on average in any year, 6% of married Hadza women in the area have non-Hadza husbands. In later chapters I show that fertility and child survival for women with a non-Hadza husband is the same as for women married to Hadza men. In this case, approximately 6% of the children in the core area (27 children) should have a non-Hadza father. This can be approximately reconciled with the 10.4% of children (47 children) in the core area attributed to a non-Hadza father by adding the 17 children of interviewed women who had returned from marriage to a non-Hadza man outside the area.

## 5.4 History of intermarriage: non-Hadza grandparents and great-grandparents of today's Hadza

Some contemporary Hadza of all ages report partial non-Hadza ancestry. We can use their reports to look back beyond the 15 years of our study to rates of intermarriage earlier in the twentieth century.

I scored the numbers of non-Hadza parents, grandparents, and great-grandparents of people for whom we have reasonable estimates of year of birth. I scored them in five-year blocks, as born in each quinquennium of the twentieth century from 1920 to 2000. To maximize the scope of the sample, I restricted it to only those seen in at least one census, thus including people seen in the Singida villages but limiting the sample to those on whom we are likely to have at least some data. Others who exist in the population register, with uncertain identity, ancestry, age, etc. are excluded. I scored the data for all the eligible individuals of any age as follows.

Suppose an index woman, "Maria," was born in 1980. She was the product of her parents' marriage, which must have been in effect in 1980. The tribal identities of her parents can be added to the count of mixed or non-mixed marriages in 1980–1985. The average age of a woman at birth of a child is nearly 29, and of a man just over 35 (Chapter 7). I assumed that Maria's father had been born in 1980 – 35 = 1945. His parents represent the marriages of 1945–1950. Maria's mother was most likely born in 1980 – 29 = 1951. Her parents are taken to represent the marriages of 1950–1955.

Maria's paternal grandfather was most likely to have been aged 35 when Maria's father was born. This would be around 1945 – 35 = 1910. Maria's paternal grandfather's parents, her paternal great-grandparents, are taken to represent the marriages of 1910–1915. The tribal identities reported for her great-grandparents were used to score the number of mixed or non-mixed marriages of the Hadza in1910–1915. The counts were restricted, for simplicity, to the male ancestors for Table 5.4 and Figure 5.1, and then repeated for just the female ancestors to give Figure 5.2. The results are, of course, very approximate. Some individuals will have been the first child of first children. Others may have been the last child of last children.

**Table 5.4** Number of individuals (born in each quinquennium) who had a non-Hadza father, or a non-Hadza grandfather or a non-Hadza great-grandfather

| Quinquennium ending in | Non-Hadza father | Hadza father | Non-Hadza grandfather | Hadza grandfather | Non-Hadza great-grandfather | Hadza great-grandfather |
|---|---|---|---|---|---|---|
| 2000 | 21 | 97 | 9 | 203 | 10 | 382 |
| 1995 | 14 | 106 | 9 | 215 | 17 | 364 |
| 1990 | 21 | 86 | 13 | 175 | 15 | 300 |
| 1985 | 11 | 106 | 13 | 205 | 21 | 302 |
| 1980 | 9 | 88 | 14 | 164 | 21 | 208 |
| 1975 | 5 | 108 | 12 | 199 | 26 | 223 |
| 1970 | 0 | 57 | 7 | 100 | 11 | 96 |
| 1965 | 1 | 76 | 6 | 130 | 9 | 115 |
| 1960 | 4 | 71 | 6 | 114 | 6 | 88 |
| 1955 | 1 | 43 | 4 | 61 | 2 | 30 |
| 1950 | 1 | 31 | 4 | 44 | 0 | 31 |
| 1945 | 0 | 35 | 2 | 38 | 1 | 16 |
| 1940 | 1 | 39 | 3 | 34 | 0 | 8 |
| 1935 | 0 | 32 | 1 | 24 | 0 | 9 |
| 1930 | 1 | 37 | 0 | 28 | 0 | 8 |
| 1925 | 0 | 18 | 0 | 6 | 0 | 0 |
| 1920 | 4 | 17 | 0 | 7 | 0 | 1 |

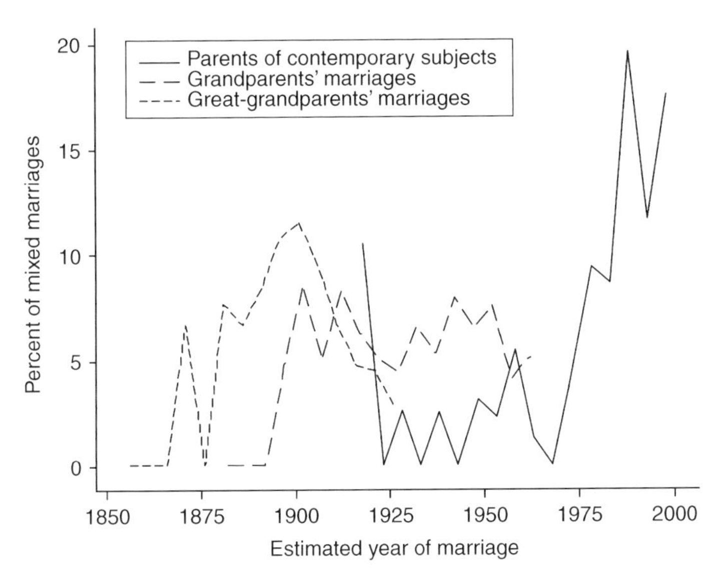


**Figure 5.1** Rates of intermarriage in the twentieth century and beyond: using percentage of non-Hadza fathers, grandfathers, and great-grandfathers to estimate rate of intermarriage in year of birth of subject, father, and grandfather, respectively and by quinquennium of subject's birth, offset by assuming average age at birth for men.

Table 5.4 shows the results. The first row concerns 118 subjects born between 1995 and 2000; 21 (17.8%) of them had a non-Hadza father. Of 212 traceable recorded grandfathers (fafa + mofa) of the subjects, 9 (4.2%) were non-Hadza. Of 392 traceable great-grandfathers, 10 (2.5%) were non-Hadza. The grandparents were assumed to represent the proportion of mixed marriages in their day. (In fact, they only represent the successful, childbearing marriages.) The great-grandparents were taken to represent their day, a generation earlier.

For the 97 subjects born in the quinquennium ending in 1980, the rate of intermarriage in 1980 was 9.3%, and in 1980 − 35 = 1945 it was 7.8%, and in 1945 − 35 = 1910 it was 9.1%. I plotted the intermarriage percentage on the Y-axis in Figure 5.1, the year of the grandparents' and great-grandparents' marriages (estimated as shown earlier) on the X-axis.

I am not trying to calculate frequency of non-Hadza genes among the Hadza, merely the proportion of mixed marriages in previous generations. Note also that I have no record of the ancestors of the non-Hadza ancestors, I assume they were all non-Hadza. By counting the number of children with non-Hadza or Hadza parents, we risk biasing the count in favor of whoever has most children who survived and whose ancestry we can trace. I show later that there are no significant differences in fertility or child mortality between Hadza women with non-Hadza husbands and Hadza women with Hadza husbands during our study period.

Field impressions are that intermarriage increased as more non-Hadza moved into Hadza country, and that it increased recently after Hadza children began to go to boarding school as young teenagers around 1990. However, there have been other periods when significant numbers of Hadza children attended school and could have met and become close to non-Hadza children, particularly at the settlement at Yaeda chini in the mid-1960s. The parentage data do not show a strong indication of intermarriage from that time. Perhaps there were still relatively few non-Hadza in Yaeda then, or perhaps more of these couples left the area and stayed away. The latter may also be true of the Hadza children who attended school in Munguli, although several "mixed" families moved from Munguli into the core study area during our study period. Bleek (1931a,b, and notebooks) indicates that a few of the Hadza she worked with were or had previously been married to Isanzus.

The data suggest that the rate of intermarriage early in the twentieth century was as high as 10%, apparently higher than during the middle 50 or so years of the twentieth century. The sharply higher data point in Figure 5.1 for parents of people born in 1920 is based on only four non-Hadza fathers in 21 cases and may belong in an earlier quinquennium; I may have underestimated the age of the oldest individuals.

## Non-Hadza mothers, grandmothers and great-grandmothers

Figure 5.2 (based on table SI 5.3) shows the same kind of computations for female ancestors. In recent decades, we know only one non-Hadza woman married to a Hadza man, a young couple who met in school and by 2000 they had yet to have a child. The data on mothers and grandmothers show this to be typical of most of the

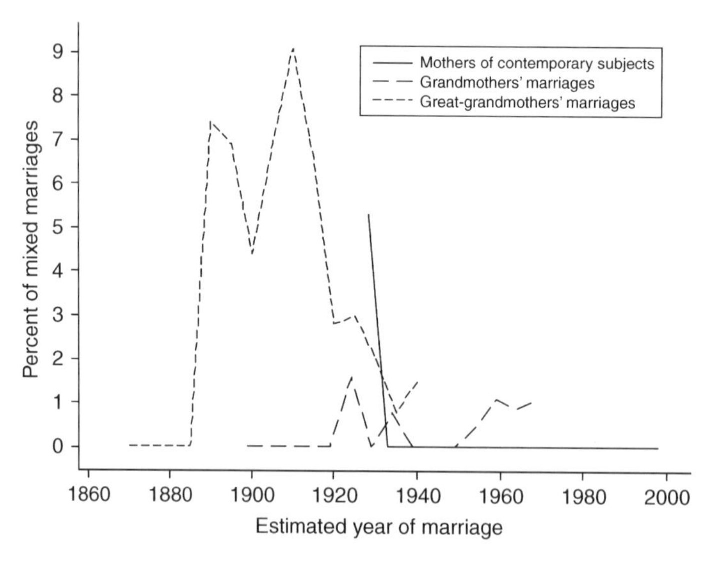


**Figure 5.2** Non-Hadza mothers and grandmothers, rates of intermarriage in the twentieth century and beyond: using percentage of mothers, grandmothers, and great-grandmothers who were non-Hadza to estimate rate of intermarriage in year of birth of subject, mother, and grandmother, respectively, and by quinquennium of subject's birth, offset by assuming average age at birth for women.

twentieth century. It was exceptional for a non-Hadza woman to marry a Hadza man. However, the data on great-grandmothers show a different picture. Early in the twentieth century, there were several non-Hadza women married to Hadza men. We seem to have found Bagshawe's Isanzu girls. Given the general pattern in which forager women marry their wealthier neighbors but neighbor women do not marry "down," we must wonder what exceptional circumstances were operating as the century began. First, the farmer life of 1900 may have been more precarious and less comfortable than the farmer life of the 1990s, thus the "general pattern" may not have been as rigid in the past. There were also the special circumstances of a series of dry years and famines that Bagshawe claims led to the Isanzu coming to the bush to live off the land among the Hadza, including several girls who married Hadza men.

### Western Hadza ancestors

I ran the program to look for western Hadza parents and grandparents of contemporary eastern Hadza. There seems to have been a roughly equal tendency between the sexes for eastern Hadza to marry western Hadza. Marriages to western Hadza formed a larger proportion of eastern Hadza marriages earlier in the twentieth century than later (SI 5.4). It is as if movement between east and west was easier for Hadza when the country was wilder and less settled by the Isanzu. The partial geneologies in Bleek's notebooks, concerning people she found at a point halfway between east and west, seem to support this. They had descendants who lived in the east and in the west. We may wonder whether separations and distinctions between western and eastern Hadza had been growing during the later part of the twentieth century. Not long before written history began, before "The Wanisanzu moved to Myadu, Mwagala and to the land of the Wakindiga [Hadza] because they were defeated by the Anankali in warfare" (Danielson, 1961), the two populations may have been contiguous.

## 5.5 Previous accounts

Our figures fit very well with the view expressed by Woodburn (1988, p. 39):

> The five hundred detailed genealogies that I have compiled ... demonstrate conclusively that immigration during the past hundred years – and probably for much longer than that – has occurred on a very small scale except during a period of a few years early this century (20th) when ... a series of famines did briefly cause a number of Isanzu farmers to come and live in the bush by hunting and gathering.... Some lived with the Hadza and a number of these, almost all men, married Hadza spouses. Many young Hadza today have one of these immigrants among their ancestors.... The number of more recent immigrants is very small indeed. Usually they only stay if they are of part-Hadza ancestry, for example the children of women who have married out. There is nothing to suggest that the proportion of immigrants is any higher than it is usually in Africa between adjacent farming communities.

Woodburn omits the Isanzu girls who married Hadza men during the famine, otherwise I have no difficulty reconciling our figures with the view expressed by Woodburn. However, I do have difficulty reconciling either Woodburn's summary or our data with the apparently higher incidence of intermarriage indicated by Barnicot *et al.* (1972b) (discussed in SI 5.5). It seems that the most likely explanation for our finding a much lower rate of intermarriage is that our sample is at least an additional 30 years removed from the time when the Isanzu sought refuge in the bush.

We should not forget that a large number of contemporary Hadza are descendants of the Hadza named and described in 1930 by Bleek (in her notebooks) and in the mid-1930s by Kohl-Larsen (1958). We should also note that the earliest authors report Hadza memories of times when Hadza women and children were taken by raiders. One of our informants recalls a few women being taken away much later by a trickster whose success she attributes to Hadza ignorance of the national situation and laws.

## 5.6 Rates of recruitment and loss by intermarriage, migration, and changed ethnic identity: implications for the Hadza population

Figures 5.1 and 5.2 suggest that the rate of intermarriage has varied greatly during the twentieth century. Rates of intermarriage have increased sharply since 1975. Furthermore, the rate before that does not appear to be constant either. If we consider both male and female non-Hadza ancestors, there is some indication of another peak in the early 1900s. Bagshawe (1925) and Obst (1912) both indicate some level of intermarriage around 1910–1920. Bagshawe mentions a famine that sent the Isanzu to live in the bush, including some girls who married Hadza men. The rate of intermarriage seems to have declined thereafter and been at its lowest during the 50 years between about 1925 and 1975. Surprisingly, we have no indication that intermarriage was increased in response to the Yaeda settlement scheme in 1964–1965. Settlements, especially at Munguli, seem to have been linked with intermarriage in the minds of Hadza informants, and we have mentioned the high rate of non-Hadza husbands of girls who grew up in Munguli and the Singida villages. Attendance at boarding school in the 1990s also seems to be linked to mixed marriages. While the non-Hadza immigrants of the early 1900s were probably mostly Isanzu, the majority in the 1990s were apparently Iraqw. This likely results merely from the greater invasion of traditional eastern Hadza country by the Walraqw. People from the Mbulu highlands have colonized and cleared most of Siponga during the past 30 years, and elsewhere in the part of traditional Hadza country that lies in the Arusha region, the Walraqw are the politically dominant majority and comprise the largest proportion of migrants to the lowlands.

What do the changes in intermarriage mean for demography? The effect depends on where the offspring end up and who they marry. At the beginning of the twentieth century, the Isanzu came and left again, except for some who stayed and parented future Hadza. But some mixed couples must have joined the Isanzu and were lost to the Hadza population. I cannot assess the size of such losses before the twentieth century or in the more remote past.

The population simulation that I use later (Chapter 9) to predict age structure from the observed fertility and mortality can be adapted to include effects of migration. The effects of including levels of migration rather greater than observed in our model are very small, probably well below the resolution of our data. Thus, it is not worth extending our effort to estimate rates of migration any more accurately. The uncertainty of estimates of the proportion who return with their children will have an effect, but it is quite small. If our estimates of migration are markedly wrong, effects should show up, such as discrepancies between predicted and observed rates of increase. Immigration is unlikely to have lasted long because it can only come from the return from settlements and from among non-Hadza, although immigration from the west has recently exceeded emigration to the west.

We are probably justified in treating the eastern Hadza as a population. Recruitment by change of ethnic identity is very rare. It is not unusual for Hadza women to marry non-Hadza men, but in the core study area, most of the children of such marriages grow up as Hadza.

Potentially, recruitment and loss by migration, intermarriage, and change in ethnic identity are problems for demographic study of many small contemporary and historical populations. Many of them are under intense pressure for land and in close contact with larger neighboring populations, and may have been for some time. It is unlikely that the fertility or mortality estimates for the !Kung or Ache, or the Agta were distorted by the way in which migration and intermarriage were treated. Early and Headland (1998) give substantial information about these processes and how they have changed during the last half of the twentieth century. Hill and Hurtado (1996) make it clear that there were almost no marriages of the Ache to non-Ache. Howell extensively discusses the effects of the arrival of Herero herders in the Dobe area and the differences between women married to Herero and to !Kung men. But for some less intensively studied populations, migration and intermarriage may make demography extremely difficult to study and to interpret.

# 6 Hadza regions: do they contain sub-populations?

Eastern Hadza describe their country as divided into four regions (Woodburn, 1968b, pp. 103–105). Are the regions so distinct that we should dissect our account of eastern Hadza demography, treating the regions as separate sub-populations? Here I attempt to quantify the description of regions.

Woodburn pointed out that the names of the regions referred to localities, but could also be used to refer to the people living in them. He suggested there was no exclusion of people from a region. Some people spent many years within one region, others moved from region to region. Among the four regions, Siponga, Mangola, Tliika, and Han!abe, the last, Han!abe, appeared to have a less certain status than the others. From his 1930s fieldwork, Kohl-Larsen (1958) seems to have been aware of a similar localization in his references to "the Matete horde" (Mangolabe) and the "Grosse-Wasser horde" (the Sanola area in Tliika). Woodburn's description seemed to apply closely to the situation we found in 1985, and to that shown in Lars Smith's 1977 census and aerial surveys. The Hadza concepts endured unchanged, although the on-the-ground realities had changed a little by the end of my data collection in mid-2000. We might wonder if there were effects of the loss of land that was associated with failed settlement schemes, the occupation and degradation of large parts of Siponga and Mangola, and the eventual arrival of tourism.

## 6.1 Regions are spatial clusters

We recorded Universal Transverse Mercator (UTM) coordinates of 243 Hadza camps all over eastern Hadza country. These represent nearly all the camps that we visited between 1985 and 2000. In our 1985 census, camps were located by compass and 1:50,000 scale maps. From late 1985, camp locations were identified by a hand-held GPS locator (an early model Magellan, then Garmin GPS12). Each camp in our records has also been given a region entry, in accord with our, and our Hadza field assistant's usage of the region names.

I fed the UTM coordinates to a cluster analysis in Minitab15. UTM coordinates measure distance from a very remote point of origin. Camps that are close to each other will have very similar UTM coordinates; camps far apart will have less similar coordinates. The options for cluster analysis in Minitab are quite limited but use of Euclidean distance as the distance measure seemed appropriate, and I allowed the Minitab software to use its default average linkage. The dendrogram (SI 6.1) was defined by requesting five clusters but results are almost identical if it is defined by

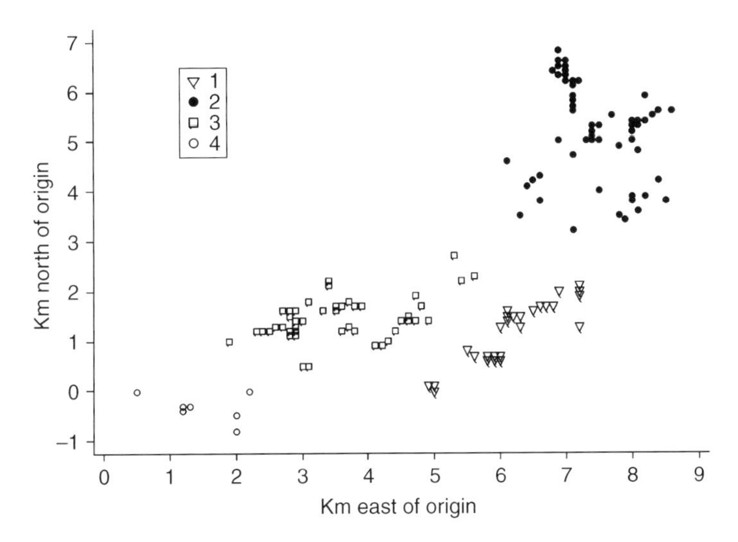


**Figure 6.1** Map of 243 camps visited between 1985 and 2000, with spatial cluster membership denoted by symbols. The spatial clusters correspond closely with the Hadza identification of the region of their camp. Cluster observations, linkage method average, distance measure Euclidean, four clusters. From late 1985, camps were mapped by Universal Transverse Mercator (UTM) coordinates. Dendrogram and data table are at figure SI 6.1. Legend shows cluster numbers.

similarity level. The cluster analysis program can save the cluster membership of each camp in the data file. I tabulated these with the regions and sub-regions noted for each camp (SI 6.1). When four clusters are requested, Mangola emerges as a single large cluster. The camp locations can be mapped, with the cluster membership denoted by symbol, as in Figure 6.1.

In Figure 6.1, cluster 1 predominantly includes camps recorded as in Siponga; cluster 2 predominantly includes camps in Mangola; and cluster 3 includes camps in Tliika. Two fuzzy results can be seen. The southern part of Tliika is included in cluster 4 along with Munguli. This is ethnographically correct; Tliika extends southwest as far as Munguli. Han!abe camps are distributed across two major regions, Mangola and Tliika. When five clusters were requested, as in SI 6.1, Mangola was divided into two regions, the southern reaching into Han!abe and close to Siponga, and the northern around Endamagha and in the Balai valley. The lack of clarity about Han!abe that Woodburn reported is supported by its geographical proximity to Mangola, Siponga, and Tliika. Cluster analyses of the spatial data for each census year give similar results (tables SI 6.2 for 38 camps in 1985, and table SI 6.3 for 1999).

## 6.2 Regions are clusters of kin

Do regions comprise clusters of related people, in which people in a region are more closely related to others in their region than to people in other regions? If so, there must be some tendency for children to stay in their parents' region; an incipient genetic separation of the regions, and a partial separation of them as distinct sub-populations. This would be important for our demographic study, and perhaps for the calculations of geneticists interested in the remote history of the Hadza. Although the regions are recognizable as clusters of kin, the separateness of the regions turns out to be quite weak.

I wrote a program which looks in the population register for each camp resident's parents, grandparents, and great-grandparents. It then looks for shared ancestors between every pair of individuals recorded in the censuses. Each ancestor is a definable number of links away from ego, and a definable number of pathways link different individuals. This allows the program to calculate the relatedness (r) between each pair of individuals, the probability of sharing a gene by common descent. The remoteness of links that the program can find is limited by the data on ancestors; we stop at great-grandparents. For a link via one great-grandparent, r=0.0156. Pairs for which we find no links get a score of r=0 (not related) (Apicella *et al.*, 2012, scored relatedness as low as 0.125, more distant kin being scored as 0). The program handles the unusual kinds of relationship that occur between Hadza, including cross-generation relationships. As children normally reside with the mother, and with the father if he is still married to her, variation in someone's success at child-rearing could tend to inflate any tendency for regions to reflect long-term reproductive isolation. Consequently, I ran the program only on people aged 15 or more at the time of the census. The program looks at each camp in a census and the relatedness between its members and all the members of each other camp in the same census. It calculates the average relatedness between members of each camp and members of each other camp and outputs these as a matrix. This matrix can be input to Minitab's cluster analysis program. The distance measure is the average relatedness between each pair of camps. The Minitab cluster analysis program, which analyzes this matrix, can save the cluster membership of each camp. I also used a second measure of relatedness between camps: the percentage of pairs with a relatedness score greater than zero.

The cluster analysis for relatedness between each camp visited in 1985 closely resembles the spatial classification of regions (Figure 6.2). These four groups represent Mangola, Mangola mountains, Siponga, and Tliika including Munguli. There was an area of expectable overlap, between camps in the northernmost end of the Kidelu range (Han!abe), camps in the hills forming the southern edge of the Balai plain, and some Siponga camps. We have seen several of the Mangola mountains people in Balai; their locations suggest vestiges of the seasonal vertical migration indicated by McDowell (1981b, see SI). The cluster analysis links the Mangola and Mangola mountain clusters at a higher level.

1. Relatedness within a camp. The mean relatedness between each pair of people in the same camp was 0.0753 (SD 0.0707, median 0.0476). The mean percentage of the people who were kin was 26.71% (SD 20.9%, median 18.2%). These compare closely with Hill *et al.*'s (2011, p. 1287) 25% for Ache and Ju/'hoansi (!Kung) bands. Mean relatedness within pre-settlement Ache bands was 0.054, very slightly lower than the Hadza figure given here (Hill *et al.*, 2011, p. 1288).

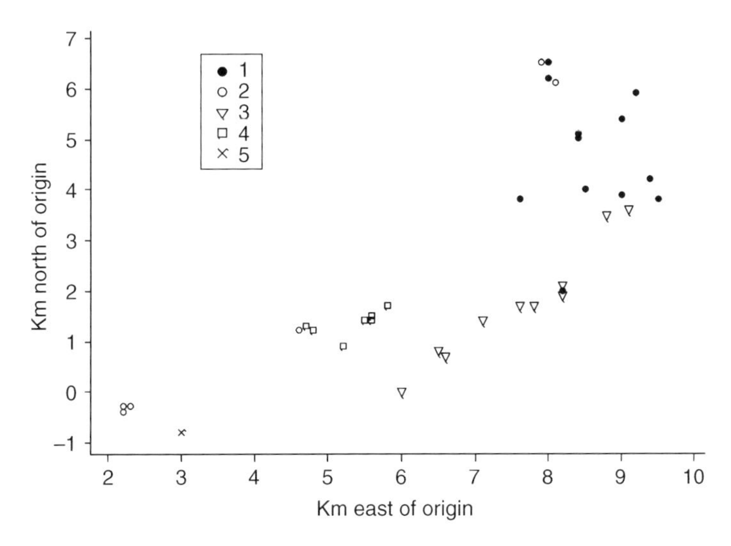


**Figure 6.2** Map of camps labeled by the cluster analysis of relatedness. Cluster membership is denoted by symbols. There is fair correspondence with the spatial clusters, with three exceptions: two camps near Endamagha with close kinship to people at Munguli, two camps in the uplands of Mangola region with affinities to Siponga, and one relatively unrelated cluster near Mwangeza. Axes are derived from camp UTM locations, measured from a nearby point of origin, as in Figures 2.1 and 2.2. The units are kilometers. Legend shows cluster numbers.

2. Relatedness between camps within a region. The 441 pairs of camps in the same region as each other were closer, and more closely related than they were to the 928 pairings with camps in other regions. Mean distance: same region, 14km; different regions, 45.7 km; $t=38.35$; $p<0.000$. Mean r: same region, 0.0256; different region, 0.0076; $t=11.37$; $p<0.000$. Mean fraction of camp members who were close kin: same region, 0.159; different region, 0.0656; $t=13.93$; $p<0.000$ (SI 6.4).
3. Relatedness across the whole study area. The average geographic distance between a pair of camps (1369 pairs) was 35 km (range 0–94 km, SD 22.9). The average r between pairs of camps was 0.01338 (range 0–0.3000, SD 0.02). The average proportion of camp members who were close kin was 0.095 (range 0–0.66, SD 0.10), or 9.5%.

The results indicate that, in 1985, the regions could also be defined as clusters of related people. People are more closely related, on average, to people in their own region. There are fewer kin, or more remote kin, in other regions. Cluster analyses of the 1999 camp-to-camp relatedness matrix (SI 6.5) did not completely reproduce the results of the 1985 analyses. The clusters of related people were located differently from those of 1985. While clusters of Mangola camps, Tliika camps, and even a ragged cluster of Siponga camps can be recognized easily; there are striking exceptions.

1. Some comprise the large number of Siponga people who stayed in the berry groves at Dumbichand near the shore of Lake Eyasi late in the year. These people were mostly seen back in Siponga in 2000. Other surprises are less easily dismissed.

2. The camp at Seikobe on the Udahaya river in Siponga had ties to people in the Mangola region. Some ancestors of a family in the Mangola region have been described to us as having come from Siponga during Kohl-Larsen's time. Camp 23, at Kidelu, which geographically clusters with Tliika, clusters by kinship with a camp at Dumbichand in Mangola region.
3. The location of people who deserted the village at Munguli may account for another "oddity." Many of the Mungulibe settled at Domanga in Tliika, but several others, went all the way north to Endamagha at the northern end of the Mangola area. The kinship links between these people (a pair of elderly sisters for example, one in each location, and each with many adult descendants) may account for the relative proximity of Domanga and one Endamagha camp in the cluster analyses.

## 6.3 Kinship and distance

We can ignore region membership and correlate the average relatedness of members of any two camps with the distance between the camps. Both average relatedness, and percentage of kin in a camp are strongly correlated with distance between camps. As in many other populations, people who live near each other are more likely to be related than people who live far apart (Figure 6.3 and SI 6.6). This was clear in 1985, and in 1999. Although, on average, camped with people 15%–30% of whom are kin, Hadza can find a few kin anywhere in Hadza country. However far they go, in a Hadza camp there will be, on average, 10% of the people with whom they share at least one great-grandparent. In a camp of 20 or so people, this would be a mean of only two individuals. While people are more closely related to people in their own region than to people in other regions, most Hadza have some chance of finding kin in each other region, convenient if they wish to go exploring another region.

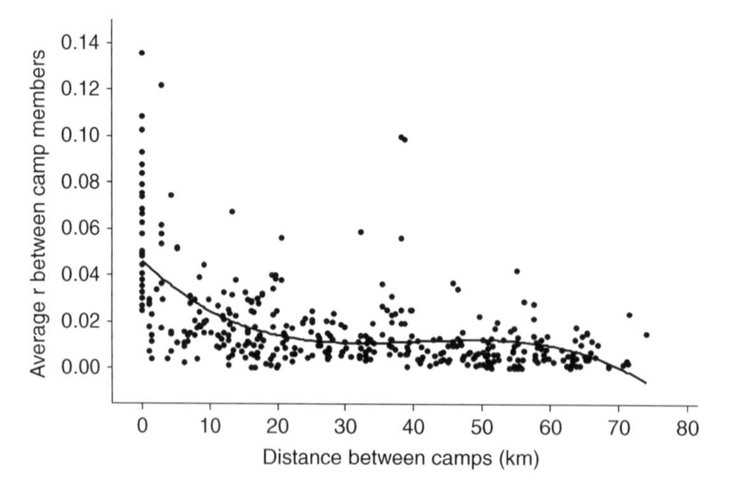


**Figure 6.3** Relatedness between members of each pair of camps by the distance between the two camps.

## 6.4 Dispersal and kinship

I illustrate people's movement from region to region by showing, in Figure 6.4, four maps of the later residence of people who were seen in each region in 1985. Each map looks like a picture of the whole of eastern Hadza country (Chapter 2, Figure 2.2). This means that, in the 15 years from 1985 to 2000, some people from each region had dispersed to all parts of eastern Hadza country. The regional structure appears not to impose firm limits on where people can go to visit or stay, just as Woodburn (1968a) suggested.

The map for people found in and near the village of Munguli in 1985 differs from the others in the concentration of its points. As noted previously, Hadza deserted Munguli during the study period, some around 1990 when they moved to the formidable settlement attempt at Mongo wa mono. A cluster of them stayed in Mongo, another cluster moved to Domanga in Tliika, and another few went to Endamagha. They stayed rather stationary in each of these places. While they described their movement as resulting from a wish to "be with my own people," they were also apparently looking for some approximation to the village life they had known since the 1970s. Because of this special history of Munguli, it is included as a separate region in the tables.

## 6.5 Long-term region-to-region movement

We can follow earlier cohorts into later censuses to quantify the impression given by Figure 6.4. Table 6.1 shows that 40% of the follow-up sightings of women (43% for

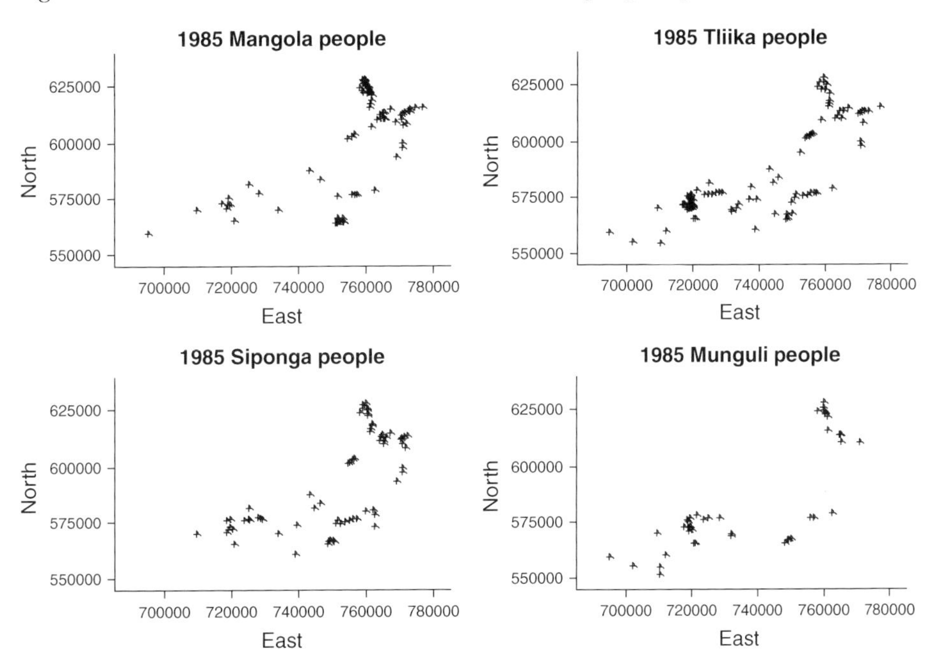


**Figure 6.4** People often move to another region. Later UTM locations of people seen in each region during the 1985 census. The distribution of their later locations resembles the distribution of the entire population shown in Figure 2.2.

men) were in a different region from the region where they were seen in 1985. The figure is similar if we follow people recorded in Lars Smith's 1977 census. On average, about 10 people of each gender change region each year. This is more than 10 times the rate of migration into or out of eastern Hadza country.

These follow-ups could not distinguish between people who were briefly visiting another region, and people who had settled there. Thus, to further assess the separateness of regions, I made two new summary scores of residence for as many individuals as possible. The first score "origin" summarizes the earliest records of where the subject was living, using Lars Smith's 1977 census and the historical interviews. The second score "nowadays" summarizes their residence during our 1985–2000 study period (see SI 6.7 for extended description of these measures).

Table 6.2 shows that the distribution of people across the regions has not changed significantly. While Munguli has lost people, no other region has markedly lost people and no region has notably gained relative to any other region. No region was being deserted between 1977 and 1985. However, the area occupied by Hadza in both the Siponga and Mangola regions have been "compressed," and even in Tliika there has been more invasion by both the Datoga and non-Hadza. These areas, once occupied only by Hadza, have been settled by more and more farmers, and the herder

**Table 6.1** Later locations of people seen in the 1985 and 1977 censuses. Individuals could be seen later in the same region as originally censused, and again in another region. Percentages are percentage of sightings

| | N at start | Died | Unaccounted | N seen again | Sightings | Same region | Different region |
|---|---|---|---|---|---|---|---|
| 1985 female | 327 | 26 | 17 | 284 | 358 | 216 (60%) | 142 (40%) |
| 1985 male | 299 | 22 | 23 | 254 | 332 | 190 (57%) | 142 (43%) |
| 1977 female | 333 | 52 | 24 | 257 | 341 | 209 (61%) | 132 (39%) |
| 1977 male | 298 | 50 | 20 | 228 | 299 | 182 (61%) | 117 (39%) |

**Table 6.2** Has the proportion of the population living in each Hadza region changed? Mangola shows a small but steady increase. Siponga and Tliika have retained about the same percentage of the population. Although Munguli was not visited in the later censuses, many Hadza left Munguli village during the study and moved into Tliika and Mangola regions (1977 from Lars Smith's census). Census counts differ because time spent on censuses varied

| | 1977 | | 1985 | | 1995 | | 1999 | | 2000 | |
|---|---|---|---|---|---|---|---|---|---|---|
| Mangola | 120 | 19.7% | 180 | 24.6% | 151 | 25.3% | 212 (305 – 93 Sipongabe) = 212 = 28.4% | | 202 | 36.4% |
| Tliika | 223 | 36.6% | 235 | 32.1% | 221 | 37.0% | 271 | 36.3% | 203 | 36.6% |
| Siponga | 158 | 25.9% | 192 | 26.2% | 177 | 29.6% | 170 | 22.8% | 150 | 27.0% |
| Munguli | 108 | 17.7% | 125 | 17.1% | – | | – | | – | |
| Census count | 610 | | 732 | | 597 | | 746 | | 555 | |

**Table 6.3** Region of origin of couples seen in censuses 1985–2000, men's regions in rows, women's in columns.

| | MA | TL | MU | Si | Totals |
|---|---|---|---|---|---|
| MA | 32 | 10 | 4 | 13 | 59 |
| TL | 17 | 42 | 17 | 31 | 107 |
| MU | 3 | 14 | 14 | 14 | 45 |
| SI | 15 | 16 | 13 | 66 | 110 |
| Total | 67 | 82 | 48 | 124 | 321 |

MA, Mangola; TL, Tliika; MU, Munguli; SI, Siponga.

For example, 32 marriages were each of a man from Mangola to a woman from Mangola; 15 were of a man from Siponga to a woman from Mangola. There were 321 marriages with information on region of origin of both spouses; 154 of the marriages had partners each from the same region, 48%

population appears to have increased although numbers of livestock did not change between 1977 and 1992 (Chapter 2).

I made the computer look at "origin" (as described earlier) for the couples listed in our household censuses (nowadays). Each couple was only used once. A few individuals could be counted twice because they featured in one census with one spouse, and in another with another spouse. These are not marriages formed per year, just married couples listed in any census. Table 6.3 shows men in rows, and their wives as columns. For example, 32 married men whose "origin" was in Mangola were married to 32 women whose origin was in Mangola. Four men whose origin was in Mangola were married to women whose origin was in Munguli. One hundred and fifty-four individuals were married to a spouse from the same region, 168 people were married to someone from a different region. There were 321 marriages, hence 48% were married to someone from the same region as themselves.

A similar analysis of parents' regions of origin (SI 6.8) showed that 42% of children had parents from the same region as each other. Sixty-six percent of children were growing up in the region where the mother grew up, 45% where the father grew up.

## 6.6 Discussion and conclusions: so much mobility

Movement from one region to another is quite extensive; in 52% of marriages, the spouses came from different regions. I will not treat the regions as sub-populations. I will treat the eastern Hadza as a relatively closed population that we can study using the stable population framework. The data are relevant to some wider issues in hunter-gatherer research: territoriality, disease transmission, ideal free distribution, the prevalence of matrilocality or patrilocality in human evolution.

The regions that Woodburn described remain clear in Hadza discourse and geographical reality. They summarize localizations of the population, which tend to have less densely inhabited areas between them. The open grassland of the Yaeda plain, which separates Tliika and Han!abe from Siponga, is little used by the Hadza. Hunting there, with no cover, is difficult, and by the end of the study there were few game animals in the open plain. There are none of the staple plant foods. The thorn scrub of the north end of the plain is used and inhabited by Hadza and comprises part of the Siponga region.

The separation between Mangola and Siponga was not based on an obvious habitat difference but may reflect change due to the invasion and habitat degradation in both regions. Higher ground, with its higher rainfall, has been more successfully colonized by farmers. In the recent past, when Hadza camps were found in the higher reaches of the Balai valley, and less of Siponga had been converted to farmland, the separation may have been merely because of altitude (Hadza often complain about the cold). The distinction may have been based also on the seasonal movement toward dry season water near the two main rivers, Balai and Udahaya. Tliika could be seen as a concentration around another major water course, the Sanola–Mugugu, and Engerao complexes of streams. In addition, small permanent water sources can be found all over the Kidelu mountains, in which Tliika and Han!abe lie. It is worth remembering, as described in Chapter 2, that each region includes patches of the major resources, soft berry patches, *Grewia* fields, baobab forest, tuber hillsides, and water. Some camps will be within easy reach of more than one resource, others will be seasonal camps, for instance among the *Cordia* berries.

There seem to be no prohibitions about using any area. One of my "filler questions" for women was "If you move to another place far away, can you find food, or is it a problem?" Answers were mostly "no problem" but one woman commented that she could follow people to see where the food was, another said she would have to ask to go with the women (Chapter 22). The extensive movement of people from one region to another supports the weakness of any territorial sentiments. People have more close relatives in the region in which they currently live but they can find quite close kin in any other region.

The regions do not denote areas of ecological specialization. Kohl-Larsen wrote of them as "hunting territories," but neither we nor Woodburn find any tendency for Hadza to exclude each other from their region. In 1985, Lars Smith remarked that he heard someone talking as if the people from region X should go back there, and said that this was the first time he had heard such a sentiment expressed. People do express belonging to an area and attachment to the area, and will refer to people in another area as people of that area.

Movement between regions occurs on a short time scale as well as the longer scale shown here. It is quite wrong to think of the Hadza population as described in Anderson and May (1991, pp. 653–654), citing Hassan (1981): "Bands of hunter-gatherers probably ranged in size from around 20 to at most 100 individuals," It is completely wrong to think of these as tiny, isolated struggling populations at constant risk of extinction. These numbers can describe camp sizes (my average camp size was 21 people) but camps are not isolated populations. In my series of censuses, which record the locality of each person in each census, we can count the number of different people that anyone was camped with during the 15 years of the study. The answer is an astonishing 69 different people (the mean for 171 women seen in four or more censuses). If we add to this the temporary visiting that goes on all the time, there is ample opportunity for transmission of contagious diseases. Camp reconstitution can include extinction, for example by 2000 there were no longer Hadza to be predictably found near Isoanakwepe in the Mangola mountains. Mzee Malele died and everyone left and they joined other camps in the region, where they and their children still live. Populations of hunter-gatherers comprise camps linked by kinship with constant visiting and reconstitution. The Hadza camps are linked in regions and the regions also linked by visiting, relocation, and marriage. Nor were hunters in a world of hunters necessarily isolated from neighbor groups of hunters who claimed different identity. Extensive trade networks have been demonstrated by archaeologists among prehistoric hunter-gatherers in many parts of the world, even in northern Tanzania as far back as 130 kya (Mabulla, 2007, p. 20). Several tragic historical accounts from around the world show that even "silent trade" can pass germs. The situation was much more favorable to maintenance of infections than Anderson and May imply. Researchers on disease transmission and demography might find Hadza situation continuing to be informative.

If Hadza are free to go wherever they please, and as people move around easily (Photographs 6.1, 6.2, 6.3), know an enormous amount about their environment, and presumably exchange information, then people might gravitate toward the richest region. If population density reduces the ease of living in a region, the process would be self-limiting. We might expect Hadza to have distributed themselves such that there is little to choose between the regions. The result would be the well-known

**Photograph 6.1** Hadza move camp frequently. Their possessions are few. Sleeping mats, cooking pots, and a plastic bottle (used for water, honey, or fat) are the bulkiest. © James F. O'Connell, 2015. Reproduced with permission.

**Photograph 6.2** A "house." On arriving at a new camp, people often build no house at first. Here you can see a sleeping mat, the ashes of the fireplace, some pots and containers, a digging stick, and some clothes. To the right of the mat is the pair of forked sticks used as a rack for arrows. © James F. O'Connell, 2015. Reproduced with permission.

**Photograph 6.3** A woman building a house. The branches are inserted into small "post holes" and wedged with small rocks; the twigs are interwoven overhead. When the structure is complete, grass is collected and used to cover the structure. The task takes little time if there is grass close by. © James F. O'Connell, 2015. Reproduced with permission.

"ideal free distribution" of Fretwell (1972). People in each region would then be expected to do equally well. We can eventually compare heights and weights, and reproductive success (RS) and other demographic parameters between the regions (Chapters 10, 13, and 16). In an "ideal free distribution," there should be no difference in RS between the regions. However, Marlowe and Berbesque (2009) found regional differences in BMI (body mass index) and percentage of body fat by region among women. Ecologists have sometimes found the ideal free distribution to be at its most informative when contradicted by the data.

The data on movement of men and women from region to region support Woodburn (1968b) in showing that men and women move with almost equal frequency, but show a slight excess of movement by men. There is a small but strongly statistically significant tendency for Hadza men to be more likely than women to change regions. Alvarez (2004) and Marlowe (2004) showed that this was the most common pattern among hunter-gatherers, particularly in the early years of a marriage. Hill *et al.* (2011) present data on observed residence (not the stated rules about residence) in 32 forager societies that support the view that residence is predominantly bilocal and flexible. Wilkins and Marlowe (2006) examined results of genetic studies that compare spatial clustering and variation in mitochondrial and Y-chromosome genes. They argue that the results of genetic studies are as expected, if patrilocality became more common with agriculture and had been preceded by a less sex-biased pattern of dispersal. This is one of several points in this book at which we see hints that men's control of events is greater in agricultural than in forager societies.

# 7 Fertility

At each large camp which I visited . . . I was also surprised at the numbers of children, and nearly half the women appeared to be pregnant.

Cooper, 1949, p. 13

Even today, a Hadza camp quickly challenges the idea that highly mobile foragers can only care for a few children. Other ideas are challenged too. Hadza children are far from helpless, gathering and processing significant amounts of food for themselves (Photographs 7.1, 7.2, 7.3), and although they are very small by western standards, babies and toddlers are usually quite fat. The children are lively, independent, curious, and often noisy. Was Cooper misled by an unusual camp, or an exceptional year? How many children are there really? In later chapters, I will examine how many children a woman keeps alive, who helps her keep them alive, and how well or poorly they grow. Here I am concerned with how many children a woman bears, and at what age she begins and ends her childbearing career. Non-anthropologists often assume that forager women would mature and begin childbearing very early in life. The !Kung and Ache studies show this is not necessarily true. However, Migliano *et al.* (2007) and Kramer (2008) show that some other forager populations have lower maternal ages at first birth than the Ache and !Kung. What do the data say about the Hadza?

I report the statistical parameters commonly used to describe fertility: age-specific fertility (ASF, births/woman/yr), marital fertility, total fertility rate (TFR, the sum of ASF from age 14 to 45), completed family size (CFS, the mean number of previous births reported by women aged 45 and older), gross reproductive rate (GRR) (total female births/woman), average age at first birth and last birth, sex ratio at birth, primary and secondary sterility, and differences between male and female patterns of fertility. I compare the results with those of Dyson (1977) to look for secular changes in Hadza fertility over the last few decades. I try to give some perspective on the results by comparing them with other populations, particularly the Ache and the !Kung.

The main analyses are based on interviews of women of childbearing age. Older women seemed to be unreliable informants, particularly prone to omit infant deaths. The interview sample comprises women at all points in their childbearing years, and I measure fertility by computing the number of women observed in each year of life and therefore at risk of a birth, and the number of those who bore a child in each of those years. The annual risk of birth gives ASF rate, and summing the ASFs gives a TFR. This "mixed longitudinal" design uses all the information we can acquire from

**Photograph 7.1** Hadza children may be "cheap" to raise. Here are two small children processing baobab for themselves. In many camps, they can also collect the pods themselves. The combination of pith and seeds gives fat, protein, and super-abundant vitamin C. © James F. O'Connell, 2015. Reproduced with permission.

**Photograph 7.2** Children can also obtain a significant amount of their food by digging shallow growing tubers such as makalita. © James F. O'Connell, 2015. Reproduced with permission.

**Photograph 7.3** Children of this age can get a third to a half of their estimated daily calorie requirements by their own efforts. © James F. O'Connell, 2015. Reproduced with permission.

this tiny population, but it could confound secular changes in fertility (if there are any) and the careers of individual women. I will address this issue in several ways.

The data were processed by more than one procedure. This account centers on "traditional" actuarial methods, as used by Howell (1979). But my account is supplemented with the use of logistic regression on year-by-year records of each woman's reproductive career. This method was used by Hill and Hurtado (1996). It has the advantage that the effects of independent variables, such as the year of observation to test for secular change, can be investigated at the same time. These data can also be examined in multilevel logistic regression to allow for the nested structure of such data, and control for the likelihood that the most fertile women are over-represented. Another advantage of multilevel regression is that we can derive a measure of each individual woman's fertility relative to her peers.

## 7.1 The reproductive history interviews

Beginning in 1992, I interviewed women of childbearing age about their children. I asked them to list all the children they had given birth to, beginning with the oldest, and giving their names. I tried to identify these children with individuals already in our population register, often unearthing alternative names for the same child. If the child was not in the population register, I added a new record for this child. I asked the women to tell us about children who had died, and tried to fit them into the sequence of births. For each child, we tried to determine where it had been born, who its father was, where it was currently. I asked several kinds of questions to determine when each child was born, as described in Chapter 4. We also asked similar questions to try to establish when the dead children had died, either by the developmental stage of the child when it died, or by linking this occurrence to other events. Occasionally, I also asked these women for similar information about the deaths of other kin. I asked whether the younger children were still suckling or were weaned. Women displayed little difficulty in deciding whether the child was "weaned." The Swahili phrase they used translates as "has rejected suckling/refused to suckle." It appears that cessation is quite sudden, no more suckling occurs. At this time, the child will have been taking other foods for many months but we have no data to show how much other food is taken in the year or so before "weaning." Sellen (2006) discusses in full the importance of the distinction between end of suckling and beginning of supplementary feeding.

I asked each woman to list her maternal siblings (children of her mother) and to give names of her parents and their parents if she could (some cannot name grandmothers, complaining that they knew her only as "Amama" or "Grandma"). I had a list of supplementary questions that were not directly about the reproductive history and were treated as optional. For example, if an interview had taken a long time to cover the children, these questions were omitted. If there were no new children to talk about, and we had covered everything about the children at a previous interview, I asked more of the supplementary questions. These questions are described in the chapters in which their data are used. They covered topics such as whether the woman or her children had ever been to a village clinic, her thoughts on sharing small game such as birds of different sizes, her access to food when her husband was away or when she moved to a new region.

I was helped by Gudo Mahiya. I used Kiswahili and Gudo translated to and from Hadzane. Many of the mothers understand and speak Kiswahili at least as well as me, and conversations took place in a mixture of both languages. None of the women spoke any English. The interviews were all conducted in or next to the woman's house and we attempted to afford her some privacy. Occasionally, husband, mother, sister, or neighbor were also present. I think their contributions were nearly always positive. Women were initially very unfamiliar with being interviewed and many needed time to build up their courage and to realize that we simply wanted knowledge and that this was part of our interest in Hadza life and in their ability to live in the bush. Those who had been in school (usually for not more than one to two years) seemed to find the procedure much less bewildering initially than those who had not.

Hadza informants were pleasant and friendly, if sometimes demanding, and nearly always willing to cooperate. However, what they say (as many previous visitors have noted) can be quite inventive. As Woodburn (1988) has pointed out, this may be part of their defense against being ensnared in subordinate debt or labor relations with their neighbors. Their multiplicity of names also has something to do with this. They can, and do, give different names to different outsiders, a demographer's nightmare as well as a device that makes individuals impossible for officials or would-be employers to track down. Repeat visits have an enormous advantage. You are no longer a stranger, you have been and gone with no evil consequences. You were interested enough to come back again. By the time we began the interviews, we had conducted three censuses and anthropometry tours, and several spells of study of subsistence activities. Many Hadza, men, women, and children, knew us very well. I gained the impression that, in addition, once the Hadza women realized that I knew a great deal about each of them (names, parents, kin, a better estimate of their age than they could manage, their children, and so on), they became much more easy to interview (SI 7.1).

I do not present any analysis of consistency or agreement about reproductive history between repeat interviews. Repeat interviews were mostly used to add to the information, to bring us up to date on new births and deaths, to fill in missing information, for instance on place of birth, to amplify the data for age estimation, and to clarify birth order (something that some women found very difficult). Later we report a variety of ways of checking our results. If our informants had been giving us severely erroneous information, we would have found a number of inconsistencies in our data.

## 7.2 How well did the sample represent the population?

I tried to interview as many women of childbearing age as I could find (and quite a few older women as well). Limits were imposed by the brevity of summer field trips, exhaustion, and logistics (primarily: how long will my supply of clean drinking water or diesel fuel last). Hadza are extremely mobile and individuals can be very difficult to find. There are women who we had seen in 1985–1991 but whom I did not meet during 1992–2000, the period of my interviews. I interviewed 91% of the women aged 15–55 who I saw in the study area between 1992 and 2000. This leaves little room for selective biases. The main influence appears to be the frequency with which I encountered them. The chance of being interviewed was greater for women seen two or more times (95%) versus 66% for women seen in only one census.

I interviewed a total of 240 women about their reproductive histories. Some of the oldest women were difficult to interview about their childbearing history, although most were able to furnish useful genealogical and historical information. The oldest women seemed to omit children who died young, and to find it difficult to give information on the relative age of their surviving children. Because of this, I limited the analysis to women born in or after 1945 (up to 55 years old in 2000), which reduced the number to 195, discarding the 45 women over age 55 who I had interviewed (SI 7.2). Another 13 of these 195 interviewed women were omitted, leaving 182 women. Eleven women were omitted because they had not given any information that allowed estimates of the age of the children, and two were omitted because during the interviews I had marked the sheet as "Do Not Use" in response to the inconsistent and confusing information provided. Of the 182 women, 127 were interviewed more than once, 76 more than twice, and 25 of them four times.

There was a risk that I might be biased toward interviewing women who had children and thus perhaps avoid interviewing those we knew to have had none. The bias could have been increased by my wish not to put very young girls (the 15–18-year-olds who we knew to be unmarried and childless) to the trouble or embarrassment of being interviewed. Instead, I added to our reproductive histories those young women who we had seen often (median number of censuses, more than five) and knew well enough to have heard if they had given birth and for the infant (dead or alive) to have entered the register. Thus, to the interviewed sample of 182 born in or after 1945, I added 45 girls and young women who we had seen frequently, and whose reproductive history we knew adequately (sometimes from their mothers as well as our census records).

We need to bear in mind that our sample was a "mixed longitudinal" sample, a combination of period approach and cohort approach, a choice arising from the small number of Hadza women of childbearing age. Being based on retrospective interviews of women of differing ages, the analysis may confound secular change (if there is any) with age difference. (I elaborate on this issue in the Discussion and in a lengthy SI referred to therein). The early births of the older women occurred as much as two decades before the early births of the younger women in the sample. Indeed, some of the older women were mothers of some of the younger women. Therefore, my computation could perhaps obscure secular changes of durations up to about 30 years. Secular change is examined later and appears to be small. Furthermore, because many of the women were young and far from completing their childbearing years, there are fewer women contributing to the estimates of fertility late in the reproductive career than early in the career. Age-specific fertility estimates at older ages are based on fewer individuals, and are likely to be subject to more error. Resampling confirms this; the 95% CI increase steadily with age.

## 7.3 Fertility of the interviewed women

The first aim was to estimate age-specific fertility (ASF) for the sample. I measure ASF as births to women of $age_x$ , divided by mid-year population of women of $age_x$ (Barclay, 1955; Hinde, 1998) (SI 7.3).

The ASF for all 227 women is shown in Figure 7.1, alongside the ASF of the 195 who had all their children by Hadza husbands. The plot of ASF shows much variance. Fitting a three-term polynomial regression to the fertility data (Figure 7.1), beginning at age 10 and ending at 50, seems to provide a reasonable smoothing of the plot ($r^2$=0.82 with two terms, and 0.84 with three terms). According to this curve, fertility peaks a little before 30 years of age, at a level close to 0.27 births per annum, then declines at a rather constant rate to the last birth in the mid-40s. The fitted curve is not symmetrical. The fertility rate climbs rapidly but declines slowly; there are many births to women in their 30s.

The TFR for the 195 women who had all their children by Hadza husbands was 6.17 (Table 7.1). The TFR based on all 227 women in the sample was 6.28 (table SI 7.4). This level of fertility is indistinguishable from the 6.12 reported from Tanzania as a whole around 1992 (Mturi and Hinde, 1994). It is lower than the CFS 6.9 found among the Datoga neighbors of the Hadza by Borgerhoff Mulder (1992). These are rates for all women in the specified category and age; they are not marital fertility rates.

**Table 7.1** Age-specific fertility rate (ASFR) (births/woman/yr) for women with Hadza husbands only. Smoothing by probability of a birth from logistic regression of women's annual hazard file: birth = woman's age + age$^2$ + age$^3$. In the total interview sample, there were 695 births, 331 females and 354 males, for a sex ratio at birth of 1.069. Gross reproductive rate (GRR) (total female births/woman) was 3.147. For women who had children with Hadza husbands only (this table) the figures were: 544 births, 275 boys and 269 girls, 103 male:100 female, GRR of 3.138

| Age | Enter | Censored | At risk | Births | ASFR | Smoothed |
|---|---|---|---|---|---|---|
| 0 | 195 | | 195 | | 0.000 | 0.000 |
| 1 | 195 | | 195 | | 0.000 | 0.000 |
| 2 | 195 | | 195 | | 0.000 | 0.000 |
| 3 | 195 | | 195 | | 0.000 | 0.000 |
| 4 | 195 | | 195 | | 0.000 | 0.000 |
| 5 | 195 | | 195 | | 0.000 | 0.000 |
| 6 | 195 | | 195 | | 0.000 | 0.000 |
| 7 | 195 | | 195 | | 0.000 | 0.000 |
| 8 | 195 | 1 | 194.5 | | 0.000 | 0.000 |
| 9 | 194 | | 194 | | 0.000 | 0.000 |
| 10 | 194 | | 194 | | 0.000 | 0.000 |
| 11 | 194 | | 194 | | 0.000 | 0.000 |
| 12 | 194 | | 194 | | 0.000 | 0.000 |
| 13 | 194 | 1 | 193.5 | | 0.000 | 0.000 |
| 14 | 193 | 2 | 192 | 3 | 0.016 | 0.033 |
| 15 | 191 | 9 | 186.5 | 5 | 0.027 | 0.075 |
| 16 | 182 | 4 | 180 | 26 | 0.144 | 0.114 |
| 17 | 178 | 7 | 174.5 | 28 | 0.160 | 0.147 |
| 18 | 171 | 12 | 165 | 32 | 0.194 | 0.177 |
| 19 | 159 | 17 | 150.5 | 38 | 0.252 | 0.202 |
| 20 | 142 | 6 | 139 | 33 | 0.237 | 0.224 |
| 21 | 136 | 2 | 135 | 30 | 0.222 | 0.242 |
| 22 | 134 | 9 | 129.5 | 31 | 0.239 | 0.256 |
| 23 | 125 | 12 | 119 | 28 | 0.235 | 0.268 |
| 24 | 113 | 2 | 112 | 35 | 0.313 | 0.276 |
| 25 | 111 | 11 | 105.5 | 28 | 0.265 | 0.281 |
| 26 | 100 | 9 | 95.5 | 26 | 0.272 | 0.283 |
| 27 | 91 | 7 | 87.5 | 30 | 0.343 | 0.283 |
| 28 | 84 | 3 | 82.5 | 18 | 0.218 | 0.281 |
| 29 | 81 | 3 | 79.5 | 23 | 0.289 | 0.276 |
| 30 | 78 | 4 | 76 | 25 | 0.329 | 0.269 |
| 31 | 74 | 1 | 73.5 | 14 | 0.190 | 0.261 |
| 32 | 73 | 2 | 72 | 14 | 0.194 | 0.251 |
| 33 | 71 | 4 | 69 | 17 | 0.246 | 0.239 |
| 34 | 67 | 10 | 62 | 14 | 0.226 | 0.226 |
| 35 | 57 | 8 | 53 | 13 | 0.245 | 0.212 |
| 36 | 49 | 5 | 46.5 | 7 | 0.151 | 0.197 |
| 37 | 44 | 7 | 40.5 | 11 | 0.272 | 0.182 |
| 38 | 37 | 1 | 36.5 | 5 | 0.137 | 0.165 |
| 39 | 36 | 4 | 34 | 3 | 0.088 | 0.149 |

**Table 7.1** (*cont.*)

| Age | Enter | Censored | At risk | Births | ASFR | Smoothed |
|---|---|---|---|---|---|---|
| 40 | 32 | 5 | 29.5 | 4 | 0.136 | 0.132 |
| 41 | 27 | 3 | 25.5 | 3 | 0.118 | 0.115 |
| 42 | 24 | 6 | 21 | 3 | 0.143 | 0.098 |
| 43 | 18 | 3 | 16.5 | 3 | 0.182 | 0.081 |
| 44 | 15 | 2 | 14 | | 0.000 | 0.065 |
| 45 | 13 | 2 | 12 | 1 | 0.083 | 0.050 |
| 46 | 11 | 1 | 10.5 | | 0.000 | 0.036 |
| 47 | 10 | 2 | 9 | | 0.000 | 0.023 |
| 48 | 8 | 1 | 7.5 | | 0.000 | 0.010 |
| 49 | 7 | 1 | 6.5 | | 0.000 | 0.000 |
| 50 | 6 | 1 | 5.5 | | 0.000 | 0.000 |
| 51 | 5 | 1 | 4.5 | | 0.000 | 0.000 |
| 52 | 4 | 2 | 3 | | 0.000 | 0.000 |
| 53 | 2 | 1 | 1.5 | | 0.000 | 0.000 |
| 54 | 1 | 1 | 0.5 | | 0.000 | 0.000 |
| 55 | | | 0 | | 0.000 | 0.000 |
| 56 | | | 0 | | 0.000 | 0.000 |
| 57 | | | 0 | | 0.000 | 0.000 |
| 58 | | | 0 | | 0.000 | 0.000 |
| 59 | | | 0 | | 0.000 | 0.000 |
| 60 | | | 0 | | 0.000 | 0.000 |

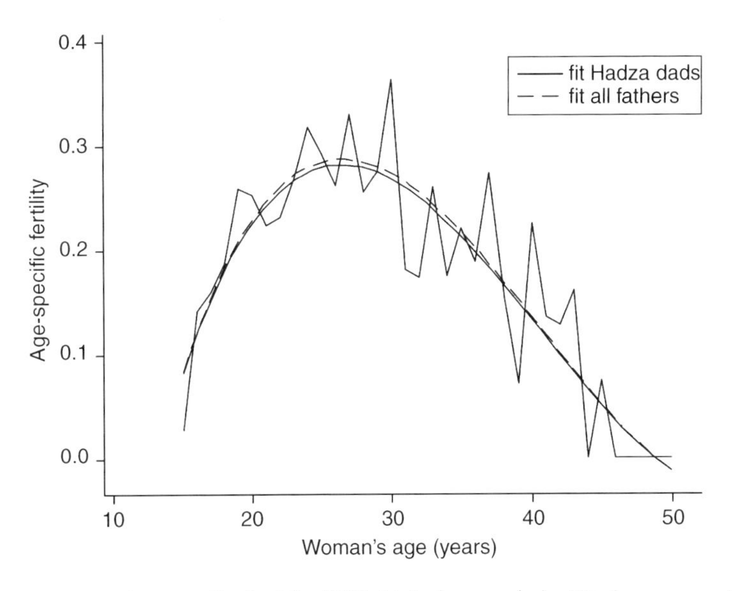


**Figure 7.1** Age-specific fertility (ASF, births/woman/yr) of Hadza women. Fitted lines for age + age$^2$ + age$^3$, women aged 15–50 only. Broken smooth line, regardless of husband ethnicity; solid smooth line, for women married only to Hadza men.

It is interesting that the tribal affiliation of the father apparently makes so little difference (SI 7.4). In comparative perspective, Hadza fertility is almost exactly at the mean (6.1) found for a sample of 70 natural fertility populations by Campbell and Wood (1988, see Wood, 1994, pp. 30–33, fig. 2.1). Hadza fertility is about halfway between the figures for the !Kung (4.7), or the central Kalahari San (4.1; Tanaka, 1980) and the Ache in the forest period (8.03).

My impression in the field was that informants were strongly biased toward reporting boys. However, the 695 births comprised 354 males and 331 females, and 10 of unrecorded sex, for a sex ratio at birth of 1.069, 107 males born for every 100 females. This is a figure in the range of many other populations. The exclusion of children with non-Hadza fathers appears to make a difference. We record 544 births to women who had children only by Hadza men, 275 boys and 269 girls; 103 boys to every 100 girls. The difference is unlikely to be significant; if all the unknowns had been male, the ratio would be 105:100, and if female, 99:100. The figure for Hadza fathers, 103:100 is within the range reported in literature for Africa. For example, James (1984) reported 104.3:100 for a sample of African countries, and Garenne (2002) reports 103.3:100 (CI 102.9–103.7). Both included Tanzanian samples.

Gross reproductive rate, the TFR of female births, was 3.15. Despite the sex ratio, this figure was higher than the corresponding figure for males (3.06) and this appeared to be because of more female births late in women's careers, when chance variation among the smaller number of cases contribute perhaps disproportionately to the TFR. There may be ways to account for a preponderance of female births late in the reproductive career but the sparse Hadza data preclude testing of these explanations.

A few Hadza adults have been described to us as a surviving twin. During the reproductive history interviews in the 1990s, we recorded three sets of twins. Between 1992 and 2000 we recorded 284 births. The three pairs give a twinning rate of 0.0106 or 11/1000 births. This is within usual figures for most human populations although rates in Africa have been reported to be rather higher (Bulmer, 1960). Two of the babies had died in infancy while their siblings survived. The third pair survived for at least 18 months, their young mother being supported by a relay team of her mother and sisters. I found no indication that twinning was seen as a context for infanticide. Indeed, my questions about infanticide met with incredulity. Informants thought it highly unlikely that any woman would ever do this.

## 7.4 Resampling

I employed a resampling method to estimate the confidence limits of the mean TFR. Resampling comprises taking repeated random samples of our sample of 227 women and calculating the TFR for each of the random samples. In each sampling, the computer randomly selects women from the original sample, until 227 cases have been included in our artificial sample. Some individuals are used more than once, some omitted altogether. It is usual to take 500–1000 such samples. A frequency table can be built up showing the number of times a run resulted in a particular TFR (table SI 7.5). Median, mean, and confidence limits can be estimated from the frequency table.

For women who had no children by non-Hadza husbands, the sample size was 195 women and 500 runs were done with these women. The mean of the 500 estimates of TFR for the 500 samples of these women was 6.07, or 6.15 if we obtain the average TFR by summing the average ASF for each year. Both are very close to the actual sample TFR of 6.17. Ninety-five percent of the estimates for women who had all their children by Hadza husbands fell between TFR 5.2 and 6.8. For all women, regardless of husband, the mean TFR was 6.18 (SD, 0.03) when calculated from the frequency distribution, and 6.24 when calculated as the sum of the average ASF at each age. The SI fig. 7.5 shows the extent of the variation in the resampling runs. Ninety-five percent of the estimates fell between TFR 5.5 and 6.7. The 67% levels were 5.6–6.4 and 5.8–6.5, respectively.

Although these resampling investigations reveal a wide range of variation, we may note that the confidence limits do not reach the TFR 4.7 estimated for the !Kung by Howell (1979), nor the TFR 8.03 estimated for the Ache in the pre-contact forest period by Hill and Hurtado (1996: table 8.1).

## 7.5 Annual probability of a birth by logistic regression

I made a file in which each record is a year in the life of an interviewed woman. The adult life of each woman is represented by a series of records, the first for the year in which she was aged 10, the next aged 11, and so on, until the year of her last interview. The variables include whether she gave birth or did not give birth in that year, her age in each year, the calendar year, and a number of other "independent" variables such as whether she was married, her husband's tribe, number of children born, and so on. These variables will be discussed in the chapters on marriage, and on helpers.

Age-specific fertility appears as the predicted probability of a birth to a woman aged x years. TFR is the sum of these over an age range, such as 15–50. The result is a smoothed prediction that should resemble the fitted graph of ASF previously calculated. An advantage of this method is that it is easy to test possible effects of the independent variables. Particularly important is the calendar year, which, if we control for mother's age, is a test for a secular trend in Hadza fertility during the period covered by the interviews. Controlling for the woman's age and age-squared, we find no significant effects of the region in which a woman lives. Women in Mangola had a slight but non-significant tendency to be more likely to bear a child. Women married to a non-Hadza man were no more likely to bear a child than women with a Hadza husband (Table 7.2).

Total fertility rate, as the sum of annual predicted probability of a birth, is sensitive to the model used and the age range over which the probabilities are summed. Predicting probability from mother's age plus age-squared gives different probabilities from the same model with age-cubed added, which gives a slightly better fit and

**Table 7.2** Region and tribe of husbands of interviewed women have no significant effect on Hadza women's fertility. Logistic regressions of the form: a birth in year x = woman's age in year x + age-squared + living in region A in year x. Sample: 2312 married years of 197 women ages 13–54, 244 woman–years with a non-Hadza husband

| | b | p | Odds ratio | 95% CI |
|---|---|---|---|---|
| Mangola | 0.201 | 0.089 | 1.22 | 0.97–1.54 |
| Tliika | -0.044 | 0.717 | 0.96 | 0.75–1.21 |
| Siponga | -0.098 | 0.395 | 0.91 | 0.72–1.14 |
| Non-Hadza husband | -0.071 | 0.648 | 0.93 | 0.69–1.26 |

a realistic plot until the high 40s and 50s in age. Summed from age 15 to 45 the last model gave a TFR of 6.02, close enough to our previous estimate. Controlling for a woman's age, calendar year was found to have no significant effect; there was no evidence of a secular change in fertility during the study period (b=-0.0062, p=0.302, OR=0.99, CI 0.98–1.01, 683 births in 4524 woman-years). A woman who was, 20 years old in 1980, for example, had the same probability of a birth in her twenty-first year as a woman who was aged 20 in 1999 had in her twenty-first year.

Because it is easy to investigate independent variables with this method, I decided to follow up a paper by Galvin and colleagues (2001) on the effect of two climatically extreme years on Maasai pastoralists in the nearby Ngorongoro region. First, in 1996–1997, the rains failed and drought was quite extreme. Maasai sold more livestock and agricultural productivity was massively reduced. Then from October 1997 into 1998 there was an El Niño year, with very high rainfall and serious flooding all over East Africa. Livestock diseases became more abundant. When I was in the field in 1997, although no Hadza complained of water shortage, people seemed under stress, and I noticed some unusual movements of people. One camp, named for a man who had spent almost his entire life in the Munguli and Tliika area, moved en masse to Mangola. They were back in their normal range in 1999 but the 1997 movement was unprecedented. This made me curious when I saw Galvin's paper.

I created two dummy variables, 1997 and 1998, to test for effects of the total failure of rains in 1997 and the unusually high rainfall of 1998, after controlling for the woman's age. The dummy variable, 1998 (scores 1) versus other years (score 0), was significant (b=-0.4643, p=0.041, OR=0.63, CI 0.40–0.98). Births were significantly scarce in 1998. The odds of a woman giving birth in 1998 were 63% of the odds of her giving birth in any other year. There was no effect of the dummy variable for 1997, nor for 1999. Many of the babies born in 1998 would have been conceived in 1997 and this finding may indicate that 1997 was as difficult a year as it appeared.

One often reads the claim that hunter-gatherers are little affected by famines, or harsh seasons and difficult years. However, Howell (1979, p. 145) commented on low fertility among the !Kung in the year after the start of a drought (1963–1964) and a similar phenomenon is reported for large-scale famines among agricultural peoples (Dyson, 1991a,b). Here we see another example, a clear effect of exceptional climate on Hadza reproduction. My curiosity was encouraged. Did the provision of food at the Mongo wa mono settlement, at its apogee in 1990, increase the probability of a birth? Again controlling for woman's age and age-squared, the year 1990 had a large and significant positive effect on births (b–0.5065, p=0.007, OR=1.66, CI 1.15–2.40). The odds of a woman giving birth in 1990 was 1.66 times the odds of her giving birth in some other year. Here is an indication that "good" years can increase fertility. Among the Hadza, it looks as if fertility is as flexibly adjustable to conditions as physiological ecologists have sometimes suggested.

## 7.6 Analysis of fertility by multilevel logistic regression

Multilevel logistic regression allows us to see how the probability of a birth varies with mother's age, while controlling for variation between mothers in fertility and in representation in the sample. An over-representation of very fertile women, or women from a more fertile era, could have biased the previous "single level" analyses.

Models were fitted to the age 15- to 50-year sub-sample, using mother's age and age-squared as predictors. All 227 women were included, regardless of whether they had Hadza or non-Hadza husbands. Adding mother's year of birth or the calendar year of each record (which stretches from about 1965 to 2000) again showed that there was no secular trend in fertility within these data. Predicted probabilities of birth in each year of life could be summed to give the TFR. The model with age + age-squared predicted a TFR of 6.1, very close to the previous estimates. Plotting the fit to this and other models suggested that age and age-squared gave the most realistic pattern of ASF. Adding age-cubed predicted a higher probability of births between ages 45 and 50 and gives a higher TFR at 6.5.

The multilevel software (Mlwin) allows one to estimate the mother residuals and store this measure of mother-to-mother variation. These measures correlate very closely (as they should) with our later estimates of individual variation in which we calculated individuals' standard scores of number of births relative to the other individuals of their age ($r=0.9454$, $p<0.0000$, adjusted $r^2=0.8932$).

## 7.7 Marital fertility

Thus far, I have reported fertility, the probability that a woman of any age and status bears a child. This is the measure we need in order to estimate rate of increase and other population parameters. Marital fertility, the probability that a married woman bears a child, can be important in other contexts. For instance, it may be considered closer to a measure of physiological fecundability. In some populations, economic and religious factors delay marriage and limit sexual relations outside marriage, and in some societies, subsistence or warfare take men away for long periods of time. Among Hadza such restrictions may not apply: (1) Fecundability is likely to be an important cause of marriage. Women's attractiveness to men probably tracks their fecundability quite closely, especially among the younger women (this is

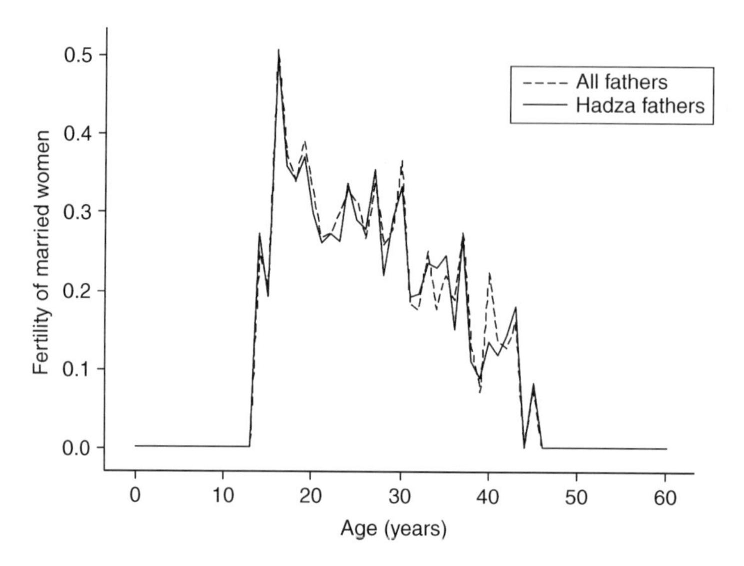


**Figure 7.2** Marital fertility computed in the same way as age-specific fertility (ASF).

examined in Part II, Chapter 15 on marriage). (2) Hadza women almost never report themselves as unmarried when conceiving a child, and marriage often follows a conception or birth. (3) My data are not well set up to demonstrate cohabitation independently of childbirth and paternity. (4) Absence is a common context for divorce (Woodburn, 1968b).

First, I estimated marital fertility the same way as I estimated fertility, with the difference that women were only at risk if they were recorded as married, and only the births to the women recorded as married were entered as births (Figure 7.2). I counted a woman as married from the first year in which we had any indication that she was married, either from birth of a child and her nomination of the father or a husband, or from census household lists.

The plot of Hadza marital fertility shows the usual result. Marital fertility climbs to a high level in the late teenage years, and declines steadily thereafter. The low marital fertility at ages 14 and 15 may indicate adolescent sub-fecundability (SI 7.6), but the sample size is very small and the age estimates are not accurate to better than a year or two. We did know one extremely young childless couple who were regarded by other Hadza, with some amusement, as married and were seen together in widely separated locations on several occasions.

The growth in numbers of women married during their late teenage years and early 20s may primarily reflect individual variation in their physical maturation. The observation that absolute fertility, and incidence of marriage, increase between ages 16 and 20, while marital fertility declines during this time, may indicate that the course of physical maturation, which determines fecundability, may in turn influence marriage and fertility. From the Hadza perspective this sounds obvious, but from the perspective of more restrictive societies commonly studied by demographers, it is probably unusual.

## 7.8 Looking for secular change in Hadza fertility

The logistic regressions showed that there was no secular trend toward higher or lower fertility between the mid-1960s and 2000, and with the brief exceptions of 1990 and 1998, no sudden transition in fertility during our study (SI 7.11). Nonetheless, if we are to use a stable population assumption to predict some features of the population, we need to have some indication of the duration for which fertility has remained at the level estimated. Are Hadza more fertile now than in former times? Some Hadza have access to farm produce, temporary wage labor and, since 1995, to tourist money. Any of these changes could have changed fertility.

There is direct evidence from the mid-twentieth century. Dyson (1977) reported on Hadza demography using data gathered by the International Biological Program (IBP) in 1966 and 1967. These data included 75 reproductive history interviews conducted by James Woodburn on women aged 20–49. These women, born between 1922 and 1947 (thus, very few could have been included in our sample of women born between 1945 and 1985), would have given birth between, roughly, 1940 and 1967. Dyson reports that the data indicate a TFR of 6.15. To judge from our resampling results, this figure cannot be significantly or reliably different from our TFR of 6.17 or any other of our estimates.

Dyson tabulates the number of births (N) to women aged 20–29, 30–39, and 40–49, and shows the mean parities for these age groups. Figure 7.3 plots these mean parities from 1967 on a similar plot of our recent interview data. The 1967 figures fall very close to the polynomial regression fitted to the 1985–2000 data. This suggests that before 1967, Hadza women were reproducing at a very similar rate and schedule to Hadza women in the 1980s and 1990s.

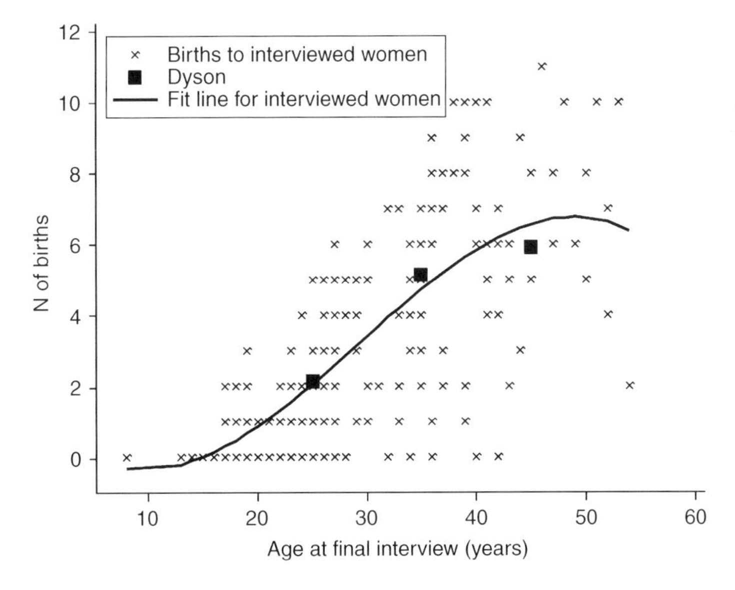


**Figure 7.3** Cumulative number of births to women of different ages. Comparing Dyson's (1977) analysis of 1967 data with our 1985–2000 data on women born 1945–1985.

## 7.9 Parity progression ratio

The parity progression ratio (PPR) is another commonly reported feature of fertility. It is the proportion of women who, after each birth, go on to have another birth. It drops to zero at the end of the childbearing career, although some women may cease childbearing at any time, which produces a decline in the PPR. One common use of the measure is as an indication of controlled fertility as opposed to "natural fertility." The idea is that it would show a woman's response to the number of children she has. If she feels that two is a good number, and has a way to lower her likelihood of becoming pregnant, then PPR might decline at two children. If many children are desired, or if she has no way to control her fertility, her PPR will remain high. We have no doubt that the Hadza are a "natural fertility" population, and given their opinions on the matter, we expect them to continue to be a natural fertility population for many years. Because PPRs are so commonly reported, I report them here. Parity progression ratios can decline without artificial contraception, and without conscious decisions about fertility control. Increased incidence of infertility or sub-fertility can arise from infection, or damage to the reproductive tract, and lower the probability of another birth.

The PPR also serves to introduce a device that we will use to examine the age at the last birth and secondary sterility Larsen's "B60." PPRs are most easily calculated for a sample of women who have completed their reproduction, but our sample is a "mixed longitudinal" sample, with many women who are still quite young. Larsen (2003) approximates failure to have another birth by counting the proportion of women, observed for at least five years (60 months) after a birth, who do not have another birth during the five years. I calculated Hadza PPRs using this device.

The SI fig. 7.7 shows Hadza parity progression calculated with the five-year waiting period, and calculated with a more conservative seven-year period. There is virtually no difference between these. Five years is evidently quite adequate. If we compare Hadza parity progression with the populations illustrated in Howell (1979, fig. 8.1), we see that the Hadza resemble the Hutterites and the Quebecois more than they resemble the !Kung. Hadza parity progression declines very little until they reach very high parities. This implies not just that we are right about absence of contraception but that there is also a fairly low incidence of premature infertility, and perhaps little effect of the added burden of an accumulating family. We will return elsewhere to the latter. In Figure 7.4, I compare the Hadza PPR with figures read from Larsen and Hollos's (2003, fig. 1) data from Pare in the Kilimanjaro region. Hadza PPR remain higher than the Pare PPR with successive parities.

## 7.10 Age at first birth

Age at first birth is an important life history variable. There is robust theory about the evolution of optimal age at first reproduction, and many other life history variables are related to age at first birth. Therefore, it is worth taking some trouble to estimate this value as well as we can.

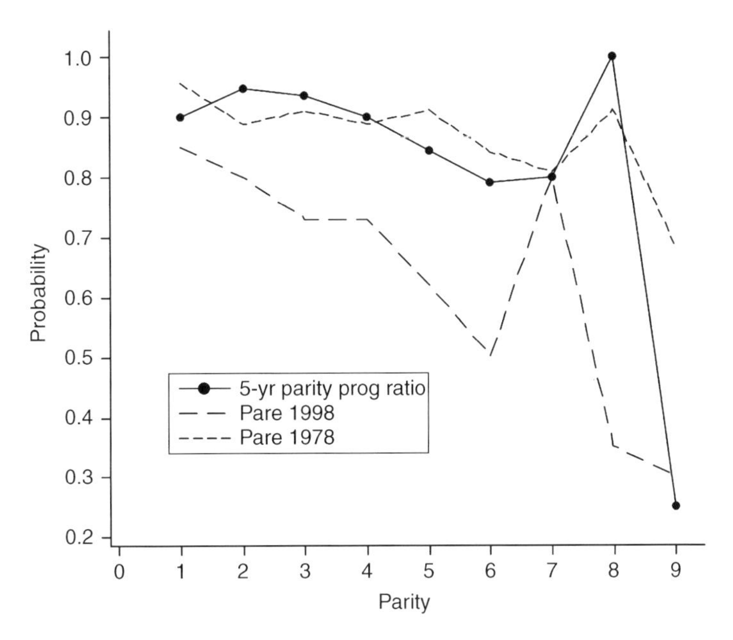


**Figure 7.4** Parity progression ratios for the Hadza, and samples of Pare women in northeastern Tanzania. Parity progression ratios for Pare women in 10-year cohorts that began their childbearing during 1989–1998, and 1969–1978 (values read from Larsen and Hollos, 2003, fig. 1). Pare women bearing children between 1969 and 1978 show a similar pattern to Hadza women who bore children between 1985 and 2000. Later cohorts of the Pare changed their pattern such that subsequent births became less likely with each increase in parity. The brief increases at high parities may reflect small sample size, or a "frailty" process in which only the most fertile women reach the highest parities.

The interview data that were used to estimate ASF are the best data available to us. We have information on each woman's reproductive history up until the last time we interviewed her. From this, a program gives us the number of women of each age on whom we have information, and the number who had their first birth at each age, the number in the study yet to have their first birth, and the number in the study who have had their first birth. The average age of mothers at the recorded first births was 18.95, the median was 18, and the mode was 17.

A better measure is the "survival function" used by Howell (1979, tables 6.3, 9.1), survival of the "yet to bear a baby" condition (Table 7.3). This takes account of the number of women entering each year of life, and their loss either by "leaving the sample" (not being interviewed at an older age), or by leaving the risk group by virtue of having borne their first child. The method is the same as for the computation of "$q_x$" and "$l_x$" for mortality in the next chapter. Figure 7.5 plots the inverse measure ($1-l_x$) for the Hadza alongside the same figures for the !Kung and Ache from Hill and Hurtado (1996, fig. 8.5). Fifty percent of the Hadza have had their first birth by age 19 whereas 50% of !Kung girls have had their first birth by age 20. Age at first birth is youngest in the Ache, then the Hadza, and oldest in the !Kung, but the differences are very small. However, both Migliano *et al.* (2007) and Kramer (2008) have reported

**Table 7.3** Age at first birth. Survival in the state of not yet having given birth, equivalent of Howell (1979, table 6.3). The number at risk in any year of age was the number of women whose record included that year of life and who had not yet had their first birth. Birth/enter ("$q_x$") is the number of first births in the year divided by the number of women at risk as they enter the year; "$l_x$" begins at 1 and is calculated as $l_{x-1} \times (1\text{-}q_x)$

| Age | At risk | First birth | Birth/enter | Survival function |
|---|---|---|---|---|
| 5 | 227 | 0 | 0.000 | 1.000 |
| 6 | 227 | 0 | 0.000 | 1.000 |
| 7 | 227 | 0 | 0.000 | 1.000 |
| 8 | 227 | 0 | 0.000 | 1.000 |
| 9 | 226 | 0 | 0.000 | 1.000 |
| 10 | 226 | 0 | 0.000 | 1.000 |
| 11 | 226 | 0 | 0.000 | 1.000 |
| 12 | 226 | 0 | 0.000 | 1.000 |
| 13 | 226 | 0 | 0.000 | 1.000 |
| 14 | 225 | 3 | 0.013 | 0.987 |
| 15 | 220 | 6 | 0.027 | 0.960 |
| 16 | 205 | 24 | 0.117 | 0.847 |
| 17 | 177 | 28 | 0.158 | 0.713 |
| 18 | 144 | 24 | 0.167 | 0.594 |
| 19 | 113 | 21 | 0.186 | 0.484 |
| 20 | 80 | 20 | 0.250 | 0.363 |
| 21 | 57 | 11 | 0.193 | 0.293 |
| 22 | 46 | 10 | 0.217 | 0.229 |
| 23 | 35 | 6 | 0.171 | 0.190 |
| 24 | 27 | 7 | 0.259 | 0.141 |
| 25 | 20 | 2 | 0.100 | 0.127 |
| 26 | 17 | 3 | 0.176 | 0.104 |
| 27 | 13 | 1 | 0.077 | 0.096 |
| 28 | 8 | 0 | 0.000 | 0.096 |
| 29 | 7 | 1 | 0.143 | 0.083 |
| 30 | 6 | 0 | 0.000 | 0.083 |
| 31 | 6 | 0 | 0.000 | 0.083 |
| 32 | 6 | 0 | 0.000 | 0.083 |
| 33 | 5 | 0 | 0.000 | 0.083 |
| 34 | 5 | 0 | 0.000 | 0.083 |
| 35 | 3 | 0 | 0.000 | 0.083 |
| 36 | 3 | 0 | 0.000 | 0.083 |
| 37 | 2 | 0 | 0.000 | 0.083 |
| 38 | 2 | 0 | 0.000 | 0.083 |
| 39 | 2 | 0 | 0.000 | 0.083 |
| 40 | 2 | 0 | 0.000 | 0.083 |

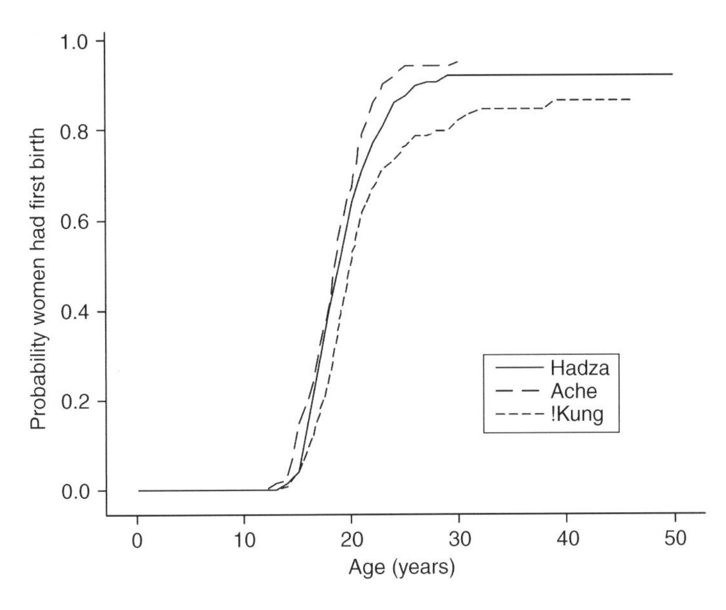


**Figure 7.5** Probability that a woman has had her first birth. Comparing the Hadza, Ache, and !Kung. Ache from Hill and Hurtado (1996, fig. 8.5), and !Kung from Howell (1979, table 9.1). Howell's 1963–1973 sample resembles the Hadza sample in including women of all ages. Howell's sample of 62 women aged 45 or older in 1968 shows the median at age 20 but climbs more steeply among the >20s, exhibiting zero primary sterility among the >30s (discussed in the text).

significantly lower ages at first birth. Among the Pume studied by Kramer, with apparently very reliable historical age markers, age at first birth was noted as 15.3–15.5, accompanied by a teenage birth rate of 195 births/1000 women aged 15–19. Kramer links this to the very high losses of first infants. Migliano *et al.*, attribute the early age at first birth of Pygmies (a collection of peoples of very short stature from Africa and southeast Asia) to high adult mortality, an explanation in line with much life history theory.

## 7.11 Average age of the mother at a birth

Median age of the mother at a birth was 28 to 29. I calculated average age at birth from Table 7.1 by multiplying the ASF by age, summing these, and dividing by the sum of the ASF (giving the TFR). The result is 28.77 for all women. For women who had children only by Hadza husbands the result is 28.75.

## 7.12 Primary infertility

The probability of a Hadza woman having had her first birth appears to climb faster and earlier than for a !Kung woman and reach a higher asymptote (Figure 7.5). The asymptote represents an estimate of the proportion of women who never bear a child. We can see that this proportion is rather lower for the Hadza (leveling off by age 30 at about 8%) than for the 1963–1973 !Kung (almost 14%).

In the 1960s to 1980s, demographers described a sub-Saharan "infertility belt" (Caldwell and Caldwell, 1983), an area covering several central African countries in which very high incidences of primary infertility could be found (Frank, 1983). The levels were strongly variable locally and tribally. The infertility has been attributed primarily to gonococcal infection resulting in occluded Fallopian tubes (Belsey, 1976), but chlamydia would seem to be as likely a candidate. Widespread use of penicillin is held to have lowered the incidence dramatically in some areas and the phenomenon apparently had been much reduced by the end of the twentieth century (Larsen, 2000, 2003). The phenomenon has, from time to time, been an issue in hunter-gatherer demography because of its possible influence on !Kung fertility (Harpending, 1994; Pennington, 2001). There have been several issues: does infection account for the low TFR of the !Kung, and for the long inter-birth intervals reported by Lee (1972) and Howell (1979)? Is it the most important factor accounting for these long intervals and low fertilities? How long has infection been a factor? Can it have been an effective limiter of hunter-gatherer populations in the more distant past?

There were 99 women in the reproductive history sample of 227 who were aged 30 or more when last interviewed. Six of them (6.1%) claimed never to have borne any children. But we could probably rely on most of the older women to be able to tell us whether they had had any children. One hundred and fifty-four (154) women in the ASF candidates file (the interviewed women) had passed age 30 by the time of their last interview. Among the older women, born before 1945, added in this count were three more who at first seemed to have had no children in our population register. However, their interviews and Lars Smith's 1977 census showed that they had borne children but these had died long ago and we had no usable information about these children. We had not added these children to the population register. Hence, in this extended sample we have six who bore no children out of 154 women (6/154=0.03896 or 3.9%).

By this count, the Hadza have a rather low rate of primary infertility, lower than the Tanzanian national figure of 10%. Intriguingly, just as among the !Kung, the older women present a very low level of infertility. Howell suggested that older infertile !Kung women are more likely to leave the area than those with many descendants to tie them. Our investigation of emigration (Chapter 5) gave no indication of such women among the Hadza. Infertility may have increased among the Hadza with the increase and encroachment of the surrounding population. Yet, sexually transmitted diseases (STDs) have been cited as one of the infectious diseases that could endure in a sparsely populated world of hunters among hunters (Anderson and May, 1991; Pennington, 2001).

Larsen (2003) uses data from married women only. If we follow this practice and exclude the two women unmarried during the period of our censuses, the rate of primary infertility drops to 4/99 (4%), (or 4/154=2.6%), not far from the level treated in the literature as a universal normal baseline level. In many African societies, infertility is given as a reason for divorce and women who are known to be infertile are not preferred marriage partners. Thus, marital fertility might be a very misleading measure of infertility in certain parts of Africa. We are aware of no strong or formalized version of this attitude among Hadza and these women's histories suggest that infertile women have not been avoided. Of the two excluded because they were unmarried at the time of our censuses, both had been married previously, thus all of the six women had been married at some time in their lives but had never borne a child. Three have been married to the same husband for many years, one has been married but has now been single for a long time, and the fifth has divorced twice and remarried three times. The sixth member of this group of infertile over 30s was married for only a brief period and is viewed by her neighbors as having a difficult temperament.

## 7.13 Age of women at last birth, and secondary sterility

My data are not well suited to determine age at last birth because I have data on relatively few women who are clearly past the childbearing years. In an unpublished study of endocrine measures of menopause in Hadza women, Phillips *et al.* found menopause arriving at an estimated age of 43, earlier than in European women (SI 7.12, and SI 19). I tried several different ways to determine age at last birth and estimate its central tendency.

There are two general classes of process that bring the reproductive career to an end. Menopause arrives, and is preceded by a period of sub-fertility, sometimes described as elongated inter-birth intervals (not the case among the Hadza, see later and Figure 7.9), sometimes as early final births. The last birth occurs some time before the endocrine indicators of menopause show an end to ovulation. Some women become infertile much earlier than this, usually assumed to be a result of infections. This was clearly an issue among younger !Kung women during the 1960s (Howell, 1979, p. 186; Pennington, 2001), and there is a large literature on the levels of early secondary infertility in central Africa, including Tanzania (Larsen, 2003; Larsen and Hollos, 2003). Even though Hadza had low primary sterility, it would be surprising if Hadza were not also affected by secondary sterility, and I attempt to assess its extent. Premature infertility will, of course, have an effect on average age at last birth and the length of the reproductive career. It would be helpful to form a rough indication of how much Hadza fertility is lost to premature sterility, and to make a rough estimate of the length of the average reproductive career.

If the ASF figures in Table 7.1 are added backward, from the oldest age to the youngest, we can get an estimate of the median age at last birth, the age at which the chance of another birth (ever) is 0.50. This age is a little over 41. This means that at age 41, half the women may have borne their final child, and half of them have not yet had their final child. I tried several other ways to check this estimated age at last birth. (Details are given in SI 7.8.) (a) I looked at the subsequent career of women in the 1985 census cohort. Most women aged up to 35 in 1985 gave birth again. The fraction giving birth declined rapidly in their late 30s and early 40s. (b) I looked at older women, even if not interviewed, by finding the time since the birth of their youngest child in our population register, and limiting this to women who we saw often, and therefore were likely to have the best records of their children. Their mean

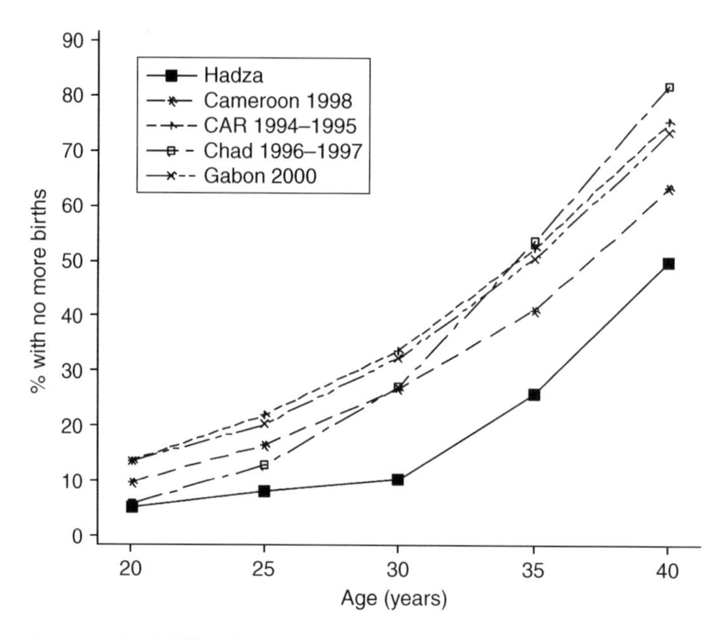


**Figure 7.6** Probability that a woman has had her final birth. Each data point marks a sample of women who were observed for another five years from age x. The score on the Y-axis is the percent of the women who did not have another birth during that follow-up period. Also plotted are data from Larsen (2003, table 3) from four West African countries.

age at last birth was 37.8 with a median of 40. (c) I employed the assumption used by Larsen in several of her studies (Figure 7.6). If observations continue for five or more years after a birth and there has been no new birth, we will count the recorded birth as the final birth. In Chapter 17, we see that most inter-birth intervals are indeed closed by five years. I also repeated the analyses using a seven-year period. The Hadza seem to have fewer prematurely infertile women than observed among other African populations (Figure 7.6). (d) I expressed these data in a survival calculation, estimating the probability of another birth at each age. According to this method, a young woman aged 19 can expect, on average, 17.8 more years of fertility.

The results show that there are many women who bear their last child in their early 40s, there are also many who bear their last child in their late 30s, and some bear their last child in their early 30s or even 20s. The Hadza are not completely spared the reproductive losses of secondary infertility. Clearly, the Hadza have their last birth later than the age 34.35 that Howell estimated for her sample of !Kung women interviewed at ages 45+. The Hadza figure is earlier than the 42.1-year mean age (median 43) at last birth that Hill and Hurtado (1996, p. 254) report for Ache women in the forest period.

The end of the childbearing career has been described as a gradual process. At a population level it certainly is. The probability of a birth in any year declines from mid-career (Table 7.1 and Figure 7.1). However, the career can look different at an individual level. Howell's data on her cohort of women aged 45+ shows closed inter-birth intervals lengthening as the women age; the Hadza data do not (Chapter 17 and

Figure 7.9). A Hadza woman tends to produce babies at a regular interval and then simply stop. The probability of closing an interval declines with age but the length of a closed interval does not increase (unless we deny the meaning of the word "interval"). Some would see this pattern as most compatible with disease-induced secondary sterility, despite its difference from the !Kung pattern. However, in Chapter 19 we see that in the Hadza case, it can also be given an energy budget interpretation, via the effect of helpers on the older woman's energy budget. The difference between the !Kung and Hadza patterns can also be seen as evidence for an influence of energy budget when we take account of the different costs of raising children of different ages in the two environments (Blurton Jones *et al.*, 1989).

## 7.14 Men's fertility

I did not interview men about their reproductive history but I have estimated their CFS, ASF, and TFR from the censuses and population register. These are sure to be underestimates, because long-dead children may never have been seen nor entered the population register, and because our interviews may not have covered all the women who bore children to these men. I took men who had been seen in three or more censuses, because these are the people we knew best and we were most likely to have recorded all their children. These men were "at risk" of a birth in each year of their lives up to the age when last seen in a census. I used the record of their children's year of birth in the population register to map out the men's reproductive histories. Most of the information on which the children's ages were based had been obtained from the mothers. Some, for older "children," came from the "children" themselves. For comparison with the men, and with the other estimates of women's fertility, I did the same with women. The sample of women seen in three or more censuses overlaps with the interviewed women, the young women, and any others for whom we have an age estimate and met three or more times during a census between 1985 and 2000.

**Age-specific fertility and total fertility rate** Figure 7.7 shows the ASF of the men and women. The difference in scheduling of births show up very clearly. Men have a childbearing career that is delayed relative to that of women. Men have few babies in their early 20s, but continue to father children throughout their 40s, 50s, and into their 60s. The population register reveals six men who fathered a child in their 60s, two of them more than once for a total of nine children. However, as will be the case in all other studies of such populations, the age estimates for the oldest people will be the least accurate. While the children were born recently and probably had their birth years quite well estimated, their fathers could be five years older or younger than estimated. Furthermore, in high mortality populations, the number of surviving old men is small. We should not make too much of population differences in age at last birth to men.

Women in this sample showed a TFR of 5.66, and men had a TFR of 5.93. As expected, the women's TFR is a little lower than that calculated from the interview sample. We can use the table of men's ASF (SI 7.9) as we did for the women, reading

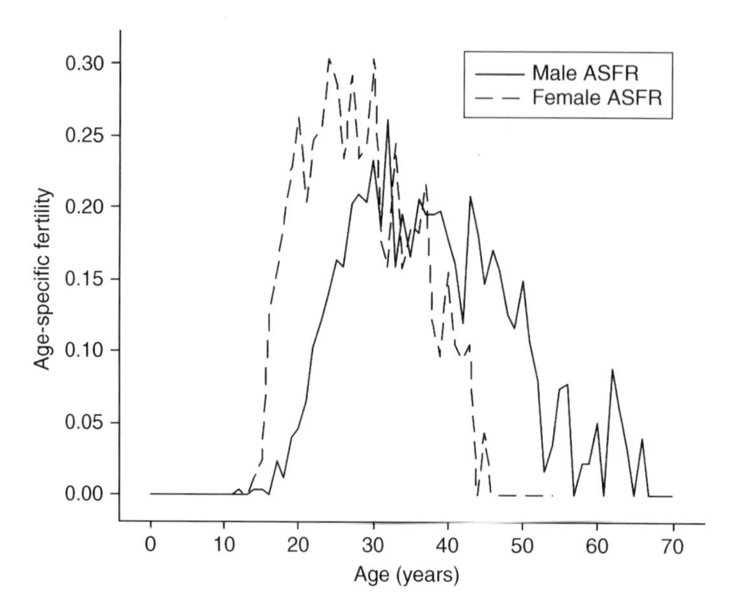


**Figure 7.7** Comparing men's and women's age-specific fertility (ASF). Data from men and women in three or more censuses and aged 45 or more at end of the observation.

the accumulated fertility at each age to read the median age at childbirth, etc. Median age at childbirth (the age when half of the TFR has accumulated) was 37. Mean age of a man at birth of a child is 33.5. At age 25 the accumulated fertility is 0.56, which I take to mean that just over half the men will have had a first child. At age 54, half the men remain to have one more child. If we sum the fitted ASF from age 40 we see that the average 40-year-old man can expect 2.4 more births, 41% of the average TFR. The average 50-year-old can expect 0.86 more births for just 14.5% of expected TFR.

**Men's parity/paternity** I also plotted the number of children born to each of these men and women (parity and paternity) against age. When I fitted a polynomial regression with three terms to these very highly variable numbers, the curves peak at a little over six for men and a little under six for women. The men's curve peaks later than the women's. The men's curve suggests that by age 25 , the average man has had one child (the mean age at first birth is 24.76). The curve levels off in their 60s, but is clearly climbing all through the 50s, suggesting the last birth is sometime in their 60s. Eleven out of 76 (11/76) men aged over 35 in 1985 had had no child by 2000.

**Men's completed family size** If we compare the 101 "census women" who have data from age 45 or more with the 68 men aged 50 or more, we can use these parity and paternity data to estimate CFS in order to examine its variance among men and women. The frequency of family sizes from zero to 17 are shown in Figure 7.8. As is well known, men can have more children than women, and more men than women have no children at all. Sixteen men had 10 or more children. Women in this sample range from zero to 11 births with a mode at six births, median of five, and mean of 4.9.

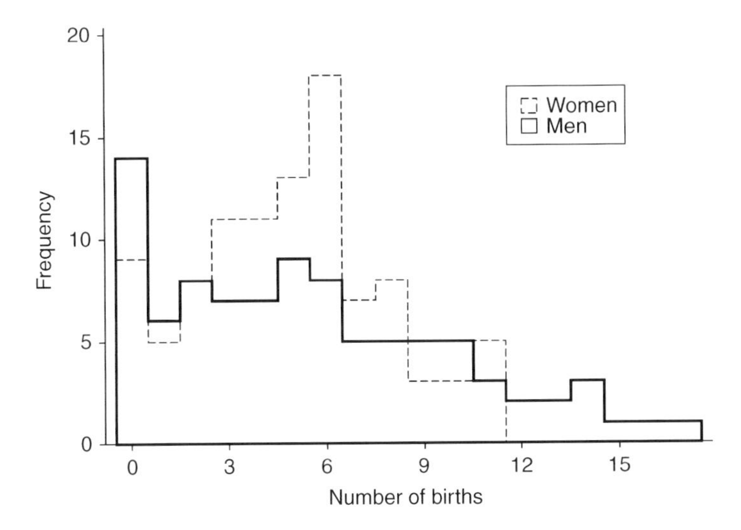


**Figure 7.8** Frequency distribution of completed family size of men and women. Men aged 50 and over and women aged 45 and over, in the population register and in three or more censuses.

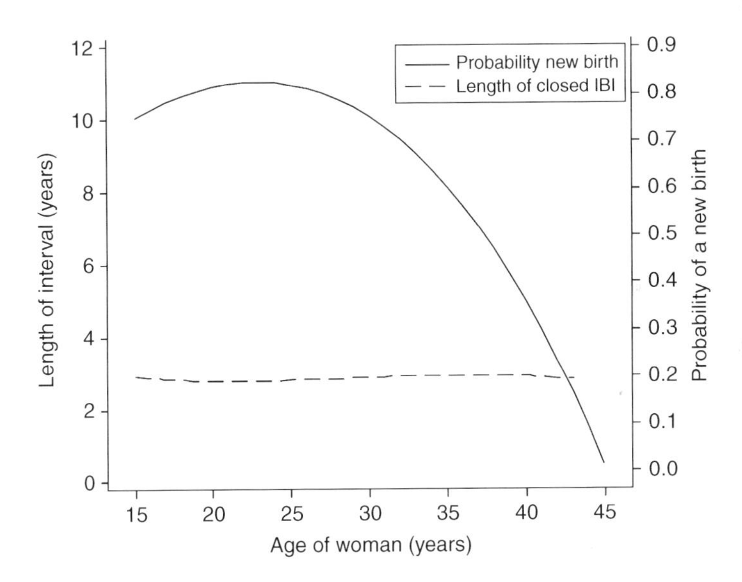


**Figure 7.9** How women's childbearing careers end. Lengthened inter-birth intervals (IBI) or not. "Probability of new birth" is the fitted probability at age x years that an interval will be closed, a birth will be followed by another birth. It declines from the age of 25. "length of closed IBI" is the length of those intervals that are closed. It does not change with age.

Men range from zero to 17 births with a mode of zero (10% of men had had no child attributable to them), median of five, and mean of 6.015. The frequency distribution for women appears roughly normal, the distribution for men shows the usual excess of zeros, and a long upper tail.

The CFS estimates for the men are slightly higher than the female estimates. Men's CFS shows great variation. The non-normal distribution among men may account for their slightly higher mean. If the difference is real, it is likely to be because of the uneven sex ratio of those Hadza aged 20–50. Genetically, if not always socially, every child has one father and one mother, therefore the total number of children produced by men and by women should be equal. However, average fertility of men and women will not be equal if the adult sex ratio is not equal, the more scarce sex would tend to have higher fertility. Thus, the higher male TFR appears to be compatible with the observation that the adult sex ratio of the Hadza in our censuses is not even; there are fewer men than women aged 20–50 (Chapter 13). But even this generalization is not safe in our case. Some Hadza women have some of their children by non-Hadza fathers, men whose reproductive histories we made no attempt to follow. Thus, we should adjust the adult sex ratio to take account of the women "lost" to non-Hadza husbands.

The difference between men and women in the age course of reproduction is interesting and important (Chapter 13) but not at all unusual. Likewise, it is interesting to see the different frequency distribution of CFS, but it has often been reported and discussed with respect to other populations. Again, the Hadza present us with a normal human demography.

## 7.15 Summary: more fertile than the !Kung, less than the Ache

Our best estimate is that Hadza TFR is just under 6.2, that the median age at the first birth is 19, and at the last birth about 37. Average age at childbirth is 28.8. Gross reproductive rate was 3.15 despite the male-biased sex ratio at birth (107:100). We recorded a twinning rate of 11/1000 births. The incidence of primary sterility is quite low, 4%, and that of secondary sterility a little higher. There is no indication of a secular change in fertility between the mid-twentieth century and 2000. Resampling showed a 95% CI of TFR 5.5–6.7, which reaches neither the !Kung at 4.7 nor the Ache at 8.0. The end of the Hadza women's childbearing careers were not preceded by lengthening inter-birth intervals, as discussed further in Chapter 19.

Women married to non-Hadza husbands had a similar ASF to that of women married to Hadza husbands. Other measures were likewise unaffected by the husband's tribal affiliation. This is surprising because non-Hadza and Hadza men pursue radically different subsistence strategies. Hadza women married to non-Hadza husbands but living within the study area (the only women in our interview and census sample with non-Hadza husbands) tend to pursue traditional Hadza subsistence. Farming is mostly men's work in this part of Tanzania. As in many other populations, men's CFS varied more widely (from zero to 17) than women's. Men fathered children between the estimated ages of 18 and 66.

Several long agonized decisions about method and exact details of calculations and statistics made almost no difference to the results. The resampling reminds us that with a sample of this size of such varied individuals, confidence limits are well outside the range of differences due to particular methods of analysis. Differences between, for example, TFR calculated by the traditional method or by logistic regression, by fits to age + age-squared with or without age-cubed, and even by cutting out some of the older informants (SI 7.2), were well within the 95% CI shown by resampling.

One methodological issue that requires further discussion is our combination of cohort and period approaches (SI 7.10). This issue has exercised demographers a great deal more than it has attracted the attention of anthropological demographers. For example, basic texts by Barclay (1958) and Hinde (1998) devote several pages to the issue, while both Howell (1979) and Hill and Hurtado (1996) hardly mention it (but see later). Anthropological demographers are mostly dealing with small populations and have little choice but to use every subject who will cooperate. In small populations, one wishes to acquire all the data possible and retrospective interviewing of as many women as one can find seems like the obvious method. Demographers of state-level societies can obtain samples of thousands of women and may have the opportunity to explore year-to-year and other kinds of variation. Their aims often include a study of change whereas anthropologists, perhaps too often, assume constancy, despite clear demonstration of changes by Howell (1979) on the !Kung, Pennington and Harpending (1993) on the Herero, Hill and Hurtado (1996) on the Ache, and Early and Headland (1998) on the Agta.

Logistic regression, as used by Hill and Hurtado (1996) and multilevel logistic regression analysis of the women's annual records largely overcome the problems of these "mixed longitudinal" data. Variation between individual women can be controlled, and calendar year can be investigated as an independent variable after controlling for women's ages. Logistic and multilevel logistic regression show no secular trend linking probability of a birth to calendar year across the period 1965–2000. This is a very long span of time, during which Hadza fertility could easily have changed. Births from this entire period contribute to our estimate of fertility of young women, aged 15 to 30, for instance. Our estimate for births after age 30 is necessarily based only on the older women in our study; the young women had not reached that age by the end of the study. We have shown that, controlling for the woman's age, calendar year (as a linear variable) had no effect on probability of a birth. There was no evidence for a secular change in fertility across the years spanned by our sample of women. We also showed Dyson's analysis of Woodburn's data on 75 women who bore children between 1940 and 1965, which yielded an identical TFR to that of our study. We probably have escaped the dangers of combining period and cohort approaches. I think, whatever the methodological intricacies, we can claim a good representation of Hadza fertility during the second half of the twentieth century.

It has long been believed that the mobility of the hunter-gatherer lifestyle restricted fertility. The data indicate otherwise. The Ache study, which found a TFR of 8.08 (Hill and Hurtado, 1996, table 8.1) in the forest period, should have confirmed the message of the comparative surveys by Campbell and Wood (1988), and by Bentley *et al.* (1993a and 1993b), and Hewlett (1991a). Hunter-gatherers do not necessarily have low fertility. The median TFR (or CFS) is almost the same for hunter-gatherers and other third-world populations. Two aspects of mobility could affect women's energy budgets and the difficulty of childcare, frequency of camp moves, and the daily walk to the food and water. Populations from which we have data vary in frequency and distance of their camp moves, and in the vigor of their daily subsistence activity. The Ache in the forest offer one extreme, a lifestyle in which they say that they moved camp almost every day, yet they achieved a TFR of 8.08. Although women provided very little of the food, their main work seems to have been transporting and caring for children. Hadza move camp every two to three weeks, more often than the !Kung as an annual average (the !Kung are anchored to a smaller number of water sources during the dry season) but the Hadza fertility is higher. These few cases suggest no association between fertility and camp moves. Kelly (2013, pp. 209–212) gives a more sophisticated account of the possible relationship of mobility and sedentism to population growth rate.

If we expect a trade-off between number and care of children, we might expect that where women have to work harder to feed their children, less energy would be available to support pregnancy, birth, and lactation. Fertility may then be slightly lowered. The data assembled by Hewlett (1991a) allows us to test whether there is a tendency for fertility to be lower where women provide more of the food. There is a small significant negative regression of TFR on percentage of food acquired by females for the 10 cases of active hunter-gatherers present in both Hewlett's table 2 and table 5. Where women gather more of the food, their fertility is slightly lower. I added the Hadza to it, using Marlowe's (2010, p. 128) 50% of food brought in by women, and my TFR 6.1. The result is unchanged; b=–0.06, CI –0.0041 to –0.113, p=0.037 (Blurton Jones, 1997). We should, of course, ask why women might vary in their allocation of energy between acquiring food and other forms of care. Perhaps the opportunity for men or others to acquire food varies among these populations. The total amount of food reaching women may be at least as important as the proportion of the food provided by their work. The topic of a trade-off between fertility and various forms of investment in children has a sizeable literature and some useful recent developments. I shall return to this in Part II.

# 8 Mortality

Howell's study of the !Kung surprised many by reporting a substantial number of older people, and a mortality schedule well above the lowest in the standard models. Hill and Hurtado's study of the Ache showed the same thing. Were our traditional beliefs about hunter-gatherers wrong, or were the !Kung and the Ache just two fortunate exceptions? Detailed information from another population should help us decide. Hence, my purpose in this chapter is to report the mortality suffered by the Hadza during our study period. I will also look at evidence about mortality at previous times in the twentieth century so that we can set Hadza mortality in the larger context of historical and geographical variation in hunter-gatherer mortality.

Commonly used measures of mortality include crude death rate (CDR), average age at death, life expectancy, age-specific probability of death, and survival. The proportion of old people in the population and the maximum lifespan are also commonly reported. Each measure has its advantages and disadvantages, and some of them can mislead the unwary. For example, life expectancy at birth ($e_0$) is heavily influenced by the mortality of infants and children, and thus is a good reflection of a widely accepted standard of life. However, $e_0$ tells us less about life expectancy at later ages ($e_x$), or about the proportion of old people in the population. Life expectancy at birth also reflects fertility (with high fertility, there are more of the vulnerable youngsters). Much of the increase in life expectancy in the industrialized nations during the past 200 years has arisen from a decrease in infant and child mortality. Only very recently has there been much change in life expectancy at age 60, for instance (Oeppen and Vaupel, 2002, and see discussion in Hawkes, 2006; Hawkes and Blurton Jones, 2005; and Gurven and Kaplan, 2007).

Population increase and decrease affect some of the commonly used measures such as average age at death. The proportion of old people in the population also cannot show us whether it signifies high or low mortality. A high proportion of old people can arise if the population is declining, as well as because old people survive quite well. By assembling as many of these measures as we can, we can offer a useful picture of mortality in any population. Sometimes patterns of mortality have been used as evidence for the importance of some particular cause of death. Occasionally, in collections of large samples of different times or populations, it is possible to see demographic effects of particular diseases or problems (Preston and Haines, 1991). We do not expect to be able to do this with the small Hadza population. However, with a relatively full picture of Hadza mortality, we can contribute to a few interesting and contentious issues such as the more remote history and pre-history of hunter-gatherer life histories.

Like Howell (1979), Hill and Hurtado (1996), and others, I need to take extreme care about methods of estimating mortality and, wherever possible, check them by using different kinds of measure. In addition, we need to take care in estimating people's ages, particularly the ages of the older people. I outlined my methods of estimating ages at length and showed my checks on the accuracy of the methods of age estimation in Chapter 4. I was fortunate in having access at various times to earlier studies, to Bleek's 1930 notebooks and to Kohl-Larsen's 1930s photographs and films. I had some willing informants who, several times, came up with unexpected and robust information such as the names and kinship of six Hadza men who joined the King's African rifles in World War II, and comments about the visits of Pastor Melander in the 1950s.

Two obvious influences on mortality must be addressed briefly, access to modern medicine, and clean water supplies and sewage disposal systems. Like the !Kung in the 1950s–1960s and the Ache in the "forest period" before contact, the Hadza have no access to piped water or sewage disposal. Hadza generally depend upon a great variety of small water holes (Photograph 8.1). The single exception is a Hadza man who has access to piped water while he is working on a European-owned farm. A couple of camps near Mangola use the spring-fed pools unique to this locality. Here and elsewhere water holes are often shared with wild animals and livestock belonging to Datoga herders. Most water sources are small, shallow, and temporary,

**Photograph 8.1** Many of the numerous small water sources are shared with wild animals and Datoga cattle. Early travelers attributed disruptive sickness to the water. Hadza have made no such comment. © James F. O'Connell, 2015. Reproduced with permission.

and may require digging each time one goes to collect water. Camps move very frequently, so disposal is less of an issue than in a more sedentary society. Household waste is swept out of the house, away from the fire and off to the side of the house or other work area (O'Connell *et al.*, 1991). Adults urinate and defecate outside camp but not far away. After a few days in a camp, a visible "shit belt" accumulates 20–40 m outside camp. Even in the "settlements," people move about locally, building new houses or finding new shelters not far from their old ones. During the long dry season, rapid dessication helps with hygiene.

During the study period, Hadza seldom made use of modern medicine. This is especially ironic, because Hadza settlement schemes had stimulated construction of three government clinics in the Eyasi area, at Yaeda chini, Munguli, and Endamagha. One of the constant arguments in favor of settlement was that modern medicine could be provided (apparently no-one foresaw the efficient traveling clinics that, by 1988, were visiting the !Kung in the Dobe area). Most Hadza currently either live at least a day's walk from each of these clinics (there is no other means of transport), or make little use of them. In southern Tanzania, Armstrong Schellenberg *et al.* (2008) found that 73% of families lived within 5 km of a clinic, and that infant mortality was higher among families who lived farther from a clinic. Furthermore, in Eyasi, although the clinic staff appear to be conscientious and well trained, they live isolated lives with unpredictable supplies and support. The problem is not unique to the Hadza. Staff, and important supplies, were also quite often missing from the clinics studied by Armstrong Schellenberg *et al.* (2008). Hadza say that visits to the clinics will be futile, that there will be no medicines, or that the medicines will be held back for more privileged patients. Nonetheless, despite the adverse, even hostile, comments about clinics by those who said they had never been to one, many Hadza report having made use of these facilities at some time in their lives. In 1999, I asked Hadza women during their reproductive history interviews whether they had ever been to a clinic for any medical problem. Forty-one percent of them (49 out of the 119 asked) had been to a clinic at some stage in their lives, and 11 (9%) had stayed in hospital at some time. The reasons for visits to the clinic were usually quite unclear. There is some indication that clinic visits were sometimes a response to supposed malaria whereas almost no deaths of adults were attributed to malaria.

As outlined in Chapter 2, Bennett *et al.* (1970, 1973) and Barnicot *et al.* (1972a,b) report the medical component of the International Biological Program (IBP) fieldwork in 1966 and 1967 (Chapter 2). Two hundred and forty two (49%) of their 491 subjects "had at some time sought western medical aid; this was so for about 75% of subjects living in the settlements, but only 40% of the nomads . . ." (Bennett *et al.*, 1973, p. 249). They suggest these figures reflect availability and the attitudes that led people to avoid the settlements. The settlements they refer to, Yaeda and Munguli, contained few Hadza by the time our study began, and even fewer by the end. It would appear that Hadza use of these facilities, which were still operating during our study period, had declined to the rate of the 1960s bush dwellers. However, when we look at the reasons given for going in search of western medicine, there seem to be some differences between 1967 and 1999, perhaps mainly in the sophistication of the informants. Three of the women we interviewed had taken their children to clinics for routine check-ups because they had heard this was a good thing to do. One showed me her child's growth chart. Nine had been to clinic or hospital in connection with a pregnancy. Several of these cases sounded as if they were quite serious. In some of those from among Bennett's high-frequency "fever," "headache" might be labeled by the informants as "malaria" or "mbu" (mosquito). Some of his "cough" and "chest troubles" might have been labeled tuberculosis (TB) by our informants. Five expressly said they went because of malaria and several of our "fever" and "headache" cases, along with three "hurt all over" among the 10 unclear reports, may have been cases of malaria.

Some of Bennett's categories were among the most frequent requests Hadza made of us. We were commonly asked if we had medicines for cough, sore eyes, sores or cuts, children's ear infections, and malaria. We were, on occasion, asked for help with scabies, venereal diseases, stomach problems, a tooth abcess, and diarrhea. None gave leprosy as a reason for clinic attendance, although at some time during our study, two victims had asked us if we had medicine for their condition. It was as if these complaints were not reason enough to try the clinics but people now knew there might be medicine for them and their conditions were troubling enough for them to ask us if we had such medicine. Our response to these requests was limited. Our basic policy was not to withhold from Hadza any intervention we would have felt competent to offer to each other. This occasionally involved medicines if we had them and would be present for long enough to supervise their use. Occasionally, we took severely sick people to local health facilities.

In 1965, many Hadza received smallpox inoculations. In recent years, we have given our limited assistance to local government measles inoculations and poliomyelitis prevention programs. Thus, Hadza have not been excluded from access to modern medicine. However, day-to-day, they live their lives without visits to clinics or doctors. Children are born in the bush with no medical involvement. Sick people mostly stay at home, resting until they recover or die. In Blurton Jones *et al.* (2002), we tried to analyze the possible impact of these factors, and of our own interventions. Numbers and an estimate of the very small effect of our interventions were given. I will discuss these issues later; our main task here is to report the analysis of Hadza mortality statistics.

Jelliffe *et al.* (1962), together with Woodburn, reported on the health of Hadza children after Woodburn had been in the field for about two years, and before the 1964–1965 major settlement schemes. Sixty-two children were examined (about a quarter of the Hadza child population at the time). Jelliffe reports complete absence of kwashiokor, nutritional marasmus, rickets, infantile scurvy, and vitamin B deficiency syndromes, and that "there were no obviously clinically underweight children and poor musculature was not seen ... no clinical evidence of anemia." Blood film examination showed a 27% malaria parasite rate. Only four of the 56 stool samples had tenia ova and three had giardia cysts. Thirty percent of the children had conjunctivitis. Jelliffe remarked on the overall vigor and good health of Hadza children, especially when compared to other populations he had experienced in Africa.

## 8.1 Causes of death

It was very difficult to get good information on causes of death. For many individuals, we have no information at all, for others some, perhaps enough to choose between disease and some kind of misadventure. We are confident that we have heard about all incidents of homicide (five during our study period), or deaths from encounters with wild animals (one from a hyena during the study period, and another from snakebite in the west). We have heard about a few accidents, perhaps fewer than the 35 reported to Bennett *et al.* (1973). The vast majority of deaths after infancy are attributed to "illness" or "old age." It was impossible to obtain very much information about the illnesses. Malaria seems to be as abundant as one might expect. Although Hadza camp well away from standing water, men sometimes hunt at water holes and are plagued by mosquitos. They say God invented the mosquito in answer to a hunter's prayers for something to keep him awake in the hunting blind. Sometimes informants had decided that the death was due to TB (five men, seven women); a few times they came up with other names. Notable among these was a measles outbreak late in 1986 that seemed to result from a very brief settlement attempt at Yaeda, and was said to have killed a significant number of children. Recognizing the extreme inadequacy of my information, I encouraged Frank Marlowe to subsequently try out a questionnaire on causes of death that had been developed for a survey in Kilimanjaro region (Setel *et al.*, 2006). Despite Frank's excellent rapport with Hadza informants, he told me that he found the questionnaire completely unusable with them.

I found myself tempted to use a broad category of "misadventure," which would include two men and one woman "poisoned by Mbulus" , another man who "died from bangi given by Mbulus," another man who died in jail for "poaching," and two who "died in hospital." Our inquiries about "mbulu poison" revealed little. Informants are firmly convinced that Mbulu (Iraqw) neighbors sometimes poison Hadza. I am not convinced. One explanation is revealing: "If you go to someone's house and he gives you food and you die, didn't he poison you?" It seems likely that a death, especially from anything resembling food poisoning, would be accounted for that way. It remains possible that Hadza "guests," insistent as they are, are sometimes given the oldest, least palatable food in the house. It is also possible that they are exposed to a new set of pathogens on their infrequent visits to Wambulu households. Some of these "misadventure" deaths probably belong in the illness category.

Five women were reported as dying in or very soon after childbirth between 1985 and 2000. During this time, we recorded 489 births. This gives a maternal mortality rate of 1022/100,000 live births, which is close to the average for sub-Saharan Africa according to UNFPA (United Nations Population Fund, formerly UN Fund for Population Activities) 2002, and a little higher than the Tanzanian national figure of 770. Olsen and colleagues (2002) report a rate of 382 deaths/100,000 births (CI 250–560) in their research at Haydom Hospital in the Mbulu highlands in 1995, which is lower than both the national average and the Hadza figure. Armon (1979) reported 80 deaths in 24,292 births between 1971 and 1977 at Kilimanjaro Christian Medical Centre, a rate of 329/100,000, very close to the Haydom figure. The Hadza figure is lower than those for 13 other African countries but higher than those for 31, according to the tables in UNFPA "Maternal Mortality Update 2002". Hill and Hurtado (1996, p. 162) give a rather lower figure for the Ache, "Between 1940 and 1970, when most births and all deaths in childbirth would have been recorded, we estimate four deaths out of 597 children born, for a maternal death rate of approximately one per 150 births." This translates to: 1/150=0.006667=667/100,000 births. Early and Headland (1998) report maternal deaths as 352/10,000 live births. At 3520/100,000 live births, it is not surprising that they comment: "The Agta level of 352 is the highest we have been able to find in the literature." We need to remember that in all these hunter-gatherer populations, we are dealing with extremely small numbers. The actual, long-term rates per 100,000 births could be some tens of a percent higher or lower.

Hadza are not noted in the literature for either an unusually high or unusually low incidence of homicide. Their relationship with their herder neighbors, the Datoga, has been described as uneasy (Bagshawe, 1925; Kohl-Larsen, 1958; Woodburn, 1972, p. 204). It is said that sometimes a Datoga youth will attempt to enhance his status by killing a Hadza instead of a lion or cattle thief. It is said that Hadza have made the occasional retaliatory killing. Murders of Hadza by Hadza are known but appear rare. Woodburn (1979) described, in some detail, the ease with which Hadza men, all lethally armed with poison arrows (silent projectile weapons), could ambush, kill, and remain undetected. However, as Lee reports for the !Kung, among the Hadza we have heard of cases where potential victims dodge and arrows miss. Nonetheless, I agree when Woodburn suggests that the potential for undetected killings is an important "equalizer" and restraint in Hadza social life.

In Blurton Jones *et al.* (2002), based on our data from 1985 to 1995 and assuming a population of 1000 Hadza, we reported a homicide rate of 40/100,000 person-years. One more person was killed between 1995 and 2000. With five people killed in 15 years in a population of about 1000, we reach an overall homicide rate of 33/100,000 person-years. Three of these people (two men and one woman) were said to have been killed by the Datoga, one for shooting cows. The remaining two, one man and one woman, were thought by the Hadza to have been killed by other Hadza. Woodburn (1979) reported another earlier case. If we recalculate our 1985–1995 estimate on the basis of the mid-study population size of 903 used, as follows, to calculate CDR, our 1985–1995 figure becomes 0.4/yr/903=44/100,000, and our 1985–2000 figure becomes 36.9/100,000. As a proportion of the 450 deaths recorded, the five homicides amount to 1.1% of the 450 deaths recorded. The murdered men comprise 3.8% of male deaths over age 20, and the five of either sex amount to 3.6% of adult deaths.

If we can believe a rate based on five cases, the Hadza suffer slightly more killings than Lee reported for the !Kung (1979, chapter 13) but considerably fewer than the 123/100,000 estimate based on a new series of interviews by Wiessner (2014). Lee gives a rate for the !Kung of 29.3/100,000 person-years. Wiessner revises this to 63/100,000 for the Botswana side of the border. All these killings were of !Kung by !Kung. Thus, compared to the !Kung, the Hadza lose about the same number of people to homicide but they kill fewer themselves. To reach the !Kung estimate, we would have to have seen 12 more homicides during our study. Hill and Hurtado describe homicide as "by far the most common cause of death to forest-dwelling Ache" (1996, p. 163) accounting for 46% of adult deaths (Hill and Hurtado, 1996, p. 171, table 5.1), much at the hands of non-Ache Paraguayans. Differences between the Ache and !Kung, or Yanomamo (Hill and Hurtado, 1996, p. 165) appear partly due to: (a) the high rate of killings by non-Ache, and (b) the practice of killing children under a variety of special circumstances. Nonetheless, killings of adult Ache by other adult Ache comprise a higher percentage of deaths than among the Hadza or !Kung. Hill *et al.* (2007) report high adult mortality among the Hiwi, 30% of adult deaths being a result of violence.

Lee (1979, chapter 13) reported that the !Kung fear inter-personal violence and recognize its seriousness in the absence of external or over-riding authority. Wiessner reports a reduced homicide rate, 34/100, 000 at Tsumkwe between 1970 and 1989, with police presence. We have witnessed two "large scale" disputes among Hadza, each reaching borderlines of deadly force. Each resulted in large delegations of people going from camp to camp, debating the issue, and apparently effecting a peaceful settlement. We have also seen communal discussion with a young woman and her suitors about the danger of her failure to chose between them (see also Marlowe, 2010). The Hadza seem not to expect any intervention from police or other authority. Nonetheless, we may wonder whether the knowledge that homicide is illegal and could bring trouble from outside could have no effect. Availability of alcohol in the twenty-first century seems to have greatly increased the number of fights (Butovskaya, 2013), and perhaps homicides, to judge from anecdotes from Marlowe and colleagues (personal communication, 2013).

Even though large animals, snakes, and predators are numerous in Hadza country, they are an extremely rare cause of death. Lions, hyenas, and leopards can be heard nearly every night. Hadza men sometimes hunt at night near water holes and tell many tales of encounters with lions, not all with harmless outcome. Hadza often scavenge from these large predators (Woodburn, in Lee and Devore, 1968, p. 342; O'Connell *et al.*, 1988a). I have been shown lion tracks in the morning passing right through a camp. In contrast, Bagshawe claims that many men had been lost to lions. Sinclair (1979) linked the early twentieth century outbreaks of man-eating lions to the shortage of large prey following the rinderpest and related epidemics. Hadza tell us that when there were rhinoceroses, they were unpredictable and dangerous. We have been told of one non-Hadza being killed by a rhinoceros long ago. Elephants are a creature much more feared by Hadza. Gudo Mahiya has described "a war between the elephants and the Datoga," which he and a friend witnessed in their youth. The elephants won decisively and gruesomely. We have heard of no Hadza killed by an elephant (nor vice versa). Woodburn, likewise (see Lee and DeVore, 1968, p. 342), comments on Hadza readiness to drive large predators from their prey, and that "within living memory only one adult Hadza has been killed by any of these animals." He adds that two children had been taken by leopards at night.

During our fieldwork, one man told of his long-ago confrontation with a lion that tried to steal "his" zebra. He tried to frighten the lion away, it leaped at him, and his arm was saved by his bow. He explained that his long-injured hand (Photograph 4.1) would have been completely severed by the lion's bite but for the strength of his bow stave. Another, presenting a badly maimed hand for treatment, told of his recent fight with a hyena. Luckily, his hunting partner was able to beat it off with a log. The garbled hand looked clean and dry so I persuaded him to leave it alone. A while later, he proudly showed me the clean and healed space where his third finger had been. Using his bow without that finger was "no problem". Another man was much less lucky with a hyena. Survivors told how, judging from the tracks, it had crept up behind him as he knelt to drink at a water hole late in the evening. The victim's camp moved first thing in the morning, talking of building a thorn fence around their next camp (they did not). People in the entire Mangola region were badly spooked by this event. Around the same time, we heard of a villager being attacked by a hyena near her hut. Yet day after day around Mangola, herds of goats browse unimpaired while guarded only by little boys.

Another older man described in detail his fight with a leopard back in the days when the Munguli area was all wild bush country. He suffered a lasting injury to one eye. Yet Hawkes, in 1985, was with three women who heard a leopard kill a small antelope; the leopard fled as they approached and they promptly cooked and ate its prey (O'Connell *et al.* (1988a, appendix case 31; and Photograph 8.2). Clearly death from predators, although always a possibility, is rare and has been for some decades. Only Bagshawe gives any indication that deaths or injury from predators was previously any more common; Kohl-Larsen and Obst do not, and they would surely

**Photograph 8.2** Women and children cooking and eating a small antelope just killed by a leopard. The leopard fled at their approach. Others have had serious injuries in fights with leopards. © James F. O'Connell, 2015. Reproduced with permission.

have relished the possibility as much as did Bagshawe. However, in Blurton Jones *et al.* (2002), we discussed the possibility that lions in particular are subdued by pastoralists, who, at least in East Africa, have a tradition of killing a lion to gain promotion from one age grade to another. This might produce a timid population of lions. Unlike East African herders, including the Datoga who live in the Eyasi basin, Herero herders in the Kalahari do not follow their cattle nor do they enclose them every night. The cattle return for water when thirsty. If any cattle are lost to lions, the Herero mount a posse to pursue a likely lion. Blurton Jones and Konner (1976) contrasted East African and Kalahari lions in the context of methods of scavenging from lions, but note that very few !Kung deaths have been attributed to predators. We have no direct data from hunter-gatherers who lack projectile weapons, which have been widely held to have been very important in human evolution, either for hunting or for protection (the literature is helpfully summarized in Zhu and Bingham, 2011). However, we should note that both !Kung and Hadza women move about in the bush far from camp, armed only with a digging stick and knife, although Hadza women often take a few bow-carrying small boys as "guards." Hadza women are, most of the time, far more concerned about encounters with Datoga than with animals.

We heard of one woman who died of snakebite during the study period and MacDowell was told of one man some years earlier. Although puff-adders, green and black mambas, and cobras are common in the area, encounters with snakes are rare. I saw two apparently harmless snakes killed when they entered a Hadza camp, and once, while out gathering with a group of Hadza, was told "let's go another way, there is no medicine for that one" when a large rustling came from a clump of grass. In 1997, some Hadza boys took us to see two very large pythons that lived in a tree near their camp; these constrictors are not poisonous and are regarded as harmless. Our field assistant showed us how to recognize the sound of one slithering by in the night. A python of the size we were shown could easily take an unattended infant.

Buffalo are regarded as dangerous. Out following the women, Hawkes found herself being pushed up a tree when buffalo were encountered. Meanwhile, the boys who were "guarding" the women ran after the buffalo. Although we have no record of a Hadza being killed by a buffalo, we do know one man who had a narrow escape. He saved himself by hiding between two tree trunks that were close together, but was severely injured. When he failed to return home, friends went to look for him and carried him home. Now he walks well but with a very pronounced limp.

Bennett *et al.* (1973) conducted medical examinations of 491 Hadza and reported 35 cases of "fairly serious injuries" resulting from falls (29 in men, six in women), mostly from trees. Only four men had been injured by wild animals (buffalo, leopard, wildebeest). The leopard case could well have been the same individual as I mentioned earlier. We have instances of falls from trees, one resulting in a death and three others in broken bones, and minor injuries from knives and thorns (Photograph 8.3).

We have one case of an injury to a woman. The elderly mother of several reproductively very successful young women suffered a badly sprained ankle and for a year or so could hardly walk. She eventually recovered completely and continued her vigorous and cantankerous old age. While following women foraging,

**Photograph 8.3** Climbing a baobab tree. Falls from baobab trees have killed or seriously injured a few men, mostly elderly. Each man makes and uses his own set of pegs for each climb. © James F. O'Connell, 2015. Reproduced with permission.

I once saw a young woman stung by a scorpion. The sting was clearly very painful but she suffered no long-term consequences. Cuts and thorn stabs can become infected and we have treated one or two quite serious infections associated with these. We know of two blind older women. One also could not walk and was being left food on occasional visits by her adult son. Women frequently complain of sore eyes, we think mostly a result of working over a fire. Occasionally, after foraging on exceptionally hot days, they complain to us of headaches.

## 8.2 The sample and methods

I have two main sources of information about deaths, the "where are they now?" interviews, and the reproductive history interviews. The latter mainly give information about deaths of children, but we also used the women's interviews to find out about year of death of older kin, husbands, ex-husbands, and others.

Even in a population such as the Hadza, that is relatively far removed from modern medicine, death is a rare event. This felicity makes mortality difficult to study. The pattern of age at death was determined using every individual in the population register for whom we had estimates of year of birth and year of death. For age-specific mortality and the life table, it is essential that our sample of people who die is drawn from the sample of people at risk. This is a more restrictive condition; some known deaths had to be excluded. Determining the at-risk sample has a much greater influence on the resulting life table than does missing the occurrence of one or two deaths. The details of sampling and procedures are described in each following section.

## 8.3 Crude death rate

Crude death rate (deaths/thousand persons/yr) gives a rough indication of the level of mortality, because it is dependent not only on mortality but on the proportions of the population who are at the most vulnerable ages. Thus, a rapidly growing population, with many births and small children, may have a higher CDR than a stationary population living under the same conditions of health facilities and nutrition. However, CDR is worth calculating, both to give us an indication of the general level of mortality of our study population, and to compare to a value we will predict by a population simulation after we have described the population.

The mobility of the Hadza made it difficult to determine the size of the population. No census was complete, but we believe that over the years we accumulated a register of every Hadzane speaker who lived at some point during our study period. We can use this register, with its records of the year of birth and death of each individual (or survival past the end of the study), to make rough estimates of the size of the population each year. A crude initial computation is as follows. We recorded 430 deaths that occurred during the 15 years between the beginning of our June 1985 census and July 2000. (A few of these cases did not have a known age at death, so these cases could not be used to estimate average age at death in the next section). Table 8.1 shows the number of deaths, and the estimated population during each year of the study. From this we can calculate a mean in a couple of different ways. The average number of deaths is 28.7/yr. The total population from which we were recording deaths could be estimated as the population size at the mid-point between 1985 and 2000. The mid-point is in late 1992, so we used the estimated population size in 1992 of 903 individuals. This gives a death rate (28.7/903) of 0.03175, or 31.75. However, the death rate varied a lot

**Table 8.1** Crude death rate (CDR) for males and females. A total of 430 deaths were recorded between 1985 and 2000. Average of annual death rates from 1985 to 2000 was 30.01. Half-year observations for the years 1985 and 2000 reduced the chance to record all the deaths in those years. Average of years 1986–1999 was 32.5 (males 33.3, females 26.2). Regression shows no significant secular trend (b=–0.00083, CI-0.00019–0.0012, p=0.6497, $r^2$=0.0194). Program counts number alive in each year from population register, with no filter for number of times seen in study area nor for % in area

| Year | Females dead | Female population | Female CDR | Males dead | Male population | Male CDR | Both sexes CDR |
|---|---|---|---|---|---|---|---|
| 1985 | 6 | 456 | 13.2 | 6 | 410 | 14.6 | 15.0 |
| 1986 | 14 | 456 | 30.7 | 30 | 403 | 74.4 | 51.2 |
| 1987 | 17 | 449 | 37.9 | 23 | 390 | 59.0 | 47.7 |
| 1988 | 22 | 443 | 49.7 | 9 | 395 | 22.8 | 37.0 |
| 1989 | 6 | 452 | 13.3 | 3 | 408 | 7.4 | 11.6 |
| 1990 | 13 | 457 | 28.4 | 10 | 420 | 23.8 | 28.5 |
| 1991 | 18 | 454 | 39.6 | 9 | 435 | 20.7 | 31.5 |
| 1992 | 9 | 465 | 19.4 | 19 | 438 | 43.4 | 32.1 |
| 1993 | 16 | 464 | 34.5 | 21 | 433 | 48.5 | 41.2 |
| 1994 | 10 | 466 | 21.5 | 15 | 438 | 34.2 | 27.7 |
| 1995 | 10 | 474 | 21.1 | 6 | 452 | 13.3 | 18.4 |
| 1996 | 16 | 476 | 33.6 | 15 | 461 | 32.5 | 33.1 |
| 1997 | 6 | 482 | 12.4 | 17 | 460 | 37.0 | 24.4 |
| 1998 | 13 | 487 | 26.7 | 18 | 456 | 39.5 | 32.9 |
| 1999 | 15 | 489 | 30.7 | 21 | 449 | 46.8 | 38.4 |
| 2000 | 3 | 506 | 5.9 | 7 | 452 | 15.5 | 10.4 |

from year to year, and if we average the annual death rates in the period 1985–2000, we get a slightly lower figure, 30.1 for the sexes combined. If we recognize 1985 and 2000 as times when only half a year's deaths were observed, and therefore take the mean for the years 1986–1999, we get an average death rate of 32.5. A general death rate for the Hadza conceals what appears to be a very large difference between male and female death rates. Even using this preliminary procedure, when we separate data for males and females (and removing seven cases of infants of unknown gender), we find, for females, 194 deaths between 1985 and 2000, and a 1992 population of 465 females gives a death rate of 27.8 (194/15=12.93; 12.93/465=27.8). For males, 229 deaths in 438 individuals, gives a death rate of 34.8.

A more reliable estimate can be obtained if we limit the analysis to the area and people we knew best, and if we look at the variability revealed by resampling. My resampling program can be set to attend only to individuals in the population register who were seen or reported (or in the case of children, whose parents were seen or reported) between 1985 and 2000 in our core study area (the area we always surveyed and the people there we knew best). These were the great majority of the eastern Hadza. Thus, the sample of deaths was limited to those known to have lived in the core area, and hence was the risk population from which they were drawn. The program resampled this population and reports means, standard deviations, and a frequency count for each estimated value. Runs of 500 resamplings produced quite a large range of variation in CDR. For females, the mean was 27.1 (median, 26.9) (95% of the results fell between 20.0 and 34.0). For males, the mean was 32.9 (median, 32.7) with 95% falling between 24.4 and 41.6. For the sexes combined, the mean was 30.3 (range 27.0–33.5, median also 30.3). Thus, our overall figure was a little lower than our first crude estimate (using a rather larger sample of deaths) but the variance is large. While female mortality rates are found to be lower than male rates in almost every population, the Hadza sex difference in CDR was quite striking and is discussed in a later section. Hadza women seem to do much better than Hadza men. The lower mortality of Hadza females shows up in every measure of mortality that we have used, and the size of the difference appears unusually large.

## 8.4 Age at death

Generally, in populations with a high mortality, many individuals die as infants and children. Relatively few survive long enough to die as old people. Thus, in general, in human populations, when mortality is high, the average age at which people die is low. When there are many births, there will be many deaths of vulnerable infants and children. Average age at death will be lowered. Within samples of high mortality populations, average age at death may be as much an indicator of fertility and rate of population change as of mortality level. Later, I discuss age at death as a reflection of age structure.

Age at death is based on the records of death; no risk group is involved in the calculation. However, the records of death may be subject to some biases. In archaeological demography, an important sampling question is "who gets buried where they will be found? . . . who gets buried in a cemetery, and who does not?" In our case, the equivalent sampling question is "whose deaths would we hear about, and whose deaths would we not hear about?". In particular, deaths reported in the "where are they now?" interviews are likely to be biased toward well-known older people. Our record will also lack the exact year of death for older people in our register, who died before 1985. Deaths of infants will be under-represented in the "where are they now?" data.

We can reduce these biases by restricting the analysis to people known to have lived in the core study area at any time between 1985 and 2000 and known to have died sometime in that period. For this restricted sample, the average age at death for females is 20.6 and males 20.9. If we resample the population, we obtain a similar average age at death, 20.7 for females (median 20.6, 95% fall between ages 17.9 and 24.3), and 21.1 for males (median 21.2, 95% between ages 17.4 and 24.6). The sex difference in age at death is not significant.

Figure 8.1 shows the percentage of these deaths that occurred in each year of age. This age pattern is a typical "non-industrialized" human pattern, with a large

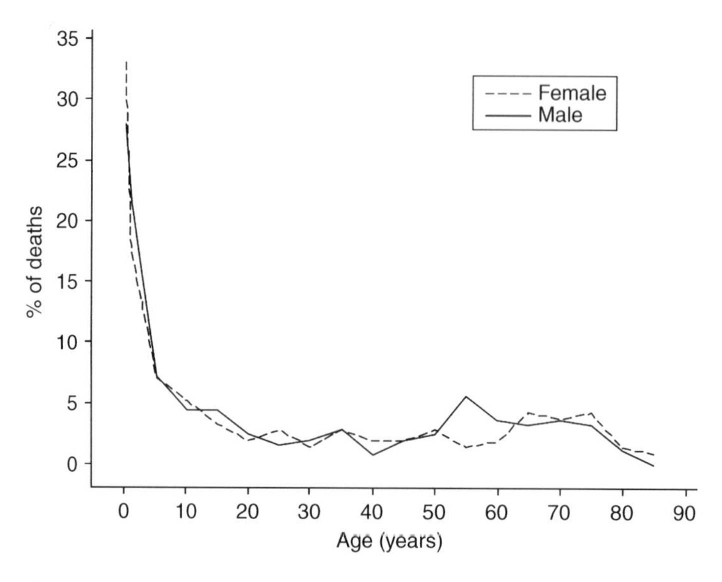


**Figure 8.1** Age at death. Deaths recorded in the first year, ages one to four years, and five-year age groups, shown as a percentage of total deaths of each sex. Average of the actual observed age at death is 20.6 for females and 20.9 for males. The slight rise among the older people described by Gurven and Kaplan (2007) can be seen and may occur earlier among Hadza men than women.

proportion of the deaths occurring in infancy and childhood, and with the remaining deaths spread very evenly through the remainder of the lifespan. This differs strikingly from some estimates made from archaeological samples, which claim high proportions of deaths to people aged 30–40, perhaps merely an extreme form of age-heaping, perhaps an indication of a "catastrophe" (Keckler, 1997; Paine, 1997,). In SI 8.7, I discuss the contradiction between demographic patterns of live, observed populations and the patterns attributed to populations known only from skeletal remains. The latter have exerted a dominant influence on discussions of human evolution, an influence which I argue is misleading.

## 8.5 Building a life table

Next I will calculate age-specific death rates ($q_x$) and a life table directly from the Hadza interview and census data. Two immediate decisions have to be made about our mortality measures. First, we might wonder whether having a non-Hadza father makes any difference to survival. I decided to present here the data for all eligible individuals, regardless of whether the father was a Hadza or a non-Hadza. We can check for effects of the father's tribe later. Second, different authors have used slightly different methods, which give slightly different results.

The $q_x$ was computed in the same way as Howell (1979, p. 81, table 4.1) did; $q_x$ was computed for each time unit as the number who die in that period divided by the number who lived through (completed) the period. For the latter, Howell used the number who started the interval minus the number "currently in the interval," those for whom it was their final year of observation. In my tables, this "at-risk group" is the number who entered the year minus the number who were censored in that year. They are censored because our observations do not allow us to tell whether they survived to the end of that final year in which they were briefly observed.

Four methods of scoring are compared in detail in SI 8.1. The range of $e_0$ given by these four methods is 32.7 to 34.15 years for the genders combined. The range for females is 35.55 to 36.75 years. This is less than the difference between Coale and Demeny (C&D) (1983) models West 7 and 8. The difference between $e_0$ by my method and by Hill and Hurtado (1996, table 6.1) is 1.8%.

## 8.5.1 Age-specific death rates ($q_x$) from interviews

The reproductive history interviews were a good source of information about children and their survivorship. I used data from the women described in the last chapter. They provided enough information for me to estimate year of birth on 690 children (I omitted five whose births we recorded after the last interview of their mother), 329 females, 351 males, and 10 dead infants whose gender I failed to ascertain. One hundred and thirty-nine died as infants (20.1%) and 103 later. When the sample is restricted to those with Hadza fathers, the figures are: 547 children; 268 females, 272 males, and seven dead infants whose gender we failed to ascertain. One hundred and seventeen died as infants (21.4%), 78 later.

A Visual Basic program calculates the probability of death in each year of life from the data on these 690 children. When a child is born it enters year zero, and it enters each year of life including its last year of life recorded, as indicated by the last interview of its mother, or until it died. A child who died aged 4.3 was recorded as entering year 0, years 1, 2, 3, and surviving each of these years, and as entering year 4 but dying during that year. A child born sometime in 1994, whose mother was last interviewed in late 1999 and who was still alive at that time was recorded as entering years 0, 1, 2, 3, 4 and surviving to the end of each of these years of life, and then entering year 5 but leaving the record during that year (censored in year 5). These records can then be treated with any of the four methods described in SI 8.1.

## 8.5.2 Age-specific death rates ($q_x$) from censuses and population register

The series of censuses give a sample of people known well enough to determine years at risk and probability of mortality. While data from interviews covers only younger individuals (children of the women who we interviewed) the census data are not restricted by age, except in so far as censuses tend to undercount any particular age group. For example, it is very likely that the censuses undercount infants.

The program that extracts the data seeks any individual ever recorded in a census, who has an estimated birth date, and for whom we either know when s/he died, or who we saw alive on a second occasion. The individual's age when first seen was calculated, and the individual was recorded as entering that year and each subsequent year until s/he was last seen, in which year it was recorded as censored, or until s/he died. Individuals that were only seen once, and were not recorded as dying, were recorded as at risk only in the year they were seen.

Among the individuals with a known age at death after mid-1985, 231 were seen in a census. For ages 1 to 20 (the oldest person whose mother was interviewed was aged 30), the probability of death produced by these data closely resemble those derived from the interview data, except during infancy (age zero to one year) where the census data give a much lower estimate of infant mortality than the interview data does (SI 8.2). Apart from the first year of life, the comparison was encouraging. It supports the view that censuses under-report infants but does not support the view that the interviewed women under-reported infant or child deaths. For comparison, we calculated $l_x$ (the probability that an individual survives to age x years) using the census data alone. The lower infant mortality has a lasting influence. If we replace the census infant mortality with an approximation to the interview value ($q_0$=0.2) (Figure 8.3) and then compute $l_x$ for the remaining census data, we get an $l_x$ curve that is quite close to that from interviews and to that from the combined data. "Where are they now?" informants evidently agreed with interviewed women.

### 8.5.3 Combined age-specific death rates and life tables for men and women

I then combined the data from interview and from censuses to create a life table for male and female Hadza and for the combined population. The interview data are probably better than census data for children but they cover only a small number of adults. Consequently, I combined these sources of data as follows. If an individual appeared in his/her mother's interviews, I used the interview data. If the individual did not appear in an interview, but had been recorded in a census, I used his/her census data in the manner described previously. The resulting $q_x$ and $l_x$ curves are shown in Figures 8.2 and 8.3, and in Tables 8.2 and 8.3. The results for the sexes combined (table SI 8.3) are similar to those presented by Blurton Jones *et al.* (2002), with an expectation of life at birth of 32.7 years, compared to 32.5 years in the 2002 paper.

The Hadza life table is much like that of any high mortality population, whether it be a third-world rural population in the mid-1900s, or a European historical population from before about 1800. The Hadza life table bears the same opportunity for understanding and misunderstanding as any other with very high infant mortality. Life expectancy at birth is very low, but it is easy to forget that life expectancy increases once the risky period of early childhood is passed, and many people live into their 60s and 70s. Very high infant mortality is followed by quite low mortality from about age 15 to the late 40s, thence climbing gradually, and eventually ascending sharply in the late 60s, 70s, and 80s. Life expectancy increases from birth to its peak, 48.1 more years for females, at five years old (Figure 8.4). A woman bearing her first child at age 19 can expect to live another

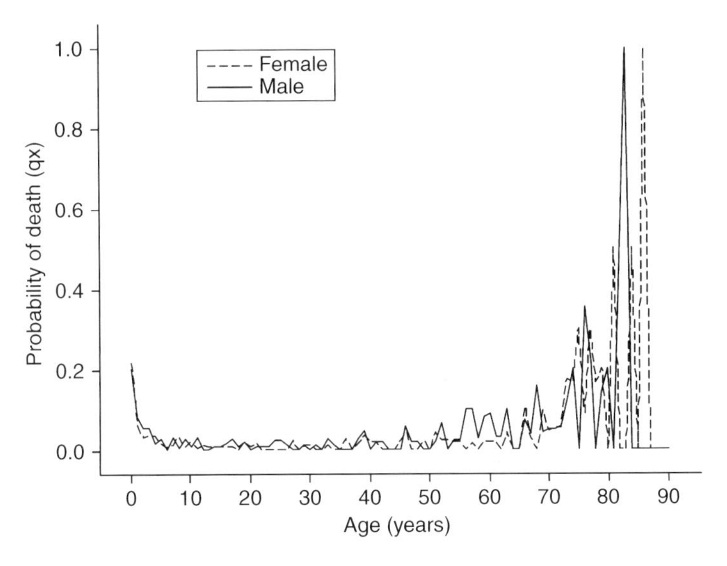


**Figure 8.2** Probability of death in each year of life ($q_x$).

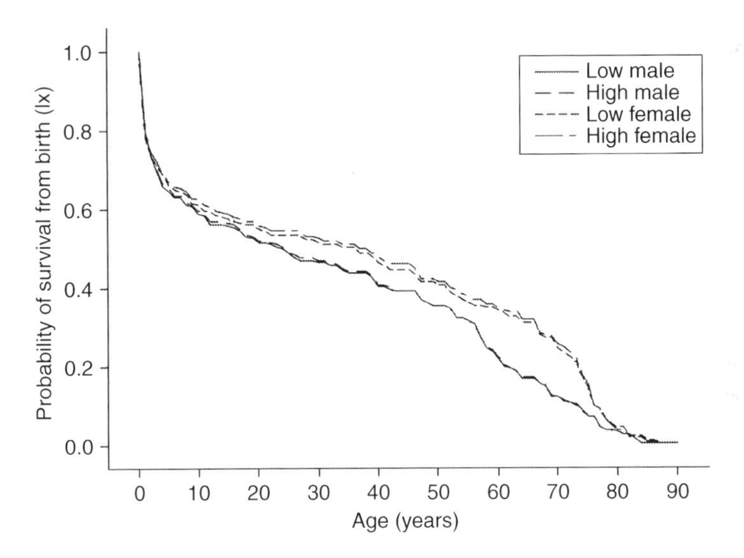


**Figure 8.3** Resampled male and female probability of survival from birth ($l_x$).

42 years and her 25-year-old spouse another 33 years (although probably not with her), thus she should sadly (but only if she reads the statistics closely) expect to outlive her newborn child. Only when her daughter has passed her first birthday can the mother expect her daughter to outlive her. When the average mother has delivered what will turn out to be her final child, at about age 37, she can still expect another 28.8 years of life. Even if her final child is born when she is 45 years old, the mother can expect to live another 23.3 years, and for many of these years she will be strong and very active. There would be no shortage of grandparents among the Hadza.

**Table 8.2** Female single-year life table. Computed following Howell (1979) and Barclay (1958). At risk group: enter – censored; $q_x$ is the number dead/number at risk; $l_{(x+1)}$ was calculated as $l_x$ $(1–q_x)$; person–years lived at each age $(L_x) = (l_x + l_{(x+1)} \div 2)$; $L_0$ was calculated as $(0.3 \times l_0) + (0.7 \times l_1)$; $L_1$ was calculated as $(0.4 \times l_1) + (0.6 \times l_2)$ (after Barclay, 1958, p. 104). Total number of years lived after each age $(T_x)$ was calculated starting from the highest ages, at $L_{86} + L_{85}$ and summing up to $T_0$. Expectation of life at $age_x$ $(e_x)$ was calculated as $T_x \div l_x$. Note: because the table results from combining interview data with census data, as described in the text, the number entering a year is sometimes larger than the number expected from the number dead and at risk in the previous year. A life table for the sexes combined can be found in SI 8.3.

| Age | Enter | Censored | Dead | At risk | $q_x$ | $l_x$ | $L_x$ | $T_x$ | $e_x$ |
|---|---|---|---|---|---|---|---|---|---|
| 0 | 330 | 34 | 64 | 296 | 0.216 | 1.000 | 0.849 | 35.55 | 35.55 |
| 1 | 235 | 11 | 14 | 224 | 0.063 | 0.784 | 0.754 | 34.70 | 44.27 |
| 2 | 215 | 19 | 6 | 196 | 0.031 | 0.735 | 0.724 | 33.95 | 46.20 |
| 3 | 194 | 8 | 7 | 186 | 0.038 | 0.712 | 0.699 | 33.22 | 46.64 |
| 4 | 185 | 12 | 6 | 173 | 0.035 | 0.685 | 0.674 | 32.52 | 47.45 |
| 5 | 172 | 12 | 3 | 160 | 0.019 | 0.662 | 0.656 | 31.85 | 48.13 |
| 6 | 162 | 7 | 1 | 155 | 0.006 | 0.649 | 0.647 | 31.19 | 48.04 |
| 7 | 158 | 11 | 2 | 147 | 0.014 | 0.645 | 0.641 | 30.55 | 47.35 |
| 8 | 151 | 8 | 5 | 143 | 0.035 | 0.636 | 0.625 | 29.91 | 47.00 |
| 9 | 146 | 5 | 1 | 141 | 0.007 | 0.614 | 0.612 | 29.28 | 47.68 |
| 10 | 152 | 14 | 3 | 138 | 0.022 | 0.610 | 0.603 | 28.67 | 47.02 |
| 11 | 148 | 10 | 1 | 138 | 0.007 | 0.596 | 0.594 | 28.07 | 47.05 |
| 12 | 146 | 9 | 2 | 137 | 0.015 | 0.592 | 0.588 | 27.47 | 46.39 |
| 13 | 142 | 11 | 1 | 131 | 0.008 | 0.584 | 0.581 | 26.88 | 46.07 |
| 14 | 132 | 6 | 1 | 126 | 0.008 | 0.579 | 0.577 | 26.30 | 45.42 |
| 15 | 135 | 11 | 1 | 124 | 0.008 | 0.574 | 0.572 | 25.73 | 44.78 |
| 16 | 126 | 8 | 1 | 118 | 0.008 | 0.570 | 0.567 | 25.15 | 44.14 |
| 17 | 122 | 11 | 1 | 111 | 0.009 | 0.565 | 0.562 | 24.59 | 43.51 |
| 18 | 120 | 15 | 0 | 105 | 0.000 | 0.560 | 0.560 | 24.02 | 42.91 |
| 19 | 113 | 13 | 2 | 100 | 0.020 | 0.560 | 0.554 | 23.46 | 41.91 |
| 20 | 109 | 7 | 1 | 102 | 0.010 | 0.549 | 0.546 | 22.91 | 41.75 |
| 21 | 109 | 3 | 2 | 106 | 0.019 | 0.543 | 0.538 | 22.36 | 41.16 |
| 22 | 111 | 10 | 0 | 101 | 0.000 | 0.533 | 0.533 | 21.83 | 40.94 |
| 23 | 109 | 10 | 0 | 99 | 0.000 | 0.533 | 0.533 | 21.29 | 39.94 |
| 24 | 105 | 9 | 0 | 96 | 0.000 | 0.533 | 0.533 | 20.76 | 38.94 |
| 25 | 106 | 8 | 0 | 98 | 0.000 | 0.533 | 0.533 | 20.23 | 37.94 |
| 26 | 107 | 17 | 0 | 90 | 0.000 | 0.533 | 0.533 | 19.69 | 36.94 |
| 27 | 99 | 12 | 2 | 87 | 0.023 | 0.533 | 0.527 | 19.16 | 35.94 |
| 28 | 91 | 3 | 0 | 88 | 0.000 | 0.521 | 0.521 | 18.63 | 35.78 |
| 29 | 92 | 6 | 1 | 86 | 0.012 | 0.521 | 0.518 | 18.11 | 34.78 |
| 30 | 94 | 7 | 1 | 87 | 0.011 | 0.515 | 0.512 | 17.59 | 34.18 |
| 31 | 93 | 1 | 0 | 92 | 0.000 | 0.509 | 0.509 | 17.08 | 33.57 |
| 32 | 96 | 5 | 0 | 91 | 0.000 | 0.509 | 0.509 | 16.57 | 32.57 |
| 33 | 96 | 7 | 1 | 89 | 0.011 | 0.509 | 0.506 | 16.06 | 31.57 |
| 34 | 93 | 12 | 0 | 81 | 0.000 | 0.503 | 0.503 | 15.56 | 30.92 |
| 35 | 85 | 7 | 0 | 78 | 0.000 | 0.503 | 0.503 | 15.06 | 29.92 |

**Table 8.2** (*cont.*)

| Age | Enter | Censored | Dead | At risk | $q_x$ | $l_x$ | $L_x$ | $T_x$ | $e_x$ |
|---|---|---|---|---|---|---|---|---|---|
| 36 | 81 | 12 | 2 | 69 | 0.029 | 0.503 | 0.496 | 14.55 | 28.92 |
| 37 | 69 | 8 | 0 | 61 | 0.000 | 0.489 | 0.489 | 14.06 | 28.77 |
| 38 | 67 | 5 | 1 | 62 | 0.016 | 0.489 | 0.485 | 13.57 | 27.77 |
| 39 | 61 | 6 | 2 | 55 | 0.036 | 0.481 | 0.472 | 13.08 | 27.22 |
| 40 | 60 | 9 | 1 | 51 | 0.020 | 0.463 | 0.459 | 12.61 | 27.23 |
| 41 | 52 | 4 | 1 | 48 | 0.021 | 0.454 | 0.449 | 12.15 | 26.76 |
| 42 | 48 | 10 | 0 | 38 | 0.000 | 0.445 | 0.445 | 11.70 | 26.32 |
| 43 | 44 | 4 | 0 | 40 | 0.000 | 0.445 | 0.445 | 11.26 | 25.32 |
| 44 | 42 | 2 | 0 | 40 | 0.000 | 0.445 | 0.445 | 10.81 | 24.32 |
| 45 | 44 | 3 | 1 | 41 | 0.024 | 0.445 | 0.439 | 10.37 | 23.32 |
| 46 | 47 | 3 | 2 | 44 | 0.045 | 0.434 | 0.424 | 9.93 | 22.89 |
| 47 | 43 | 1 | 0 | 42 | 0.000 | 0.414 | 0.414 | 9.51 | 22.96 |
| 48 | 46 | 4 | 0 | 42 | 0.000 | 0.414 | 0.414 | 9.09 | 21.96 |
| 49 | 44 | 2 | 1 | 42 | 0.024 | 0.414 | 0.409 | 8.68 | 20.96 |
| 50 | 46 | 0 | 0 | 46 | 0.000 | 0.404 | 0.404 | 8.27 | 20.46 |
| 51 | 49 | 4 | 2 | 45 | 0.044 | 0.404 | 0.395 | 7.86 | 19.46 |
| 52 | 45 | 3 | 1 | 42 | 0.024 | 0.386 | 0.382 | 7.47 | 19.34 |
| 53 | 46 | 2 | 1 | 44 | 0.023 | 0.377 | 0.373 | 7.09 | 18.80 |
| 54 | 47 | 2 | 1 | 45 | 0.022 | 0.369 | 0.364 | 6.72 | 18.22 |
| 55 | 56 | 3 | 1 | 53 | 0.019 | 0.360 | 0.357 | 6.35 | 17.63 |
| 56 | 55 | 1 | 0 | 54 | 0.000 | 0.354 | 0.354 | 5.99 | 16.95 |
| 57 | 56 | 1 | 1 | 55 | 0.018 | 0.354 | 0.350 | 5.64 | 15.95 |
| 58 | 55 | 4 | 0 | 51 | 0.000 | 0.347 | 0.347 | 5.29 | 15.24 |
| 59 | 52 | 0 | 1 | 52 | 0.019 | 0.347 | 0.344 | 4.94 | 14.24 |
| 60 | 54 | 4 | 1 | 50 | 0.020 | 0.340 | 0.337 | 4.60 | 13.51 |
| 61 | 50 | 8 | 1 | 42 | 0.024 | 0.334 | 0.330 | 4.26 | 12.78 |
| 62 | 44 | 1 | 0 | 43 | 0.000 | 0.326 | 0.326 | 3.93 | 12.08 |
| 63 | 47 | 2 | 2 | 45 | 0.044 | 0.326 | 0.318 | 3.61 | 11.08 |
| 64 | 43 | 4 | 0 | 39 | 0.000 | 0.311 | 0.311 | 3.29 | 10.57 |
| 65 | 44 | 6 | 0 | 38 | 0.000 | 0.311 | 0.311 | 2.98 | 9.57 |
| 66 | 38 | 1 | 4 | 37 | 0.108 | 0.311 | 0.294 | 2.67 | 8.57 |
| 67 | 37 | 1 | 1 | 36 | 0.028 | 0.278 | 0.274 | 2.37 | 8.54 |
| 68 | 35 | 5 | 0 | 30 | 0.000 | 0.270 | 0.270 | 2.10 | 7.77 |
| 69 | 31 | 0 | 3 | 31 | 0.097 | 0.270 | 0.257 | 1.83 | 6.77 |
| 70 | 30 | 9 | 1 | 21 | 0.048 | 0.244 | 0.238 | 1.57 | 6.45 |
| 71 | 20 | 1 | 1 | 19 | 0.053 | 0.232 | 0.226 | 1.33 | 5.74 |
| 72 | 20 | 2 | 1 | 18 | 0.056 | 0.220 | 0.214 | 1.11 | 5.04 |
| 73 | 18 | 1 | 3 | 17 | 0.176 | 0.208 | 0.189 | 0.89 | 4.30 |
| 74 | 14 | 2 | 2 | 12 | 0.167 | 0.171 | 0.157 | 0.70 | 4.12 |
| 75 | 11 | 1 | 3 | 10 | 0.300 | 0.143 | 0.121 | 0.55 | 3.84 |
| 76 | 11 | 0 | 1 | 11 | 0.091 | 0.100 | 0.095 | 0.43 | 4.27 |
| 77 | 10 | 0 | 3 | 10 | 0.300 | 0.091 | 0.077 | 0.33 | 3.65 |
| 78 | 7 | 1 | 1 | 6 | 0.167 | 0.063 | 0.058 | 0.25 | 4.00 |
| 79 | 5 | 0 | 1 | 5 | 0.200 | 0.053 | 0.048 | 0.20 | 3.70 |

**Table 8.2** (*cont.*)

| Age | Enter | Censored | Dead | At risk | $q_x$ | $l_x$ | $L_x$ | $T_x$ | $e_x$ |
|---|---|---|---|---|---|---|---|---|---|
| 80 | 4 | 0 | 0 | 4 | 0.000 | 0.042 | 0.042 | 0.15 | 3.50 |
| 81 | 4 | 0 | 2 | 4 | 0.500 | 0.042 | 0.032 | 0.11 | 2.50 |
| 82 | 2 | 0 | 0 | 2 | 0.000 | 0.021 | 0.021 | 0.07 | 3.50 |
| 83 | 2 | 0 | 0 | 2 | 0.000 | 0.021 | 0.021 | 0.05 | 2.50 |
| 84 | 2 | 0 | 1 | 2 | 0.500 | 0.021 | 0.016 | 0.03 | 1.50 |
| 85 | 1 | 0 | 0 | 1 | 0.000 | 0.011 | 0.011 | 0.02 | 1.50 |
| 86 | 1 | 0 | 1 | 1 | 1.000 | 0.011 | 0.005 | 0.01 | 0.50 |
| 87 | 0 | 0 | 0 | 0 | 0.000 | 0.000 | 0.000 | 0.00 | 0.00 |
| 88 | 0 | 0 | 0 | 0 | 0.000 | 0.000 | 0.000 | 0.00 | 0.00 |
| 89 | 0 | 0 | 0 | 0 | 0.000 | 0.000 | 0.000 | 0.00 | 0.00 |
| 90 | 0 | 0 | 0 | 0 | 0.000 | 0.000 | 0.000 | 0.00 | 0.00 |

## 8.6 Estimating confidence intervals for the mortality estimates by resampling

I used resampling (SI 8.4) to give an impression of the likely sampling error associated with the results. I added a new routine to the mortality program to choose individuals from the original sample at random, making an artificial sample from our original sample. In each run, some cases were omitted, others were used twice, or more times, until we reached the original sample size. Then the mortality of this artificial sample was calculated exactly as before. This was repeated 1000 times and the median, percentiles, and mean and standard deviation of $q_x$, $l_x$, $e_0$ were calculated and reported. This method gives an excellent summary of the extent to which our initial result depends upon the membership of the sample.

Observed $e_0$ was 35.55 years for females, 30.8 years for males (Tables 8.2 and 8.3) (for sexes combined see SI 8.3). The resampling closely matched these observed means (females 35.5 and males 30.8 years). Confidence limits are indicated as follows: 95% of the runs fell between $e_0$=35.4 and 35.9 years for females, and 30.7 and 31.1 years for males. The scatter of $e_0$ is very small (SI 8.6).

Resampling shows only that calculating the life table with slightly different samples of the individuals in the Hadza population makes rather little difference to our result, including the sex difference in $e_0$. This is useful to know because of the large potential effect of choice of the sample "at risk." Resampling does not address other conceivable sources of error. It does not protect us from, nor measure the extent of, errors in measurement, except in so far as these vary from subject to subject. If there were a systematic bias in our knowledge of subjects, so that a large number with a different pattern of mortality had been missed, no number of resampling runs could address the problem. Nor does resampling protect us from systematic errors in informant report (such as if the Hadza tended to underreport deaths of women), or computation. I employ other checks on these estimates as best I can.

**Table 8.3** Male single-year life table (Details as for Table 8.2.)

| Age | Enter | Censored | Dead | At risk | $q_x$ | $l_x$ | $L_x$ | $T_x$ | $e_x$ |
|---|---|---|---|---|---|---|---|---|---|
| 0 | 357 | 16 | 69 | 341 | 0.202 | 1.000 | 0.858 | 30.81 | 30.81 |
| 1 | 281 | 10 | 22 | 271 | 0.081 | 0.798 | 0.759 | 29.95 | 37.55 |
| 2 | 255 | 15 | 13 | 240 | 0.054 | 0.733 | 0.713 | 29.19 | 39.83 |
| 3 | 231 | 10 | 12 | 221 | 0.054 | 0.693 | 0.674 | 28.48 | 41.09 |
| 4 | 211 | 20 | 3 | 191 | 0.016 | 0.656 | 0.650 | 27.81 | 42.42 |
| 5 | 191 | 15 | 5 | 176 | 0.028 | 0.645 | 0.636 | 27.16 | 42.08 |
| 6 | 176 | 9 | 0 | 167 | 0.000 | 0.627 | 0.627 | 26.52 | 42.30 |
| 7 | 173 | 13 | 5 | 160 | 0.031 | 0.627 | 0.617 | 25.89 | 41.30 |
| 8 | 160 | 14 | 1 | 146 | 0.007 | 0.607 | 0.605 | 25.28 | 41.62 |
| 9 | 149 | 16 | 4 | 133 | 0.030 | 0.603 | 0.594 | 24.67 | 40.90 |
| 10 | 137 | 11 | 1 | 126 | 0.008 | 0.585 | 0.583 | 24.08 | 41.15 |
| 11 | 132 | 12 | 4 | 120 | 0.033 | 0.580 | 0.571 | 23.49 | 40.48 |
| 12 | 120 | 8 | 0 | 112 | 0.000 | 0.561 | 0.561 | 22.92 | 40.86 |
| 13 | 119 | 8 | 0 | 111 | 0.000 | 0.561 | 0.561 | 22.36 | 39.86 |
| 14 | 120 | 3 | 1 | 117 | 0.009 | 0.561 | 0.559 | 21.80 | 38.86 |
| 15 | 122 | 5 | 1 | 117 | 0.009 | 0.556 | 0.554 | 21.24 | 38.19 |
| 16 | 120 | 7 | 2 | 113 | 0.018 | 0.552 | 0.547 | 20.69 | 37.51 |
| 17 | 119 | 9 | 3 | 110 | 0.027 | 0.542 | 0.534 | 20.14 | 37.18 |
| 18 | 111 | 5 | 1 | 106 | 0.009 | 0.527 | 0.524 | 19.61 | 37.21 |
| 19 | 108 | 5 | 2 | 103 | 0.019 | 0.522 | 0.517 | 19.08 | 36.56 |
| 20 | 108 | 11 | 0 | 97 | 0.000 | 0.512 | 0.512 | 18.57 | 36.27 |
| 21 | 102 | 8 | 1 | 94 | 0.011 | 0.512 | 0.509 | 18.05 | 35.27 |
| 22 | 100 | 7 | 1 | 93 | 0.011 | 0.506 | 0.504 | 17.54 | 34.64 |
| 23 | 97 | 7 | 1 | 90 | 0.011 | 0.501 | 0.498 | 17.04 | 34.02 |
| 24 | 96 | 9 | 2 | 87 | 0.023 | 0.495 | 0.490 | 16.54 | 33.39 |
| 25 | 95 | 10 | 2 | 85 | 0.024 | 0.484 | 0.478 | 16.05 | 33.17 |
| 26 | 92 | 6 | 1 | 86 | 0.012 | 0.473 | 0.470 | 15.57 | 32.95 |
| 27 | 92 | 10 | 0 | 82 | 0.000 | 0.467 | 0.467 | 15.10 | 32.34 |
| 28 | 88 | 9 | 0 | 79 | 0.000 | 0.467 | 0.467 | 14.64 | 31.34 |
| 29 | 86 | 8 | 1 | 78 | 0.013 | 0.467 | 0.464 | 14.17 | 30.34 |
| 30 | 84 | 4 | 0 | 80 | 0.000 | 0.461 | 0.461 | 13.71 | 29.72 |
| 31 | 82 | 4 | 1 | 78 | 0.013 | 0.461 | 0.458 | 13.25 | 28.72 |
| 32 | 84 | 5 | 0 | 79 | 0.000 | 0.455 | 0.455 | 12.79 | 28.09 |
| 33 | 83 | 9 | 2 | 74 | 0.027 | 0.455 | 0.449 | 12.33 | 27.09 |
| 34 | 74 | 5 | 1 | 69 | 0.014 | 0.443 | 0.440 | 11.88 | 26.83 |
| 35 | 72 | 4 | 0 | 68 | 0.000 | 0.437 | 0.437 | 11.44 | 26.22 |
| 36 | 71 | 4 | 0 | 67 | 0.000 | 0.437 | 0.437 | 11.01 | 25.22 |
| 37 | 71 | 3 | 0 | 68 | 0.000 | 0.437 | 0.437 | 10.57 | 24.22 |
| 38 | 74 | 8 | 2 | 66 | 0.030 | 0.437 | 0.430 | 10.13 | 23.22 |
| 39 | 68 | 5 | 3 | 63 | 0.048 | 0.423 | 0.413 | 9.70 | 22.92 |
| 40 | 65 | 10 | 0 | 55 | 0.000 | 0.403 | 0.403 | 9.29 | 23.05 |
| 41 | 58 | 6 | 1 | 52 | 0.019 | 0.403 | 0.399 | 8.89 | 22.05 |
| 42 | 56 | 4 | 1 | 52 | 0.019 | 0.395 | 0.392 | 8.49 | 21.47 |
| 43 | 52 | 2 | 0 | 50 | 0.000 | 0.388 | 0.388 | 8.10 | 20.88 |
| 44 | 52 | 2 | 0 | 50 | 0.000 | 0.388 | 0.388 | 7.71 | 19.88 |
| 45 | 59 | 7 | 0 | 52 | 0.000 | 0.388 | 0.388 | 7.32 | 18.88 |

**Table 8.3** (*cont.*)

| Age | Enter | Censored | Dead | At risk | $q_x$ | $l_x$ | $L_x$ | $T_x$ | $e_x$ |
|---|---|---|---|---|---|---|---|---|---|
| 46 | 55 | 3 | 3 | 52 | 0.058 | 0.388 | 0.377 | 6.93 | 17.88 |
| 47 | 54 | 9 | 1 | 45 | 0.022 | 0.365 | 0.361 | 6.56 | 17.94 |
| 48 | 47 | 3 | 1 | 44 | 0.023 | 0.357 | 0.353 | 6.20 | 17.34 |
| 49 | 43 | 1 | 0 | 42 | 0.000 | 0.349 | 0.349 | 5.84 | 16.73 |
| 50 | 51 | 2 | 0 | 49 | 0.000 | 0.349 | 0.349 | 5.49 | 15.73 |
| 51 | 50 | 2 | 1 | 48 | 0.021 | 0.349 | 0.346 | 5.14 | 14.73 |
| 52 | 50 | 4 | 3 | 46 | 0.065 | 0.342 | 0.331 | 4.80 | 14.03 |
| 53 | 43 | 3 | 0 | 40 | 0.000 | 0.320 | 0.320 | 4.47 | 13.98 |
| 54 | 43 | 1 | 1 | 42 | 0.024 | 0.320 | 0.316 | 4.15 | 12.98 |
| 55 | 47 | 6 | 1 | 41 | 0.024 | 0.312 | 0.308 | 3.83 | 12.28 |
| 56 | 41 | 2 | 4 | 39 | 0.103 | 0.304 | 0.289 | 3.52 | 11.58 |
| 57 | 42 | 2 | 4 | 40 | 0.100 | 0.273 | 0.259 | 3.24 | 11.84 |
| 58 | 37 | 1 | 1 | 36 | 0.028 | 0.246 | 0.242 | 2.98 | 12.10 |
| 59 | 36 | 0 | 3 | 36 | 0.083 | 0.239 | 0.229 | 2.73 | 11.44 |
| 60 | 39 | 6 | 3 | 33 | 0.091 | 0.219 | 0.209 | 2.50 | 11.43 |
| 61 | 32 | 0 | 1 | 32 | 0.031 | 0.199 | 0.196 | 2.30 | 11.52 |
| 62 | 31 | 1 | 1 | 30 | 0.033 | 0.193 | 0.190 | 2.10 | 10.88 |
| 63 | 31 | 1 | 3 | 30 | 0.100 | 0.187 | 0.177 | 1.91 | 10.24 |
| 64 | 28 | 3 | 0 | 25 | 0.000 | 0.168 | 0.168 | 1.73 | 10.32 |
| 65 | 30 | 3 | 0 | 27 | 0.000 | 0.168 | 0.168 | 1.56 | 9.32 |
| 66 | 27 | 0 | 2 | 27 | 0.074 | 0.168 | 0.162 | 1.40 | 8.32 |
| 67 | 28 | 2 | 1 | 26 | 0.038 | 0.155 | 0.152 | 1.23 | 7.94 |
| 68 | 25 | 0 | 4 | 25 | 0.160 | 0.149 | 0.137 | 1.08 | 7.24 |
| 69 | 23 | 1 | 1 | 22 | 0.045 | 0.126 | 0.123 | 0.94 | 7.53 |
| 70 | 24 | 4 | 1 | 20 | 0.050 | 0.120 | 0.117 | 0.82 | 6.86 |
| 71 | 20 | 1 | 1 | 19 | 0.053 | 0.114 | 0.111 | 0.71 | 6.20 |
| 72 | 19 | 2 | 1 | 17 | 0.059 | 0.108 | 0.105 | 0.59 | 5.51 |
| 73 | 16 | 0 | 2 | 16 | 0.125 | 0.102 | 0.095 | 0.49 | 4.83 |
| 74 | 15 | 0 | 3 | 15 | 0.200 | 0.089 | 0.080 | 0.39 | 4.44 |
| 75 | 14 | 0 | 0 | 14 | 0.000 | 0.071 | 0.071 | 0.31 | 4.43 |
| 76 | 14 | 0 | 5 | 14 | 0.357 | 0.071 | 0.058 | 0.24 | 3.43 |
| 77 | 9 | 0 | 2 | 9 | 0.222 | 0.046 | 0.041 | 0.19 | 4.06 |
| 78 | 7 | 0 | 0 | 7 | 0.000 | 0.036 | 0.036 | 0.14 | 4.07 |
| 79 | 7 | 0 | 1 | 7 | 0.143 | 0.036 | 0.033 | 0.11 | 3.07 |
| 80 | 6 | 1 | 1 | 5 | 0.200 | 0.030 | 0.027 | 0.08 | 2.50 |
| 81 | 4 | 2 | 0 | 2 | 0.000 | 0.024 | 0.024 | 0.05 | 2.00 |
| 82 | 2 | 0 | 1 | 2 | 0.500 | 0.024 | 0.018 | 0.02 | 1.00 |
| 83 | 1 | 0 | 1 | 1 | 1.000 | 0.012 | 0.006 | 0.01 | 0.50 |
| 84 | 0 | 0 | 0 | 0 | 0.000 | 0.000 | 0.000 | 0 | 0 |
| 85 | 0 | 0 | 0 | 0 | 0.000 | 0.000 | 0.000 | 0 | 0 |
| 86 | 0 | 0 | 0 | 0 | 0.000 | 0.000 | 0.000 | 0 | 0 |
| 87 | 0 | 0 | 0 | 0 | 0.000 | 0.000 | 0.000 | 0 | 0 |
| 88 | 0 | 0 | 0 | 0 | 0.000 | 0.000 | 0.000 | 0 | 0 |
| 89 | 0 | 0 | 0 | 0 | 0.000 | 0.000 | 0.000 | 0 | 0 |
| 90 | 0 | 0 | 0 | 0 | 0.000 | 0.000 | 0.000 | 0 | 0 |

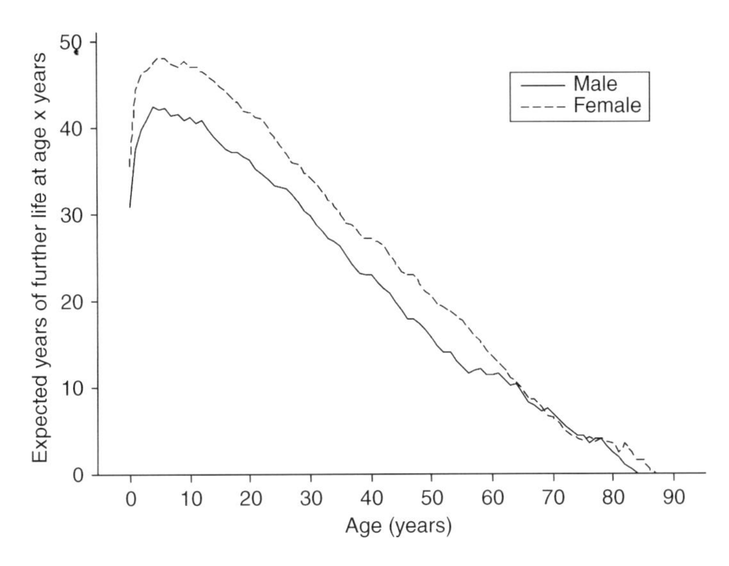


**Figure 8.4** Life expectancy from birth to old age. Note that life expectancy increases until age six years. Those who survive the perils of early childhood have a greater life expectancy than they had at birth. Until their mid-60s, Hadza males have shorter life expectancy than females. If males survive an apparent challenge in their 50s, they have the same old age life expectancy as women.

## 8.7 Checking the risk group and the mortality estimates: follow-up of the 1985 census cohort

Like Howell (1979. p. 73), I recognize that while deaths are relatively easy to record, obtaining a correct count of the population at risk is much more difficult in a small and mobile population. The size of the correct "population at risk," the population at risk of death, that contributes the deaths in the record, exerts a large influence on the $q_x$ measure. Very minor variations are produced by the methodological issues, such as whether censored individuals are counted as being at risk for a portion of the year in which they were last seen. A more serious issue for our study was the decision to include only children who were offspring of interviewed women, or had themselves been seen in a census between 1985 and 2000. One important decision concerned the many people who left the sample when last observed but whom our final "where are they now?" interviews suggested might have been still alive a year or so later. Should those extra years of likely survival have been added to their number of years at risk, perhaps lowering our mortality measures?

Here I conducted a check, not just of the computations and calculations that gave the mortality estimates, but more importantly, a test of whether I used an adequate "risk group," because now we derive our risk group for the check by a quite different method. Instead of finding the years at risk (years under observation) from our repeat censuses, we take just one cohort of people, those seen in the 1985 census, and see whether the number who died by 2000 matches the number we would predict from our life table. In this test (similar to that used in Blurton Jones *et al.*, 1992),

**Table 8.4** Mortality and survival of the 1985 census cohort. Comparing number of deaths observed by 2000, and number predicted, by applying the life table to the people in the 1985 census on whom there was follow-up data. See also Figure 8.5. The observed number are close to the predicted figures, and are between the 15% boundaries (predictions from 15%-higher or -lower mortality). If the life tables had been based on a misleading "at-risk group," using the known 1985 risk group would give a poor match between observed and predicted deaths

| 1985–2000 deaths | Females | Males |
|---|---|---|
| Predicted by $q_x$+15% | 73.7 | 96.9 |
| Predicted | 66.6 | 83.7 |
| Observed | 65 | 88 |
| Predicted by $q_x$−15% | 59 | 77.8 |

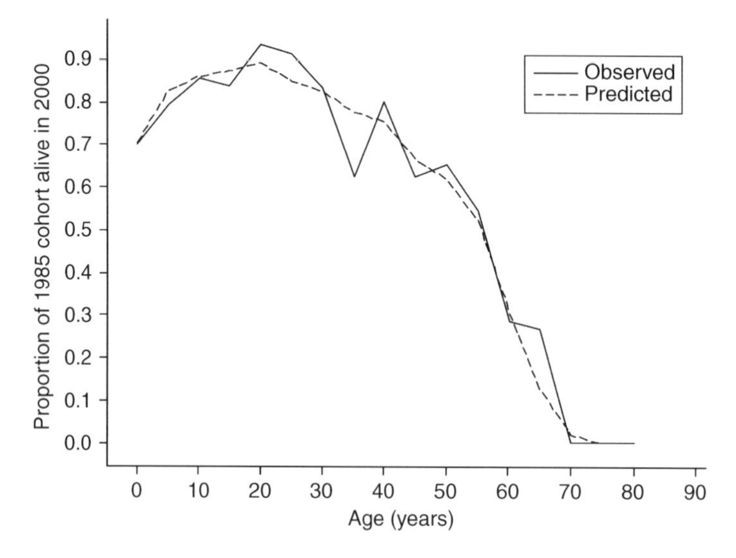


**Figure 8.5** Survival of 1985 cohort to 2000. Sexes combined. (Data table in SI 8.8.)

each one-year sub-cohort in the 1985 census suffers the $q_x$ for its age, and the survivors go on into the next year. Then they suffer the $q_x$ for that year ($q_{x+1}$), and so on until 2000. Table 8.4 compares the number of male and female deaths predicted and observed, and Figure 8.5 shows predicted and observed survivorship of five-year age groups. Visually, the match in Figure 8.5 is quite close. In Table 8.4, we show the number of male and female deaths observed and predicted but we also show the number of deaths that would be predicted if mortality were 15% higher, or 15% lower, than estimated. The match between observed and predicted values is very close indeed, the numbers predicted by higher or lower mortality levels (while perhaps not significantly different) are noticeably different from those observed. In SI 8.5, I show similar data for the 1977–1985 period. These results add to our confidence in our estimates of mortality. The selection and computation of our risk group, a procedure to which mortality estimates are most sensitive, must have been quite good.

## 8.8 Looking for secular change in mortality

I have now described the level and pattern of mortality seen during the 15 years between 1985 and 2000. We may wonder how well these data represent other periods. If we are going to report the 1985–2000 mortality data as representing the Hadza condition during a longer period, and if we are going to use them as input to any kind of stable population model, or generalize them in any way, it would be useful to know whether the mortality we have observed is a stable representation of a longer period, or merely a snapshot of a period of transition.

We cannot define any clear-cut transitions in Hadza life such as are denoted by Hill and Hurtado's (1996) "Forest," "Contact," and "Settlement" periods, or Early and Headland's (1998) similar categories. Nor are there changes as apparently transformational as the settlement of significant numbers of Herero in the Dobe area. However, 1985 to 2000 was a time during which many things happened around the Hadza, and during which interventions in their life were at least as frequent as previously. Did these events affect mortality? Has Hadza mortality declined as the outside world brought different resources? Has mortality increased as newcomers degraded the environment, and accelerated disease transmission? The estimates of CDR during 1985 to 2000 (described previously and shown in Table 8.1) showed great annual variation, but regression showed no significant trend toward increasing or decreasing mortality within our study period.

We can look for recent secular change in mortality using the children's annual hazard file, more fully described in Part II where we investigate several factors influencing child survival. In logistic regression controlling for child age + age-squared, we see a non-significant and very small secular increase in probability of child deaths (calendar year, b=0.0127, p=0.173, OR=1.01, CI 0.99–1.03). We can also use this file to see if there are regional differences in child mortality, and whether the 112 children of non-Hadza fathers were any more or less likely to stay alive than the 583 children of Hadza fathers. Father's tribe had no significant effect on child survival (deadalive = age, $age^2$ $mom_{age}$ $dad_{trb}$; b=–0.1456, p=0.442, OR=0.86, CI 0.60–1.25).

In the case of the Hadza, we are fortunate to have the account of Hadza demography by Dyson (1977). Dyson based his analysis on reproductive history interviews of 75 women, conducted by Woodburn in the 1960s (we did not use these in any way; furthermore, few if any of the women were alive and could have been interviewed during our study), and on census data and age estimates conducted during the IBP fieldwork in 1966 and 1967. Dyson reports that he noted 64 deaths of infants out of 294 live births, and based on this proportion, selected from the C&D tables, a level 6 mortality. Using the age structure and parity data available to him, he selected North 6 with a rate of increase of 13.9/1000/yr, total fertility rate (TFR) of 6.15 live births, and CDR of 32.8/1000/yr. This death rate is 7% higher than our estimate (30.3/1000/yr) for 1985 to 2000, and falls at the 88th percentile of our resampled death rates. Dyson reports that the model indicates an $e_0$ of 31.0 years for the sexes combined. This is 5% lower than our estimate of 32.7 years for the sexes combined.

On the one hand, we may say that, given the indirect methods used by Dyson and the considerable passage of time since the period his estimates represent, these estimates are remarkably similar. On the other hand, the comparisons point in the same direction, toward a very slightly higher mortality in the 1960s.

## 8.9 Regional differences in mortality 1985–2000

The 141 children born in Mangola region were significantly more likely to die than the remaining 554 children (b=0.3843, p=0.009, OR=1.47, CI 1.10–1.96, after controlling for child age and age-squared). The finding was quite robust; it showed in Kaplan–Meier and "Cox regression" analyses. Children born in Munguli were less likely to die than others, but this effect was not significant (b=–0.4527, p=0.144, OR=0.64, CI 0.35–1.17). Although both the Mangolabe and Mungulibe tend to live nearer health services than others, they differ from the overall sample in opposite ways. The Hadza in Mangola have much more contact with a large, mixed, and mobile non-Hadza population, and much more access to alcohol (and after 1995, the money from tourists to purchase it), which may produce more inter-personal violence and some child neglect. Tourists did not go to Munguli. Since 1995, many went to Mangola, and their money was often quickly spent on alcohol. We have witnessed plenty of drunkenness among the Hadza close to Mangola village but only minimal indications of violence. Nor do we have direct evidence of child neglect. The nearest indication of this we have is Frank Marlowe's comment that in one camp where he stayed in 1995–1996, frequently visited by tourists and very close to the village where adults quickly went to spend the tourist money, the several small children were being fed primarily by a 12-year-old boy. Nonetheless, the higher mortality in the Mangola region pre-dates the 1995 upsurge in tourist money. In the twenty-first century, the alcohol problem became much worse and spread throughout Hadza country (Marlowe personal communication; Crittenden personal communication). Butovskaya (2013) describes the effect of alcohol on Hadza violence.

## 8.10 Sex differences in mortality

Both the CDR and $e_0$ show a large difference between Hadza males and females ($e_0$ is 35.5 years for females, 30.8 years for males). Hadza males appear to suffer a greater probability of death than females during early childhood, adolescence to young adulthood, and most strikingly, in late middle age (50s and 60s). A similar observation is reported for many other populations, but among the Hadza the difference among the 50- to 60-year-olds appears unusually large. I was initially doubtful about the size of this difference because age assessments for males were made a little differently from age assessments for females. The data on males comprises fewer individual reports of age relative to the interviewed person, and more partial rankings of series of individuals by teams of two informants. A detailed reinvestigation left me much more confident of the result (SI 8.6). Figure 8.3 compares Hadza male and female $l_x$, and SI fig. 8.6b shows the difference, female $l_x$ minus male $l_x$.

The difference appears to accelerate during three periods when males fare less well: early childhood (1–4 years of age); adolescence–young adulthood (16–27 years of age); late middle age (56–64 years of age). Middle childhood (5–16 years of age) and adulthood (28–55 years of age) appear to be periods of relative equality in the survival prospects of the two sexes.

The excess of male deaths in adolescence and late middle age (shown by steep upward slopes in SI fig. 8.6b) could plausibly be attributed to accidents and misadventure. Sixteen of the 48 male deaths that had an attributed cause were by homicide (5), accidents (3), or misadventure (8). If we exclude deaths in childbirth, six of the 43 female deaths with an attributed cause were caused by homicide (2), accident (2), and misadventure (2). Tested as a $2 \times 2$ chi-squared analysis, this difference between the sexes is significant at $p<0.05$. As reported earlier, Bennett *et al.* (1973) inquired about injury from accidents and found 29 attributed to men and six to women.

Males in many societies expose themselves to more environmental and interpersonal hazards, and the Hadza seem to be no exception. Many human populations show better survival of females than males (Kruger and Nesse, 2006 reports a recent example). This can be seen in all of the C&D models, but the Hadza data show a larger difference in 55- to 65-year-old mortality. Resampling indicates that the Hadza sex difference in $l_x$ significantly exceeds that of the C&D models for all ages after 20, and would not be much affected by errors in the age estimates for men over 40.

Among hunter-gatherers, the Hiwi seem to show a similarly striking divergence of mortality between men and women aged 40–60 (Hill *et al.*, 2007, fig. 1b). The Ache $l_x$ curve remains higher for males than females until age 40. After that age, as among Hadza and Hiwi, males die faster than females. The !Kung data in the study by Howell (1979, fig. 4.4) resemble that of the Hadza in showing the $l_x$ curves diverging during middle childhood and the teenage years. Howell's data were derived from the interviews of women who had completed their reproductive careers. The source thus resembles that of the Hadza data, and mostly pre-dates the increasing interaction with cattle owners since the late 1950s. The hunting and gathering way of life might expose males to hazards that men succumb to more easily as they age. Falls from baobab trees could be such a hazard among the Hadza; we came across two memorable elderly examples.

Hiernaux and Hartono (1980), writing on the anthropometric data gathered by the IBP fieldworkers in 1966–1967, commented that when compared to the men, Hadza women seemed to be nutritionally and physically better off than many African women. Their greater subcutaneous fat and superior body mass index (BMI) may indicate that, despite their short stature, they are relatively well fed and have been since childhood (Chapter 16). This advantage might be showing itself in survivorship. Alternatively, contacts with surrounding populations are more abundant for men and might be more risky for them, either due to exposure to disease, or rarely to violence. Even if the adult data that impressed Hiernaux is the result of the severe filtering in a high mortality population, it still indicates a relative advantage of those women who survive to adulthood. Alternatively, it could result from absence of a parental preference for one sex over the other. In some populations, intervals following the birth of a boy are longer than those following the birth of a girl. This is not the case among the Hadza (Chapter 17). Nor do weight or height differences between Hadza boys and girls show up until the teenage years (Chapter 16). These relative advantages of Hadza women could contribute to their relative advantage in survival in adulthood and late middle age.

## 8.11 Summary and discussion: long adult life expectancy again

Life expectancy at birth was 30.8 years for men, 35.5 years for women, and 32.7 years for the sexes combined. We found a mortality level very similar to that proposed by Dyson, based on the 1966–1967 IBP survey. The gender difference was quite large among Hadza, larger than C&D models would have suggested. As in many rural developing country populations and historical demography populations, the biggest portion of the deaths fall on the under-five-year-olds.

I found no strong evidence for secular change in Hadza mortality. However, Dyson's estimated $e_0$ (31.0 years with sexes combined) was sufficiently shorter than mine (32.7 years) to raise some doubt. Although just outside our 95% CI (31.4–33.6) from resampling the observations, the difference is less than that between adjacent C&D models. In a population of 900 people, for example, the difference in CDR amounts to a difference of 1.5 deaths per year, undoubtedly well below the resolution of our data or our methods in this small and very mobile population.

Hadza $e_0$ was intermediate between that of the !Kung (30.0 years for females, 27.7 years for males) and the forest Ache (37.1 years for females, 37.8 years for males). These estimates do not change the comparative position of the Hadza reported by Gurven and Kaplan (2007), in which they used my earlier published estimates. But there are other recent studies of hunter-gatherers, some of them omitted from the detailed review by Gurven and Kaplan (2007), that report rather lower life expectancies. The Hiwi $e_0$ is 27 years (Hill *et al*, 2007), the Agta 24.9 years (Early and Headland, 1998), and Migliano *et al.* (2007) collected several more cases, apparently with an $e_0$ in the low and mid-twenties. Most of these cases are also people of very short stature (males $<155$ cm), contrasted with the 160 cm of the Hadza, the !Kung (Howell, 2010, fig. 3.19), and the Ache (Hill and Hurtado, 1996, fig. 7.2). Migliano shows that the high mortality populations also tend toward early ages at first birth, and she suggests that these populations show a life history adaptation to high mortality (challenged on demographic and other grounds by Becker *et al.*, 2010). Migliano also points out that most of the Pygmy peoples live in tropical forest. We may note Dunn's (1968, table 1, p. 227) observation that people living in tropical forest were exposed to a greater number of infectious and parasitic organisms. The association of high adult mortality, early age at first birth, and short stature seems to be found in tropical forest in several continents, including Africa. I will discuss this in Chapter 22 in connection with Terashima's (1980) suggestions about differences related to forest and savanna.

None of the mortality schedules found among modern hunter-gatherers resemble the estimates of paleodemographers. This is important for thinking about the evolution of human life history. In SI 8.7, I discuss the several possible reasons for the difference. Among them we may note that populations with written records show life histories that resemble modern hunter-gatherers. Populations assessed from bones do not, even those that date from more recent times than the earliest written population records. Skeletal collections from populations that had contemporary written records show errors in age estimations from bones of the elderly (Molleson *et al.*, 1993), and differential preservation of the bones of old and young (Walker *et al.*, 1988). Comparative evidence, finding primate lifespans (oldest age recorded) to be strongly predicted by body weight and brain size, supports the view that long lifespans characterized not only *Homo sapiens* but also *Homo erectus* (Weiss, 1981; Hammer and Foley, 1996; Judge and Carey, 2000). Newer methods may be expected (Finch and Stanford, 2004).

Hill *et al.* (2007) and Gurven and Kaplan (2007) discuss the relative importance of violence and disease as causes of mortality in simple societies. Black *et al.* (1974) and Bennett *et al.* (1970) show that even "remote" populations had been exposed to modern crowd diseases by the time they had been "contacted" and studied. However, with the exception of the 1986 measles outbreak, these diseases apparently account for few of the Hadza deaths. The Hiwi $e_0$ reported by Hill was 27 years, about halfway between that of the Agta and of the !Kung. It is not clear to me what the $e_0$ would be if the deaths by violence were subtracted. The effect of removing killings of girls aged 5–10 from the Ache data would be quite large; it appears to account for the unusual excess of male survival over female survival shown at ages five and 10 years (Hill and Hurtado, 1996, figs. 6.2, 6.3, and table 6.1). Effects of violence on population increase may be quite large, or very small, depending partly on who the victims are, as discussed in Chapters 10 and 11.

Hill *et al.* (2007) point out that the savanna African hunter-gatherers have low rates of violence compared to the Ache and Hiwi. They indicate that this is true even when external attacks are excluded, and suggest a cultural difference resulting from dissimilarities in colonial and post-colonial intervention. The colonial influences in South America and in Tanzania seem to have been radically different. The Hadza have received no genocidal raids from settlers. In Tanzania, colonial powers gradually reduced the nineteenth century inter-tribal violence that had centered on control of the slave and ivory trades, and replacing tribe with nation has been the permanent goal of Tanzanian governments since independence. In Chapter 10, I discuss the extent to which we should regard the colonial experience as pacification of the Hadza or of their neighbors. Hadza share with other Tanzanians the propensity to "talk a problem into the ground," seemingly endless discussions dwindle away into some kind of peace by default.

# 9 Testing the estimates of fertility and mortality

If fertility and mortality persist at a given level for several decades, they generate a population with a predictable age structure and rate of increase. Demographic measures have an essential inter-relatedness that derives from this phenomenon, commonly referred to as stable population theory. In general, if you know two of these measures, and have reason to believe that fertility and mortality have been approximately stable, you can predict the other measures. I want to exploit this opportunity for triangulation to see how firmly we should believe our estimates of Hadza fertility and mortality, and how well we have represented average Hadza demography during the second half of the twentieth century. I will show whether the age structure predicted by our estimates of fertility and mortality is matched by the age structure we observed in the field. In addition to age structure, several other measures can be predicted for a stable population and can be compared to field observations.

A stable population, one in which fertility and mortality remain constant for some decades, and in which migration is negligible, is something of an abstract ideal. In reality, fertility and mortality vary year to year, age structures are jagged (Figure 10.2), migration likewise varies and may occur in quite irregular bouts responding to some climatic or social opportunity or hardship. In small populations such as the Hadza, these variations become especially influential. Nonetheless, demographers of modern states, anthropological demographers, and historical demographers continue to find stable population models to be extremely useful. At the very least they can show when apparent results are inconsistent and require re-examination. Sometimes demographers have used stable population theory as a solution to the practical problems of measurement in less accessible populations. Especially useful have been the tables developed by Coale and Demeny (1983), and the Weiss (1973) models for anthropological populations have been widely cited. If one has data that allow one to select one of the models, then many other demographic parameters can be simply read off the model, or compared to further field observations to verify the choice of model. In the example best known to anthropologists, Howell (1979) used the Coale and Demeny (C&D) tables to obtain initial estimates of !Kung mortality and age structure. Based on the rate of deaths in infancy, and the proportion of live individuals known to be under five years old and under age 10 (a highly abridged age structure), she selected C&D model West 5 as a first approximation to the !Kung population. She used the predicted age structure to help in estimating ages of all the other individuals in her population. The model also predicts life expectancy at birth ($e_0$), and subsequently, the proportion of people aged over 60, and so on. Howell interviewed women of completed childbearing age, which allowed her to match completed family size (CFS) and gross reproductive rate (GRR) to her chosen model. Comparison of the model predictions and relevant observations showed that fertility and mortality prior to the 1960s conformed rather well to this model, and that since that time, mortality had greatly declined.

Bagnall and Frier (1994), working with papyri containing partial census records from Egypt between AD 11 and 257, made careful use of the C&D models to estimate demographic parameters of the Egyptian population under Roman rule. Arguing that the long-term records imply a long-term, stationary or only slowly increasing population during their study period, they were able to match the female age structure (for ages 5–80) to model West, level 2. Model expectations about fertility could be supported with information of mean age at childbearing and other data to give a plausible picture of fertility. Bagnall and Freier arrived at a very supportable picture of Egyptian demography during the Roman period. The population, although with higher mortality, conforms in most ways to the picture of third-world and historical populations, including the presence of old people in their 70s.

The C&D models were also used in earlier studies of Hadza demography. Thus, when Dyson analyzed the data collected on the Hadza as part of the International Biological Program (IBP) in 1966–1967, he used data on infant mortality from reproductive history interviews conducted by James Woodburn to select a mortality level from Coale and Demeny. Having chosen C&D model North 6, he compared its predicted age structure to the observed age structure, concluding that the fit was adequate. The reproductive history data suggested a mean age at childbirth of 29, which the model suggests is given by a total fertility rate (TFR) of 6.15 and GRR just over 3.0. Observations on the number of births by age of woman interviewed (Dyson's table 2) supported these estimates. The model predicts a crude birth rate (CBR, births/1000 woman/yr) of 46.7, crude death rate (CDR) 32.8, $e_0$ of 31.0 years and rate of increase of 13.9/1000/yr. Thus, from a relatively short-term study, by using the C&D models, Dyson was able to offer estimates of most of the important demographic parameters of the Hadza.

In our 1992 paper on Hadza demography, we used data gathered in a single field season (1985) in which we could measure age structure, and a rough estimate of mortality from survival of individuals since Lars Smith's previous censuses in 1977–1978. These were used to match to the C&D model North 6 chosen by Dyson. We found that these, and the numbers of live children of women of different ages, and an estimate of the rate of increase using a "capture-recapture" method, also fitted well with model North level 6. From these rather fragmentary data, we could support the match to model North 6. The model could then be used to extrapolate any other demographic measure one wished to see, such as fertility (TFR, GRR) and mortality (life expectancy at any age, average age at death), and so on.

However, some anthropologists have argued that such studies may be more dependent on the use of the models than is wise (Hill and Hurtado, 1996; Gage, 1998). Two reservations about using the existing life tables (such as those of C&D or

Weiss) should be distinguished. Howell's use of the models to help estimate ages was criticized by Hill and Hurtado. The slight risk of circularity would seem best avoided. Gage (1998) among others criticized the use of life table models for its tendency to impose the common pattern on what may be unique cases. To use the models with remote, important and little known, thus potentially unusual populations, such as surviving hunter-gatherers, reduces the opportunity to find cases that do not follow the usual patterns, and may reduce the value of such studies. Howell has argued that the models provide the most logical "null hypothesis" for the population parameters. Anyone who would argue that their study population is unusual can try to show a deviation from the models, and perhaps even test the significance of the deviation, for instance by resampling their observations.

In this study, as in Hill and Hurtado (1996) (but in contrast to Dyson, 1977 and Blurton Jones *et al.*, 1992), I have gone to some lengths to avoid any dependence on the existing models. I do not assume that the Hadza population resembles any other population summarized by Coale and Demeny (or by Weiss, 1973), nor do our data and measurements depend on those models in any way. I have much more direct data about fertility and mortality and I try to let the Hadza data speak for itself. Instead, I use my fertility and mortality estimates as the "input" or generators of predictions from a population simulation that simply mimics the natural process of reproduction and death across many simulated years (SI 9.1). The simulation runs until long after the arrival of a stable population structure with its unvarying rate of increase and other parameters. These can be measured directly from the simulated population and compared to the values observed in the field. Nonetheless, our predictions do depend on the assumption that the Hadza population has been stable; in other words, that fertility and mortality did not change significantly during the last half of the twentieth century. In Chapters 7 and 8, I offered evidence for stability of Hadza fertility and mortality during this period.

## 9.1 Requirements for the tests

I will predict measures such as age structure, rate of increase, average age at death, and CDR, and compare the predicted values with our directly observed estimates. If the observations match the predictions, we gain confidence in our results; more or less different kinds of data are found to fit together in a predictable manner. However, there are limitations. The observations are taken in the short term, to test predictions about long-term averages. This renders the enterprise vulnerable to the short-term fluctuations that may occur in any small population. However, the observed values could turn out to be strongly incompatible with the estimated fertility and mortality. If there is such a mismatch, then something is seriously wrong. A mismatch could arise because our estimates of fertility or mortality are incorrect. Alternatively, a mismatch could arise because our estimates were correct but they had not endured for enough time without major disruptions in fertility or mortality, or major migrations, to generate the observed age structure and associated measures. A mismatch might also arise because the data on which the observed age structure was based are unreliable or biased. For example, censuses, such as our household lists, are notoriously liable to undercount children. This would badly distort the observed age structure. We will see, as some readers will realize, that these are not very sensitive tests. Nonetheless, especially given the difficulties of demographic study of small, mobile populations such as the Hadza, almost any test is worthwhile.

Our tests raise another problem. We are trying to see whether our observations match our predictions. How do we test for a match? Will we be trying to prove the null hypothesis? I will report three kinds of comparison: (1) compare the predicted value with the observed value, and with the 95th percentile ranges of each, obtained by resampling (see Figure 9.1, Tables 9.1, 9.2; and SI 9.2 for the sexes combined); (2) compare the observations with values predicted by adding or subtracting 5%, 10%, or 15% to the values of age-specific fertility (ASF) or mortality that are fed to the population simulation (SI 9.3); (3) compare the Hadza observations to predicted values for the Ache and the !Kung.

Resampling statistics can show how the predictions and observations vary due to the chance composition of the samples from which both predictor variables (fertility and mortality) and test variables (age structure, etc.) were derived. Resampling to generate the probable range of variation in the predictions was done as follows. Individuals were chosen at random from the population that created the estimates of ASF and age-specific death rates ($q_x$). Some individuals were omitted, others used more than once. The runs provide a range of estimates of ASF and $q_x$ that became the input to the population simulation. Predicted values are then gathered from the results of 500 such resampling runs of the population simulation, each running for 300 "years," ample time to generate stable predictions. Variation in the observed values was generated by resampling the sample of people eligible to be measured; seen in the core study area on more than 70% of their observations.

In addition, I made the computer run the population simulation with different levels of fertility and mortality (SI 9.3), and with different assumptions about migration (SI 9.4), to show how much variation we could expect in the test variables. We can say, therefore, whether our observations fall outside the 95% CI of the predictions from estimates lower or higher than our best estimates of fertility or mortality. We can compare the predicted levels with the observed 95th percentiles and say whether, for example, there is a 5% chance that the real fertility was as high as 10% higher than we estimated. We may be able to conclude that on only two to three out of 500 resamples, the observed measure falls at a level that would arise from a 10% higher fertility. I thus attempt to show that an observed value is significantly different from the value predicted by fertilities and mortalities that differ from our estimates.

A couple more issues must be discussed before I report the results: which demographic measures are the best ones to predict, and what is the appropriate sample from which to collect them?

A measure of age structure is the classical third side of the triangle but there are other measures that vary with fertility, mortality, and structure, such as rate of increase, and average age at death. The most sensitive tests will be on the measures that vary by the greatest extent as fertility or mortality vary. I chose to predict and test measures that seemed most sensitive to variation in age structure, mortality, or fertility in the C&D tables. The most meaningful measures will be based on data that were gathered independently from, or with least dependence upon, the data used to estimate fertility and mortality. Two measures should be particularly meaningful:

1. Age structure ascertained from the sample of people who turned up to be measured, and which is a sample of people very much unrelated to the interviews with women that determine fertility and much of the mortality data. Age structure can be presented as an age pyramid, as a percentage below each age, or summarized by average age of the population, or percentage below age 20 years.
2. Age at death can be ascertained from the "where are they now?" interviews, and completely independent of the computation of the live age structure and of the denominator for the mortality measures – the sample at risk of mortality. Estimation of mortality is quite sensitive to the methods of determining the risk group. Age at death can also be represented as a percentage of deaths below each age, or summarized as mean age at death of the population. However, in the range of mortality levels to be expected with remote high mortality populations, age at death distribution is not very sensitive to level of mortality. It is at least as sensitive a reflection of age structure, fertility, and rate of increase as it is of mortality level, so it will be of relatively little use in testing our mortality estimates. Nonetheless, it will provide an excellent check on our fertility estimates, a test based on data gathered independently from our test of live age structure.

The test is strengthened by matching observations to several different variables that are predicted from an enduring schedule of fertility and mortality, such as average age at death, rate of population increase, CDR, and birth rate. Other measures could be checked against the population register but they would be more closely tied to the data that generated the fertility and mortality estimates. I use the population register data to compute observed rate of increase, CBR, births per woman aged 15–44, and CDR. These measures are not derived solely from the interviewed women from whom ASF was derived.

The population that I use to check the predictions should resemble, as closely as possible, the population that I used to estimate fertility and mortality. One important way in which the "test sample" and the "original sample" can resemble each other is in the chance that we know much about the individuals in the sample. For example, we know very little about births to women who lived mainly in the west. Among the people we saw in 1985 in Munguli, there were some who we often saw, and whose fate we were able to follow. There were others who we never saw again, and we do not know how many more children they had, nor whether or when they died. We should exclude such people from our "test sample." Each of my eight census visits covered a core area that we have called "the study area" (described in Chapter 4). Other locations were sometimes visited, or we made household lists by talking to individuals who had

**Table 9.1** Female demographic variables predicted by observed fertility and mortality, compared to observed values. Rate of increase (sexes combined) was assessed by several methods discussed in the text; results ranged from 0.012 to 0.018. Observed percentage of population aged <20 and average age of population are from the anthropometry sample

| Variable number | Variable name | Females, predicted median and 95% CI | Females, observed median of resampled observations and 95% CI |
|---|---|---|---|
| 1 | Rate of increase 1987–2000 | 0.0162 (0.0124–0.0195) | 0.0176 (0.0119–0.0244) |
| | 1985–2000 | | 0.0123 (0.0075–0.0175) |
| 2 | Crude death rate | 27.0 (26.7–27.4) | 27.1 (20.0–34.0) |
| 3 | Crude birth rate | 43.3 (39.2–46.7) | 41.8 (30.9–49.0) |
| 4 | Female births per woman aged 15–44 | 0.102 (0.091–0.111) | 0.101 (0.075–0.124) |
| 5 | Average age at death | 20.3 (18.1–23.3) | 20.2 (16.5–24.4) |
| 6 | Average age at death >5 | 43.1 (40.9–45.6) | 40.9 (37.2–45.5) |
| 7 | % of population aged <20 | 48.9 (45.8–51.5) | 47.1 (43.0–51.3) |
| 8 | Average age of population | 24.5 (23.2–26.0) | 25.7 (23.3–27.9) |

just come from those locations. We think such lists are much less reliable than our regular house lists, and specially under-represent children. Consequently (except where specified otherwise), I derived the "test measures" from all individuals for whom >70% of the records we have of them came from within our study area. For the two measures of age structure: percentage below 20 years old and average age of the sample, I used all individuals who presented themselves for measurement.

Tables 9.1 and 9.2 show for females and males, respectively, and SI 9.2 for the sexes combined, the predicted values and observed values of the eight variables that change with age structure, fertility, or mortality. In these tables, the predicted values and their 95th percentiles are listed in the left columns, and the observed values are shown on the right with the median and 95th percentiles from resampling. These results are illustrated in Figure 9.1. SI 9.3 shows results when predictions are drawn from higher or lower fertility or mortality than observed. SI 9.4 shows results when migration is included in the simulation. Later I summarize and discuss these results. Details of the observation measurements are also given here.

## 9.2 Rate of increase

The combination of recruitment and loss determines whether a population increases, decreases, or remains stationary. If I had grossly underestimated mortality or overestimated fertility, my predicted rate of increase would be higher than the observed rate. Of course, errors in opposite directions could tend to cancel each other out and might be harder to identify. The mobility of the Hadza make estimating the total population size and the observed rate of increase quite difficult. Several methods are

**Table 9.2** Male demographic variables predicted by observed fertility and mortality, compared to observed values. Observed percentage of population aged <20 and average age of population are from the anthropometry sample

| Variable number | Variable name | Males, predicted median and 95% CI | Males, observed median of resampled observations and 95% CI |
|---|---|---|---|
| 1 | Rate of increase, both sexes, as in Table 9.1 | 0.0162 (0.0124-0.0195) | 0.0176 (0.0119-0.0244) |
| 2 | Crude death rate | 31.8 (31.3–32.2) | 32.9 (24.4–41.6) |
| 3 | Crude birth rate | 48.0 (43.8–51.4) | 46.3 (34.4–56.8) |
| 4 | Male births per woman aged 15–44 | 0.109 (.098–0.119) | 0.1080 (0.077–0.14) |
| 5 | Average age at death | 18.5 (16.6–21) | 21.2 (16.9–24.5) |
| 6 | Average age at death >5 | 39.3 (37.6–41.6) | 40.8 (37.2–44.3) |
| 7 | % of population aged <20 | 52.5 (49.6–54.9) | 53.7 (49.2–57.9) |
| 8 | Average age of population | 22.4 (21.3–23.7) | 23.2 (20.9–25.4) |

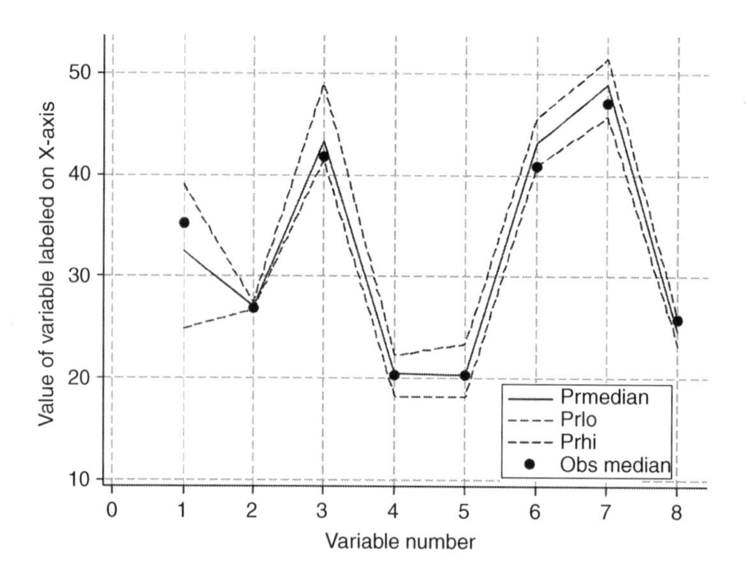


**Figure 9.1** Illustrating Table 9.1. Predicted female values and their 95% CI compared to observed values of the eight variables. The solid line "prmedian" links median predicted score on variables 1 to 8. The broken lines link the high "prhi" and low "prlo" 95% CI of the predictions, based on resampling. The filled black circles mark the observed values. To fit on the graph, population increase was multiplied by 2000, and births per woman aged 15–44 was multiplied by 200. Crude death rate is very little affected by changes in fertility but much more affected by adjustments to mortality (SI 9.3).

compared here. Rate of increase was predicted from the simulated population and two other methods.

Predicted rate of increase was calculated from the total simulated population at the end of year x / total simulated population at end of year x − 1. The predicted rate of increase was 0.0156 (15.6/1000) with the CI 0.0112–0.0195.

The crude rate of natural increase is calculated from the difference between female birth rate and death rate. Crude birth rate 41.8 minus CDR 26.9=14.9/1000 or 0.0149.

Rate of increase can also be calculated from net reproductive rate and mean age at childbearing (Howell, 1979, p. 214). Net reproductive rate was 1.58 and mean age at childbearing was 28.34, which gives a rate of increase of 0.0162.

Ideally, the actual observed rate of increase would be measured by two complete censuses, taken some years apart. However, our censuses varied greatly in their completeness and were not a good choice for trying to assess increase or decrease in the total population. The population size at the time of the June 1985 census was estimated as the number of people in the register who had been born before mid-1985 and were still alive at that time, and had 70% or more of their observations in the study area. The population size in 2000 was estimated from the population register in the same way; the number of people who had been born by mid-2000 and were still alive at that time. In resampling the observations, the program randomly selected from these eligible people in 1985 and in 2000 and gave quite a wide range of variation. I calculated rate of increase for both sexes together.

The two populations thus estimated were 684 in 1985 and 822 in 2000. The actual eastern Hadza population was rather larger. The observed annual rate of increase from 1985 to 2000 was 0.0123, or 12.3/1000. Removing the 70% in area selection, thus including more of the people seen in Munguli and neighboring Singida villages in 1985, gave a slightly lower rate of increase, 0.0118. These people had had little chance that their children would be represented in our censuses, interviews, or population register.

The observed rate of increase between 1985 and 2000 was clearly at the low end of the predicted 95th percentiles. I wondered whether this was caused by the measles outbreak during the very brief settlement attempt at Yaeda late in 1986. Although the deaths we recorded at this time have contributed to our estimates of mortality, their concentration in a short period of time may have had a visible but brief effect on population size. This seems to be true. If we calculate the rate of increase from 1987 to 2000, we get a much higher figure, 0.0176 (using either selection criterion specified earlier), which is a little above the predicted value of 0.0156.

While we cannot claim an exact match between observed and expected values, we can note that the 95th percentile ranges of both observation and prediction overlap substantially and give us little grounds for believing that the population actually increased substantially more or less than predicted. The observed values also encompass Dyson's (1977) estimate of 13.9/1000. But how well do they support our estimates of fertility and mortality? The observed values, assessed by the various methods, range from 0.012 to 0.018 (12–18/1000), which correspond approximately (SI 9.3) to the values predicted when 10% is added to or subtracted from mortality or fertility.

## 9.3 Crude death rate

Crude death rates strongly reflect mortality level but also vary with age structure. Especially when infant and child mortality is high, death rate can increase with high fertility and with population increase.

The observed CDR was reported in Chapter 8. Deaths recorded in the population register were counted for each year between 1985 and 2000, and expressed as a proportion of the number of people alive in each year as indicated by the population register. Results are shown in Table 9.1, line 2, for female deaths per female population; and in Table 9.2, line 2, for male deaths per male population; and SI 9.2, line 2, for sexes combined as deaths of either sex/total population.

For females, the observed value (27.1 deaths/1000 females/yr) is almost identical to the predicted value (27.0). For males, the observed value (32.9) is higher than predicted (31.8), falling above the very narrow 95% CI of the predicted value. However, the much wider confidence limits of the observations easily encompass the predicted value. This measure is one of the least independent of the original data used to estimate fertility and mortality; it differs from the age-specific mortality estimation only in having required no age estimation but resembles it in its dependence on the estimate of the "at-risk" population. SI 9.3 shows that predicted CDR is quite sensitive to changes in age-specific mortality. If 10% is added to age-specific mortality, predicted CDR rises to 31 for females and 36.7 for males, both well above the observed figures although within the 95% CI of the observations. If 10% is subtracted from age-specific mortality, predicted CDR falls to 25.4 for females and 29.9 for males, well below the observed values of 27.1 and 32.9, respectively.

## 9.4 Crude birth rate

Observed CBR was computed from a slightly wider sample than was used to estimate fertility, the births entered in the population register. In contrast to ASF, CBR does not depend on age estimations, only on the computation of whether an individual was alive during the study period.

Observations included individuals or their parents in the population register who were recorded in the core area for ≧70% of their records. For children under age 15, this qualification included the presence of their mother in the study area. Number of births and population size for each year are counted, and summed across years, and an average birth rate for the study period was calculated. Births in year 2000 were pro-rated by the number of months of that year covered by the study.

The results are shown in line 3 of Tables 9.1 and 9.2 and in SI 9.2. At 41.8 observed births/1000 women/yr, the observed and predicted values match quite closely for each gender. Genders indicate female births per woman, male births per male, all births per total population.

## 9.5 Birth rate of women aged 15–44 years

This measure of birth rate uses the same births as previously. It takes some account of age structure, by limiting the denominator to the population of reproductive age. Hence, it depends less on the accuracy of age estimates than does ASF. At any mortality level it varies, of course, strongly with fertility and rate of increase. I computed this measure for female births to females aged 15–44, male births to females aged 15–44, male birth to males aged 15–44, and births of either sex to women aged 15–44. Line 4 in Tables 9.1 and 9.2 shows female births per woman aged 15–44, and male births per woman aged 15–44, respectively. The match between observed and expected values was very close.

## 9.6 Average age at death

Average age at death is useful in that while it uses the same records of deaths as do the mortality estimates, it is independent of the data used to derive the "at-risk group." The "at-risk group" is required as the denominator in calculations of CDR and age-specific mortality but is not needed to calculate average age at death. Average age at death is very much affected by fertility and strongly associated with age structure, and it appears to be quite insensitive to changes in mortality within the range of Hadza-like populations. Observed and expected figures, for males and for females, are close and well within the confidence limits of either predicted or observed values. While we can say that observed and expected were close, the observed figures could also be compatible with rather higher, or rather lower, levels of mortality (SI 9.3).

## 9.7 Average age at death over age five years

By excluding infant and child mortality, this measure is more indicative of adult mortality but it is very insensitive to changes in mortality level among Hadza-like populations. There was a surprisingly small sex difference in the observed figures. Among males, the observed and predicted figures matched very closely. The match for females was less close, although the 95% CI for the observations easily encompassed the predicted value.

## 9.8 Percentage of population under age 20 years

I chose this measure as one of the most sensitive single figure summaries representing age structure. Any age between 20 and 25 would do as well. However, much older ages vary less because, for example, most of the population are aged less than 60, therefore their number can vary little.

The predicted value was the percentage of the final simulated population (after the simulation runs for 300 "years"), who are aged less than 20 (up to age 19.99). The figure is calculated and stored for every run, and when all runs are completed, for the median and percentiles scored.

The observed value was estimated from those who came to be measured in our anthropometry sessions. The numbers in each year of age were summed over all censuses. Thus, values for the two-year-olds measured in 1990 were added to those of the two-year-olds measured in 1991, to which values for two-year-olds measured in 1992, 1995, 1997, 1999, and 2000 were added. This smooths out some of the year-to-year variation in the age distribution. It removes information on year-to-year fluctuations in births or deaths, but hopefully gives a more general picture of Hadza age structure. This anthropometry sample was free of the errors in our household lists (such as undercounting children) but probably subject to its own different set of biases. For example, a handful of toddlers refused to be measured but were recorded, and one or two of the oldest men also declined to be measured.

Observed medians were 47.1% for females below age 20, 53.7% for males, 50.2% for both. The observed female value is a little lower than the predicted 48.9%, the observed male value a bit higher, but both observed values are well within the 95% limits of the predicted values.

Observed females had slightly fewer under-20-year-olds, an older age structure than predicted. The same tendency is seen if we estimate age structure from other samples. Observed differ from predicted values in a direction that suggests we saw fewer than expected young females and more than expected young males. We can show that our census household lists counted fewer children and infants than arrived to be measured, and fewer than were listed in interviews. However, this does not explain the sex difference as we find it both in anthropometry data and in structure estimated from the population register. If emigration was confined to females under age 20, we would expect a lowering of the proportion aged under 20. But SI 9.4 shows the negligible effects of various levels of emigration of females aged 15–30.

The age structure reported by Blurton Jones *et al.* (1992, table 1) was based on the 1985 census, probably our most complete census. We reported the sexes combined; 48.6 below age 20, close to Dyson's 47.6, a little lower than the 50.2 we predict for both sexes combined but well within the confidence limits of the prediction. Thus, these older observations were close to the predicted values.

## 9.9 Average age of the population

The average age of members of a population provides another single number summary of the age structure; it shows whether we are dealing with a relatively young population, as in a rapidly growing population, with many children per adult, or an old population, perhaps stationary or declining, with fewer children. The population simulation predicted an average age of 24.5 for females, with 95% of the predictions falling above 23.2 and below 26.0 (Table 9.1, line 8). The observed average age from the resampled anthropometry population was 25.7 for females, 23.2 for males. These were close to the predicted values and within the 95% limits.

## 9.10 What can we conclude from these tests?

In general, agreements between predicted and observed values are encouraging. The results for the sexes combined (table SI 9.2) are very close to the predicted values. Mean values are close, means of observations fall within 95% CI of predictions, mean predicted values fall within 95% CI of observations.

Among females, all the observed values fall within the 95% CI of the predicted value. The worst match is CDR of males, for which the observed value lies outside the very narrow range of the predicted value. Observed average age at death for males is at the upper 95% CI of the predictions. For all measures, the 95% CI of the observed values are wider and enclose the predicted value and its confidence limits.

We can conclude that Hadza fertility and mortality remained stable during most of the second half of the twentieth century at levels close to my estimated values. However, these are not very sensitive tests. For example, the confidence intervals suggest there is a chance of five in 100 that our observed age structure could result from fertility as much as 10% lower (or 5% higher) than I estimated. Nonetheless, because the logic of the triangulation is so firm, I think the tests were well worth while, especially given the difficulty of demographic study of small, mobile populations such as the Hadza.

I also examined the match between observed and predicted age structure defined as the percentage of the population below age x years, as featured in the C&D tables (SI 9.5). I did the same with age at death structure (the age distribution of deaths). Age at death structure (SI 9.6) is interesting because it is the only direct demographic information that archaeologists can obtain on ancient populations. The results do not alter the picture given by the single measures of structure, average age of the population, average age at death. They do add a clear indication that Hadza age structure and its 95% CI lie between those expected for the Ache (by running the population simulation on the mortality and fertility data in Hill and Hurtado, 1996), and for the !Kung represented by C&D model West 5. Closer examination of the age structure can show slight mismatches, and draws attention back to the jaggedness that is visible in Figure 10.2. The jaggedness is likely to arise from a mixture of age heaping (that is, errors in age estimation, especially among the oldest) and genuine year-to-year variation, some random, some meaningful. These will be discussed in the next chapter.

In SI 9.7, I report a comparison of the observed and predicted number of living siblings for women of each age. The prediction is given by a simulation that keeps track of kin. The simulation is described in SI 9.7 and used for various purposes described in Chapters 11 and 19.

If the results had shown severe discrepancies between predicted and observed values, how might we have accounted for them? There are several possible reasons for mismatches. Some would be methodological, such as the well-known tendency for household censuses to undercount children. Our household lists do undercount children when compared to numbers arriving to be measured. Age estimates could be wrong. Age heaping would decrease the match to all models. Systematic overestimation or underestimation of ages should generate differences between observed and predicted measures of structure. Other mismatches could signal an unsuspected level of migration or interesting historical changes in fertility or mortality.

Among the Hadza, there is a small amount of emigration, which I attempted to measure in Chapter 5. The departure of some young women, and loss of their offspring in every generation, could have mimicked the effect of lower fertility by everyone. The population model is able to include the observed emigration of young women and return of older women (SI 9.4). At the observed rate (young women leaving at about 0.5% per year, and returning later at 0.25%), the match of predicted to observed structure is almost imperceptibly improved. At a rate of emigration of 1% per year, the matches to observed fertility and mortality are improved but still to a trivial extent.

There must always be variation in fertility and mortality from year to year and even over strings of years. But there might also be a steady trend toward increase or decrease in fertility or mortality, in which case the assumption of a stable population would be wrong. I could see no clear evidence for such trends in the Hadza data. The mean ages of the populations in 1967 and 1977, as reported by Dyson (1977, on data from 1967) and Lars Smith (1977 data), do not suggest any such trend. There is no support for either a gradual change, nor for sudden and massive changes such as the high mortality that the Ache experienced on settlement (Hill and Hurtado, 1996), or the Herero age structure showed even some 80 years after they were expelled from German Southwest Africa in 1905 (Pennington and Harpending, 1993). Two potentially interesting deviations in the age structure will be examined in the next chapter. There I also compare my Hadza data with the C&D models, to see whether there is anything unusual about Hadza demography, or as seems much more likely, if they support the "uniformitarian assumption" (Howell, 1976) that all human populations follow a general pattern with which demographers are well familiar.

# 10 Hadza demography: a normal human demography sustained by hunting and gathering in sub-Saharan savanna

During the twentieth century, the eastern Hadza comprised a nearly closed, stable, but quite rapidly increasing population. The great majority of Hadza earned their living during this time as hunter-gatherers. Even in the late twentieth century, they were getting about 95% of their food from wild animals and plants (Marlowe, 2010, p. 36). As the size of the population increased, the area inhabited decreased slightly and the number of people occupying it with other economies and other languages increased.

In the last chapter, I reported that the observations fit with predictions drawn from a stable population that has the schedules of fertility and mortality that I estimated for the Hadza. However, the 95% CI of the observations enclose levels of fertility that could be as much as 10% higher or lower, and levels of mortality that could be more than 10% lower.

Marlowe (2010, p. 256; table 10.1) described the Hadza as the median hunter-gatherer. With regard to demography, I can support the claim. At a total fertility rate (TFR) of 6.1 of births and life expectancy at birth ($e_0$) of 32.7 years (genders combined), Hadza fertility and mortality were close to the median for hunter-gatherer populations. They demonstrate that a very normal human demography can be maintained in a predominantly hunter-gatherer economy in sub-Saharan savanna (but at a low density and small body weight).

Because Campbell and Wood (1988) and Bentley and colleagues (1993a,b) found only small differences between hunter-gatherer fertility and other traditional subsistence populations in developing countries, we can regard Hadza fertility as close to the central tendency for this wider array of ecologies. Populations described by historical demographers also overlap Hadza fertility and mortality. These include mortality of European populations before about 1800 (Laslett, 1995), Roman Egypt (Bagnall and Frier, 1994), and adult $e_0$ in rural China AD 0 to 1749 (Zhao, 1997, table 3, average $e_{20}$ was 24.0 years). Chinese families with written records between 1365 and 1849 (Yuan, 1931) also showed similar adult life expectancies. Fertility in the classic historical case of Colyton in Devon was 6.4 for married women below age 30 (Wrigley, 1966), and fertility in a variety of locations in Britain averaged 7.23 between 1600 and 1799 (Wilson, 1984). A more comprehensive set of examples can be seen in the tables in Paine and Boldsen (2006). Twenty-first century European and North American populations fall outside this range by their very low fertility and mortality. Some developing country populations exceed the range by their continuing high fertility accompanying declining mortality to give very high growth rates;

Tanzania has had almost double the rate of increase of the Hadza. If there is a trade-off between fertility, survivorship, and individual growth (Migliano *et al.*, 2007; Walker *et al.*, 2008), then Hadza achieve their high population growth rate at some expense to their individual size, for like other hunter-gatherers, they are quite small in stature (Chapter 16). Additionally, despite the high growth rate, the Hadza population is distributed at very low density (0.24/km$^2$; Blurton Jones *et al.*, 1992, but see following) compared to agricultural populations.

## 10.1 Summary: a nearly closed, stable, but increasing population with average fertility and mortality

### Migration

Migration in and out of eastern Hadza country was nearly balanced at around one woman a year. Women leave to live among non-Hadza at a rate of less than one per year (0.5%/yr) (Chapter 5), but many of them return later and bring most of their children back to be raised as Hadza. These women tend to leave young and return older and impose a small loss of reproductive capacity on the population. Hadza move between east and west at a low rate (about 0.8 females/yr each way) (Chapter 5).

### Intermarriage

Six percent of the married women in our censuses had non-Hadza husbands. Their fertility and child survivorship was the same as for women married to Hadza men. I investigated the history of intermarriage by looking at non-Hadza parents, grandparents, and great-grandparents of contemporary Hadza of different ages. This suggested that intermarriage is currently higher than previously but also shows a brief peak near the beginning of the century, as indicated by early reports. Mid-century saw a very low rate of intermarriage. In the longer term, the rate of intermarriage has been very low (Tishkoff *et al.*, 2007).

### Regions and the ideal free distribution

Hadza describe their country as divided into east and west, and in the east into three main regions: Mangola, Siponga, and Tliika. Spatial clusters of camps corresponded to these, and people tend to have more close relatives in their region than in other regions. However, in 52% of couples, the spouses are from different regions. There were no significant differences between regions in fertility. Mangola had higher fertility (but not significantly), but significantly higher child mortality, perhaps caused by greater contact with mobile members of the larger national (onion farmers and truckers) and global (tourist) populations. Munguli had non-significantly lower child mortality (perhaps due to greater access to farm foods), but I did not attempt to predict or test for differences in age structure, etc. in these tiny regional samples. Despite this, differences in reproductive success (RS) (live children controlled for the woman's age, Chapter 13) between the regions were small and not significant. Had there been significant differences in RS between the regions, we should have had to ask why people had not moved away from the least hospitable region. Hadza appear to conform to an ideal free distribution.

### Fertility

Median age at first birth is 19 and at last birth 38. Average age at childbirth is 28.8 for women and 33.5 for men. The TFR was 6.17 births. Childlessness was very low, about 4%. Men had longer reproductive careers than women; more variance in completed family size (CFS); and 10% of men fathered no children.

### Mortality

Infant mortality was high at 21.8%. Thereafter, mortality falls rapidly, and child mortality (aged one to four years) takes probability of survival from birth (survivorship, $l_x$) from 0.7817 at age one year to 0.6433 at age five years. Survivorship from birth to age 15 ($l_{15}$) is 0.5565. Life expectancy at birth is 35.5 years for Hadza females, 30.8 years for males, and 32.7 years for both sexes combined. Highest $e_0$ is 48.1 at age five years for females, and 42.42 at age four years for males. Adult life expectancy ($e_{20}$) was 41.7 years for women and 36.3 years for men. Life expectancy at age 45 was 23.3 for women and 18.9 for men. The oldest woman was estimated to be aged 86, the oldest man 83. Both could have been older.

### Age structure and tests

Age structure and related measures were predicted from a population simulation that assumed our observed fertility and mortality had been stable for much of the twentieth century. The observed age structure and other measures were used as tests of the stability assumption and the fertility and mortality measures. Observed values were 47.1% of females below age 20, and 53.7% of males. Observed crude birth rate (CBR) was 44.2, crude death rate (CDR) 30.5 for the sexes combined (SI 9.2). Average age at death was 20.3. Observed values almost all fell within the 95% CI obtained by resampling the individuals who give rise to age-specific fertility (ASF) and mortality data that enter the simulation. Predicted values fall within the 95th percentile limits estimated by resampling the observations.

### Rate of increase

The Hadza population was increasing. The population simulation gave a rate of increase at r=0.0162 (0.0124–0.0195) or 16.2/1000/yr (12.4–19.5). The difference between CBR and CDR gave r=0.0149. Net reproductive rate gave r=0.0162.

### Density

In Blurton Jones *et al.* (1992), we estimated the density of the roughly 750 eastern Hadza as 0.24/km$^2$ (if those living in the Singida villages were excluded, and 0.30/km$^2$ if included). If we use the same estimated area (10 × 10 km squares marked on the 1:250,000 scale map), 2500 km$^2$, then the estimated population in the year 2000, which was 822, gives a density of 0.33/km$^2$ (Singida villages included). However, the clearing of land by non-Hadza immigrants continued throughout the study period. The area available to the Hadza thus decreased, and the estimate of density could be somewhat greater. I also changed my method of estimating population size. In Blurton Jones *et al.* (1992), the population was the number in the very thorough 1985 census. The re-estimates discussed earlier obtain the number of live eastern Hadza from the population register. The new method takes people seen in three or more censuses, and on >70% of occasions, seen in the study area. The two populations thus estimated were 684 in 1985 and 822 in 2000, 0.27/km$^2$ in 1985, and 0.33/km$^2$ in 2000, a 22% increase in density during my study.

All these population parameters are very close to the estimates by Dyson (1977), based on 1966–1967 data gathered by the International Biological Program (IBP) group, which supports the view that the Hadza population has been stable but increasing during much of the twentieth century. This medium-term stability of fertility and mortality comprises a smoothing of inevitable year-to-year variation. For example, the total population, the CBR, and the CDR vary from year to year. Some of this variation may be meaningful. For example, in the fertility chapter we showed that the total failure of the rains in early 1997 was followed by significantly fewer births in 1998. Year-to-year differences or trends can be relevant to population dynamics, as discussed later.

## 10.2 Comparing Hadza demography with Coale and Demeny models

All the previously discussed estimates were made with no reference to or dependence on standard life tables. My reason for avoiding the use of existing population models was to allow the Hadza data to show us whether they were unusual in any respect. Thus far, they seem like "the median forager" (Marlowe, 2010) or even "the median human" (if we exclude the industrialized nations). Let us look more closely at how they fit among the Coale and Demeny (C&D) (1983) model life tables. The C&D tables were built around 326 demographic studies of human populations in many countries, and show a great deal of uniformity in the life course of fertility and mortality.

The regional variants in C&D reflected apparent geographic variation in mortality patterns. "East" refers to eastern Europe, showing unusually high mortality in infancy, and over age 50 (N=31 studies). "North" refers to northern Europe, especially Scandinavia, at a time when mortality was relatively low in infancy and above age 45–50 (Norway, Sweden, Iceland; N=9 cases). "South" refers to Spain, Portugal, and southern Italy, 22 studies finding higher than usual mortality under age five years, lower than usual mortality in ages 40–60 , but higher than usual mortality after age 65.

The most general pattern was in the residual group, called the "West" series of models. This series was built from the remaining 264 studies of large, well-recorded populations, mostly from western Europe, but including studies from several continents (United States, Australia, Canada, Israel, New Zealand, Taiwan, South Africa) (Coale and Demeny, 1983, p. 12).

It is unlikely that our sample of fewer than a thousand Hadza can be definitively matched to any single C&D model. We may not even be able to discriminate among a sizeable range of models. Choosing a model requires that one begins with any two of several parameters. I chose $e_0$, which I estimated at 35.5 years for women and 31.7 years for men, and gross reproductive rate (GRR), which I matched to 3.0 in preference to the C&D tables adjacent 2.5 and 3.5. In workings shown in SI 10.1, I tabulated the figures given by a likely range of C&D models and compared them to the 95% CI of: (a) my predictions from Hadza fertility and mortality, and (b) the Hadza observations for the eight variables reported in Chapter 9. I counted the number of times the C&D values fell outside these confidence intervals. Note that in previous chapters I showed C&D model West level 5 as representative of the !Kung to be outside the 95% CI for the Hadza.

The sex difference in $e_0$ seemed to be unusually large in my Hadza data. While women ($e_0$=35.5, CI 35.4–35.9) were close to West level 7 with $e_0$ 35.0, men ($e_0$=30.8, CI 30.7–31.1) were closer to West 6 at 30.07 than to level 7 at 32.48. I compared the eight variables examined in Chapter 9 for these four models. West 6 and 7 fit equally well but rather worse than North 6 or 7. If the North models, either level 7 or level 6 selected by Dyson, are indeed the best fit for the Hadza, it implies that Hadza may have low mortality in infancy and late middle age (over age 45–50). Plotting Hadza $l_x$ and its confidence intervals with $l_x$ from the C&D models suggest North 7 is a good fit for women and that more Hadza women survive through the years 40–60 than expected from the West models (less evident for men). If there is anything unusual about the Hadza mortality pattern, it is that survival in the ages 40–60, especially among women, was better than even in the North series of models. The first of three possible reasons that immediately come to mind is errors in age estimation. I have discussed this in previous chapters, and could find no way to adjust the ages of men or women in this age range. Second, the C&D data come from a period and from locations where industrialization had become prevalent. Diseases of the industrial revolution among adults were a dominant public health challenge while deficiency diseases of crowding and unbalanced nutrition were major issues in child health. Both challenges were absent among the Hadza. Third, Hadza aged 40–65 are survivors of a harsh early life which, given the high infant and child mortality, only the most robust (least frail) survive.

C&D East and South models at any level of mortality fitted poorly. The Hadza do not share the peculiarities of either of these groups. For their mortality level, Hadza do not suffer higher than usual infant mortality, child mortality, or middle-aged and later mortality. There was little difference between fits to North and West models. I have not made a systematic comparison to the Weiss (1973) models, nor to the United Nations models for populations in developing countries. I believe that the

Hadza data would show a greatly exaggerated departure from the Weiss models in the direction of better survival among older adults.

Because my estimates were independent of models such as the C&D model, if they conform to the models, they strengthen the uniformitarian assumption, the view that the C&D models represent universal features of our species (Howell, 1976). None of my estimates depend upon the existing models. Instead, I made the computer mimic the stable population process and, for example, produce the age structure that would result if the fertility and mortality that we have reported has endured for some time without severe interruptions. For example, in Chapter 9, I compared this predicted age structure with the age structure observed in our censuses, as a test of the validity of our estimates of mortality and fertility. The Hadza represent a relatively normal human population; they give no reason for thinking that a hunting and gathering way of life generates unusual demographic parameters or a peculiar life history.

## 10.3 The role of model life tables

The model life tables provide the only way of making any demographic reality out of less well-studied samples and the fragmentary reports of early anthropologists or colonial administrators. The C&D models seem preferable to the Weiss (1973) models for two reasons: the large number of detailed direct demographic studies of living populations that they summarize; and the relative ease of working with increasing or decreasing populations. Weiss provides a table of conversion coefficients that allow the user to produce their own tables for stable increasing or decreasing populations. The main set of tables assume stationary populations (a sign of the times; in the 1960s and 1970s, we assumed that "anthropological populations" were all long standing, stable, stationary, and likely to be in exquisite balance with their environment). Users may have been tempted to assume their samples were from stationary populations just for ease of choosing a matching table.

I could have used other methods but did not. Using the C&D life tables would have enabled a much faster study. Both Dyson (1977) and I (Blurton Jones *et al.*, 1992) were able to report what turn out to be rather good estimates relatively quickly by using the model life tables. However, we made no test of the uniformitarian assumption, and could have conducted fewer independent tests of the estimates. The two-census method such as described by Gage *et al.* (1986) and successfully employed by Gage *et al.* (1984) on the Trio, and Fix (1989) on Semai would also appear to be much more economical of field time. I could have tried to apply this method to any pair of my censuses, but given the mobility and elusiveness of the Hadza, it seemed inappropriate. Each of a pair of censuses would be a different sampling of the population. The censuses differed quite a lot in the time taken and number of localities visited. Asking about absent people may be adequate for determining whether they are alive or dead, and roughly where they have been living, but among the Hadza it would be a very poor way to collect data on births.

Some populations are much more difficult to study than others. For example, Ernestine Coast (2001) made a large-scale study of Maasai demography in Tanzania and Kenya. The study was intensively planned, and enumerators were local people carefully trained. However, the Maasai will not talk about dead people. No amount of wrangling the numbers could overcome this problem. It might be that a smaller-scale, longer-term, more "anthropological" study, would eventually break through the subjects' resistance. The Hadza showed no such reluctance but are far from easy about names. Being confident about people's identities was one of the reasons for my setting filters for the individuals; for example, limiting an analysis to people seen in three or more censuses, or to people seen in the core study area.

Recent studies have used a combination of census and interview similar to mine. For example, Hill and colleagues report on seven years of intermittent fieldwork among the Hiwi using census and interview (Hill *et al.*, 2007). The Tsimane (Gurven and Kaplan, 2007) and Pume (Kramer and Greaves, 2007) studies have also been quite long-term studies. Repeat visits over a significant number of years improve rapport with research subjects and field assistants, and give one a chance to get at least a feel for year-to-year variation and how the observed period fits into the history of the population. Perhaps the most rapid successful study was by Borgerhoff Mulder (1992) and colleagues on the Datoga, neighbors of the Hadza. Even this entailed three field visits, in 1987, 1988, 1989, with household censuses and reproductive history interviews, and the publication is entitled *Preliminary Data on the Datoga of Tanzania*. None of these studies used Weiss or C&D life tables. But if they added a comparison of their results with selected models, our confidence in their results might be strengthened, at least to the extent that nothing unexpected shows up. In some cases (perhaps the Hiwi would be an example judged by Hill *et al.*'s description of the history), it might be wrong to assume stability in the population, in which case stable population models would be inappropriate.

## 10.4 Population history shown in age structures

Age structures can be useful as a way to trace historical events (Paine and Boldsen, 2006). We might wish to test for an effect of a major change in lifestyle, such as a lasting settlement, where people give up a mobile forager lifestyle for a less mobile life with a mixed economy. Thus, Hill and Hurtado (1996, p. 149, fig. 4.10) can show the effects of high mortality after contact on subsequent Ache age pyramids. Penningon and Harpending (1993, p. 51) suggest that Herero age structure shows the persistence across at least two generations of effects of the Herero expulsion from German Southwest Africa (Namibia) in 1904–1905. The age structure of the !Kung could be examined for effects of the apparent decrease in mortality during the 1960s. Paine (2000) shows that, in medieval Europe, the plague left traces in the population structure for up to 50 years. Paine and Boldsen (2006) also show that the frequency of epidemics affected details of age structure during the last 800 years or so of European history.

In the case of the Hadza, we have no abrupt change in the last 50–100 years that affected the whole population. There are no massive anomalies in the Hadza abridged age pyramid (Figure 10.1), such as can be seen in the Ache and Herero age pyramids.

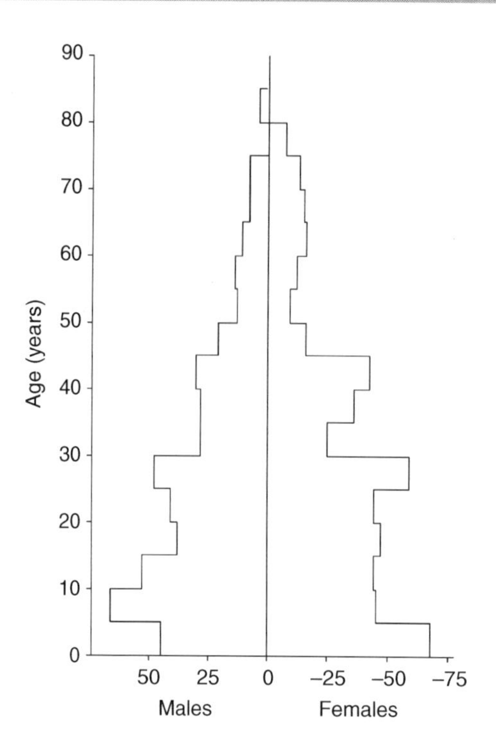


**Figure 10.1** Abridged age structure derived from the population register. Number of males and females in each age group (0–4.99, 5–9.99, etc.) in the register and alive in the year 2000. Table in SI 10.4.

There is no consistent informant's or visitor's report to suggest that there should be; there is no indication of a major perturbation during the twentieth century. The Maasai wars and Isanzu slave raids late in the nineteenth century were too long ago to show a direct effect on the age structure. Only if they approached the severity of the Herero expulsion from Southwest Africa, would we expect to see a "second generation" effect. However, in the Hadza age structure, there are many lesser fluctuations (Figure 10.2). We must ignore the fluctuations among the oldest people, the samples are small, and obvious age heaping can be seen for age 60 upward. Nonetheless, we might wonder whether other factors discussed in Chapter 3, such as the series of settlement attempts, the gradual encroachment by people with other economies, and the variable rainfall, left small but persisting traces in the age structure.

There are indications of a shortage of people, especially women aged 40–50 in 2000 visible in Figure 10.1. I mentioned it previously as a shortage visible in Dyson's age structure (the people missing from Dyson's age structure would be aged 43–54 in 2000). Post hoc explanations can be offered. These people were estimated to have been born between 1946 and 1957. There was a widespread and severe drought in 1949 that decreased [by sale or death] the cattle in Sukumaland by more than a half (Brooke, 1967; Baker, 1974, p. 174), although it does not show in the Mbulu rainfall records of Meindertsma and Kessler (1997, fig. 1.8). Perhaps this drought led to a decrease in the number of births (as in 1998, after the total failure of rain in 1997),

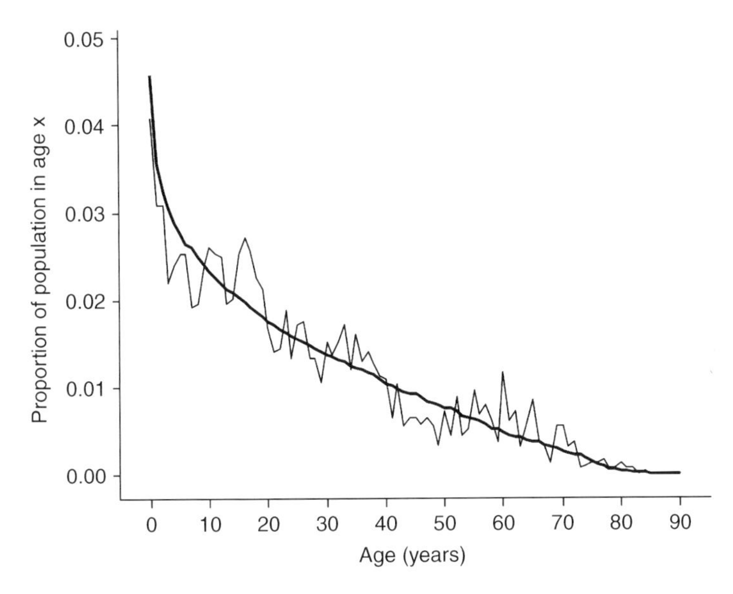


**Figure 10.2** Observed and predicted age structure. Proportion of the population in year of age, observed at anthropometry and predicted by population simulation that uses fertility and mortality estimated in Chapters 7 and 8. Number in each year of age is averaged over the seven censuses in which people were measured. A similar figure, for females only, is given in SI 10.4.

and/or an increase in deaths. A shortfall of people for any particular year in the age structure may mean fewer were born that year but it could reflect events that befell their birth cohort at any time between birth and observation. In view of the association between high rainfall and malaria (Chapter 2, and SI 10.2), we might wonder about the reportedly very high rainfall of 1961. The missing people were aged 4–15 in 1961. Children are very vulnerable to malaria but it would be surprising if the effect showed over such a wide age range. Another explanation would note that these people were aged 8–18 at the time of the Yaeda and Munguli settlements.

## 10.5 Settlements

The shortfall of people born between 1946 and 1957 can be seen making its way through all the age structures up to 2000. These people were aged 8–18 during the peak years of settlement at Yaeda and Munguli in the 1960s. The shortfall is more evident among women than men, and this may imply that a larger than usual number of Hadza girls married non-Hadza men and left Hadza country at that time. They could have been relatively unlikely to have entered our population register, but the interviews about location of siblings of the interviewed women did not indicate a large number of such people (Chapter 5).

Because informants had told us that the time of the 1964 settlement at Yaeda was a time of many child deaths, I looked for a dip in the age structure that might reflect these losses. I have seen none, in fact it is possible to argue there is a slight excess of people born at that time (ages around 35 in Figure 10.2). To judge from the Mono wa mono settlement in 1990 (Chapter 7), there may well have been a burst of births as the free Yaeda food supplies took effect on women's energy budgets. Without altering mortality rates, this would have increased the absolute number of child deaths that informants had witnessed.

The brief settlement attempt in late 1986 led to a measles outbreak in which a number of children died. This shows quite strikingly in the age structure. The abnormally high number of deaths can be tested by logistic regression in the child annual survival file. The year 1986 stands out as associated with a high probability of deaths (b=1.2634, p=0.000, OR=3.54, CI 2.13–5.88). This year also seems to have been accompanied by a brief drop in fertility.

The lower than expected number of young children in the 2000 age structure could have resulted partly from the deaths in the 1986 measles epidemic, and partly from a decline in the birth rate in the late 1990s. Logistic regression on the annual record of births of the interviewed women shows that the years 1996–2000 had a significantly lower probability of a birth (b=–0.2258, p=0.04, OR=0.80, CI 0.64–0.99). No comparable length previous period (index years) was significantly different from the remaining (non-index) years. Marlowe and his students may be able to show us whether this heralded a lasting decline in the fertility and rate of increase of the Hadza population.

I looked for an effect of settlements on fertility in the fertility annual hazard file. Crucial years of settlements can be added as independent variables, such as 1990, and 1964 and 1965 when Yaeda, Munguli, and Endamagha were started. Their effect on probability of a birth to any woman in the record can be examined, first controlling for her age (age + $age^2$, and $age^3$). There was no significant effect of 1964 or 1965. If these years were combined with 1990, there was still a small significant positive effect on fertility (OR=1.44, CI1.07–1.94). However, when 1986 was scored together with these as a "settlement year," there was no effect of settlement years on fertility. Obviously, this depends on what happens at a settlement. In 1986, it quickly led to a measles epidemic and was abandoned. The earlier settlements had included food supplements and no clear account of an epidemic. In the last half of the twentieth century, the overall effect of the settlements on Hadza fertility and population increase seems to have been small.

## 10.6 Hadza population history and raids upon the Hadza

My best estimates imply that the eastern Hadza population increased at about 0.016 per year (16/1000/yr) (0.012–0.017) for most of the twentieth century. If we compute the population size backward in time, the eastern Hadza numbered around 200 people in 1900, rather more than Obst's guess that there were only about 100 of them in 1911. Around 1920 there would have been around 250 Hadza, rather fewer than Bagshawe's (1925, p. 119) estimate that there were "not many of them, probably not more than five or six hundred." At around 375 in 1935, my extrapolation passes closer to Kohl-Larsen's estimate of 450 people in the 1930s (1958, p. 53 of translation), even closer if we attend to the region he knew best (SI 10.3). My extrapolation is only a little higher than Woodburn's estimated 400 Hadza in about 1960 (Woodburn, 1968a).

In Blurton Jones *et al.* (1992), I suggested that the Hadza were recovering from losses inflicted by the Maasai, which came to an end with the rinderpest epidemic in 1890 and German colonial pacification. A closer review of the historical information leaves a more complex picture. In Chapter 3, I reviewed some of the literature on the Maasai expansion, and the disasters that fell upon them late in the nineteenth century. I also reviewed the evidence from informants of Obst (1912), Bagshawe (1925), Bleek (1930), Kohl-Larsen (1958), and other sources that Hadza were under pressure of slave raids and warfare from their neighbors between about 1860 and 1910, which may have greatly reduced their population. The slave trade may have been more important than the Maasai raids, especially as it is described as taking women and children, far more demographically important than men. These raids ceased after German pacification, and the twentieth century apparently saw a steady increase in the Hadza population.

Pacification by colonial powers and the post-independence government has been discussed in the context of African populations in general, and with respect to the Hadza in connection with their low rate of homicide. The colonial powers eventually, but not until 1915, suppressed the slave trade and thereby the wars over its control and over the access to foreign trade goods. Historians seem to agree on the whole that these wars reached an unprecedented level during the nineteenth century and that we cannot assume that the same frequency or extent characterized previous centuries. Pacification removed pressure on the Hadza from the Maasai and from Isanzu slave traders. Eventually, it may have contributed to the massive national population increase that led to encroachment on Hadzaland late in the twentieth century. However, whether the Hadza themselves were pacified is an even more complex question.

Hadza reported extensively to Bleek and Kohl-Larsen on the German pacification of the Maasai. They clearly knew about the potential power of the colonial governments. At the beginning of our fieldwork, the Hadza regarded the government with fear, mainly because of the settlement schemes. They were afraid that they would be "rounded up" again. They recognized that the government was powerful, if capricious. It is far from clear that they viewed their internal affairs as subject to government control or retribution. The situation is complicated. Nonetheless, it is possible to argue that the Hadza did not regard homicide as anything but an internal affair until 1995. A homicide occurred in that year, which was of great concern to the people close to the participants. They seemed relieved when a retired local official suggested this should be taken to district authorities to decide what would happen. No such discussion had accompanied previous homicides, nor a subsequent homicide, as far as I could ascertain. It was as if pacification affected Hadza external but not internal relations. On the many occasions when I witnessed Hadza complaining to each other about intrusions on their land, the potential solution was always the government, "why doesn't the government fix this?" or "help us ask the government to put this right." There was no indication that the Hadza ever thought about taking violent action against the intruders. They seemed to fully appreciate their weakness in numbers and weaponry.

Yet the reach of legal authority was quite short during our study period. Police and judicial authorities were unlikely to make the journey into Hadzaland; distance and lack of vehicles, fuel, facilities, and infrastructure made it forbidding and expensive. They would be at least a day and a half away (a day to notify them, half a day to come if they had a functioning vehicle ready). One anecdote, whether strictly accurate or not, efficiently illustrates the situation. A while after the 1986 economic reforms, traveling rural markets began to appear in surrounding villages. Usually these were peaceful and festive occasions. I was always reminded of the livestock fair that occurred annually across the road from my grandparents' house in Glastonbury, England, complete with escaping livestock. On the occasion at issue some trouble broke out. Some herder youths, it is said, got into an argument with a stall holder, refusing to pay for the meat they had "bought" from him. They left the market, collected their weapons, stashed routinely at the gate, and began firing arrows into the crowd. The story continues with the rapid departure of the people's militia (practical uniforms, and in those days, lots of drills but no weapons). The trouble continued and someone decided that with the police inaccessible, the solution was to fetch a Hadza with his poisoned arrows. One was found and had apparently no hesitation putting a stop to the trouble. The herder youths wisely fled as best they could. Everyone was happy to see peace and quiet restored. In what sense are the Hadza pacified? The recent history of the Datoga also suggests limits to the reach or scale of "pacification" (Ndagala, 1991; Hussein *et al.*, 1999). While the Hadza have been encapsulated, the Datoga have been "chivvied," repeatedly moved on, first by the Maasai expansion, then by expanding farmland, and last by industrial-scale agriculture (Lane, 1996). They have occasionally resisted.

## 10.7 A "bottleneck" in Hadza population history

Foreshadowed by Harpending *et al.* (1993, p. 489), Henn *et al.* (2011, p. 5160) report for their Hadza sample "Extremely elevated LD, increased runs of homozygosity, and very low HLA and haplotype heterozygosity estimates in comparison with other sub-Saharan African populations ... all indicate a severe population bottleneck in the Hadza. We estimate their $N_e$ to be only 2500 individuals, which corresponds to an ancestral population size of 15,000 (given the estimated six-fold bottleneck) ... more than twice the current census size of the Hadza, suggesting that the bottleneck is ongoing and that historically the Hadza had a much larger population size." Several issues arise. First, how can the Hadza have reversed course from being a declining population to increasing? Second, as the Maasai can hardly have killed 14,000 Hadza in the space of a few years, what else could account for the decline? Even the slave trade is unlikely to have removed so many of these elusive, widely scattered people with any rapidity. Initial exposure to modern crowd diseases, spreading from the trade caravans early in the nineteenth century could have been important. However, we have no mention of such by any author or any informants. We probably should not identify the geneticists' bottleneck with the low population attributed to raiding late in the nineteenth century. The genetic bottleneck cannot be timed very precisely and its effect seems much too large. The Bantu expansion, leading to extensive change of ethnic identity by intermarriage, is a better example, perhaps more compatible both in size and timing. The Hadza of the Singida villages illustrate the likely process, gradual but, on the long-time scale, quite rapid exclusion of large numbers of Hadza from the marriage pool from which today's eastern Hadza are descended.

Kaare and Woodburn (1999) noted the flexible ethnicity of some Hadza, which is especially true of people at Munguli. There was a high rate of intermarriage between Hadza and Isanzu, and an influx of Isanzu to the village and surrounding area. Country that, in the lifetime of some of our informants, was wild bush country, inhabited only by Hadza and perhaps linking the eastern and western Hadza, is now heavily populated farm country. In Chapter 5, I wrote: "However, if we look at ... people seen in the core study area on less than 70% of their records,... we find a much higher percentage with a Swahili father. There were 48 with a Swahili father out of the sample of 111 children who had been most often seen outside the core area, 43.2%." Many of these were children of women who grew up in the Singida villages, married a non-Hadza from that area, and lived there for most of the study period. Their children were less often seen in the core study area. Among the children of these mixed marriages are many on whom we have no information, they appear lost to the Hadzane-speaking population. They probably will not marry a Hadza, and their children will probably not speak Hadzane. These people probably represent the most common process (other than death) by which people of Hadza descent leave Hadza society. They will not figure in any later demographic or genetic study of the Hadza. It is also possible that, in the past, the children of the mixed marriages were at a disadvantage, less able to accumulate property and find wives, and more likely to have been sold as slaves. Then they would have left few genetic traces among the absorbing Bantu populations.

We know very little about the former distribution of the Hadza. The extensive range of the western Hadza implied by Baumann's (1894b) map, and the kind of country that the Hadza thrive in, may give us some clues. Woodburn (1968a) suggested there were perhaps about 250 western Hadza. In 1978, Lars Smith found 231 western Hadza, clustered around Paji and a few other localities. In 1992, I found all western Hadza confined to the quite small area of their "village" (humiliatingly referred to as Iramba ndogo) in the Meatu district, where they remained, living partly as the dependents of the government and the Cullman and Hurt Foundation. In contrast, Baumann's (1894b) map implies that they ranged widely across the whole area northwest of Lake Eyasi, including country that Hadza sometimes traverse even today when going from Mangola to the west. Their population could have been spread even wider over the very suitable country of the Maswa Game Controlled Area and between there and equally suitable country south of the present-day eastern Hadza range. Not only do older Hadza recall a time when much of Isanzu country was wild bush used only by Hadza but much more of Iramba quite resembles the baobab-strewn rocky hills of eastern Hadza country. Eastern Hadza country included more of the hill country around Siponga, and higher up, the Balai valley than today but these amount to only a few 100 km$^2$. Expansion to the north or east seems unlikely because of the highland forest cover and the cold. The areas to the south and west are all predominantly occupied by the Iramba/Isanzu, Sukuma, and Nyamwezi, all part of the Bantu expansion from Lake Victoria that began about 2000 years ago. If this had been the 45,000km$^2$ that 15,000 Hadza would fill at the latest density (0.33 people/km$^2$), we should expect some archaeological traces in that area, and perhaps genetic traces of the Hadza in the Isanzu, Iramba, and Sukuma populations. Tishkoff *et al.* (2007) found evidence of flow of mitochondrial genes (via females) from the Hadza to the Sukuma, and flow of Y-chromosomes from the Sukuma to the Hadza. They have yet to collect a sample from the Isanzu and Iramba.

It is easy to picture two processes. First, the larger, denser populations of farmers bring diseases to which the foragers have not been exposed, and from which they likely suffer badly. This effect could be as passing as the Ache population crash upon settlement (Hill and Hurtado, 1996, fig. 3.1), or as massive and lasting as the consequences of European invasion of North America (Thornton, 1987). Second, some of the surviving Hadza women marry these farmers, raising their children as Bantu speakers and farmers and losing their ties with Hadzane speakers. The in-marrying population of Hadza quickly becomes smaller. We have to differentiate between potentially large losses due to absorption into other populations, and the condition of the surviving population of people who identify themselves as Hadza, speak the language, and predominantly marry other Hadza speakers (from a shrinking pool of marriage partners), and live by hunting and gathering in the area where they have been observed during the twentieth century. These "surviving" people are clearly increasing. However, on a longer, wider scale, perhaps since the Bantu expansion, the number of Hadza has apparently decreased greatly.

## 10.8 Surrounded, encapsulated, invaded, but not impoverished

The non-Hadza populations in the areas around the Eyasi basin have increased about twice as fast as the Hadza during the later years of the twentieth century (Table 10.1). As argued in Chapter 3, early in the twentieth century non-forager neighbors of the Hadza were present in all directions, but at some distance away and at low numbers and density. As their numbers increased, the Hadza became not just surrounded with trading partners, who could be reached when needed, but truly "encapsulated" in a

**Table 10.1** Populations of three wards overlapping Hadza country, showing the dramatic increase in two of them. Hadza are included in national censuses but as they number only about one thousand, the numbers shown are almost entirely non-Hadza. Data from Tanzania National Census Bureau

| | 1978 | 2002 |
|---|---|---|
| Mangola ward | 6846 | 16568 |
| Yaeda ward | 9293 | 5960 |
| Mwangeza ward | 2445 | 12414 |

tight circle of much more numerous and "developed" peoples. Finally, these people began to move into Hadza country and change the landscape, particularly around Mangola and Munguli, and in the higher parts of Siponga. Herders began to live in every part of Hadza country, and it seems likely that they have modified the landscape in several ways. By 2000, there was still no strong indication that the invasion had affected Hadza population growth, although our Hadza census counts for the Mangola region were very similar for 1985 and 2000, perhaps implying less growth in that region. It has been clear that since the 1950s, there was nowhere for the Hadza to expand. By 2000, it began to look as if their habitable range was beginning to contract. As previously mentioned, the newcomers may have impaired hunting and foraging but they offered new trading opportunities in exchange. Researchers who have worked with the Hadza since 2000 may be able to show us whether Hadza population growth has begun to slow, or continues a relentless increase.

The historical and demographic data imply continuity of Hadza lifeways and a growing population despite encroachment by neighboring populations. Anthropometry might suggest a different picture. It could provide a more sensitive indication of hardship. Anthropologists tend to think of height and weight as an enduring characteristic of their study population. In contrast, public health researchers are impressed with rapid change in size, as between the 1700s and the 1900s in northern Europe, or between migrants and their home-bound kin. Growth and adult physique are widely regarded as indicators of the conditions under which people live and grew up. Measurements have been found to respond quite rapidly to changes in conditions. Height is regarded as reflecting longer-term conditions while weight and fatness respond more rapidly to short-term change in conditions.

Is there evidence of secular change in Hadza height and weight? The IBP fieldwork in 1966–1967, reported by Hiernaux and Hartono (1980), and Lars Smith's fieldwork in 1977 give us substantial samples to compare with our data collected between 1985 and 2000 (Chapter 16). Marlowe's recent publication (2010, pp. 141–149, and table 6.1) of adult means and standard deviations add the early 2000s to the comparison. Table 10.2 shows there are no significant differences in height between any of these samples. Weight and skinfold thickness may tell a slightly different story. Weights of males and females were lower in the 1990s and 2000s than in 1977 or 1967, and the differences were close to or below the 0.05 significance level. Women were 1.36 kg lighter and men 2.16 kg lighter in the 1990s.

Men's and women's upper arm circumferences were also significantly lower ($t=-3.08$) in the 1990s, whereas women's triceps skinfolds were significantly larger in the 1990s ($t=2.52$) than reported by Hiernaux. I use the word "significantly" in the merely statistical sense; the differences are very small, and the overall impression of these data is of a remarkable constancy. Hadza in the 1990s and 2000s were much the same size as Hadza in 1967. There was no support for the view that they were now starving, nor that they were currently over-fed and under-exercised. Body mass index (BMI) differs very little between these samples. Thus, the figures in Table 10.2 give little support for the view that Hadza living conditions (availability of food and incidence of disease) changed sufficiently between

**Table 10.2** Comparison of height and weight across five decades. Adults aged 20–60, based on my age estimates. Hiernaux and Hartono identified adults in 1967 as those between the age at which height and weight leveled off and before any decrease of old age began. The table shows means in column 3, t-tests in columns 4–6. The 5% probability level for these t-tests would be 1.96 (10% at 1.64, 1% at t=2.58). Significant t marked with asterisks * (5%) and ** (1%)

| Males | N in sample | Height (cm) | 1977 | 1990s | 2000s |
|---|---|---|---|---|---|
| 1967 | 125 | 160.95 | -0.47 | 0.64 | 1.54 |
| 1977 | 96 | 160.6 | | 0.83 | 1.46 |
| 1990s | 174 | 161.4 | | | 0.95 |
| 2000s | 253 | 162.0 | | | |
| **Males** | **N in sample** | **Weight (kg)** | **1977** | **1990s** | **2000s** |
| 1967 | 126 | 54.26 | -0.63 | -3.44 ** | -1.91 |
| 1977 | 96 | 53.8 | | -1.81 | -0.96 |
| 1990s | 174 | 52.1 | | | 1.79 |
| 2000s | 252 | 53.04 | | | |
| **Females** | **N in sample** | **Height (cm)** | **1977** | **1990s** | **2000s** |
| 1967 | 109 | 150.37 | 0.38 | 1.49 | 0.16 |
| 1977 | 108 | 150.7 | | 0.86 | -0.28 |
| 1990s | 204 | 151.4 | | | -1.55 |
| 2000s | 238 | 150.48 | | | |
| **Females** | **N in sample** | **Weight (kg)** | **1977** | **1990s** | **2000s** |
| 1967 | 110 | 48.26 | 0.3569 | -1.91 | -2.59 ** |
| 1977 | 108 | 48.58 | | -2.20 * | -3.06 ** |
| 1990s | 204 | 46.9 | | | -0.96 |
| 2000s | 238 | 46.33 | | | |

1966–1967 and the early 2000s to show a consistent and substantial change in height or weight.

The earliest visitors to the Hadza also reported a few measurements. Obst (1912) reported the average height of a small number of Hadza men as 161.2 cm and women as 150.2 cm, strikingly close to our measurements, after so much recent history, closer contact with neighbors, and several resisted settlement schemes instigated by free food. Kohl-Larsen (1958) measured Hadza near Mangola in the 1930s. Men averaged 158 cm (147–164 cm) and women averaged 145 cm (113–155 cm). He comments on the exceptionally small stature of one apparently adult women measured at 113 cm and she must have lowered the mean of his small sample from about 146.5 cm. Bleek (1930, 1931a) also gave a few measurements and included people with Isanzu ancestry, notably taller than those without. Fosbrooke (1956), writing about his 1950 field visit, reported "in the matter of stature, my measurements over a small group tally closely with those recorded by Bleek (1931) and

Reche (1944 [sic, probably 1914]), as quoted by Huntingford (1953), namely the men averaging about 5 ft 4 in (160 cm) and the women just under 5ft (150 cm)."

On a short time scale, living conditions for the Hadza have varied over the years, with the dramatic but short-lived settlement schemes (Chapter 3, Table 3.1; and SI to Chapter 3). In 1967, when the data reported by Hiernaux and Hartono were collected, many Hadza still lived at Yaeda and Munguli, where food may have been provided by the government, especially to the children attending school. Blurton Jones and Marlowe (2002) reported that Hadza children in school during 1990–1995 were heavier than children who did not attend school. In 1977, fewer Hadza lived at Yaeda or Munguli and food aid had probably ceased (except in school). Even fewer Hadza were at these locations during our 1985–2000 study period. These changes imply a decrease in use of government food from 1967 to 1977, and to the 1985–2000 period. During this period, the Hadza have also lost significant portions of their land, compressing an increasing population into a shrinking area. A slight decrease in average adult weight would not be difficult to believe, indeed we may be more surprised that weight and fatness measures have not changed more. Nonetheless, these comparisons are limited. The data were collected by different teams, using different equipment. In view of the possible data collection differences, the degree of constancy, over so much time and after so many events, is more striking.

As Howell (2010) points out, constancy of size across time and space (evident on a much longer time scale in southern Africa, Pfeiffer, 2012) raises fundamental questions about what controls human physique. If in future we combine the individual anthropometric data from the different research groups to link children's growth to their eventual adult size, we may be able to test some of the ideas about this control.

## 10.9 Comparison with the nation and the neighbors: parallel, isolated, pacified, or integrated?

In earlier chapters I argued that the eastern Hadza were a sufficiently self-contained population to allow demographic study of "the Hadza population" in isolation from the surrounding countryside. Now that we have completed the description of the eastern Hadza population, we should compare it with the neighboring and the national populations. We may wonder whether the Hadza population has or has not been influenced by or entrained to neighboring populations or to the wider population (Hammel and Howell, 1987; Headland and Reid, 1989). Populations could affect each other but they could also react in parallel to the same aspects of the environment, like annual variation in rainfall or longer-term climate change. The Hadza and their Datoga neighbors share some aspects of their involvement with the local, regional, and national economy, such as the state of the tourist industry, but not others, such as the price of cattle. Available demographic data on the neighbors are summarized in Table 10.3. The national population and its history can be described from the data of the Tanzanian Bureau of Statistics, its national censuses, and some specialist studies such as the historical study by Walters (2008), and others such as Mturi and Hinde (1994), and their colleagues.

**Table 10.3** Some population parameters of the Hadza compared to neighbors and the nation. Datoga from Borgerhoff Mulder (1992), Maasai from Coast (2001). Maasai infant mortality from Meegan *et al.* (1994). Mbulu district and Tanzania from the national census (Tanzania Bureau of Statistics). No cases of HIV were identified among Hadza, 1985–2000. HIV incidence in adjacent parts of Manyara and Singida regions during 2003–2004 were low, at 2%, well below the national estimate of 8% (Yahya-Malima *et al.*, 2006). Hadza age at marriage is reported in Chapter 15

| | Hadza | Datoga | Maasai in Tanzania | Tanzania 2002 |
|---|---|---|---|---|
| Total population | 950 | | | 33,461,849 |
| % increase/yr | 1.6 | 2.2–2.8 | 3.9 | 2.9 |
| TFR | 6.17 | 6.9 | 6.4 | 5.5 |
| Infant mortality | 216.2 (f) | 200.3 (f) | 186 | 100.3 |
| | 202.3 (m) | 205.8 (m) | | |
| Male life expectancy at birth | 30.81 | | | 52.4 |
| Female life expectancy at birth | 35.55 | 42.5 | 55 | 55.1 |
| Women's age at first birth | 19.0 | | | 18.2–19.2 |
| Women's age when first married | 18.5 | 17.5 | | 16.6–18.4 |
| Childlessness | 3.9 | 3.0 | 1.4 | 10.7–12.0 |
| % below age 15 | 40 | | 53.1 | 44 |
| Crude birth rate | 43.3 | | | 40.2 |
| Crude death rate | 27.0 | | | 12.9 |
| HIV | None | | | 2–8 |
| IBI (mo) | 27-33 | 33.68 | | 33.3/33.4 |

Compared to the general Tanzanian population, the Hadza have higher mortality, less childlessness, and less change in fertility and mortality. Hadza mortality is quite a bit higher, even than the mortality of their closest neighbors, the pastoralist Datoga. I have argued that Hadza fertility and mortality have changed very little during the twentieth century, and are observed to have changed very little between 1966–1967 and 2000. In contrast, the national population has shown declining mortality since about 1940, particularly of infants and children under five years old, accompanied by increasing and then slowly declining fertility (Walters, 2008).

The national population shows a much higher incidence of childlessness than the Hadza (10.7% versus 3.9%). Walters, (2008, fig. 5.19) shows that the incidence of childlessness in Tanzania may have been declining slowly. As well as isolation from medical care, and from propaganda about birth control and birth spacing, Hadza have apparently been relatively spared from disease sources of infertility. However, the isolation from sexually transmitted diseases (STDs) is far from complete. The STDs exist among the Hadza, were reported by Bennett *et al.* (1973) from 1966–1967, and by the time of Tishkoff's fieldwork (in 2002), a few HIV-positive cases were identified near Mangola (the first case among non-Hadza having been reported there in 1991). Despite its more than two-fold greater incidence of childlessness, the national population has a growth rate double that of the Hadza.

## 10.10 Reasons for neighbor and national population increases

I attributed the growth of the Hadza population in the twentieth century to the "Pax Germanica" claimed by the informants of Obst (1912), Bleek (1930, 1931a,b), and Kohl-Larsen (1958). Raids by Maasai and others ceased. The Hadza no longer suffered the associated (but quite limited) losses. Clearly "pacification" made a difference to the Hadza at this particular time. Does this mean that warfare was important in limiting hunter-gatherer populations in the more distant past? Some of the early colonial administrators, and historians, have argued that warfare was a significant brake on population and economic development all over Africa in past centuries. Most of their examples concern agricultural and pastoral societies, and their claims have been contested on other grounds. I explore this issue in the next chapter, but the population history of some of the Hadza neighbors, and of the nation itself, may be instructive.

Colonial officials in the early decades of the 1900s claimed that the Tanzanian population was decreasing rapidly. In some locations, it may have been. The Maasai were badly hit by the rinderpest epidemic in 1890. Other populations had suffered from the slave trade, the wars for control of the trade, the German scorched earth suppression of rebellions, and the enlistment of large numbers of men as bearers during World War I. Some populations were apparently doing much better. For example, Borjeson (2004) points out the extreme difference in Baumann's description of the condition of the Maasai in Ngorongoro, and the Iraqw near Mbulu, just a few months apart. While Baumann encountered starving Maasai in March 1892, he describes and illustrates with enthusiasm the successful farming practiced by the Iraqw in January 1893.

Later authors attributed the survival and eventual increase of the population to a variety of factors. Among them were pacification, end of the slave trade, health services, modern medicine (effective antibacterial sulphur drugs like the famous M&B693 became available in the 1940s), modern crops, new cattle medicines, generally higher rainfall following exceptionally high rainfall in 1961, or the end of the most vigorous colonial depredations. It is likely that each was important in some areas and unimportant in others; the Iraqw may have been especially equipped to benefit from new crops, and they benefited from colonialist help with extending their farmlands northward.

The total size of the Iraqw population provides an interesting comparison. When first recorded by Europeans, the Iraqw were concentrated in a small area "Iraqw'ar Da/aw," also known as Mama Issara, southeast of Mbulu. Their oral history claimed they had moved to this enclave from further south to escape cattle raiding. Their 1890 population has been estimated at 26,000 people by Winter and Molyneaux (1963), based on extrapolation from later rates of change (1948–1957), and on Werther's (1898) reports from 1892–1893 and Jaeger's (1911) from 1906–1907. Their population size has increased, forming a large fraction of the present-day 348,000 population of the Mbulu district in 1995, perhaps exceeding Winter and Molyneaux's predicted 288,000 people by 1990. Borjesen (2004) points out that their society has been inclusive, absorbing people from other identities. In contrast to the Hadza, they have greatly extended the area in which they live. Winter and Molyneaux (1963) argue that the population of Mama Issara remained roughly constant but that people spread west across the remainder of the highlands, and north into the Karatu–Mbulumbulu area (as also described by Fosbrooke, 1972; Lawi, 1999; and others). Hadza foragers, in contrast, had no suitable available habitat into which to expand. However, both populations and other neighbors increased throughout the twentieth century, the Iraqw almost twice as fast as the Hadza.

In Chapter 2, I suggested that Hadza were not isolated from diseases that require a large population to persist (measles, smallpox, etc.). Measles has appeared among Hadza as the occasional destructive epidemic. Tuberculosis (TB) seems a constant, and perhaps ancient (Gutierrez *et al.*, 2005; Stone *et al.*, 2009) threat. Dunn (1968) showed the number of disease organisms with which forest dwellers are infected exceeds the number with which desert dwellers are infected. Bennett *et al.* (1970) showed that Hadza were intermediate in this respect. This result could be a feature of the slightly richer wooded savanna environment, or it could be a demonstration of the greater disease contact between the Hadza and the large, dense, and less mobile populations of "mainstream" Tanzania. A dubious implication of this could be that, in a world of hunters among hunters, a population such as the Hadza would be exposed to fewer life-threatening diseases than today. This implication relies on the untested belief that mortality levels reflect number of diseases. It may be that the most significant causes of mortality among the Hadza are the inconspicuous infections of the pulmonary and digestive systems of infants and children, many caused by common but little-studied viruses. The association of number of infectious disease organisms and climate may mean that the abundance of diseases has varied extensively during Hadza history; more important may have been the association between malaria and rainfall. Another possibility is the ceaseless struggle between humans and their virus parasites. While such a large proportion of the deaths are a result of common respiratory and intestinal infections, the possibility of rapidly evolving new forms must remain a candidate for a long-term and significant restraint on hunter-gatherer populations.

## 10.11 How do hunter-gatherers disappear?

In many instances, hunter-gatherer populations have not disappeared but have added to or changed their economy and endured as a distinct population with their own language and many of their earlier traditions. The Hadza case seems different. Hadza have repeatedly resisted economic change, for reasons both mysterious and provocative (Woodburn, 1988; Blurton Jones, 2015). The geneticists tell us, however, that they are much fewer today than not long ago. What could have happened?

In the twentieth century, the eastern Hadza were a viable, self-contained, and increasing population. Yet they are a very small population occupying a very small area. The evidence of a genetic bottleneck implies that they were once much more numerous. It has been widely assumed that any hunter-gatherer population is a dwindling remnant, soon to disappear. The contemporary demographic parameters of the Ache and Hadza appear to contradict this. The Hadza have been increasing quite rapidly, at least throughout the second half of the twentieth century. The Ache were increasing even more rapidly during the forest period, only to dramatically crash at contact and settling, soon to increase almost as dramatically again. Only the !Kung approximated a steady population (based on Howell's fertility and mortality estimates for the time before the arrival of many Herero herders in the 1950s). My earlier discussion of the bottleneck (Henn *et al.*, 2011) implies a dual process. Those who call themselves Hadza and have been marrying other Hadza are increasing. However, the number of those who call themselves Hadza is small today compared to the past. Hunter-gatherer populations disappear, not because they cannot support themselves or do not have a viable life history but because other lifeways attract them, limit their options, and brought new diseases in the remoter past. Land is taken, even when neighbors and states intend to help by settlement schemes and food handouts, intermarriage increases, people change their ethnic identity (which changes the label they are given by anthropologists and molecular genetic historians). Entering these other societies at the lowest level (as seems often to be the case) may ensure the rapid disappearance of the forager genes. Changing their identity changes who they marry, and for the men, whether there is anyone left who will marry them.

The Hadza population history gives us another puzzle. If the Hadza population is recovering from losses around the year 1900, for how much longer will it increase? Is it returning to some pre-nineteenth century density at which it will level off? Will it continue to increase until some disaster strikes? These questions lead us to important general questions about hunter-gatherer population regulation to be discussed in the next chapter.

# 11 The Hadza and hunter-gatherer population dynamics

## 11.1 The "forager population paradox"

During the 1960s and 1970s, many of us believed that hunter-gatherers, living "in a state of nature," would have largely stationary populations, closely regulated by density dependent factors. In contrast to agricultural and industrial societies, they had not exhausted, eroded, or polluted their world. Their populations were small and sparse. We assumed they had always been that way. We may have been totally wrong.

In those days, debate centered on the processes that restrained populations. Infanticide, senilicide, birth spacing, famine, drought, and warfare were discussed. A stationary population was thought to be consonant with Deevey's (1960) and Hassan's (1978) widely cited calculations of the slow pace of increase during the Pleistocene. Hassan's papers comprise a substantial and scholarly integration of demography and archaeology that deserves more than the usual fleeting citation. In collecting and estimating the range of parameters of hunter-gatherer reproduction and mortality, Hassan (1973, p. 540) notes the incompatibility of observed hunter-gatherer fertility and mortality with a long-term stationary population. Birdsell (for example, 1968) had also noted the incompatibility of hunter-gatherer fertility and mortality with stationary populations, and he used this as one of his arguments for his view that infanticide had been a population regulation mechanism.

"Control mechanisms" of hunter-gatherer populations were proposed as society level restraints, only occasionally as results of individual practical decisions. Caldwell and Caldwell (2003) review the early phases of this history from the perspectives of demographers and anthropologists. My perspective is unashamedly biased by close exposure to the parallel debates in biology that preceded them. In biology, the theory that there were societal level restraints, mainly attributed to Wynne-Edwards (1962), was quite quickly discarded. It was discarded because natural selection would be expected to favor individuals who eschewed the restraint, who would quickly outnumber those who volunteered to restrain the number of their descendants. The short-lived debate over societal restraints was replaced by a lasting debate about whether limits were imposed by increasing population density, or by random processes such as unpredictable climatic extremes not caused by population density (although in reality, the size of their effects could be correlated with population density). In the history of animal population studies, density dependence was championed by David Lack (1954), and extrinsic factors were best known from the writings of Andrewartha and Birch (1982). In modern studies of animal populations, rapid increases and rapid declines are noted, their causes intensively investigated, and the time lags of responses to density (Turchin, 2009) and the shape of the relationship of population growth rate to population density elaborated (Winterhalder, 1993; Godfray and Rees, 2002; Sibly *et al.*, 2002, 2005).

Studies of hunter-gatherer populations that included any time depth and careful examination of rates of increase or decrease were few. Howell's study of the !Kung was far more careful and thorough, and based on more, and more direct, evidence than any hunter-gatherer population study that had preceded it. For many readers, the results of Howell's demographic work on the !Kung would have seemed like confirmation that hunter-gatherer populations were close to stationary.

Subsequently, Hill and Hurtado's Ache study (1996) brought attention forcibly back to the contradiction noticed by Hassan (1973) and Birdsell (1968). The Ache in the forest, before settlement, in quite extreme isolation from neighboring populations and from farm produce, with no access to modern medical care, were increasing very rapidly. There were historical data compatible with the idea that the Ache were recovering from a previous era of extreme persecution from other populations. Hadza demographic history tells the same story. However, instead of assuming that the Ache would increase back to some density dependent stationary population, Hill and Hurtado remarked on the basic paradox in human life history (I will refer to it as "the forager population paradox"). Hill and Hurtado suggested that to generate a stationary Ache population would require greater mortality than we have ever observed in any human population under any conditions. Yet over the long run, such as during approximately the last 80,000 to 40,000 years preceding agriculture (Cox *et al.*, 2009; Zheng *et al.*, 2012; Aime *et al.*, 2013), human populations increased at far below the rates of increase observed for recent hunters and gatherers, as Hassan had pointed out (1973, p. 540). The rapid increase observed in a hunter-gatherer ecology caused Hill and Hurtado to speculate that human populations may have had a long history of rapid rise and even more rapid crashes. Pennington (2001), reviewing hunter-gatherer demography, also saw the conflict between the evidence of rapid increase and the inevitability of much lower long-term rates of increase. She argued that high incidence of infertility resulting from sexually transmitted diseases (STDs) might comprise one possible resolution. This reminds us that we should consider decreases in rate of recruitment by births as well as increases in loss by mortality.

Keckler (1997) followed up Hill and Hurtado's "sawtooth" population idea. He outlined the contradiction between Hill and Hurtado's Ache observation and Hassan's argument that before agriculture, human populations had increased at best at a very low rate over a very large time span. Keckler modeled the effects on three hunter-gatherer-like populations of crashes that killed a substantial percentage of every age group in the population. As reported by Boone (2002, fig. 2), Keckler found that a long-term stationary population could result if a population with an intrinsic rate of increase of 0.007/1000/yr (similar to the !Kung's rate of 0.005) experienced crashes in which 25% of the people died, and which occurred at random intervals with an average interval of 50 years. The result was, not surprisingly, quite sensitive to the size of the crash. Keckler's main goal was understanding the very strange "age-at-death distributions" reported for several archaeological assemblages (SI 8.7). Yet Keckler's crashes offer a potential solution to the "forager population paradox". What causes of population crashes are likely? Epidemics due to the well-known, all too destructive diseases of modern populations (measles, plague, smallpox, whooping cough) are thought to be unlikely in a world of hunters among hunters. Warfare remains a possibility (Hill *et al.*, 2007), and one that might be provoked by rapid population increase, and thus be density dependent (Kelly, 2013, pp. 207–250). Nonetheless, we should explore other possible sources of greatly increased mortality, or much lower fertility, than we have seen in the Ache, !Kung, or Hadza environments.

## 11.2 Hadza demography displays the "forager population paradox"

The Hadza have sustained a quite rapid population increase during most of the twentieth century while living almost entirely off wild foods and with limited access to modern medicine. The Hadza resemble the Ache in their high rate of increase. Even the lower end of the 95% CI (0.0123–0.0176) does not take the Hadza remotely near to the realm of stationary populations. The Hadza rate of increase would take them from a few hundred to 32 million in a thousand years. The Hadza exemplify the forager population paradox. Like the Ache, the Hadza history suggests a reduction in population about 100 years ago. In both the Ache and the Hadza cases, the cause of the hypothetical "crashes" from which the populations were "recovering" were captures and homicides. Some form of warfare or raiding would undoubtedly be a prime candidate for a cause of crashes.

I used my simple population simulation to find out what changes in fertility or mortality, or both, would reduce Hadza population growth to zero. Table 11.1 summarizes the results of these and subsequent simulations reported in this chapter. To render the Hadza population stationary, we have to add a little over half to the observed Hadza mortality, and make the age-specific death rates $q_x = q_x + (q_x \times 0.5925)$ in the simulation program (or multiply $[q_x]$ by 1.5925 at every age) (Table 11.1, line b). This is equivalent to changing the Hadza mortality level from the Coale and Demeny (C&D) model level 6 to level 1, the highest mortality that Coale and Demeny modeled, with a life expectancy at birth ($e_0$) reduced from 32 to 20 years. Thus, while Hill and Hurtado are probably correct to say that preventing the Ache from increasing would take a level of mortality never observed for our species, the rather less fertile Hadza offer less of a challenge. Notwithstanding, could we imagine them with West level 1 mortality? In my effort to quantify the effect of modern influences on Hadza mortality (Blurton Jones *et al.*, 2002), I raised the estimate of Hadza mortality to about West 5 – similar to Howell's !Kung data, with a female $e_0$ of 30 years. Pennington (2001, table 7.6) shows that to render the Ache population stationary, mortality would have to rise to an $e_0$ of 19 years, worse than West 1. The Hadza are close to Hill and Hurtado's formulation: to offset their fertility, a mortality level as high as ever observed is needed to render their population constant (Figure 11.1, Table 11.1 line b).

**Table 11.1** Summary of results of population simulations discussed in the text. Changes that could make the Hadza population stationary (achieve zero population growth [zpg]). Row c: gives an $l_x$ (survival) curve very similar to the C&D model West 1. Row f: annual risk of infertility; percentage of women permanently removed from the breeding pool at each age over 18. They remain in the population to be counted

| | Type of change | Change | % change to achieve zpg | Equivalent to TFR, $e_0$, etc. |
|---|---|---|---|---|
| a | Enduring | Reduced fertility | ASF – (ASF × 0.35125) | TFR 4.0 |
| b | Enduring | Increased mortality | All ages $q_x \times 1.5925$ | $e_0$=20.0 |
| c | Enduring | Increased child mortality | $q_{1\text{-}7} \times 1.75$ | |
| d | Enduring | Fertility decrease and mortality increase | See Figure 11.1 | |
| e | STD enduring secondary infertility | Age at last birth | 32 years See Figure 11.2 | Last birth at 32 |
| f | STD enduring risk | Annual risk of becoming sterile | 0.0385/yr See Figure 11.3 | TFR 3.87 |
| g | Occasional crash or raid | All ages | See Figure 11.4 | |
| h | Occasional crash or raid | Children killed | $q_x \times 10$ for crashes, 1 in 10 years | |
| i | Occasional wars or raids | Women stolen | Approx. 25% stolen every 10 years | |
| j | Occasional wars or raids | Children killed, women stolen | 22% every 10 years | |

The Hadza population is slightly more sensitive to career-long changes in fertility. Subtracting 35% of ASF at each age gives zero population growth and a total fertility rate (TFR) of 4.00 (ASF = ASF – [ASF × 0.35125]) (Table 11.1, line a). This is a TFR similar to the 4.2 estimated for the central Kalahari San by Tanaka (1980, table 18). It is close to the potential TFR (3.99) displayed by the low fertility of Hadza in 1998 after the exceptional drought of 1997. If fertility dropped this low for a substantial period, the Hadza population would be stationary even if there were no change in mortality. In Figure 11.1, we can see that the Hadza population would be stationary (at zero population growth) if mortality increased by 30% and fertility decreased by 20%. In Pennington's (2001) simulations, lowering Ache fertility to a TFR of 3.7 renders the population stationary. Pennington points out that this level of TFR has been observed not infrequently, whereas mortality with an $e_0$ of 19 years was, at the time she was writing, almost unknown. She views low fertility, a common result of STDs, as a more likely target for resolving the hunter-gatherer population paradox. In support of her view of the importance of STDs, Pennington suggests that lactational amenorrhea could seldom elongate inter-birth intervals beyond about two years. However, subsequent discoveries about the role of women's energy budget,

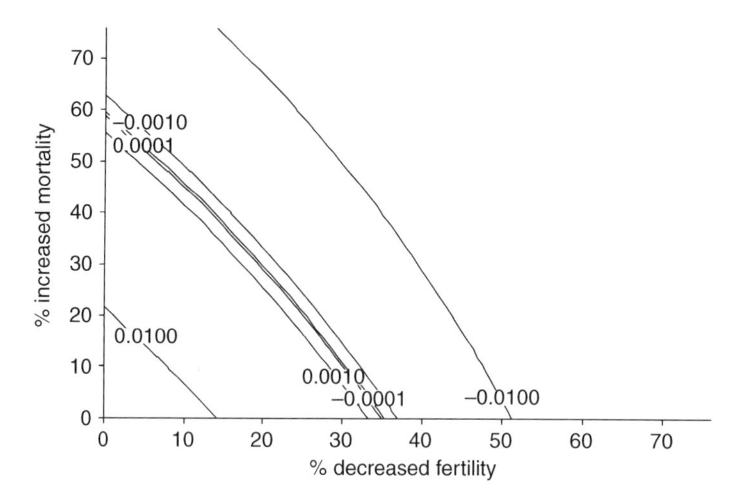


**Figure 11.1** Contour plot showing effects on simulated Hadza population growth from combinations of increased mortality (Y-axis) and decreased fertility (X-axis). Y-axis shows percentage added to $q_x$ each year of age. X-axis shows percentage subtracted from age-specific fertility each year. The closest pair of lines mark area of zero population growth (zpg). Populations to the right of the line decrease; populations to the left increase. For example, a 30% increase in mortality, coincident with a 20% decrease in fertility would result in a stationary population (zpg).

summarized in Ellison's "metabolic load model," suggest that we cannot dismiss the possibility that severe living conditions can greatly reduce fertility. For example, Ellison (2001, and elsewhere) cites the three-year inter-birth intervals of Lese women, who wean infants by age one year. This and many other lines of evidence suggest that infection is not the only possible route toward very low fertility.

## 11.3 Density dependent regulation of human populations

The dominant view of animal populations has been that upon colonizing a new locality, population increase followed a sigmoid curve, gathering pace while at low density, then slowing and becoming stationary as density reached "carrying capacity." In biology, carrying capacity was defined as the level at which population increase slowed and ceased. In anthropology, it was more often taken as an empirical economic measure, as the population calculated to be supportable by measured technology in a measured environment. Much discussion concerned carrying capacity and the observation that populations (briefly observed, and assumed to be stationary) were at levels well below calculated carrying capacity. The anthropogists' usage encouraged the idea that traditional human societies had adopted restraints that maintained their standard of life and gave long-term security to their economy. Most calculations of carrying capacity ignored the time required to acquire and process apparently abundant food. Populations believed to be stationary and below carrying capacity may have been neither.

Theoretical modeling of density dependent population regulation using models of hunter-gatherer populations were developed by Belovsky (1988), Winterhalder *et al.* (1988; Winterhalder, 1993), and Rogers (1992). Each of these authors attended primarily to hunting and its effects on prey populations, and assumed that shortage of prey would limit a forager population. This was probably a reasonable assumption for the boreal foragers considered by Belovsky and by Winterhalder (and for others like the Ache, where men apparently acquired most of the food). The most important result of these studies was that large oscillations of population can, in theory, result from interaction between predator and prey, and that effects depend in part on the kind of contests between predators. Rogers distinguishes scramble competition from contest competition. Scramble competition most resembles what we see in cultures such as the Hadza and !Kung, where anyone who has little is likely to get something from someone who has more (Woodburn, 1998; Howell, 2010; Marlowe, 2010, pp. 229–252). In this circumstance, crashes are, in theory, more devastating – everyone runs short at the same time and the population crashes massively. In contest competition, some keep more resources than others, and while the poor may die off rapidly, the rich tend to survive. However, as Rogers repeatedly reminds us, these are intentionally simplified models. In reality, we find individual differences in reproductive success (RS) in even the most egalitarian societies, and should expect that a few individuals survive to give us the new population and whose descendants we eventually observe. Furthermore, if we are thinking about people, as the economics of contests change, so may the behavior.

Relationships between foragers and plant foods might easily have some of the same properties as relationships between predator and prey; but there are differences. Harvesting nuts and fruit does not kill the tree, it merely slows its rate of reproduction by a very small amount. Even storage organs, which require the forager to dig up part of the tree or vine, do not kill the plant but may impair its subsequent growth and storage capacity (see for example Stewart, 2009). Effects would depend on rate of regeneration of the plant. For instance, a patch of makalita at /ilo camp in 1986 was nearly wiped out by the children, but when I passed by this site in 1989, a few makalita plants were growing there again.

Direct evidence of density dependent regulation in human populations has been reported by Wood and Smouse (1982) on Gainj, Ronald Lee on medieval Europe (1987), and Lutz and Qiang (2002) globally. These authors demonstrate density dependent responses but have not shown that these responses resulted in a stationary population. The density dependent effects (a damping of survival of children and old people) among the Gainj were seen during a period of population growth following a decline of 6.5% resulting from a flu outbreak. The interaction between fertility, density, and wages that Lee demonstrated for historical data from Europe only reveal themselves in "de-trended" population parameters during a period of substantial increase. Lutz and Qiang find, in a study of 187 national populations, "the relationship [with population density] is much stronger and more statistically significant in the case of fertility as the dependent variable, although the signs are consistently negative for both fertility and the growth rate."

Turchin (2009) reviews his work on long-term population cycles in state-level agrarian societies. His central contention is that increasing population density does not necessarily have an effect either early or rapidly (as Winterhalder's models also showed). The result would be long-term oscillation, of the order of two to 300 years, which Turchin finds is matched by the several available long-term historical censuses in Europe, China, and Vietnam. He traces the social unrest and disorganization that relates to the peaks and declines. It is not difficult to believe that civil war, unemployment, and internal migration lead to higher infant and child mortality and lowered fertility. He seems to suggest that some of the process applies to non-state societies, commenting on the limitation of cultivation when organized defense is not available. His comments from historical Europe are reminiscent of the literature on the consequences of the slave trade in rural Africa. Turchin discusses "contests of the elites" and other components of societal breakdown. The atmosphere of the Hadza in 1997 provokes thoughts of civil breakdown. Things could have become worse, but by 1999 and 2000, everything seemed back to normal. What could "getting worse" amount to in a society with "minimal politics" (Woodburn, 1979)? Crowding into the handful of localities with abundant water? Tiny camps, later marriage, slower remarriage? Each can have small demographic consequences. Concentration around a smaller number of water sources would result in faster depletion of plant foods, and longer foraging and hunting trips. As the ecology approached that of the !Kung or G/wi, the demography may be expected to follow, including a decrease in fertility.

History suggests another possibility, which may preserve the idea that hunter-gatherers were under density dependent regulation. Perhaps the living hunter-gatherer populations that show evidence of rapid increase are those that have been through the same filter, a recent bottleneck. Expanding agricultural populations were converting hunter-gatherers everywhere they went, or exterminating them by violence, disease, or habitat destruction. Then the colonial era arrived, in time for some of these groups to be described, and some of them, including the Hadza and the Ache, to be rescued by colonial pacification of the invasive herding and agricultural peoples. With few exceptions, perhaps in the Kalahari and the high Arctic, the hunter-gatherer populations that still existed had been depleted well below any density dependent limit. Consequently, anthropologists have been presented with a very biased sample, not just the ones who lived on land unsuitable for agriculture (Foley, 1982), but those whose populations had been greatly reduced and had become free to grow again. This history is an oversimplification but it may be useful to bear in mind whenever early demographic data are examined.

## 11.4 Is there direct evidence for density dependent regulation of the Hadza population?

Do the Hadza data encourage us to preserve the 1960s' view that hunter-gatherer populations had been living in rough balance with their environment, subject to density dependent regulation? Unlike the Ache, who settled in the 1970s and have remained settled, the Hadza may give us the chance to look for an eventual leveling of the population, slowing of the population growth rate, or changes in fertility or mortality. There may be density dependent limits that only take effect at a higher density than yet observed. If we wait, we will see the Hadza population growth slowing, if it has not already started to slow. But the Hadza have been under observation for nearly a century, and up to 2000, we had seen no sign of their population growth slowing down.

One sign that the population is below carrying capacity and hence may later reduce its rate of growth might be an unexpectedly low density for the habitat. Marlowe (2010, p. 263, fig. 10.4) shows a plot of warm climate forager density against NPP (net plant productivity). In Porter and Marlowe (2006), we see NPP in Hadzaland at 607 g/m$^2$/yr. In Marlowe's fig. 10.4, this level NPP is matched to a smoothed density around 0.35 persons/km$^2$. Observed density for the Hadza was 0.27 persons/km$^2$ in 1985 and 0.33 persons/km$^2$ in 2000, quite close to the expectable level. Given that no confidence level is shown, we cannot safely claim that Hadza density in 2000 was any different from what would be expected from the net primary productivity of their environment. Both measures of NPP and measures of density have problems. It is difficult to know how to calculate density when foragers, and their resources, are patchily distributed. It is not clear how well NPP reflects the availability of water to people, nor of edibility and accessibility of the biomass (which Porter and Marlowe discuss in relation to forest habitats). Kelly (1995, table 6–4; 2013, table 7.3) shows Hadza density at 0.15 persons /km$^2$ derived from Woodburn's (1968a) rough estimates of 400 Hadza in 1000 square miles. Woodburn would be the first to caution that his estimates were approximate. Our evidence of increasing population and decreasing habitat should support the idea that Hadza density has been increasing. Perhaps we should expect some slowing in eastern Hadza population growth in the early years of the twenty-first century.

Are fertility or mortality or population growth responding to population size? There was no significant relationship of population growth in the following year to current population size (a potential test for density dependence, r=0.0402, p=0.8869). The sharp recovery between 1988 and 1992 after the 1986 measles epidemic may have obscured any such relationship. One can see a decline in the rate of increase after 1995, but the period is too brief to say whether this is real. There has been no consistent decrease in fertility or increase in mortality during our study period, nor since Dyson's (1977) analysis of the 1960s data.

Are there population adjustments to the environment? Having seen a response of fertility to the drought of 1997, I wondered whether earlier periods of drought had a similar effect. There is a trough in the age structure that had mystified me until I found references to severe famine and drought in 1949 (Baker, 1974). Brooke (1967) lists this as severe for the Mbulu and Singida. The coincidence suggests a similar episode of extreme drought being associated with either a drop in fertility, or an increase in infant and child mortality, or both. In Chapter 10, I suggested that this trough might have resulted from departure of young women who married non-Hadza men. Furthermore, there are other troughs that I have generally assumed to be random or effects of errors in age estimation. It might be too easy to find such associations by chance alone. The effects of the occasional drought year are much too small to prevent the Hadza population increasing. Nonetheless, a climate change that led to the average ASF resembling that of 1998, giving a completed family size (CFS) of about 4.0, close to that of the central Kalahari G/wi (a 33% reduction in fertility), could have a significant effect on the population (Table 11.1, line a).

If the rate of increase is not slowing, nor its contributory fertility and mortality changing, we cannot say the Hadza have reached carrying capacity in the biologists' sense. What about carrying capacity in the anthropologists' sense? We could calculate the number of people who could be fed each year by the tubers and numerous baobab trees in eastern Hadza country (1977, aerial survey). Annie Vincent's record of the abundance of tubers gives us a start. Vincent (1985b, p. 136, table 2) shows the enormous supply of tubers in the Mangola area. The means of five varieties that she lists total close to 70,000/kg/hectare (which converts to 7 million kg/km$^2$). At 75 kcal/100 gm, a conservative mean calorie value of these tubers (Chapter 2), this amounts to a standing crop of 5250 million calories/km$^2$. At a generous 2000 cals/d, this would provide the calorie requirements of 7000 Hadza for a year. However, this rather approximate calculation (which takes Vincent's startlingly high figures at face value and assumes these tubers can be regenerated in a year, perhaps too short a time) would neglect a much more important restraint. The amount of time taken to collect and process each resource is likely to be a much more striking immediate limit to the number of people who can be fed by a working woman and her kin. Especially important are the resources used at the most difficult time of year (discussed in Marlowe and Berbesque, 2009). Limits for the Dobe !Kung likely come from the time taken to reach the nut groves (determined by the location of dry season water sources) and process the nuts, not from the size of the nut groves. The number of people who could be fed by the nut grove if some benevolent machine collected, transported (some use donkeys now), and processed all the nuts, would be as unthinkably large as for the Hadza and the tubers. The neglect of processing costs is not unique to anthropologists. When Lars Smith and I visited a wealthy Datoga elder, he asked Lars for some ideas on what he could do with his money. Lars suggested he buy a maize grinding machine, a sound business plan for rural Tanzania in the 1980s. The elder pointed to the elegant young women here and there around his house and replied "but I have three of those already."

The Hadza population is increasing, therefore, women must be successfully processing sufficient food to raise enough children to generate the increase. If this continues long enough, there would eventually be enough Hadza to eat all the tubers and baobab fruit that the habitat can produce. Thus, there may be a calculable carrying capacity (sensu anthropology) but it would take quite a long time for the growing population to reach it. If food were sufficiently depleted, such that women had to walk further to the baobab trees or the nut groves, or dig for longer to get the same amount of tuber, the number of children their energy budget allows might decrease toward replacement level. Anthropologists encounter populations that are smaller than they think they could be because of effects of depletion on acquisition and processing costs, not on the abundance of the environment. Omitting these costs makes it inevitable that anthropologists will record populations as below carrying capacity.

The Eyasi environment, and the Hadza population, seemed to absorb the influx of the Isanzu and Datoga refugees who came to live as foragers early in the twentieth century, a period of low rainfall. The Isanzu women who married Hadza men presumably added to the subsequent Hadza population. If the Hadza had been limited at that time by the availability of bush foods, we might have heard a different account of these events. Woodburn (Lee and DeVore, 1968, discussion, p. 244) is easy to agree with: "I think the population could be a good deal larger than it is without causing hardship for the Hadza or damage to the environment."

For the Hadza, even with loss of berries from the Balai delta, and the loss of a chunk of Siponga, food is abundant. However, it is hard to get and hard to process, and the limits are work time and exhaustion. Other density dependent factors may be more urgent. There may be higher infant, child, and elder mortality as population growth causes an increase in the number of individuals infected with a communicable disease. If food acquisition becomes more exhausting and takes more time, social relations may become stressed and fights and fissions begin. In the field, I gained the impression that one motivator of camp moves was hungrier children when they had depleted the easy, nearby child plant foods. Elsewhere, I have emphasized the contribution of Hadza children's foraging to their mothers' work budgets (Blurton Jones *et al.*, 1989). The depletion of children's more accessible plant foods might sometimes be a significant influence, not just on the frequency of camp moves, but also on women's work and energy budgets and thus their fertility.

Sometimes even famously abundant plant foods can be exhausted. In 1988, the Dobe Mongongo nut grove was quite sparse when Draper, Hawkes, and I measured foraging return rates in a variety of habitats. It was so depleted that other habitats, where berries or roots were the best targets, were just as rewarding. To estimate return rates from the Dobe grove in better years, we used the higher collecting rates from the much more distant groves (Blurton Jones *et al.*, 1994).

## 11.5 Disease

Perhaps we have been wrong about epidemics in a world of hunters among hunters. Anderson and May (1991, pp. 653–656) briefly discuss the role of disease in the history of human populations in light of their extensive modeling and model testing on infectious diseases. They rely on population estimates by Hassan (1981) and by Deevey (1960), including the view that "Bands of hunter-gatherers probably ranged in size from around 20 to at most 100 individuals." They suggest that such populations could maintain diseases with "long periods of infectiousness or ... with asymptomatic carriers ... [or] ... with a normal habitat in a host other than man ... [or] ... able to survive and multiply in soil." They continue "but the directly transmitted microparasites responsible for much mortality in historical times – smallpox, measles, cholera, and the like – have very high host threshold densities, and could not have been present in the pre-agricultural era." Investigations of genetic history of disease organisms are tending to support the recent origin of these diseases but reveal others of great antiquity, including tuberculosis (TB) (Chapter 2).

To take "band sizes" of 20–100 people as measures of population is to misunderstand anthropologists' observations. Camp sizes and "residence groups" of 20–100 people are not isolated populations, they are notoriously variable and fluid. In cases such as the !Kung and the Hadza, people come and go all the time between camps within the 800 or so "dialect group," which may correspond to a biological population. Furthermore, these populations were not necessarily isolated from visiting with other such populations, as the archaeological evidence from Eyasi shows (Chapter 3). From Australia, Birdsell (1968, p. 237) cites Tindale as reporting a rate of 14% intermarriage between neighboring Aboriginal language groups. The "meta-population" of interacting individuals in Australia, for example, and in sub-Saharan savanna Africa could have been very large, although spread very thinly. But given the population sizes estimated to be needed to keep a disease such as measles from extinction, it is easy to agree with Anderson and May's conclusion. However, Sattenspiel and colleagues' (2000) model of the 1918 flu epidemic in aboriginal Canadian hunter-fur-trapper society, with interacting populations only a little more concentrated than the Hadza, suggests that modeling disease flow in a world of hunter-gatherer meta-populations might be rewarding.

While some diseases may generate density dependent mortality (more people, more transmission; less well-fed people, more vulnerable), others may arrive unpredictably. Thus, another speculation concerns unknown zoonoses. One would think hunter-gatherers were especially likely to contract animal diseases; in fact, searches for new diseases concentrate on active bushmeat procurers (Wolfe *et al.*, 2004). The research is also concentrated on forest areas, where the number of primate species (the most likely source for zoonoses) is high. In the wooded savanna of Hadzaland, baboons and vervet monkeys are the only non-human primates seen, and these are very locally distributed. Most bushmeat hunters have recently intensified their involvement in hunting, which makes them useful for showing us current disease threats. People like the Hadza may have been exposed to any disease that infected any of their prey species (which does occasionally include baboons) during the last several tens of thousands of years. Primates may be hunted more intensively when other prey become scarce. Each zoonosis may have initially caused severe population crashes, followed by recovery. There might be little superficial indication of this succession of "extinct" diseases. Fieldwork such as Bennett *et al.*'s could only test for well-known diseases for which field tests were available. However, 5%–8% of the human genome is virus in origin (Mayer and Meese, 2005). Some of this material dates from common primate ancestors 30 million years ago, some dates from since the chimpanzee–human split. The material is much modified, and arrives in the germ line by re-infection from somatic cells (Belshaw *et al.*, 2004). The exact history of these ancient diseases may never be worked out (the remarkable work of Marchi *et al.*, 2013 on Neanderthal and Denisovan retroviruses may imply that my pessimism is unwarranted), but the history is likely to have included severe outbreaks that could have a serious impact on small populations.

Today, the Hadza (like the "isolated" South American populations described by Black *et al.*, 1974) are exposed to "post-agricultural" diseases, like measles, and some decades ago, smallpox. Deaths as a result of these are included in our estimates of mortality level. Without the occasional arrival of such diseases, Hadza mortality would presumably have been lower and rate of increase higher, exacerbating the forager population paradox. However, it is probably important to note that most deaths under Hadza conditions arise from "coughs and colds" and diarrhea. These diseases, trivial under "western conditions," are still major causes of death in the third world. The absence of modern diseases may have made little difference to hunter-gatherer mortality. While the modern world has added such diseases to the Hadza experience, it has also added very slightly better access to modern medicine since the mid-1960s. I have shown how little use Hadza make of these facilities, but I could have underestimated the demographic effect of even limited use. Determining the effects of modern medicine in the African countryside is not straightforward. Orubuloye and Caldwell (1975) showed the startlingly lower child mortality rates in a Nigerian village that had a hospital as contrasted with an otherwise closely matched village with no hospital. Morley (1963) had reported a similarly striking improvement in child survival during the five years following the establishment of an "under-fives clinic" in another Nigerian village. But Orubuloye and Caldwell comment on the observation that clinics and hospitals have very small catchment areas. In Tanzania and Kenya, where nationwide networks of local health clinics were set up soon after independence, the effects seem confined to a 5 km radius (Armstrong Schellenberg *et al.*, 2008; Moisi *et al.*, 2010). Despite the relative inaccessibility (much more than 5 km for most Hadza, and with no public transport), perhaps Hadza mortality was affected by the clinics set up at Munguli, Endamagha, and Yaeda. Perhaps mortality was so high before this that the Hadza population was scarcely increasing. Dyson's account contradicts this possibility. His mortality estimate is only very slightly higher than ours. The data on child mortality that he used came from reproductive history interviews conducted by Woodburn in 1966–1967 that covered the reproductive lives of 75 women who were bearing children between the 1920s and 1967. The clinics were built after 1965, that at Munguli in or after 1967, and at Endamagha after 1971.

## 11.6 Sexually transmitted diseases

Pennington (2001) suggests that infertility caused by STDs played a significant role in restricting human population growth during our history as hunter-gatherers. She comments that "Many authors believe the diseases affecting fertility are too new to have greatly affected reproduction in our species history. But even using the lowest hunter-gatherer survival rates, we cannot account for low growth rates without them."

At the time Pennington wrote this, there were few molecular evolution studies of disease history. While genetic history has been employed in the "syphilis from the New World" debate (Harper *et al.*, 2008), there appears to have been little done on the more remote history of organisms that cause diseases of the reproductive tract (mostly sexually transmitted), other than HIV. As has been frequently stated, these can have substantial demographic effects. Could they have played the role in forager population regulation that Pennington (2001) suggested? Anderson and May (1991, p. 228), with reference to *Neisseria gonorrheae* and *Chlamydia trachomatis*, say "This suite of properties – virtual absence of a threshold density of hosts, long-lived carriers of infection, absence of lasting immunity – adds up to microparasites that can be well adapted to persisting in small, low density aggregations of humans." Pennington remarks that STDs are widespread among other animals, and they sometimes affect fertility (reviews include those by Longbottom and Coulter, 2003, on chlamydioses; by Knell and Webberley, 2004, on insects; by Sheldon, 1993, on birds; and by Nunn *et al.*, 2000, on primates; and relationships between STD rates and mating systems were investigated by Nunn *et al.*, 2000, and Altizer *et al.*, 2003). Reed *et al.* (2011) summarize information on diversity and history of several sexually transmitted infectious agents. Anderson and Seifert (2011, and elsewhere) suggest that *N. gonorrheae*, for which humans are the only natural host, has a "long and exclusive evolutionary history with its human host." One other member of the genus, *N. meningitidi*, causes meningitis particularly in infants, but several others are regarded as harmless commensals. The evolutionary history of bacteria is presumably rather difficult to study given their ability to transfer genetic material from other species. Indeed, Anderson and Seifert (2011) found that some *N. gonorrheae* strains have incorporated a human gene. They date this as "recent," since the divergence of *N. gonorrheae* and *N. meningitidi*. Thus, it is probably not yet clear whether this particular STD pre-dates the invention of agriculture and ensuing population increase, but Anderson and Seifert's papers seem to imply that it probably does. I have not found a time depth offered in the literature for chlamydia as a sexually transmitted infection of humans, although the differentiation between the species that target different tissues has been described as resulting from an evolutionary "arms race" with the human host immune system (Nunes *et al.*, 2008).

If the effects of infection on reproductive performance vary, for instance with nutritional level and presence of other infections, environmental stressors might produce sporadic dire fertility effects. If such were the case, then there would be a density dependent relationship to STD. Increases would follow in availability of infected partners and the rate of spread would increase. All disease organisms are co-evolving with their hosts' immune systems. The process may sometimes lead to periods of stability, sometimes to instability and rapid outbreaks of disease. Other possible dynamics, including epidemic waxing and waning, and density dependence of STDs, should be explored if one wants to fully understand the population dynamics of hunter-gatherers in a world of hunter-gatherers.

I tried some crude models of the effects on a Hadza-like population from primary and secondary sterility. One preliminary model showed, obviously, that the effect of

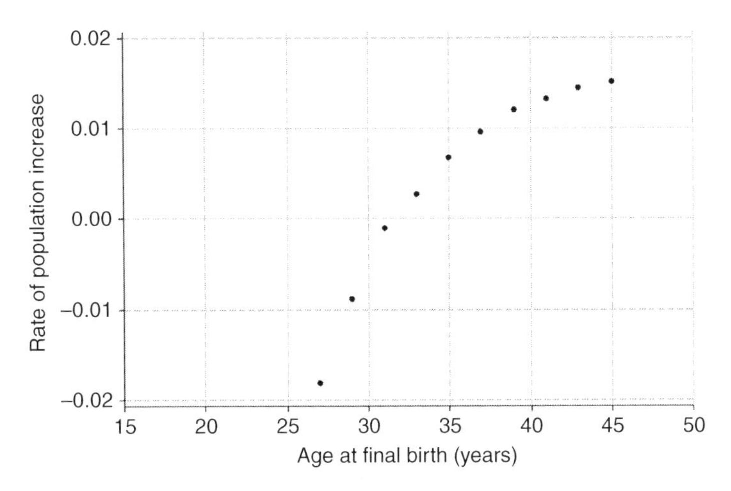


**Figure 11.2** Effect on population increase, in a model of the Hadza population, from curtailing fertility at different ages. A crude simulation of effects of secondary infertility. Births cease at $age_X$. Until $age_X$, births occur at observed age-specific fertility.

secondary sterility on fertility and population growth rate was greater when sterility began at earlier ages. Less obviously, it showed that this was not a linear effect, lowering the age at which the women became infertile from age 40 to age 35 had much less effect than lowering the age from 35 to 30 (Figure 11.2, and SI 11.1). Zero population growth arose when fertility reached zero at age 32 (similar to the !Kung but lower than the Hadza estimated age of 37).

More realistic was a model of primary and secondary sterility together. Fertile women were at a constant risk of becoming infertile in each year of their reproductive career. Having become infertile, they stayed infertile. An annual risk of 0.0385 gives close to zero population increase (r=0.0001), and 0.039 gives population decline at r=−0.00008. Thus, starting with Hadza ASF and mortality, an annual risk of 3.9% of fertile women becoming infertile gives a declining population. With TFR 3.87, this is a close match to Pennington's (2001, table 7.6) estimate of the TFR required to give a stationary population at the observed Ache mortality. In this simulation, the resulting median age at last birth is 35 instead of the observed median near 40 (at this median age half the women are expected to have a subsequent birth). An interesting side result was a change in the shape of the ASF (Figure 11.3). By slightly increasing the hazard of sterility, one can mimic the differences between the Ache (most symmetrical – 30–39-year-olds as fertile as 20–29-year-olds), the Hadza (intermediate), and the !Kung, among whom fertility was estimated by Howell (1979) to decline rapidly from age 30. The experimental 3.9% of fertile women becoming sterile each year results in 44.5% of women being sterile by age 30, a very high figure compared to the Hadza 4%–6% (Chapter 7).

These calculations were based on sterility. They do not show us infection rates. Not all infected women become sterile. Sterility is said to develop in 10%–30% of infected women (clearly, it must differ from one disease to another). Therefore, my simulated

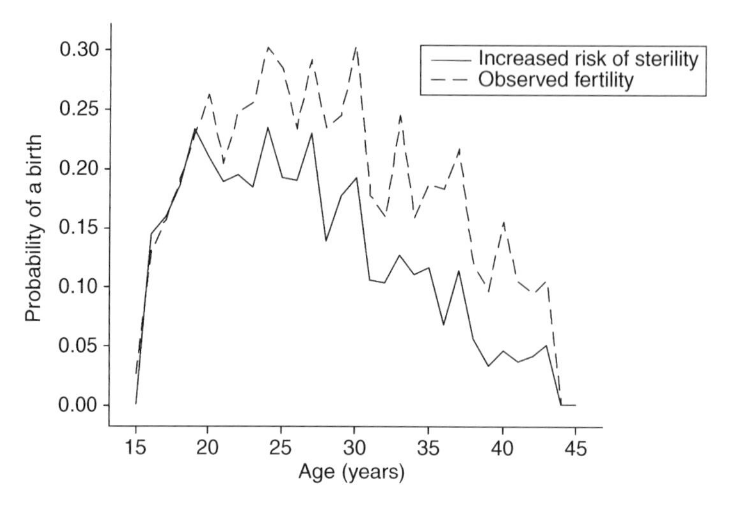


**Figure 11.3** Change in age pattern of age-specific fertility in a simulated Hadza population owing to simulated annual risk of infertility sufficient to give a stationary population.

4% of women becoming sterile per annum represents a much higher level of infection. In studies of "remote" populations in Australia using modern techniques of identification of infection, up to 15% of women were found to be infected with gonorrhea. In a study in rural antenatal clinics in Tanzania, 8% had gonorrhea and/or chlamydia infections (Mayaud *et al.*, 1995). By self-selection, it was an antenatal clinic; these women had presumably not been rendered infertile by the infection. I found no figures on how long it takes for infection to produce sterility, which results from lasting inflammation and tissue damage, particularly of the Fallopian tubes. I have not been able to make a realistic estimate of the annual rate of new infections that would result in 4% of women becoming infertile each year; but it must be rather high.

These models from the Hadza agree with Pennington (2001) that some of the higher recorded levels of primary and secondary infertility could possibly generate a stationary population. Human evolutionary biologists should follow the literature on molecular evolution of disease organisms, as well as that on primate social systems, STDs, and immune systems. Another reservation about STD as a resolution to the forager population paradox derives from the earlier literature. Bennett *et al.* (1973, p. 268) comment, "The Hadza are now getting gonorrhea from neighboring tribes; this is most obviously so among the western Hadza who visit beer-drinking establishments during the cotton sales. This disease may lead to sterility in Hadza women and would be difficult to treat if it became prevalent among the nomads." Tanaka (1980, p. 90) writes "Gonorrhea is prevalent in Botswana but has not penetrated into the Kade area. Venereal syphilis was not found (Jenkins and Botha, personal communication), but there were many cases of non-venereal syphilis" (see also Nurse *et al.*, 1973). Such comments can be found in the early literature on other populations but it is easy to suspect that some reflect the "pristine hunter-gatherer" image more than any direct evidence. Bennett implies that sterility was not prevalent among nomadic eastern Hadza, and it still was not by 2000. Dyson (1977) does not report the number of women infertile in Woodburn's reproductive histories of 75 Hadza women. Tanaka reports on the central Kalahari San "women surviving until menopause have given birth to an average of four children." Tanaka's 1967–1972 study appears to give us the lowest TFR recorded for hunter-gatherers with evidence of no serious STD or primary infertility. The central Kalahari reserve, where they live, had no permanent water accessible to them, an average annual rainfall of 392 mm over a 11-year period, extreme temperatures (–10°C to >40°C), and may represent the harshest of habitable savanna. Tanaka's subjects would appear to be a better fit to Ellison's "metabolic load model" than to an STD interpetation. If the central Kalahari situation is a solution to the hunter-gatherer population paradox, it would have to represent a chronic situation, such as would result from significant climate change.

## 11.7 Evidence about population crashes

I ran my population simulation with crashes of different sizes (decimations of fixed percent of all ages) and different intervals. When I scheduled the crashes at regular intervals, the effects were not as large as reported by Keckler or Boone; it needed a larger increase in mortality, or a shorter interval between crashes, to generate a stationary population. When the crashes occur at random intervals (as in Keckler's model), you can sometimes have two or more crashes in quick succession, which has a more decisive effect on population. Many other explorations of Keckler's modeling enterprise could be imagined and might be productive. However, the overall conclusion, that substantial decimations occurring on average as rarely as once every 50 or 100 years could give long-term zero population growth, is easily supported (Figure 11.4). Apart from the genetic evidence for a recent bottleneck (Chapter 10), do we have any evidence that crashes such as Keckler modeled have happened to the Hadza, or any indications about the kind of factors that might produce a population crash? For example, the kind of factors that produce random crashes, including the occasional occurrence of closely repeated crashes, might be different from the causes of more regularly spaced crashes. Density dependent factors would seem more likely to generate the latter than the former.

### 11.7.1 Waiting 100 years for the next crash

The Hadza have been observed by anthropologists frequently, almost continuously from 1959–2010, a period of 50 years, and before that in the 1930s, around 1920, and in 1911. Thus far, we have seen no crash of any kind nor heard talk of any such disaster from the Hadza (other than the historical accounts of wars and raiders before 1911). While we have awaited a crash in vain, we might wonder if the special circumstances of the modern world prevented such an event.

Settlement attempts (with food) were not at times of special hardship nor in response to food shortage or disease, they were results of external political or

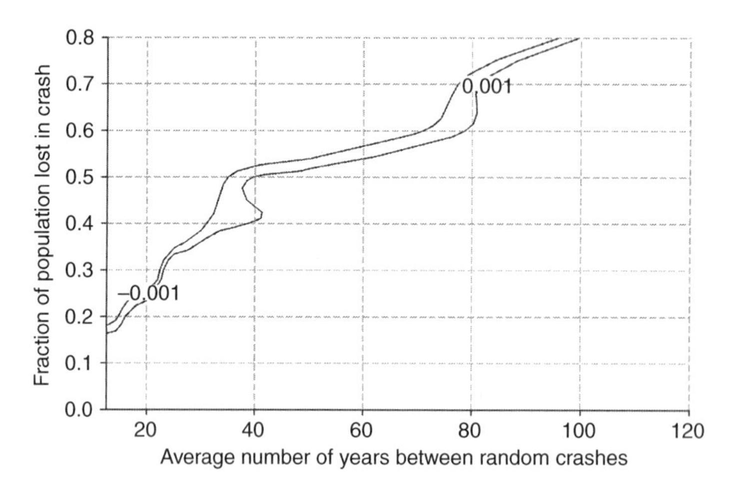


**Figure 11.4** Contour plot to illustrate relationship between frequency and size of crashes. X-axis is average interval between randomly timed crashes. Y-axis is proportion of population removed in each crash. Populations in top left decrease, those to the bottom right increase.

religious ideology. They appear to have had quite limited and inconsistent effects on fertility (Chapter 7). Three of them (Yaeda, 1964, "many children died"; Munguli "epidemic of sleeping sickness"; Yaeda, 1986, measles outbreak) may even have caused disease outbreaks. Disease outbreaks upon settling have been observed in several continents including South America (Hill and Hurtado, 1996; Garcia-Moro *et al*., 1997) and Australia. Another potentially important modern influence is Hadza knowledge of the legal system and the size and political power of their intruding neighbors. Without this knowledge, to judge by the extent of the intrusions, and the occasional noisy discussions among Hadza about the intrusions, there might well have been violence between the Hadza and the newcomers, with catastrophic consequences for the Hadza.

## 11.7.2 Famines

A general impression of the literature is that hunter-gatherers are the first to complain of hunger. "I haven't got anything and I'm starving to death" could be a generic greeting among foragers. But when asking about past serious food shortages (famines), every anthropologist is told there have been none. Famines happen to farmers. If a crop succeeds, it barely lasts until the next harvest. If a crop fails, what do they eat? African farmers' famines arise from failed rains, rainfall at the lower tail of the wide normal distribution of rainfall. If densities are high enough, farmers might soon run out of bush foods. The Isanzu came to the bush to survive a 1917 famine. There was apparently wild food enough for the Hadza and a lot more people. The Hadza population was well below "carrying capacity" (in either sense) at that time. During our extended period of observation, Hadza required no food intervention. They have once or twice received a few bags of maize (one or two bags reaching each Hadza region) when there have been nation-wide handouts.

Effects of historical famines on mortality have been measured (Dyson, 1991a,b; Kinoshita, 1998). Kinoshita reports from records from a farming village in pre-industrial Japan between 1760 and 1870. Death rate varied greatly from year to year, with no obvious secular trend. Kinoshita selected, as mortality crises, those years with 50% higher mortality than the moving 25-year average. The village experienced 14 crisis years by Kinoshita's criterion, out of the 110 years of records. Mortality crises were associated with a flood, harvest failures, and epidemics (smallpox, measles, influenza, cholera). In the crisis years, mortality was increased in all age groups (0–65 + years) but most of all among children under 10 years old with the five- to nine-year-olds suffering more than the two- to four-year-olds. (Village registration documents omit the infants who died and record newborns as aged 1 year; therefore, these ages equate to one- to three- and four- to eight-year-olds). What are the results if we try to match Kinoshita's figures in the Hadza population simulation? If we enter a rate of crashes, mean interval 7.86 years, (annual probability 0.125) and a size of crash as four times the observed mortality for under five-year-olds, and 1.999 multipied by mortality for others, we get a slowly declining population. If frequency of crashes is reduced to one every 10 years on average, the population is slowly increasing. These simulations include a reduction in fertility during the crash years (a phenomenon much debated in the historical demography of Japan). Dyson found a quite large effect on fertility that lagged the price of food by about a year in south Asian famines of the late nineteenth century (as I found Hadza fertility declined in the year following the failed rains of 1997). Hence, it is just possible that "farmer-sized" famines could have a serious impact on a population like the Hadza, if they were to occur one year in 10 or more frequently. Such events give very ragged age structure, with no stable percentage of people aged below 20, and unpredictable population histories but apparently quite normal age at death distributions, not a close match to the Hadza data. Much depends on how the dice fall; if the random process results in a string of closely spaced famines, the population can crash precipitously. We have no evidence of foragers suffering "farmer-sized" famines. Furthermore, Dyson shows that much of the mortality in the South Asian famines was related to disease epidemics, including outbreaks of malaria associated with the resumption of rain after a drought and a period of high food costs, as discussed in Chapter 2 for East Africa.

### 11.7.3 Prey crashes?

Large mammals such as those hunted by the Hadza suffer occasional population crashes of their own. Young (1994) summarized records of such crashes. His account was criticized on the grounds that a few of the records turned out to be duplicate reports of a single crash, and that his claims about the frequency distribution of the size of crashes suffered statistical errors (Erb and Boyce, 1999). Nonetheless, his central claim seems well accepted (Elkan *et al.*, 2009). Few if any of the observations included populations that were subject to human predation (but it is not clear how many of the crashes began with disease outbreaks among local domestic stock). If such crashes are a normal feature of the Hadza prey, it might have significant effects on the Hadza population or subsistence strategy; but crashes do not seem to synchronize among species. Thus, Prins and Weyerhaeuser (1987) found that, in Manyara Park, impala crashes were unrelated to buffalo crashes, and to some extent the biomass of species balanced out, overall biomass was more constant than biomass of a single species (Prins and Douglas-Hamilton, 1990). If the Hadza find no buffalo, there may be plenty of impala, or vice versa. Impala are much more often taken by Hadza and their population oscillates quite rapidly, often due to anthrax, from which the impala population recovers quickly. However, unsynchronized, random crashes could coincide in rare years, producing a drastic reduction in game animals available to the Hadza. However, I find it hard to believe that this would have much effect on the Hadza population. After grumbling a while, men would turn their attention to plant foods, honey, birds, and smaller game. If they did not, then since men aged over 18 acquire 43% and married men 50% of the calories arriving in Hadza camps (Marlowe, 2010, table 5.5), there could be a substantial effect on women's work, women's fertility, and small children's survival.

### 11.7.4 Plant food crashes

I have found little information on temporal variability in Hadza plant foods. Ecologists do study what they label "Large infrequent disturbances" in historical environmental data (Gillson, 2006 on the period 1883–1902 in East Africa). They define these as periods in which environmental variables fall beyond two standard deviations from usual annual variation. We observed one year in which baobab seemed very scarce. One would expect that chance strings of droughts would affect many plant species. It is also standard horticultural lore that plants produce more reproductive organs (fruit, seeds) when stressed, and in wet years accomplish relatively more vegetative growth. Wheelwright (1986) showed a negative correlation between vegetative growth and fruiting in his seven-year sample of 22 wild species.

Individual plant species might suffer disease epidemics. There are reports of a new disease of baobab trees in southern Africa (Patrut *et al.*, 2007). If this were to spread to Hadzaland and kill large numbers of these slow-growing trees, it would make Hadza life much more difficult for a very long time.

In times before agriculture, locusts may have specialized on wild plants of importance to humans. I have heard of no hunter-gatherer difficulties associated with locusts (which are quite edible). It can probably be assumed that plagues of locusts and *Quelea* are dependent on the agriculture to which they can be so devastating. The small seeds preferred by *Quelea* and other small birds are of no significance to the Hadza (or the !Kung) but may have been to peoples foraging in more arid environments, for example in central Australia (O'Connell and Hawkes, 1981).

### 11.7.5 Droughts

Rainfall records have been kept quite systematically at many locations in Tanzania for close to 100 years (Chapter 2). Annual variance in rainfall is very high indeed

(30% for an annual mean of around 500 mm, and greater at lower mean rainfall) and one might think that random processes would generate runs of severe drought years. Periods in which rainfall is below average for more than a year are not scarce, and are, of course, very dangerous to farmers. However, longer runs of extreme drought years seem quite unusual. Previously, I reported our two pieces of evidence about drought and the Hadza: the observed low fertility in 1998, a year after the severe drought of 1997, and the under-representation in the age structure of people born around 1949. The fertility of Hadza aged 15–44 in 1998 translates to a TFR of 3.99, similar to the central Kalahari San (Tanaka, 1980) at an average rainfall of 390 mm/yr. A string of such droughts might reduce the Hadza population to near zero population growth. However, we might also expect "under-burdened" women to quickly increase their birth rate as soon as normal rainfall resumes.

### 11.7.6 Climate changes

Long-term changes in climate are superimposed on the annual variation in already sparse rainfall. The extent of forest, savanna, and desert has varied through the history of modern *Homo sapiens* in Africa (Potts, 1998). These must have had major effects in many locations (contrast the desertification and depopulation of the Sahara and the long history of habitation in the Eyasi basin). As I discussed in Chapter 2, the extent and discontinuity of the habitat variation may have been exaggerated (Cerling *et al.*, 2011). The supposed challenge to hunting and gathering by modern *Homo sapiens* from the climate variation may also have been exaggerated. Given that *Homo sapiens* became a prime example of Pott's flexible organisms, the environmental changes that resulted from variation in temperature and rainfall may have offered little challenge. At the high rainfall extremes, montane forests may have expanded over the highlands at least sufficiently to separate the Hadza and the Sandawe, or during lower rainfall periods contracted sufficiently (as when Lake Victoria dried up 20 kya) to allow the Hadza and Sandawe to meet (Chapter 3). However, it seems unlikely (especially in the view of Cerling *et al.*, 2011) that the semi-arid and arid lands would have departed from the Zambesian phytochorion and its associated fauna. Mabulla (2007) indicates the presence of a Hadza-like suite of prey during much of the last 100 thousand years in the Eyasi basin. A more probable threat to Hadza-like populations would be extremes of aridity that went beyond those experienced by the !Kung in Ngamiland and approximated the levels in the contemporary central Kalahari. At such a time, the Hadza could simply follow their preferred habitat uphill, or cluster around the mountain-fed springs at Mangola. There are many locations in East Africa where vertical migration could have allowed a similar response.

Long-term climate changes imply a scenario with crashes spaced by thousands of years. Some of the climate changes are described as rapid, perhaps as short as 100 years. Although each such crash might be devastating (despite likely broadening of diet to include, for example, foods that the !Kung take but that the Hadza know about but pass by, such as grass corms), the population in the interim could have grown to enormous size (the Hadza rate of increase can take us from a few hundred people to 32 million in 1000 years, even !Kung-like rates of increase generate an unsupportably large population in 1000 years). Thus, the major climate changes cannot solve the "forager population paradox," they have been neither severe enough, nor long enough, nor frequent enough. The same argument applies to other very large-scale natural events such as volcanic eruptions. Suppose the Eyasi basin were covered in ash (some of it has been), its small surviving population would leave and try to settle elsewhere. Once the flora regenerated, and the fauna returned, a few people might move in, and within 1000 years they could have increased to 32 million. Major eruptions are insufficiently frequent to solve the forager population paradox.

## 11.8 Demographic consequences of raids and wars

Raiding and warfare is one possible source of population crashes, or some form of ragged, even density dependent population limit (Kelly, 2013, p. 208). Wiessner (2014) reports that !Kung homicide rates were substantially higher than reported by Lee. Hill *et al.* (2007) report relatively high young adult mortality among the Hiwi. Much of it was caused by violence, and these authors (like Kelly, 2000) suggest that homicide may once have been a larger contribution to mortality than among today's "pacified" populations. Were raids and wars large enough and frequent enough to resolve the forager population paradox?

Recent surveys and analyses (Kelly, 2000; Otterbein, 2004) are weighted toward populations outside Africa. The situation in Africa is less clear. Reading Burton (1859), Iliffe (1995), Tippoo Tib (Brode, 1907), or the Iramba oral history, which abounds in clan warfare, can leave one readily believing that warfare was abundant and important in the history of agricultural and pastoral societies in Africa. However, many historians would argue, as I did in Chapter 3, that this situation was relatively recent, and strongly tied to the coastal and overseas trade in slaves and ivory. We have much less information about warfare in the history and archaeology of African foragers. Hadza told Kohl-Larsen (but no other investigator) that they used to fight with men in other camps but that long ago they got together, had a feast, and decided that eating was better than fighting. To me this reads more like an effort to account for the men's epeme feast than a historical report. I am not arguing that the Hadza are devoid of aggression. The oral histories give ample indication of their readiness to retaliate against attacks. Even today they show concern over the possibility of homicidal conflict arising from internal disputes (Butovskaya, 2013).

The literature agrees that warfare and raiding are less common and less effective among foragers than among farmers (Wrangham *et al.*, 2006), and the archaeological evidence of warfare (such as fortifications, burned settlements, numbers of skeletons with injuries) is more evident among farmers than foragers (Flannery and Marcus, 2003). Nonetheless, war and raiding are by no means absent among foragers (Ember, 1978). Recent literature pays close attention to the conditions that differ between foragers among whom war or raiding are frequent and those among whom war and raiding are infrequent or absent (Kelly, 2000; Otterbein, 2004). Features such as population density, mobility, nature and abundance of resources, and social organization tend to be associated with each other. This makes it more difficult to pinpoint a crucial difference between the "warlike" and "peaceful." Kelly emphasizes the difference between segmented and non-segmented egalitarian societies. Otterbein notes the importance of fraternal interest groups that can cooperate in a warlike enterprise. This, in turn, may depend on patrilocal residence patterns. Some of us fieldworkers have been impressed by the frequent failure of Hadza and !Kung men to cooperate in groups, most especially in their relations to outsiders.

War and the characteristics of resources have received attention from others. The incidence of warfare differs not only between farmers and foragers, but also between the denser, more sedentary hunter-gatherer populations (with "alienable" resources) and the low-density, more mobile populations with less "alienable" resources (Knauft, 1991; Manson and Wrangham, 1991; Wrangham *et al.*, 2006). The nature of resources might best be approached as an issue of economic defensibility, as set out by Dyson-Hudson and Smith (1978). This principle is also applied to variation within farming groups, for instance in Vayda's (1969) interpretation of warfare among swidden horticulturalists. The possibility that warfare is a response to density, and has effects related to population size, density, or resource competition has also been debated (Durham, 1976; Kelly, 2013).

Marlowe (2010, p. 264 and fig. 10.5) discusses the relationship between hunter-gatherer camp sizes and net primary productivity of their environment. He points out that if warfare were important, we should expect people to live in large groups to better defend themselves. The opportunity to live in large groups should be greater in richer environments, yet he did not find such a signal in the relationship between group size and NPP.

We know that Hadza have been at the receiving end of attacks by the Maasai, and raids to take women and children. They attempted some retaliations. We know that Hadza used to respond to the arrival of strangers by running away and hiding. Camps can be extremely difficult to find, even if your Hadza guide has been told where they are by a recent visitor. We do not know how long such raids had been going on. We do not know whether forager neighbors had been involved in raids upon the Hadza.

The key issue for us, seeking to resolve the forager population paradox, is the size of the demographic impact of wars and raids. Raids and wars exert a variety of costs with demographic consequences. Fear of raids could restrain movement, limit home range (Kelly, 2005), and impede foraging, as when Kohl-Larsen's informant described Hadza as moving nearer to the Isanzu to avoid raids by the Maasai.

We could rank in order of ascending demographic influence: men killing men, killing of small children, killing or capturing women. A population that loses men loses only their food production, their genes can easily be replaced by those of other men. In societies where men produce most of the food, their loss can be a serious influence, lasting until boys grow up to become effective at subsistence. The loss of men's labor due to slaving and new diseases arriving with trade caravans was probably very important to rural Tanzanian farming societies throughout the nineteenth century and during the early colonial period (Iliffe, 1979). If weaned children are killed, given the common sensitivity of populations of long-lived mammals to infant and child mortality (Stearns, 1992), the demographic effects will be severe and direct, if slightly delayed. If women are taken away, their reproductive production is lost to the donor population, but at least partly gained by the winners, with little net effect on regional population. This consequence of fitness maximizing female victims has been interestingly approached by Sugiyama (2014).

I looked at some of the approximate effects by adjusting my population simulation. This is a simple simulation that makes no allowance for effects of helpers (unrealistic but typical of population simulations), and assumes no scarcity of willing males (realistic in the Hadza case and many others). The baseline condition gave a rate of increase, r=0.0156 and this and other measures (for example, TFR of 6.16) show it is a good replica of the Hadza population.

Frequent raids in which children are killed or stolen could suppress population growth in the raided population. If mortality of children below age seven years is doubled, the population declines rapidly (r=–0.0035). If child mortality is multiplied by 1.75 each year, the population is almost stationary (r=0.001) (Table 11.1, line c). Hill and Hurtado's (1996) point is demonstrated, a massive increase in mortality of the most important age group is needed to counter an average human fertility. If we look at crashes or wars that occur, once every 10 years for example, only a much higher child mortality rate will produce much effect. When mortality of under-seven-year-olds is multiplied by 10 ($q_x = q_x \times 10$) in the crash years, r is lowered to.003 (Table 11.1, line h). Traces of the crashes can be seen in the later age structure. The percentage aged under 20 varies strikingly from year to year. At this level, all the newborns die during crashes. The relative ineffectiveness of crashes in child survival seems to have to do with the brevity of the crashes (each crash lasts only one year). When each crash lasts for two years, multiplying child mortality by 7.5 reduces r to 0.001. The winners gain nothing in this scenario, unless they steal women, including the mothers of these children. If children are taken away, they are presumably costly to the host population until the girls become reproductively mature and the boys become able to work.

If 10% of women aged 15–45 are removed each year and they and their reproduction are lost to the population, the population decreases very rapidly (r=–0.035). If a raid capturing 10% of the women only occurs every 10 years on average, then the population continues to increase, at r=0.010. If the raids take 25% of the women, the population declines rapidly with r=–0.07 (Table 11.1, line i).

Thus, raids have to be both frequent, and quite destructive to cut long-term population growth to zero. Indirect costs, such as reluctance to use all the terrain and resources available, for fear of being captured, may be greater. The profitability of raids to the raiders is quite a different calculation. It has to balance the reproduction gained by the men who make the raids, against their loss of usable land if their enemies retaliate. The net demographic effect of raiding to capture women, when the two warring populations are summed, may be a slight decrease in growth rate commensurate with the extent to which the transferred women, as slaves or second wives in their new home, fail to reproduce at their expected capacity. Such raids, between two warring forager populations, in so far as they lead only to relocation of women, do not help us solve the forager population paradox. (SI table 11.2 sets out likely demographic and fitness consequences of different targets of raids.)

Raids that combine capturing women and killing or capturing children have a larger effect on the losing population. With raids every 10 years, when women are captured and children under seven years are killed, a take of 22% every 10 years produces a stationary victim population (Table 11.1, line j). Suppose raids are on a small scale, one raiding party succeeds in finding and raiding one camp and takes all the women and all the young children. If one camp out of every five were raided, the roughly 20% capture rate is matched. This happens on average every 10 years. This translates to any single camp getting raided (with all the women taken) once in 50 years. It is conceivable that the raids described by Obst's and Bagshawe's informants reached or exceeded this rate. These raids, however, were conducted by more numerous and organized farmer neighbors and are unlikely to have great antiquity. Raiding between forager populations, with likely retaliation, is likely to have had weaker demographic effects. Its value for resolving the forager population paradox is unproven.

Raids in which women of one population are captured and taken away to another population, or raped and left *in situ* by men raiding from another population, would produce considerable gene flow between the warring groups. If population mixture of this kind had been frequent and extensive enough to have significant demographic effects, it might have slowed the development of genetic differences between nearby groups, and would have obscured genetic differences between previously distinct populations. Yet genetic historians find many distinguishable populations with long histories of very low rates of interbreeding.

## 11.9 Summary and conclusions

The limits on growth of hunter-gatherer populations remain unclear. Population simulation of the effect of raids suggests that the Isanzu raids that took women and children may have had a larger demographic effect than the "wars" with the Maasai. Raiding could solve the "forager population paradox" only if quite severe, vigorously reciprocated, and women and children killed not captured.

The idea of population increase as a prime mover of history fell into disfavor when "intensification" was found to be not an inevitable consequence of population growth. It seems likely that the disfavor was premature. We have little evidence that human populations stabilize in a density dependent manner. Instead, they disperse and migrate if there is anywhere to go, and conquer and expand if the surrounding country is occupied. Sometimes they broaden their diet, or intensify their subsistence methods. If we link the tendency for populations to increase, with the tendencies of males to seek ways to stand out among their peers, we may have still more ammunition with which to account for history. The tendency of human populations to increase seems to weaken the need for special stimuli to explain expansion of *Homo sapiens* out of Africa. The greater puzzle is why it did not happen earlier. The rapidity of the expansion, once out of a continent of hunters among hunters, if it implies faster population growth than within Africa, would be evidence for some kind of density restriction on rate of increase of the ancestral African forager populations.

Consequences of the "sawtooth idea" for life history evolution have been followed up by Boone and Kessler (1999) and Boone (2002) who suggest that such a history might select for strategies that buffer individuals or their progeny from succumbing to the next crash. On the other hand, we might expect hunter-gatherers, having passed through many periods of rapid increase, to have been selected for fast and early reproduction, almost the opposite of Boone's suggestion. If populations alternate periods of rapid increase with periods of sharp decline, would our expectations about life histories be different? For example, it has been suggested that selection favors earlier reproduction in increasing populations, and later reproduction in declining populations. Much life history theory in biology (heavily borrowed by anthropologists in the last 20 years) has been based on the assumption of long-term stationary population. This is almost a necessary assumption, for in the long term, no surviving organism can have grown at commonly observed rates for long, nor declined severely enough or for long enough to become extinct. There are many studies of animal populations that fluctuate but endure, and on whom life history theory seems to make good predictions.

In discussing the bizarre age-at-death distribution of the Libben sample, Howell (1982) argues that the population the sample had appeared to represent would suffer from a distorted dependent:producer ratio. There would be too few adults to support the implied number of children, many of whom would be orphans. Howell's subsequent work on !Kung support systems (Howell, 2010) makes her suggestion about Libben even more persuasive. Population models, such as the one I used to replicate Keckler, and models that use Leslie matrix methods, generally omit direct attention to parental care and effects of helpers. Their net effects are implicit in the child mortality measures, but the models do not allow, for instance, for increased deaths of adults to increase child mortality. If there are such effects, and few would doubt that they exist, the effects of population crashes may be larger and a little longer lasting than thus far proposed. I briefly looked at this issue with a population model that keeps track of individuals and allows for effects of helpers. The effects seemed to be small. Simulated populations that include helper effects took longer to recover from a single crash, and it appeared that long-term zero population growth could arise from crashes that were slightly less severe and slightly less frequent than in the model with no helpers. Recovery from a raid that took women of reproductive age, modeled to increase mortality of the orphans, was slower than recovery from an epidemic that killed children and old people.

# Part II

# Applying the demographic data to interpreting Hadza behavior and biology

# 12 Introduction

In Part I, I reported Hadza population parameters, the central tendencies of fertility, mortality, age structure, and population growth. My focus was on doing this to a knowable level of accuracy, and independently from model populations such as those of Coale and Demeny (1983) and Weiss (1973). I wanted to allow the Hadza to be as different as they might be from other known populations, to be able to assess the degree of accuracy of my estimates, and to look for evidence about change during the second half of the twentieth century. I answered my main question: have we got the main parameters right? With "Yes. Measures collected independently of each other, yet linked in a stable population, fit well enough with predictions derived from our estimates of fertility and mortality that we are unlikely to be wildly wrong." Details of Hadza demography support Marlowe's (2010, chapter 10) characterization of them as "the median foragers," and their demography is close to that of many historical and third-world populations. A normal human demography can be supported by hunting and gathering in sub-Saharan savanna. I finished by discussing the puzzles of hunter-gatherer population dynamics. Simulations suggested that, in a population where helpers were important, the effects of sporadic population crashes could be magnified.

In the second part of this book, I leave the anthropological version of classical demography and try to use the variation in our data, looking at individual differences in the whole population, to address issues our team has long pondered. Issues such as: Are grandmothers effective helpers? How big are the effects of helpers? Do inter-birth intervals influence child mortality? Does the reproductive system successfully manage the allocation of resources between fertility and childcare? Is big game hunting a paternal investment, the way men help their children survive, or a display, a way to gather good neighbors, or what? What are the origins of marriage?

This takes me into another group of disciplines, evolutionary ecology, behavioral ecology, and sociobiology, which have successfully been applied to studies of people in a variety of contexts. Excellent introductory accounts of behavioral ecology in biology are by Davies *et al.* (2012) and Krebs and Davies in *Behavioural Ecology: An Evolutionary Approach* (1978, and subsequent editions), which first included a chapter on humans in its 1991 edition (Borgerhoff Mulder, 1991). Landmark volumes devoted entirely to behavioral ecology studies of humans were by Winterhalder and Smith (1981) and Smith and Winterhalder (1992). Newer brief introductions include those by Smith *et al.* (2001). I could have called this part of the book "Sociobiology," which would be the shortest label, but it can be a very controversial one. Roughly, the label stands for the view that the social behavior of any animal is a product of evolution and can be studied as such. This has been a long-standing and mainstream view in biology. The idea was expressed by Darwin; in animal behavior, a large step toward studying behavior as a product of evolution was taken by Konrad Lorenz's (1941) use of behavior in taxonomy. Founding figures were David Lack, E.O. Wilson, Robert MacArthur, Gordon Orians, E.E. Pianka, George Williams, and W.D. Hamilton.

Among biologists, the evolutionary ecology, behavioral ecology, or sociobiology approach is seen as emphasizing one of the four levels of explanation offered by biology. In principle, we can explain behavior as a result of its immediate causes (including the physiological processes underlying it); or by the processes by which the observed responses developed in the individual. Behavioral ecologists study the next kind of explanation: the consequences of behavior that natural selection can act upon, and thereby eventually alter the frequencies of genes that promote this behavior or channel its development. A fourth layer of explanation concerns evolutionary history and the phylogenetic constraints on the solutions to adaptive problems that confront the animal. Earlier steps in evolution may have excluded some, promoted others. Most mammals cannot fly, birds and many insects can. This fact influences the ways they can solve problems of predation, searching for food, and so on. This is a crude summary of what are often known as Tinbergen's (1963) "four why questions." Much of the time it pays to think of them separately. However, in principle they are linked and may sometimes be used to guide each other. For instance, it may be easier to find out how a mechanism works if you know what it is built to achieve. Furthermore, it may be pointless trying to add to one's model of selection, a distinction that physiologists can tell us the animal has no equipment to make.

The evolutionary ecology approach began to be applied to humans in the 1970s. Leading examples include work by Daly and Wilson (1978), Alexander (1979), Chagnon and Irons (1979), and their contributing authors. The more polemical and high profile suggestion by E.O Wilson (1975) that biology was about to invade the social sciences initially attracted intense hostility. After all, its central assumption, that people tend to perform adaptively in the biologist's sense, was at first glance completely contrary to most of what we believed about ourselves. Its saving grace, for scientists, was its ability to generate, from basic principles of natural selection, predictions that could be tested. While economists (almost alone among social scientists) had long been making predictions from theory and testing them, the ultimate base of their theory was not clear. Nonetheless, economists share with evolutionary ecologists a working assumption that access to resources is limited, and that allocation of the resources between different outcomes is a key problem for the organism. In other social sciences, "theory" was a label for relatively rootless and cryptic ideas, most of them entirely post hoc, and generalizations or summaries more than theories. Fortunately, this did not prevent the growth of empirical social science, which has accumulated substantial data on phenomena like marriage.

Striking, some might say "strident," claims such as Wilson's made life very difficult for those of us who were in the field at the beginning. Somehow the field survived, and somehow the sound and fury has subsided (a few exceptions persist, such as those cited by Daly and Wilson, 2008). The growing acceptance of sociobiology may owe something to the success of molecular genetics. Some of the doubts about sociobiology arise from data, but the more forceful doubts arise from the "naturalistic fallacy" and the history of disastrous uses of ideas about human biology in politics. The dangers of the naturalistic fallacy should not be forgotten. Just because something is "natural," or has been favored by natural selection, does not mean we have to choose to like it. Opposition to sociobiology has diminished so much that younger academics in this field seem, to us who were there at the beginning and struggled to maintain an academic career, dangerously innocent in their total faith in their paradigm. To some extent their innocence may arise from the solidity of their conviction that knowledge about adaptation is not a source for personal values.

## 12.1 Predictions and tests in behavioral ecology

The success of this field is measured by its ability to derive testable predictions, which is done by applying knowledge of local conditions to simple first principles of evolutionary theory. The success of the field can be further assessed by the number of predictions that come true (a surprisingly large number), and by the way in which research is pursued when predictions fail (a test of any regular science). The response to failure consists of: (a) re-examining the way in which the predictions were derived from theory, (b) trying to see what other costs and benefits were ignored in the initial study, and (c) most important, seeking ways to test the added ideas, and actually conducting the tests. The new ideas ideally allow new predictions beyond the failed predictions they were created to encompass. (Lakatos, 1970, and Lakatos and Zahar, 1975, p. 369, are among philosophers who have argued for this view of science.)

We expect individuals to adopt behavior that increases their fitness. There is an underlying assumption in human behavioral ecology that people are very flexible optimizers. Somehow, we manage to perform in ways that maximize fitness under a very wide array of circumstances. The circumstances, by their consequences for fitness, can help us account for the variety in behavior. The local conditions set the costs and benefits that result from different courses of action. The organism "chooses" a course of action that "optimizes" the costs and benefits, finding the outcome that maximizes fitness. The costs and benefits are ultimately measured in evolutionary fitness, or measures that have a good chance to be proxies for fitness. Combinations of behavior and compromises between outcomes are expected, and they are presumed to lead to "optimal" behavior. "Optimal" is used as shorthand for "fitness maximizing," it carries no implication that we (or even the subject?) should like the behavior. "Optimal" might be better thought of as "the best of a bad job." We expect individuals to adopt behavior that increases their fitness and this expectation guides our development of hypotheses. There are exceptions, of course, and two other branches of the evolutionary study of human behavior may be better equipped to deal with such instances, Darwinian psychology and cultural transmission theory (SI 12.1). The exceptions may indicate weaknesses in our understanding of the selection pressures acting on individuals, probably most commonly by underestimating conflicts of interest between individuals or groups of individuals.

Testing the predictions helps us choose between opposing hypotheses. For example, if we follow the traditional line of thought about male provisioning and pair formation, we might expect an effect of the father's presence on child growth and survival, and perhaps a tendency for men to avoid caring for another man's children. If we follow some of the other ideas about men's reproductive strategies, we might look at readiness to marry younger women, and expect this to be unrelated to her burden of small children. These expectations can be tested with our demographic data. With these data, it may not be possible to test every elaboration of either view; a different kind of data, or just more data, may be required. For example, males might less often desert the family when the cost to their offspring would be severe. We might find we had too few data to discriminate according to the vulnerability of the offspring.

With animals or plants, we can test hypotheses by experimentally manipulating circumstances. An example would be testing the hypothesis that male birds gain fitness by feeding nestlings. The experimenter removes males and records the growth or survival of their offspring and the work rate of their mates and of newly arrived males. When studying people, or animals that cannot be manipulated, we measure whether individuals who show one form of behavior have higher fitness (or score higher in some proxy measure) than individuals who show some other kind of behavior. The comparison is often between the most common or average form of behavior and less common variants. Problems follow from the assumption that these other forms of behavior are "mistakes." There may be some unobserved difference in individual capacities or opportunities, such that individuals are actually employing behavior that maximizes their fitness, even if it would not maximize the fitness of the majority of individuals. Thus, any population may comprise individuals following a variety of strategies, and much attention is directed toward understanding how this may come about and how proposals about individual variation can be tested. In addition, the optimal behavior may sometimes be determined by what others are doing, and how many of them are doing it. Such situations are analyzed using theoretical games. I will not be pursuing any such cases in any depth, although perhaps I should. I will pay some attention to individual differences in strategy, especially to the idea that individual differences in access to resources, or differences in individual frailty (easier to think of as its inverse: vigor), may account for some of the observations. In particular, such differences could obscure expected trade-offs between, for example, number and survival or size of offspring. It is also important to remember that not all individuals in a population have the same interests; their fitness may not benefit equally from the same outcomes. Then one individual, one sex, one set of individuals, one alliance, may wholly or partly prevail, and the processes by which this outcome is shaped may be important for understanding what we see.

In general in evolutionary ecology, we look for variations that predict reproductive success (RS) or some of its obvious contributors, such as fertility or offspring survival. Thus, we ask of the data whether children who have a living grandmother live longer than children who have none. We ask whether longer intervals between births go with better child survival, and if so, whether there is a compromise between child survivorship and maternal fertility, between numbers and survival, and whether there is a length of interval between births that maximizes the rate at which a woman produces live descendants. We ask whether men who are well known as successful hunters have more children than other men, and if so, how this comes about, and especially whether they keep more of their children alive. We ask how various features of marriage, divorce, and remarriage relate to RS, fertility, and offspring survival. I look at some aspects of height and weight as predictors of adult reproduction or as indicators of children's prospective fitness. Along the way, I describe some general aspects of Hadza patterns of marriage, reputations, and growth.

## 12.2 Hunter-gatherers

In Part II, I am building on the field in general, but specifically on earlier work on hunter-gatherers by Eric Smith (including the classic by Dyson-Hudson and Smith, 1978), by Bruce Winterhalder (Winterhalder and Smith, 1981), and by Kristen Hawkes, and her students (1982) Kim Hill, Magdi Hurtado, Hillard Kaplan, and their students; and by Strassmann and Gillespie (2002) and Strassmann (2011) on farmers and Borgerhoff Mulder (1988, 1992, 2009) on herders and farmers. I am, of course, paying most attention to previous ideas and research on the Hadza discussed by Hawkes, O'Connell, me, Frank Marlowe (2010), and Frank's students. I use these previous works as guides to where to look, what factors I should or can test for their effects on survival, fertility, and fitness. As I said in the introduction, papers by James Woodburn are a constant source of insight, and sometimes support for our independently derived ideas, but mostly they are a confirmation of the continuity of what we have seen. The Harvard Kalahari Project was a model for all of us, although much less narrowly driven by evolutionary theory than by the authors just mentioned. The books by Richard Lee (1979) and Nancy Howell (1979) are known to every anthropologist, but a wealth of other leading projects grew from Irv DeVore's drive, imagination, and inspiration. For example, Lee and DeVore (1976) features early papers by graduate and postdoctoral students who became major figures in anthropology, including Yellen, Harpending, Draper, and Konner. Mel Konner's (1972) work on !Kung infancy was especially influential on me, as was the chance to join him and Marjorie Shostak in the field for a brief visit in 1971. From those early beginnings, Konner's work has culminated in his recent, monumental, multidisciplinary work, *The Evolution of Childhood* (Konner, 2010).

The assumption that hunter-gatherers are closest to the circumstances that contain the selection pressures that evolved our species should be examined. Hunter-gatherers have lived in every continent and in many ecologies from tropical savanna, tropical forest, to boreal forests, arctic tundra, and coastal habitats worldwide. Each may have exerted its own blend of selection pressures, and generated population differences in traditions and in gene frequencies. Each tells us about the ways people with a hunter-gatherer economy respond to changes in circumstances. The environment that has exerted its influence over the greatest time depth is sub-Saharan African savanna. There are extensive data on how changeable this environment has been (much of it summarized by Potts, 1998, Spinage, 2012, and discussed in Chapter 2). Selection pressures may be shaped by the environment but they are not the same as the environment. Aspects of selection are also shaped by demography, anatomy, and life history and its rules of variation. Moreover, understanding the selection pressures on modern *Homo sapiens* in sub-Saharan savanna is difficult enough. Trying to understand selection much further back in time may be much too ambitious, although the more we understand about the opportunities and obstacles in savanna environments, the better are our chances for success.

Rainfall and temperature vary. Plants, animals, and access to water vary. How these change selection pressures, intensity, and direction is less clear, and needs ethnographic knowledge. Nuts of a different species are still nuts – they need to be opened (with greater or lesser difficulty), tend to be abundant, and are edible for a long season, but they fail in the occasional year. Tubers are still tubers, but some are shallow and some are deep, some grow in flats and pans, some on hillsides among rocks, some are poisonous, most are not. The ubiquitous baobab may be numerous or sparse. Some species of *Grewia* or other fruit bush will be everywhere, but the lengths of their seasons differ. Food and water may be far apart, or close together. Animals may be easily stalked (if there are trees, bushes, or undulations) or impossible (in the rare open plains), or hunted at night near water or trails. The amount of groundcover may affect the ease of tracking. Some important factors may be due to geology, and within broad limits, little affected by rainfall and temperature (but the broad limits include loss of water and total desertification). Thus, in Blurton Jones *et al.* (1994, 1996) we discussed the inevitable distance between food and dry season water in much of !Kung habitat, and contrasted this with the many small water holes all over Hadza country.

Our questions are intimately tied up with ideas about human origins. In fact, some of my colleagues, much more than I, have been interested in what we can learn from the Hadza that can improve our interpretation of evidence from the past (O'Connell *et al.*, 1988a,b, 1990, 1991, 1992, and see Hawkes *et al.*, 1998, and Monahan, 1998). However, all the topics we have investigated, even if primarily aimed at understanding contemporary behavior, have some relevance to the remote past.

## 12.3 Sensitivity, selection, and the life history

Because evolutionary ecology draws its predictions from evolutionary theory, it is intimately tied to demography. Having completed the demographic study, we are in a relatively strong position. Fitness is the rate of spread of a gene that tends to develop some phenotypic character. The rate of spread has been equated to the rate of growth of a population of bearers of the original mutation. The rate of growth of any such population depends on which part of the life history the character affects. The classic example is that a character that improves survival of very old individuals, unlikely to

**Photograph 12.1** Population growth is most responsive to survival of children aged zero to five years. © James F. O'Connell, 2015. Reproduced with permission.

reproduce again, would fail to spread. Using a simple population simulation, I computed the increases in population growth rate resulting from percentage increases in fertility or decreases in mortality at different parts of the life history.

The rate of population increase responds more to a percentage change in child survival (age group one to five years) than to a percentage change in old people's survival. This helps us focus our attention on the possible proxy measures and on the most rewarding targets for helpers (Photograph 12.1). We can give special attention to child survival, with some confidence that differences will be reflected in differences in fitness (rate of increase of descendants, r). The very low sensitivity of population increase to changed survival of women near the end of their reproductive career focuses our attention on the special issues surrounding the evolution of post-reproductive life. In general, according to Stearns (1992), among large long-lived mammals, r is most sensitive to changes in early survival, the age at which reproduction begins, and the fertility of young females (Photograph 12.2). Stearns' and Crow's previous generalizations were recently explored in a sample of 142 mammal species by Oli and Dobson (2003). James Jones (2009) confirmed Stearns' generalization for several human populations. I add a crude approximation of the female Hadza example in Figure 12.1. Children under five years old are clearly rewarding targets for any helper. It is more difficult to test the sensitivity of changes in male life

**Photograph 12.2** Population growth is next most responsive to fertility of young women. Here is a baby born when her mother was a little below the median age at first birth. Age at first birth has a stronger influence on the eventual reproductive success of women than of men. 

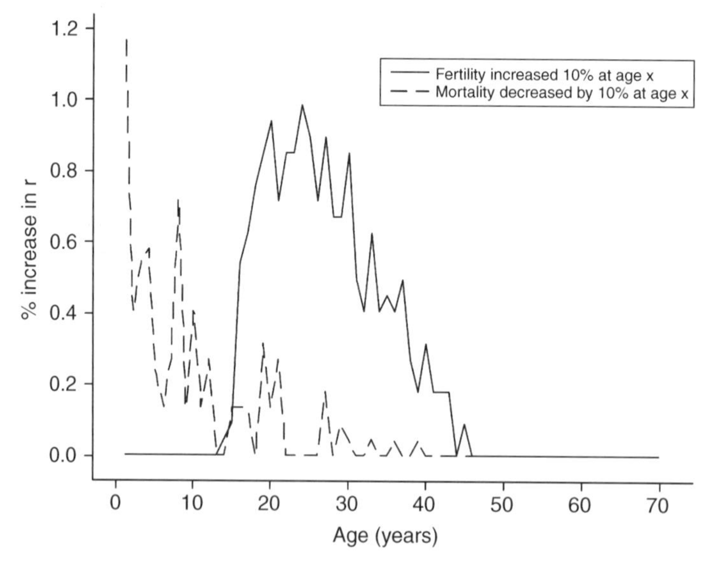


**Figure 12.1** Response to selection on different parts of the life history. Percentage increase in the rate of increase of a Hadza population that would be produced by a 10% decrease in mortality (dashed line) suffered by the people in a single year of age. Continuous line marks response to 10% increases in fertility. The program uses observed Hadza age-specific mortality and fertility reported in Chapters 7 and 8. Baseline rate of increase (r) is calculated using Stearns (1992, p. 38, appendix A). The program then increases fertility by 10% at a single year of age and recalculates r. Repetitions continue until the effect of the changes have been explored at each age. This process is repeated for mortality. Mortality is lowered 10% at each age x years by subtracting ($q_x \times 0.1$) from the observed age-specific death rates ($q_x$). The program stores and reports the rate of increase that results from each change and compares it with the baseline. (Data values in SI 12.3.)

history. In a population simulation that uses observed Hadza male mortality and fertility (SI 12.3), survival of the under-five-year-olds comes out as the strongest influence on population growth, followed by the later fertility of males. The fertility of 40–50-year-old men appears to be an important feature. Nonetheless, I doubt the meaning of such a model of "parthenogenic" males. Male fertility is so dependent on success in competition for access to females that a single sex model for males seems likely to be misleading. A single sex model for females seems much more meaningful; it assumes only that there is no shortage of males willing to mate.

## 12.4 Juvenile dependence

The questions addressed in Part II all cluster around the consequences of the economic dependence of children, and the overlapping, accumulating family that mothers and others have to care for. Who reaps the rich rewards of affecting early child survival, and how do they do this? The route to answers is lined with alternative hypotheses and pitfalls, as will become evident.

Age at first reproduction is considerably later in people than among our closest genetic relatives (chimpanzees); as a result, the length of the juvenile period (weaning to first reproduction) is longer. Hawkes *et al.* (1998) linked this to low adult mortality that resulted from selection to elongate the female lifespan because of the fitness rewards from surviving to be a helpful grandmother. The length of the human juvenile period fits quite closely on plots of mammalian and primate adult lifespan, size, and age at first reproduction (Hawkes, 2006). Thus, there is nothing special about its length. Needless to say, not everyone agrees; some point out, for example, that we have yet to show any demographic effect of hard-working Hadza grandmothers (examined in Chapter 18). However, two things do seem to be different from the juvenile life of our closest primate relatives: early weaning and juvenile dependence.

The age at which human offspring are weaned seems to be lower than among our closest relatives, although one recent paper argues that this age is no younger than we should expect from a carnivore of our size (Psouni *et al.*, 2012). There follows a vulnerable period, not clearly evident among other Great Apes, which become self-sufficient soon after weaning, when children are highly dependent on adults for food. The sensitivity analysis suggests children of this age should be a specially rewarding target for related helpers. Psouni *et al.*'s result on carnivores may arise because many carnivores have non-parent helpers, perhaps responsible for a parallel evolution of earlier weaning.

Attempts to explain the dependence of human juveniles can be roughly classified into two kinds: ecological or trends related to increasing brain size. The ecological scenarios attend to expected differences in the ecology of foraging in savanna and forest (Hawkes *et al.*, 1997; Kaplan, 1997; and Chapter 2). Foods available in the savanna are plentiful and densely nutritious, but much more difficult to access than food taken by our closest primate relatives in tropical forest, where fruit is available around the year. Some foods, like tubers, baobab seeds, and nuts need tools, knowledge, and strength to acquire or process. They are therefore difficult for newly weaned juveniles to acquire, hence the economic dependence of weaned children. Blurton Jones *et al.*, (1989, 1997), Hawkes *et al.* (1995a), Crittenden (2009), and Crittenden *et al.* (2013) have tried to provide some of the details. In Blurton Jones *et al.* (1994), we document the economics of foraging by !Kung children during their long dry season. A woman's children get more food if the children do not attempt to follow her on the inevitably long walk between dry season camp near water and nut groves on dunes far away. If the older children stay home they can be more help cracking nuts than by attempting the strenuous gathering trips. In the much richer Hadza environment, children offer a contrast and perhaps help illustrate the argument even better. Children aged five years and older can get some food for themselves. Short-seasoned soft fruit can be picked by all of them, as can the longer-seasoned unappetizing-looking *Grewia* fruit. Baobab fruit can be processed by small children using their unsophisticated wet process. Some tubers can be dug but the returns are low for the smallest children. Older children can dig shallow or deep tubers and process baobab more rapidly but still cannot support themselves completely, especially outside the brief soft fruit seasons.

Whatever the cause of the lengthy dependence of human juveniles, it involves periods of total or partial dependence on adults for food. This means that the number of dependents that a woman tends will increase with time. In contrast to almost all other mammals, who suckle a baby or litter until weaning, whereupon the offspring become self-sufficient, competent food gatherers, a human family builds up in layers. Unlike other primates, where mothers feed one at a time, a human mother is feeding the latest baby, the last surviving weanling, a child or two, and later on a teenager or two. Children are competing for mother's resources, and the resources of other helpers, in a way other juvenile mammals are not. The mother's allocation of resources may take a very different form to that of other mammals. As Hrdy (2009) has documented at length, this both sets a social task for infants, and opens an opportunity for helpers, whether father, grandmother, older siblings, or other adults and children.

## 12.5 Helpers and "cooperative breeding"

There is now much reference to the idea of humans as "cooperative breeders." Despite having seen too many labels masquerade as explanations in the behavioral sciences, I welcome the attention this brings to some important features of human behavior. Humans are contrasted with other primates in the number of individuals who direct time and resources toward infants and juveniles, and by the number of years of childhood during which this care is offered. While most young primates acquire their own food soon after weaning, children receive many years of provisioning, guarding, and direction. However, the Hadza, and probably other human populations, differ from classical cooperative breeders in the ease and low cost with which individuals can change groups.

Four chapters are aimed specifically at potential helpers; grandmothers in Chapters 18 and 19; older children, siblings, aunts, and uncles in Chapter 20;

and in Chapter 21, fathers. I look for their effects on children's growth and survival and on women's fertility.

Not least among the pitfalls in these studies is an aspect of studying helpers (whether husbands or grandmothers, or others) to which Hill and Hurtado (1996) drew attention. Helpers can cancel out their own effects by directing help where it is needed (if, we must predict, this coincides with directing help where it is most effective for the helper's own fitness).

## 12.6 Men and their reproductive strategies

It is widely believed that the long dependency of human children increased the payoff for provisioning by males. Savanna habitats were thought to have given increased opportunities for scavenging or hunting, further promoting this tendency. Only when O'Connell *et al.* (1988a) measured Hadza scavenging did we realize two important limitations: if you cannot get to the kill rapidly, and if you cannot chase away the other large predators and scavengers (lions and hyenas), you get very little food by scavenging. For many years, the dominant view in human evolution has been that by hunting or scavenging, men were able to provide food for their wives and children and thus made life in the savanna possible. Men's "paternal investment" was held to have become important very early in our evolution and to be the ultimate reason for the evolution of pair formation, and several consequent features of our species.

A wider range of explanations for the evolution of monogamy is available in the animal literature but these have received less attention from human behavioral ecologists. Research on birds (and almost as much on insects) has led the way in showing us the possibilities. Experiments showed male care was essential in a few species but not in others, equally pair bonded. Females seemed able to cope without a male if he were removed. Ideas like mate guarding, male coercion (Smuts and Smuts, 1993; Muller and Wrangham, 2009), resource brokering, and helpful coercion (Gowaty, 1996) became current. Studies of primate mating systems showed a minority of monogamous species. Both male coercion and infanticide by males were observed. However, so were "consortships" in some of the group-living primates with great differences in power between males.

Early in our project, Hawkes began to think of Hadza big game hunting as something different from paternal care. Her 1991 experiment seemed to show that trapping and shooting smaller game produced a daily supply of meat adequate for a family. Hunting big game, even when the meat was shared as widely as it is among the Hadza, left many stretches of meatless days. This seemed like a less effective way to provide for one's children than a near daily meal from small game. Hawkes suggested other outcomes of hunting large game. The immense but rare supply of meat was obvious to everyone. Perhaps all would do better to keep the successful hunters in camp. Concessions to good hunters were thus not necessarily a matter of reciprocation but simply of self-interest. Eventually, the idea became crystallized as "costly signaling" (Smith and Bliege Bird, 2000). Big game hunting might be an accurate and honest (hard to fake) sign of a man's physical and mental qualities. It could signal "good genes." Some feel the idea has become too influential and the pendulum has swung too far from paternal investment and the idea of human reproduction as a cooperation with division of labor (Gurven and Hill, 2009; Wood and Marlowe, 2013, 2014). Nonetheless, it is worth keeping in mind that hunting may benefit men, and women, in more ways than originally supposed (Bliege Bird and Bird, 2008). The benefits may include relationships with other men, and a reputation that makes a man a good neighbor. Both ideas suggested data to collect and analyses that I should conduct on the available data.

Men may have multiple strategies (Gangestad and Simpson, 2000). In some populations, men may specialize more in one route to fitness than in another population. Within any population, some individual men may specialize in one strategy, others in another. I make a case for such variation among Hadza men in Chapters 15 and 21. This view seems to be becoming more acceptable to those who work in this field (Winking and Gurven, 2011; von Rueden *et al.*, 2014). The recent papers display the complexity of opportunities involved in reproductive strategies, and should encourage us to make the most of the behavioral ecology paradigm in its ability to predict variation in behavior with ecological and demographic circumstances. Any piece of the complex web of findings from one population may be different in another population. Additionally, I get the impression from some of the debate that we all find it very difficult to visualize each other's field situation and the key parts of it that may shape the behavior of the people who we study. If one of our aims is understanding the evolutionary history and the selection pressures that favored behavior such as marriage, we will do better to understand causes of its variation and see how these might map onto what is known about ancestral anatomy and environment.

Men may gain fitness by fathering more children or by keeping more of them alive. Many authors have pointed out that some males have more extensive opportunity than females to increase fitness by increasing the number of conceptions (and this is likely to increase the variance in male RS). The origins of this situation have been discussed recently by Kokko and Jennions (2008); Kokko *et al.* (2012), Liker *et al.* (2014), and Szekely *et al.* (2014), who find inadequacies in Trivers' (1972) formulation tracing the difference to size of gametes, and find more promise in attending to adult sex ratios. For our purposes here, we can take the physiological scheduling limitations on mammalian females as given. Even so, we have to consider several routes by which males may increase the number of conceptions and survival of their children, or both at once. Infanticide by males is quite widespread among mammals and especially so among primates. Males can increase the survival of their offspring by defending them against infanticide. The horrifying accounts in Hill and Hurtado (1996, pp. 434–439) suggest that protecting their children from death at the hands of other men is a major contribution of Ache men to their fitness. Van Schaik and Kappeler (2003) argue that comparative study implies that social monogamy in primates has arisen independently several times from infanticide, recently supported by Opie *et al.* (2013). They suggest that male care of offspring may have evolved secondary to this social monogamy. Alonzo (2011), and subsequently Gavrilets (2012), have shown that in evolutionary models, if females can choose their mates, then male care can more readily evolve. An interesting feature of human societies is the existence of step-fathers, males who (most of the time, and perhaps not quite so well, Daly and Wilson, 1987; Marlowe, 1998) appear to care for their new wife's children by a previous husband (Rohwer *et al.*, 1999).

## 12.7 Women and their reproductive strategies

Females, especially Hadza females, are not passive spectators of the mating game. While some of the factors that influence their RS are fairly clear, and relatively easy to study, others are much more subtle and are difficult to investigate. It is extremely difficult to know when one is studying the outcome of male decisions, or female decisions. If females are able to choose effectively between mates, a wide array of consequences may follow, including selection for male help (Alonzo, 2011), and including male efforts to deprive them of their choice (Gowaty, 1996; Muller and Wrangham, 2009). In the present case, the uncertainties are easy to see. When a couple divorces, I cannot tell who is responsible. Did he desert, or did she throw him out? What the participants say is not much help. Men say they don't know why she threw him out, only to later reveal they had been seeing another woman (Marlowe, 2010). Women will seldom say "he left me" or "he found someone else." "He went crazy" is as near as she gets to a reason why she threw him out. The difficulty of separating effects of one upon the other makes studying mating systems the most complicated thing we attempt. Readers should be alert for one-sided interpretations of the observations about couples.

In Chapter 13, I look at variance in male and female RS and notice that while male variance is lower in human forager populations than in some other mammals, it is still higher than among females. In Chapter 14, I look especially at men's reputations among women as good hunters or traders, qualities that both Woodburn and Marlowe have indicated as criteria for choosing and keeping a husband. In Chapter 15, I report on Hadza marriage and especially on some features that bear on ideas about men as providers. In Chapter 15, I also offer some data on variation between different categories of men: dads, cads, show-offs, and "regular Joes" (the average man), and the six or so Hadza who earned some kind of wage during the study period. In other chapters, the role of men as helpers enters as a continuing question, and in Chapter 21, this is addressed directly. I look for effects of men's presence and reputations on children's growth and survival.

Chapter 16 describes aspects of growth and physique of Hadza (anthropometry). Children's growth is an indication of their well-being and has often been suggested as a predictor of their fitness, and of their contribution to their parents' fitness. This chapter is limited to more speculative issues about control of growth, descriptive data being given in SI. Chapter 17 deals with inter-birth intervals. The effect of birth spacing on health and survival of mothers and children is a central target of numerous studies and interventions in the developing countries. However, it has been a contended issue in hunter-gatherer studies. Hadza birth spacing, much shorter than described for the !Kung by Lee (1972) and Howell (1979), clearly needs close examination.

## 12.8 Statistical issues: irreplaceable subjects, false positives, and false negatives

My questions and analyses vary between the strongly predicted and the almost exploratory. I asked very few questions of the data that were not motivated by some degree of evolutionary cost–benefit analysis, even if much of it is merely intuitive. At both ends of the spectrum between strong predictions and exploration, my questions are driven by the proposition that people tend to behave in ways that would be favored by natural selection under the circumstances in which the people live. In the case of the Hadza, these circumstances include the savanna habitat, an ecology that varies and changes, but perhaps less boundlessly than some have argued (Chapter 2).

The strong predictions come from long-lasting theory, especially "the grandmother hypothesis," and the paternal investment theory of marriage and male reproductive strategies. Weaker predictions concern children as helpers; there is extensive literature suggesting they may be important but much less developed theory about the ways in which they would increase their fitness by helping their younger siblings or mothers.

These distinctions are important when we look at the abundance of my statistical analyses. By conducting many analyses, I run the risk that some of the answers will be significant merely by chance ("false positives"). Most of the results to which I give any attention are significant at the 1% (0.01) level or lower. There will be fewer false positives at these levels than at a 5% level. I also usually report series of results, and in most of these, the whole series have similarly small p-values. This is important for it was sometimes easy to think of a variety of possible variations on the models or subsets of data. In this situation, the danger of attending to false positives is heightened. I have only reported analyses in which the results were robust across such minor variations. There is quite an extensive literature discussing and debating the issue of false positives. One important point made is that adjustments to alpha values that reduce the number of false positives lower statistical power and increase the number of false negatives.

A classical statistical recommendation is to do very few analyses on each sample. This is acceptable when there are limitless subjects (laboratory mice, psychology undergraduates) who can be recruited for new studies. This approach would be a scandalous waste of the scarce, unique, global treasure (the Hadza) that we are trying to understand. There were no more Hadza to be recruited during our study. We are describing almost the entire population. A new generation is growing up, and are studied by Marlowe and his students, but we know some aspects of their circumstances have changed, and any attempted replications will have to be thought through carefully. For example, the trough in the age structure described in

Chapter 10 is currently working its way through the grandmothering ages. I expect a shortage of grandmothers, and others will fill the gap or suffer the consequences.

In the behavioral ecology context, driven by a large body of theory, the context for interpreting results is very different from fields where, for example, a large number of questionnaire items of unclear theoretical provenance comprise the dependent variables. To choose the occasional significant result in such a study, no matter how adjusted the probability levels, must be doubtful, and requires replication. Nonetheless, behavioral ecologists run the risk of false positives just as does anyone. But, when suites of coherent predictions are consistently supported, or consistently refuted, it is difficult to believe that the whole series are false positives. It is worth keeping the possibility of false positives in perspective – in any body of research there may be some false positives. However, this does not mean that the results are ALL false positives.

False positives are not the only danger in statistical analyses. False negatives may be just as misleading. In fact, Hill and Kintigh (2009) suggest that our analysis of Hadza food sharing (Hawkes *et al.*, 2001) is an example of false negatives. Our data may have lacked the power to find significant relationships, for instance between the amount given by man A to man B and the amount reciprocated. There is a strong view in the statistical literature that "post hoc power tests" are worthless, adding no information to that already given by the p-values and confidence intervals. Power tests have a clear place in planning studies where the researcher has a choice of the sample size and where granting agencies do not want to pay for the study of an unnecessarily large number of subjects. Nonetheless, power tests depend on a choice of alpha level (p-value), and a choice about expected effect size. Only sometimes is there a rational basis for predicting effect size.

Power is much less of a problem in the present study than false positives. Only when the sample is subdivided excessively does lack of power become a risk. Readily available power calculators suggest that most of my regressions have very high power, although I have looked crudely at contributors to power; for instance, to see whether a particular sub-sample has more than a handful of deaths or births as dependent variables, or missing grandmothers, siblings, fathers, or step-fathers as independent variables.

I introduced multilevel regressions in Chapter 8, but they will be used rather more in this second part of the book. This technique, especially when it can work on logistic regression, helps us distinguish between effects that are because of individual variation and effects due to differences in circumstances. This is especially important when, for instance, we want to see whether effects of helpers are effects of the presence or absence of a helper, or are effects of some, perhaps heritable, personal vigor of grandmothers, mothers, or others. The technique allows us to "control for" individual differences when we try to look for effects of circumstances that may be confounded with individual differences. For example, a sample of inter-birth intervals may show a relationship to child survival because short intervals are dangerous to children, or because impoverished or less well-treated women have frequent babies and give less effective childcare.

My philosophy of presenting results of regression analyses may be worth describing. I have tried to present the results as they follow my thinking. In general, this is: first, is there "a case to answer" on behalf of my primary prediction or question? Given some obviously needed controls, such as age, is there an effect of, for example, the grandmother's presence on child death? If there is a "case to answer," then I proceed to see whether I can remove the effect by including other possibly important variables, like mother's age, or the presence of other potential helpers, or the nested nature of the data. At a third stage, I look at other details, such as whether the effect is primarily due to the mother's mother, or any grandmother, and so on. This can lead to a fairly lengthy series of regressions and can make for tortuous reading. Consequently, I have tried to relegate details of some of the lesser steps to SI and give a more readable report. An alternative plan might have been to report one final model in each chapter. If our theories and predictions were more focused than they are or could be, this might be ideal. To my mind, there are two disadvantages. First, it is always possible to render an effect statistically non-significant by giving away degrees of freedom by adding more and more barely relevant independent variables (SI 12.2). Second, there will be more than enough independent variables that might explain away the result and that are absent from the final model. The skeptical reader (all readers should be skeptical) is left wondering about these, unclear as to whether they have been considered. I think my approach is close to the statistician's common exhortation to "get to know your data." However, it may, at the same time, risk accusations of "over-analysis," and produce a text that is more difficult to follow.

In the next chapter, the beginning of Part II, I describe and justify my key dependent variables, measures of individual RS, and its components, fertility and mortality. I then explore the variation of these measures between individuals, genders, and ages.

# 13 The outcome variables: fertility, child survival, and reproductive success

In the following chapters, I assess effects of helpers, marital status, growth, child spacing, and reputations on reproductive success (RS) of Hadza men and women. The first task of this chapter is to introduce the dependent variables: fertility, child survival, and an approximate indicator of Hadza lifetime RS. Having established our measures of RS, I compare features of the RS of men and women, reviewing aspects of the data that imply differences in the ways Hadza men and women may maximize their fitness.

## 13.1 Fertility

The childbearing careers of the 227 interviewed women were set out as a file in which each record represents a year in the woman's life, beginning at age 10 and ending in the year in which she was last interviewed. The total record lists 695 births in 4524 woman-years. Each record includes the woman's estimated age, decimal year of birth, whether or not she had a birth during the year, whether any of her children died during the year, her accumulated number of births, and number of living children. Other variables were added as the data processing proceeded, such as her marital status in each year, the identity of her husband, whether he was present or absent (divorced), and so on. The file was designed for analysis by logistic regression, predicting birth or no birth under various conditions. The best fit of probability of a birth to mother's age was to age + $age^2$ + $age^3$ (Chapter 7). Because women are represented for different lengths of time, some of the analyses use multilevel regression. This method also allows derivation of individual differences in fertility, as reported in Chapter 7 and as follows. For a few analyses, I used another summary measure, the number of births during each quinquennium of the woman's adult life.

## 13.2 Child survival

The lives of the 695 children of the interviewed women were set out as a file in which each record was a year in the life of a child. There were 5650 child-years of data, and 242 children died. Each record included the child's ID, its mother's ID, its year of birth, age, whether it survived the year or died during that year, and whether its data were censored, as in the final year in which it was observed. This file can be used for logistic regression and multilevel logistic regression. Independent variables are added for each chapter, such as whether the father is still married to the mother, whether a step-father is present, or whether the mother is living alone. The mortality of these children was partly reported in Chapter 8. Children are much more likely to die during the early years, especially during the first year of life. For most of the analyses in Part II, where we test the effect of independent variables such as grandmother's presence, it is important to note that the best fit predictor of child death is age + $age^2$ + $age^3$, and we should enter these variables first in any regression model. The mother's age at the birth of the child is also an important predictor and is included in most regression models (SI 13.1).

## 13.3 Measuring reproductive success

Fertility and offspring survival undoubtedly contribute to RS and to fitness. However, if we expect organisms to face problems of allocation of limited resources, we should expect some trade-off between outcomes, such as between the number of offspring and the survival of these offspring. Then simply showing effects on child survival, or only on fertility, give us a limited reflection of fitness. Reproductive success ideally combines these, measuring the number of offspring that survive to adulthood, and lifetime reproductive success (LRS) measures the number of such descendants produced in the parent's lifetime.

In studies of small populations in remote locations, we often are unable to collect data on completed reproductive careers for more than a few individuals. Direct measures of LRS of large samples are out of reach. In my study, relatively few individuals of either sex had clearly completed their reproductive careers and left reliable data (only 39/173 interviewed women were aged 40 or more; only 44/188 men with adequate data were aged 50+). If one required another five years, for example, to assess the survival of the final child, sample sizes would be even smaller. Like others before me (Borgerhoff Mulder, 1988, and Strassmann and Gillespie 2002, 2003), I try to use measures from incomplete reproductive careers as indicators of eventual lifetime RS.

At any age, some individuals have accomplished more reproduction than others. If success at an early age predicts final success, then we can use the early success to give each individual a measure of relative RS. We could score each individual by his/her number of live children, controlled for their age when they were last interviewed or observed. As long as we have established the best fit to age, we could control for age in each analysis, use residuals, or standardized residuals of number of live children on age of parent, or express these as standard scores for parental age. We can also use multilevel regression to derive individual scores. For most purposes, I chose to use standard scores, standardized within each sex.

I computed the number of living children (and their means and standard deviations) for adults of each age, and expressed each individual's reproductive accomplishment at a given age as a standard score. I took an individual's standard score at the latest age at which s/he was observed ("end-age") as his/her final, single measure of RS. I followed the same procedure for the number of children ever born, and for the proportion of those born who remained alive at observation. I did this separately for men and for women.

Thus, my usual measures of an individual man's or woman's RS were the standard score (ss) of the number of children (N) who were still alive (sslive) when we last saw the individual (RS = [N – mean]/SD). Two other measures were: the standard score of the number s/he had borne up to this time (ssborn); and for offspring survival, the standard score of the proportion of those born who remained alive at the final observation(or who had survived) (sssurv).

The sample of women was restricted to those with whom I had conducted reproductive history interviews. The representativeness of this sample was discussed in Chapter 7. Women were slightly more likely to have been interviewed if we had encountered them more often. I set no criterion child age for child survival. At each age of the mother, it is possible that one of the children will be less than a year old, and thus at quite some risk of failing to survive. At younger parental ages, this vulnerable latest infant will comprise a larger portion of the still small, accumulated surviving family.

I conducted no reproductive history interviews with men, so the sample of men was limited to those men on whom adequate data were likely to have been obtained and entered in the population register. These were the 188 men seen in three or more censuses, and on 60% or more of these occasions, seen in the core study area and still alive in 2000. This results in a smaller sample than I used in Chapter 7, particularly excluding several prolific older men who died during the 1990s. These restrictions were designed to maximize the chance of using only men on whom I had relatively full information, such as men for whom we had some record of all of their wives. As among the women, the reproductive history was extracted from the population register. Children's fathers had been entered in the population register and I used this to reconstruct men's reproductive histories (as described in Chapter 7). A computer program did the scoring, as described in SI 13.2. A file was written for each gender, listing the totals, means, standard deviations, and standard scores for number of live children, number born, and percentage surviving for each year of age of adult reproductive lives. The standard scores are independent of age at last observation (SI 13.3). The standard scores are, not surprisingly, very highly correlated with the residuals, and the standardized residuals of number of live children by parent age.

## 13.4 Testing the validity of the standard scores: do early scores predict later scores?

We need to show whether the standard score at the last observation is a reasonable indicator of final RS. Using this measure implies that I believe that the final observation of a 25-year-old is as good as, and can be combined with, data on a 50-year-old who has completed her childbearing years. The measure implies that I believe that a 25-year-old with more children than others of her age will end up with more children than others. Thus, I assume that reproductive careers tend to find a trajectory and stick to it, people who start poorly continue poorly; people who start productively continue productively.

In addition to the "random disasters" that can befall any family (SI 13.4), there are reasons to expect that early reproduction might be a poor indicator of later reproduction. Individuals who fall behind at one age might catch up at a later age, after observations ended. Some researchers have argued that we should expect to see variation in the timing of allocation of effort, with trade-offs between early and late reproduction overshadowing individual variation in total output. Some expect early depletion of resources for reproduction to cause fast early reproducers to become slow late reproducers. For some species, resources are accumulated before reproduction and rapidly exhausted by reproduction. In others, resources are clearly renewable. Reproductive effort might also be anticipated to increase with age. As expected years of future reproduction diminish, it pays less to conserve resources for future reproduction, and it pays to invest more heavily in reproduction while the opportunity is available ("terminal investment"). The payoff from conserving resources to expend in parental care or grandparental care would complicate the picture.

The principles involved help account for features of some plant and animal life history phenomena (Stearns, 1992, pp. 82–85). However, attempts to demonstrate increased effort at greater age in humans have failed to find it (Sear *et al.*, 2003b). Effects of reproductive effort on the very low adult human mortality would be hard to show. Nonetheless, it is easy to point to possible risks. For example, although at a very low rate in any population, Hadza women do sometimes die in childbirth (Chapter 8), and too many early births might jeopardize the later years of reproduction.

Similar possibilities can be imagined for men. Competition for wives can be costly. In any population, men can get into fights over women, and homicide rates are not negligible (Chagnon, 1968; Lee, 1979; Daly and Wilson, 1988; Hill *et al.*, 2007; and Chapter 8). The frequency of fights may vary with age. Older men may fight less than young men (old men may have more allies, or may be more skilled at negotiating conflicts of interest without violence, or simply have less to gain or lose by changes in their situation). Additionally, among the Hadza, some men's reproduction continues long past the age at which women bear their last child.

I test the validity of my assumption that reproductive careers stay on trajectory by looking at the retrospective information on the early careers of the people on whom we have later information. The later we look at them, the smaller our sample size. Therefore, I looked at intermediate ages as well. Thus, I used the early career of people who were observed at an older age to give us some indication of whether the scores of people who were observed only at a young age could be taken as an indication of the young person's eventual RS. Families accumulate as time goes by. A child born to a 20-year-old woman does not disappear from her total number of births; a child who survives the first few years of life becomes ever less likely to die. Time "wasted" with no birth cannot be regained; a child lost may be followed by a shorter birth-interval but the time from conception to its death is still lost time. These are important and

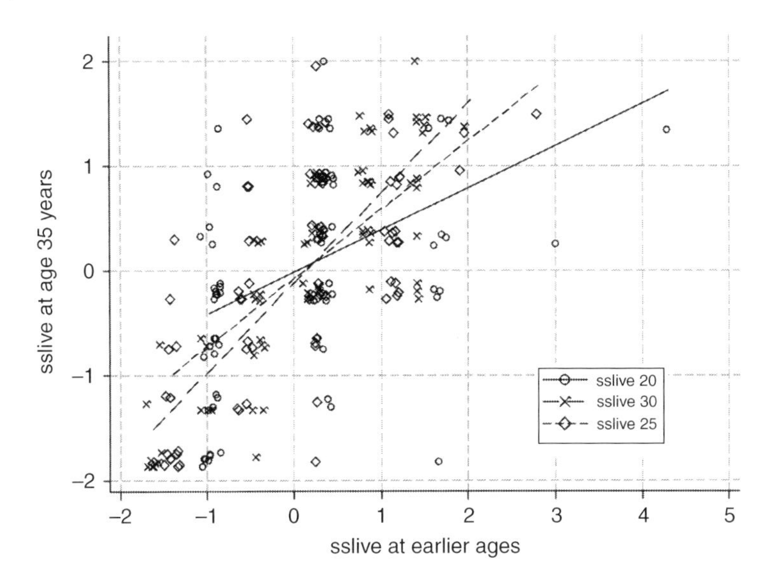


**Figure 13.1** Reproductive success (standard score of number of living children) of women aged 35 predicted by their standard scores at earlier ages. Fitted regression lines show that the prediction from age 15–20 is weaker than the prediction from the next two quinquennia.

real limits on women's reproduction that work in favor of our finding that early reproduction predicts final accomplishment.

Tables in SI 13.5 show correlations between individuals' standard scores at one five-year age point and the next, and each subsequent five-year point (all are strongly positive). Figure 13.1 shows an example, the relationship between women's standard score at age 35 and the earlier scores for the same individuals. SI 13.6 shows regressions of the standard score at the last observation on previous standard scores. I also ran principal components analyses on the data matrices. SI 13.7 shows correlations between the number of births in each quinquennium and the next and all subsequent quinquennia (most are positive and significant).

There is a consistent pattern of positive and significant correlation between standard scores of achieved reproduction at any age and end-age (S1 13.6). Regressions predicting the score at end-age can give a "prediction, $r^2$," which is only slightly lower than the adjusted $r^2$. The pattern is similar for the standard score of the number of births, number of live children, and percentage survival. The relationships are slightly stronger for women than for men. An individual's reproductive achievement at any age was a good predictor of his/her accumulated achievement five years later, 10 years later, and even longer.

There were only weak hints of resource-budgeting, or terminal investment. For both men and women, the earliest five-year period is the worst indicator of later reproductive achievement. For women, this is the age group 15–19; for men, it is 20–24. This is not surprising; by 19 years of age, the average woman will have had only one birth (and about half of them, no births), and mortality of infants and very young children is quite high. Few men aged 20 are married, and if they are, their wives, on average almost five years younger, can have had few children, and early marriages are especially vulnerable. Thus, in several of the analyses in the following chapters, I use only samples of women aged 20 or more, and men aged 25 or more. Not surprisingly, in view of the correlations just reported, principal components analysis (SI 13.5) showed strong first components for each sex (66% of the variance for men, 73% for women) with even the youngest ages loading positively on the first "fast–slow reproducers" component.

## 13.5 "Person residuals" from multilevel analysis as individual reproductive rates

The annual hazard files comprise a record for each year for each individual in the sample. The annual records are nested within individuals. Multilevel analysis (introduced and briefly discussed in Chapter 4; Goldstein, 1987,;Rasbash *et al.*, 2003; and Twisk, 2006, is a helpful introduction) can give us measures of individuals, reporting the individual residuals, departures, for instance, from a regression of women's fertility on age + $age^2$ + $age^3$. I compared these measures of women, with the standard scores discussed earlier. Keep in mind that these are merely slightly different ways of looking at and scoring exactly the same raw data.

Multilevel logistic regression allows us to see how the probability of a birth varies with the mother's age (from 15 years to end-age), while controlling for variation between mothers in fertility and in representation in the sample. An over-representation of very fertile women, or women from a more fertile era, could bias "single level" analyses. The multilevel software (Mlwin) allows one to estimate the mother residuals and store this measure of individual, mother-to-mother variation. Then, we can see which women were least and most fertile. The measures of the 227 individual women correlate very closely with our individual standard scores ($r=0.9454$, $p<0.00001$, adjusted $r^2$ 89%). Some women are more fertile than others.

I did the same thing with the child survival file. The records are nested within child, and children are nested within mother. Multilevel logistic regression predicting death or life from child age + $age^2$ + $age^3$ (the best fit to childrens' probability of death by age) allows one to save the mother residuals. I linked them to women's standard score for child survival, the measures are strongly associated ($r=-0.8154$, $p<0.0000$, adjusted $r^2$ 66%, N=147.) The association is negative because sssurv scores survival and the logistic regression is predicting probability of death. Offspring of some mothers consistently survive better than the offspring of others.

The analyses discussed here suggest that the standard score of live children (sslive) is a worthwhile measure of the RS of individual men and women in our samples. Being established separately for men and women, they reduce the apparent variation between the sexes, and in some later analyses, where I compare men and women, the raw number of live children, controlled for parent's age may be a more useful measure. In the remainder of this chapter, I will not be using the standard scores but will look at variance in the numbers of live children, numbers of births, and percentage surviving within specified age groups.

## 13.6 Variation in reproductive success: pointers toward reproductive strategies?

Sensitivity analysis (Chapter 12) pointed us toward the high selection pressure on the survival of young children (ages zero to five years) and the fertility of young women, and perhaps the fertility of older men. We can get more pointers about potential differences between reproductive strategies of men and women by looking at sex differences in variance of reproductive parameters. We can also look at the opportunity for sexual selection given the current demography of Hadza men and women.

In Chapter 7, we saw that men's and women's childbearing careers differed. Men's completed family size (CFS) (final number of births) ranged from zero to 17, with 10% of men fathering no babies. Men's age-specific fertility (ASF) climbed more slowly, from age 20 to 30, and remained high during their 50s, with a few births to men in their 60s. Women's ASF dropped to zero in their early 40s. Women's CFS followed something more like a normal distribution; women clustered around a central tendency near the total fertility rate (TFR) of 6. This suggests that the Hadza may fit a common pattern in which men have a higher variance in RS than women. If we plot the number of live children among the 38 women aged 40 or more, we find a near normal distribution with a mean of 4.05 (CI: 3.37–4.73) and variance of 4.366. The standardized variance in live children for these women (variance/mean$^2$) is $4.37/4.05^2=0.2665$. The exact score varies slightly with the age group measured. The percentage of their children surviving also follows a roughly normal distribution with a mean of 67.7%.

The 44 men aged 50+ at the last record had a mean of 4.2 live children (median three children), from 5.9 births (median five births). However, the distributions departed far from normal, with an excess of men having a very low RS (Anderson-Darling tests for departure from a normal distribution, p=0.005 for live children, and 0.028 for births). The variance for the number of live children was 10.80, for a standardized variance (variance/mean$^2$) of $10.80/4.2^2=0.6122$. The percentage survival of the men's children showed a more normal-looking distribution except for an excess of men with a high survival of their children. This appears to be due to a small cluster of men who had their first child late, probably with an older woman (who tend to keep more of their children alive than younger women), as shown in later chapters.

The small difference between the sexes in mean RS is not statistically significant (if the adult sex ratio [ASR] is one, and our informants attributed parentage correctly, there should be no difference, for every child has one mother and one father), but the difference in variance is significant (Levene statistic 9.69, p=0.002). The high variance in men's RS implies that some men gain much more fitness than others. The five men (11% of the sample) who raised nine or more surviving children have a success that is more than double the mean RS and three times the median. The equivalent figures for women are much lower. The four women (10% of the sample) who raised eight or more live children have a success that is just below double the mean, and exactly double the median. Some men can achieve exceptional success relative to others; women less so. Nonetheless, as Schacht *et al.* (2014) and others have pointed out, this difference in variance does not tell us the means by which some men excel over others.

Standardized variance in RS has been offered (Crow, 1958; Wade and Arnold, 1980; Arnold and Wade, 1984a,b; Clutton-Brock, 1988, pp. 2–3, and chapters therein; Shuster and Wade, 2003) as an index of the opportunity for selection, and the ratio of male score to female score as an index of opportunity for sexual selection. The Hadza index of opportunity for sexual selection was $(I_m/I_f)=2.3$. While these computations support common expectations about differences in the ways selection acts on men and women, they have weaknesses that I will try to illustrate. There are important arguments about what these measurements do or do not tell us and how useful they are for testing the adaptiveness of particular aspects of structure or behavior. We should take them as rough pointers, to be compared with "selection gradients," and to be re-examined as we learn about how Hadza reproductive strategies work (SI 13.8). Useful discussions of this difficult topic are in Brown (1988) and Grafen (1988).

Because the sample of men and women at older ages is small, I repeated the calculations at all younger ages. Among the youngest, variances are inflated, probably because some individuals have had one birth, and others none. From a little over age 20 in women and 25 in men, the numbers begin to stabilize. There are two important observations: (1) At all ages, men's standardized variance in the number of children alive and in the number born is almost twice the value for women. (2) Among women, the standardized variance in child survival is about the same as the standardized variance for births. For men of all ages, the standardized variance for births is much greater than that for child survival. The "opportunity for selection" among men is much greater for number of births than for child survival. This is not so among women (Figure 13.2).

The Hadza are just one small human population and we should look at these estimates of variance in RS for some other populations before generalizing. Table 13.1 is one such compendium (SI 13.9). There are larger compendia recently published, all with minor and not so minor differences (Brown *et al.*, 2009; Betzig, 2012; Henrich *et al.*, 2012). Comparable measures of RS are not universally available; for example, some reports only show fertility. Howell (1979, tables 16.1, 16.2) gives figures that show standardized variance for the !Kung RS as 0.3227 for men and 0.2147 for women. !Kung women appear to have a similar standardized variance to Hadza women but !Kung men have a lower score than the Hadza men. Reproductive success is more evenly spread among !Kung men than among Hadza men. Hill and Hurtado (1996, p. 411, fig. 12.5) show mean RS and SD for three cohorts of Ache men and women in the forest period. I calculated the mean of these cohorts, then the standardized variance (variance/mean$^2$). The Ache women scored 0.1885 and the men 0.5347, almost the same as the Hadza men. Borgerhoff Mulder (1988) reports the values for three cohorts of Kipsigis farmer-herders. Women's standardized variances were 0.28, 0.30, and 0.17 (the mean of these is 0.25, Brown *et al.*, 2009, report 0.1354), while the cohorts of men ranged from 0.32 to 0.79 (Brown *et al.*, 1988, p. 425, but table 26.1 presents values of 0.51, 0.23, 0.97, 0.28; mean of which is

**Table 13.1** Standardized variance in RS, and "opportunity for sexual selection" in some mammals and birds compared to people. The studies shown here were conducted without genetic identification of paternity. The human samples are adults; pre-adult mortality was not included. I tried to select animal cases that matched these criteria. Studies of large human populations may neglect young men who moved to cities and may or may not return. Under-counting peripheral males may also be a problem in studies of primates, and possibly other animals. Sources for most animals are chapters in Clutton-Brock (1988). Detailed sources of human data in SI

| | Males | Females | $I_m/I_f$ |
|---|---|---|---|
| Northern elephant seal | 21.77 | 5.23 | 4.16 |
| Red deer | 2.51 | 0.36 | 6.97 |
| | | | |
| House martin | 0.76 | 0.46 | 1.65 |
| Bewick's swan | 1.26 | 1.24 | 1.02 |
| Kittiwake | 0.83 | 0.69 | 1.20 |
| Song sparrow | 2.41 | 1.51 | 1.60 |
| Great tit | 1.89 | 2.53 | 0.75 |
| | | | |
| Hadza | 0.6122 | 0.2665 | 2.30 |
| !Kung | 0.3227 | 0.2147 | 1.50 |
| Ache | 0.5347 | 0.1885 | 2.84 |
| Dogon | 0.2835 | 0.2210 | 1.28 |
| Kipsigis | 0.23–0.97 | 0.17–0.30 | 1.99 |

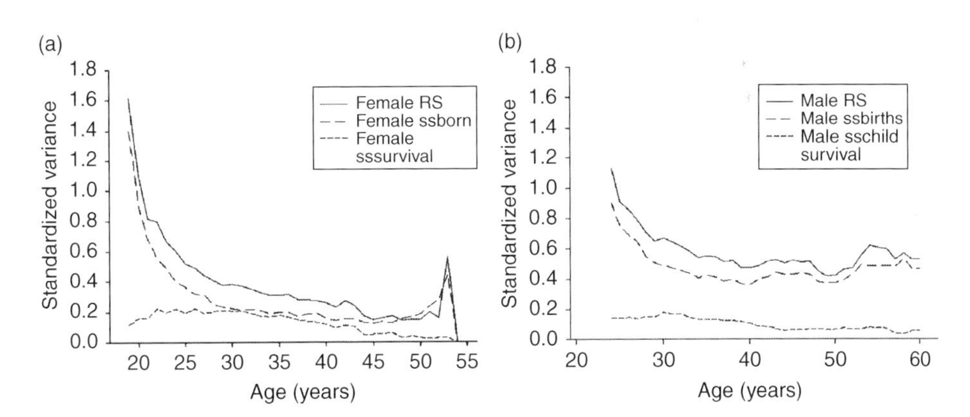


**Figure 13.2** Standardized variance (opportunity for selection) on RS, births, and offspring survival by age among women and among men. Note that while standardized variance in child survival is similar for men and women, variance in RS and fertility is higher in men than in women, during most of the career and especially at higher ages. (a) Standardized variance (opportunity for selection) on RS, births, and offspring survival by age among women. (b) Standardized variance (opportunity for selection) on RS, births, and offspring survival by age among men.

0.4975; Brown *et al.* estimate 0.2054; Schacht *et al.*, 2014, show about 0.65 on their fig. 2). Clearly it is not easy to replicate these measures very closely. Schacht *et al.* (2014) attempted a compendium with standard age restrictions.

Values in some mammals and birds are shown in Table 13.1. The human values vary little, and are quite low among mammals, and even quite low for birds. Marriage may be the primary correlate of the low human male index. Brown *et al.* (2009) comment that low $I_m/I_f$ seem to accompany monogamy and higher figures are found under polygamy and serial monogamy. Betzig (2012) and Henrich *et al.* (2012) show the same result from samples that include more "large scale" and historical populations. We should wonder whether this means that marriage, as has often been suggested, is an outcome of contests among males.

While standardized variance in human male RS is low compared to other mammals, the ratio $I_m/I_f$ is not (Table 13.1). This measure, "opportunity for sexual selection," for the Hadza and Ache is higher than among monogamous birds, and approaches the figures in Packer *et al.* (1988, fig. 23.2) for lions. It is important to remember that the measure is only a measure of opportunity. For sexual selection to occur, there needs to be some heritable factor associated with higher (or lower) than average RS. For our purposes, it is perhaps more important that the opportunity can also be taken by phenotypic processes; some men manage to excel over others to a larger degree than women can excel over other women. As Schacht *et al.* (2014) point out, how men reach the top end of the distribution is a topic for research rather than instant assumptions. The existence of these differences in variance do not tell us the means by which men compete and by which some attain much RS while others little. It especially does not tell us much about patterns of violence toward other men, or toward women. Men may deal with competition by alliances (Rodseth, 2012). They may maintain marriage partners by compliance (particularly where female choice is possible [Alonzo, 2011; Gavrilets, 2012]), or by duress (Muller and Wrangham, 2009). The strength differences between men and women (and in many societies, the economic differences) may sometimes make female choice impossible. At other times, female choice may be facilitated when it comprises tipping the balance between competing men.

Reproductive success is affected by the number of births, the survival of children, and the length of the reproductive career. Men and women may maximize their RS by different combinations of these variables. Figure 13.3 suggests one difference. Hadza women with the highest fertility (number of births) do not raise more surviving children than those with slightly fewer births. Strassmann and Gillespie (2002) found the same among Dogon farmers. Pennington and Harpending (1988) had a different result from their wide survey of !Kung women. Among Hadza men, those with the greatest number of births also had the greatest number of live children. It is as if there is a trade-off between numbers and survival for Hadza women but not for Hadza men. This will be followed up in greater detail in Chapter 17.

The Hadza data (Figures 13.2 and 13.3) suggest that men can excel most by achieving a large number of births, but we need to find out more about how they achieve a higher number of births. Women can excel about equally by ensuring high

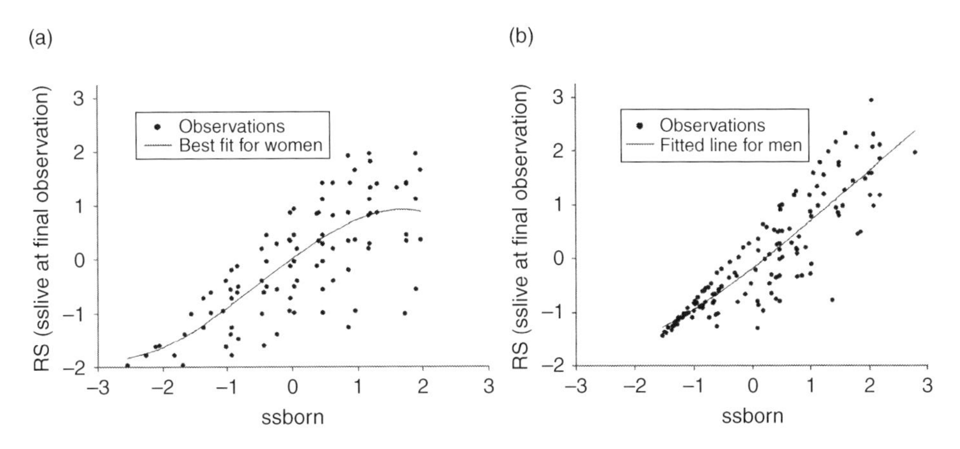


**Figure 13.3** Relationship between RS (sslive) and fertility (ssborn) in men and women. For women, the added terms are each significant and make a small contribution to $r^2$. For men, the added terms are not significant and make no improvement in $r^2$. (a) Women. (b) Men.

fertility, and high survivorship of their children. In SI 13.10, this potential for gender difference in reproductive strategies is estimated using stepwise regression, and using the method described by Brown (1988) to overcome limitations on the use of regression for analyzing multiplicative predictor variables. Both methods support the view that Hadza women may enhance their RS more than men by promoting the survival of their children.

## 13.7 Length of the reproductive career and reproductive success

In many animals the length of reproductive career is a major contributor to lifetime RS (Stearns, 1992; Oli and Dobson, 2003). This is often because adult mortality cuts off the reproductive career. Among humans, adult mortality is very low but there is still some variation in length of career. People have their first baby at different ages, and they have their last baby at perhaps even more widely differing ages. We can look at the relationship between RS (sslive) and age at first birth for those who have had a first birth (Figure 13.4, and SI 13.11). The relationship is linear, negative, and significant for both men and women but the regression coefficient for men ($r=0.0847$, $p=0.000$, $r^2=16.6\%$) is about half that for women ($r=-0.1480$, $p=0.000$, $r^2=18.9\%$). In both cases end-age (the age observations ended) had no statistical effect.

It is more difficult to look at the end of the career in my data. I noted earlier that Hadza men continued to father children to the end of their 50s and into their 60s, unlike women, few of whom had births during their 40s (Chapter 7). I have 38 women in the RS file observed at more than age 40, and 44 men observed at older than age 50 (and 89 aged 40+). Births after age 40 comprised 4.6% of the women's total number of births, and 41% of the men's. Continued reproduction accounted for a

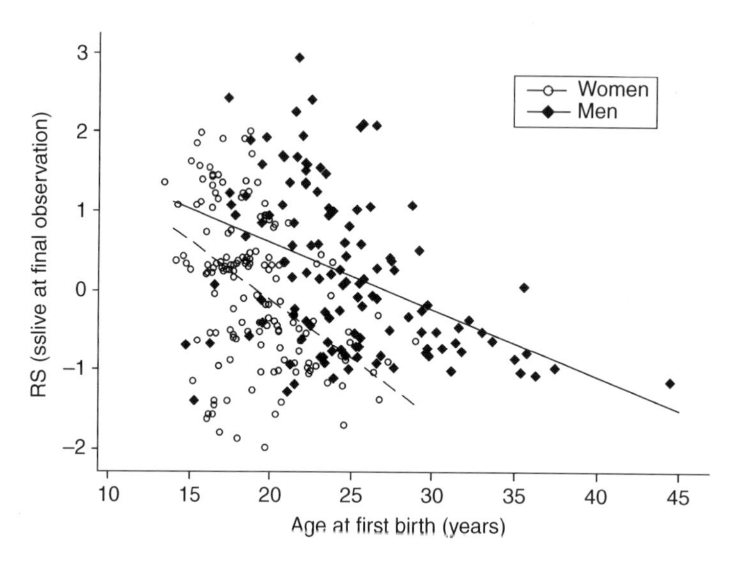


**Figure 13.4** Relationship between RS (sslive at final observation) and age at first birth for men and for women. Lines are from linear regressions. Regression analyses were controlled for age at last observation.

bigger proportion of lifetime RS in men than in women. In both sexes, continued reproduction is partly predicted by a high score for early reproduction.

Because women's reproductive careers are physiologically limited, the age at which they begin is more important than it is for men. For the same reason, the later phase of the career makes more difference to men's RS than to women's. This may remind us of the sensitivity tests of the life history where we saw that the effect on fitness from increases in fertility of older women was much lower than for younger women. The same was seen, in a very similar analysis in Chapter 11, simulating population growth rate with childbearing ceasing at various ages from 45 downward. The effect was very small at ages above 40. These patterns may go some way toward accounting for gender differences in marriage to be reported in Chapter 15. A delay may pay off in fitness for men if it enables them to increase skills, reputation, and end up with a wider choice of wives during a longer portion of their lifespan. Not surprisingly, even though male mortality is greater at all ages than female mortality, and this reduces the effect of males' continued fertility, men's reproductive value (RV, contribution to population growth) peaks later than women's and is higher than women's RV during the 40s and 50s (SI 13.12). Note that the classical calculations of RV ignore effects of post-reproductive women on the fitness of their children.

## 13.8 Adult sex ratio and operational sex ratio

Adult sex ratios have recently received much attention in our most difficult sub-field, the study of sexual selection and mating systems (Kokko and Jennions, 2008; Kokko *et al.*, 2012; Schacht *et al.*, 2014; Szekeley *et al.*, 2014). Studying ASRs is itself not simple. There are methodological as well as conceptual difficulties.

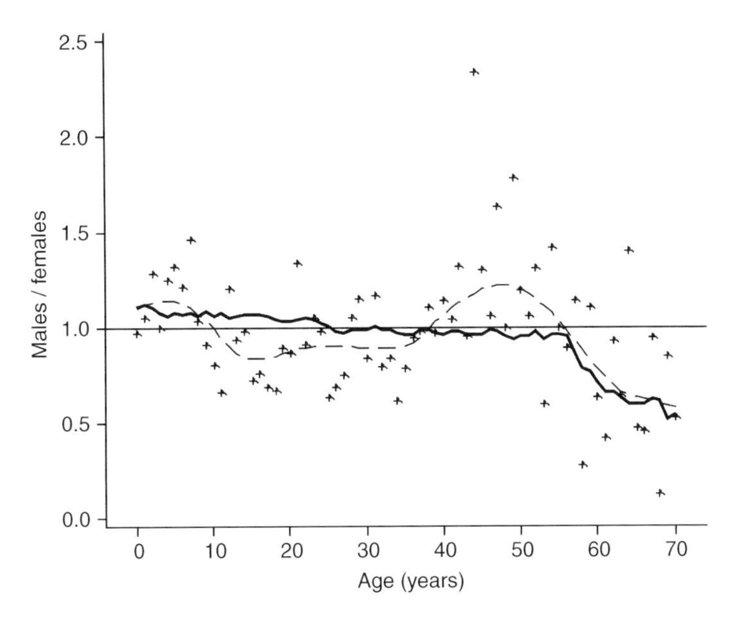


**Figure 13.5** The Hadza sex ratio (number of males divided by number of females). Solid line shows the value predicted from estimated fertility and mortality. Crosses mark observed counts for each year of age from the population register. The broken line is a Lowess fit to the observed values. Short-term variations can be striking.

The population simulations used in Chapters 9 to 11 can be used to predict the long-term average sex ratio at any age. Figure 13.5 shows the predicted value, based on observed fertility, mortality of boys and girls, and sex ratio at birth. The life course of the predicted sex ratio is very much as expected in any human population. A slight excess of boys at birth dwindles to a slight deficit of males in adult life and a striking shortage of elderly males. The Coale and Demeny (C&D) life tables show us the generality of such a pattern. They also show that, presumably because males are generally more vulnerable to survival threats, the ASR is consistently higher in lower mortality populations.

One of the conceptual difficulties was made clear to me when Ryan Schacht and Monique Borgerhoff Mulder asked me and other anthropologists for sex ratio data. They gave us two options: their standard measure, the number of males aged 18–65/ females aged 15–50; and the anthropologist's preferred "ethnologically appropriate" measure. For the latter, I used males aged 25–55 (the ages at which Hadza men are most likely to become fathers) and 15–45 for females, which covers the span of the childbearing years. Having quite intensively read the literature that Schacht and Borgerhoff Mulder were responding to, I believe that my "ethnologically appropriate" age choices were circular; they incorporated some of the results that sexual selection researchers are trying to account for by variation in ASR. Hadza men aged 18–25 could, physiologically (and sometimes do), father children. Why do they not have children more often? Hadza men aged 55–65 sometimes do, but why is it unusual? The answer is very likely to be that they cannot outcompete the "prime" aged men.

Either women do not like them as much, or their kin do not, or other men edge them out, by means that vary greatly from culture to culture. We need to distinguish between the competitive situation, the extent to which fertile women are generally a scarce resource for fertile men, and the various ways in which individuals of either sex may respond. We need to be careful with measures such as my "ethnographically correct" measure because it includes some of the results that it was intended to help explain. Other responses to the competitive situation may also modify both the situation and our attempted measures. In many rural areas, young men migrate to the city, believing they can earn enough money to come home and buy fields and livestock. However, by departing they modify the situation of those who remain. They lower the number of men in what the ethnographer in the rural "homeland" may label "the mating pool." Similarly, juvenile primates, by fleeing the huge dominant males and joining a bachelor troop, or simply wandering and risking predation, change the "mating pool" for their elders. Moreover, we do not know how much of the increased male mortality in Coale and Demeny's high mortality populations would be because of interactions between competing males.

The ASR determined in a small population cannot be very reliable (SI 13.13). The adult population will be about half the total population; divide that by sex and we are comparing one quarter of the population with another quarter. Errors in age estimation and year-to-year variation in fertility and mortality can make the definitions of "adult" more influential than we would wish. My population register gives a larger count than any individual census. Figure 13.5 shows the counts from the population register for each age. The broken line in the figure shows a Lowess smoothing of the markedly varying observations. This shows a shortage of males aged 15–40 (age mates of the women of childbearing age), and a surprising excess of men aged 40–55. The latter reflects the shortage of women of that age noted in Chapter 10. Sex ratio recorded at a specific time is evidently vulnerable to historical fluctuations.

Operational sex ratio (OSR) has long attracted interest as a measure of potential accessibility of mates, attempting to combine sex differences in number with sex differences in "time out/dry time" determined by the evolved propensities of each sex. Thus, for mammals we may be excused, initially, from having to explain why females gestate and only females lactate, while males have very short "refractory periods" after intercourse. Different authors have employed different measures of OSR. In its simplest form, it measures the number of males of reproductive age to the number of females of reproductive age (like, for example, Schacht's standard measure). We may choose to specify the built-in circularity that I discussed earlier by following Clutton Brock and Isvaran (2007). They defined a "duration of effective breeding" (DEB) as the ages between which ASF, fitted to age + age$^2$, exceeds 25% of fitted maximum ASF. The quadratic fit curtails the upper age limit for Hadza men, the cubic fit better represents the number of men who father children in their 50s and 60s.

For my Hadza data, the DEB encloses the ages 19–57 for men and 15–44 for women, and gives a mean OSR of 0.840 for the eight census years (slightly lower than in Blurton Jones *et al.*, 2000). This appears lower than the 1.14 that Marlowe reported (2010, p. 157). When I use the age criteria that Marlowe used (females aged 16–49, males 18–60), my result is 0.946, which is not significantly different from his figure. More sophisticated measures of OSR have been proposed by Mitani *et al.* (1996), allowing for an estimate of female "time out," and by Hurtado and Hill's (1992) fertility units per male. It is worth noting that, in most computations, the criteria for females are based on a physiologically rather fixed criterion while for males they are identified by criteria (ages when we see some men fathering children) that are not based on a physiological given. The male figures are based on the outcomes of the competition between males that they are used to explain. This detail criticism does not negate the fact pointed out by Marlowe (2000), Marlowe and Berbesque (2012) and others, that the physiological difference between men and women implies a high OSR, with more males competing for reproductive females than in many other species in which survival and reproduction age at the same rate.

Schacht *et al.* (2014) argued against a now traditional expectation of those who study human behavior and evolution that an excess number of males predict more promiscuity, more divorce, more aggression between men and by men toward women, and other components of the active promiscuous male/coy chaste females stereotypy. Notably they point out that when women are scarce, men have to treat them better, and when women are more numerous, divorce is more frequent, observations also reported by Liker *et al.* (2014, p. 880). Scarce females may be guarded more intensively, have a stronger bargaining position, and be able to elicit more help from males. Our (Blurton Jones *et al.*, 2000) reported association among four hunter-gatherer societies between divorce rates and fertility units (number of females multiplied by TFR) per male supports this suggestion. Those who write about wife beating may contend this, some discuss beating as a form of mate guarding, but they should perhaps look again at its relationship to sex ratios. The main point is: precisely how the competitive situation translates into behavior cannot be simply assumed.

## 13.9 Patterns of reproductive success among Hadza men and women

Individuals who are reproducing fast at one age tend to be reproducing fast at a later age. Standard scores of RS at the age of the last observation are predicted by standard scores at earlier ages. This is true of sslive (RS), standard score of the number born, and of the percentage born that are still alive (sssurv). There is only weak support (in the second component in principal components analysis, SI 13.7) for the idea that slow early reproduction is compensated by faster late reproduction or vice versa. Not surprisingly, the standard scores correlate strongly with individual residuals derived from multilevel regression. There are individuals who are successful reproducers and others who are less successful reproducers. The standard score of the number of live children at the last observation seems to be a good proxy for lifetime RS among the Hadza. The data give little support to the idea of a trade-off between early and late reproductive effort.

Hadza women show a stronger effect on RS from age at first reproduction and from child survival. Men have a weaker effect of age at first reproduction, and child survival and a strong effect of late fertility; some men continue to father children

**Photograph 13.1** Six of these "paanakwete" (senior men) aged 41–57 are in the reproductive history file. Between them, they had had 17 wives, on average 8.8 children born and 5.5 surviving. At the time of the photograph all their wives were younger than 45, one was in her 20s and another in her early 30s. Men's reproductive success varies widely, and successful older men such as these may be significant competition for other men in search of a share of available paternities. © James F. O'Connell, 2015. Reproduced with permission.

throughout their 50s and into their 60s. Standardized variance in RS, and in births, is greater for Hadza men than for women. The ratio, index of opportunity for sexual selection, for the Hadza and for the Ache, is similar to that for lions and higher than for several species of socially monogamous birds. The index for the !Kung is lower, and similar to that for the monogamous birds. Precise comparison with recently published compendia is difficult, partly because results for small populations vary, with small differences with age and year-to-year variation.

The ratio of male to female standardized variance in RS shows not just the current opportunity for sexual selection on any genes that may underlie the individual differences in RS, but it summarizes the opportunity for some individuals to gain or lose in competition with others. The chart of completed family size (Figure 7.8) shows that some men can gain more than ten-fold lifetime fertility over others. There will be a great fitness payoff for those who can be in the top 10% or so, and a great loss to those who fall into the bottom tail of the distribution. As there is limited paternity to be acquired, one man's gain is another's loss (Photograph 13.1). Competition is involved, and this is the usual definition of male–male competition. But we cannot say how much of this is direct competition, such as for access to females, and how much is due to factors not shaped by other men, such as an accident of low sperm count, and how much is because of female preferences. Standardized variance can be taken as an indication of potential for gains from successful competition between individuals. There will be gains in fitness from any action that pushes a man further up the distribution. However, action that increases the number of conceptions, for example, could be offset if those actions decreased the survival of resulting offspring. We have seen that among Hadza men, increases in births do not seem to be accompanied by a decrease in offspring survival (because RS increases linearly with fertility, Figure 13.3), whereas for women, RS begins to level off at high fertilities, implying poorer survival of offspring. In Chapter 17 we will see that, among women, offspring survival is highest at intermediate frequencies of births.

Among mammals, the index of opportunity for sexual selection is high among harem-holding species such as the elephant seal and red deer, among which much of the male–male competition is aggression over access to females. The index is low among monogamous species, such as many birds, but these are not devoid of aggressive contests between males. Wherever there is competition among males, there must be the potential for aggressive, even lethal contests, but we cannot take the standardized variances as direct indications of aggression. Among hunter-gatherer men, with not insignificant observed homicide rates, even in those populations with a relatively high standardized variance, men are in peaceful contact most of the time, despite the underlying potential competition. Indeed, one of the earliest quantitative direct observation studies of hunter-gatherers, by Bailey and colleagues (Bailey and Aunger, 1989a, 1990) showed that Efe men spent more time in the company of other men than with any other class of people except for their wives. Among Hadza, where men gather at "the men's place" whenever in camp, they may even spend more daylight time in the company of other men than in the company of their wives. While alliances between males have been closely studied in other primates, Bailey's study represents almost the only exception to the startling neglect of relationships between men since the lively attention to the topic by Tiger (1969) very early in the origins of the evolutionary approaches to human behavior. Some of the early literature linked men's relationships with other men to cooperation during hunting. Modern savanna hunter-gatherers, with good projectile weapons, can hunt alone successfully. In forests (Terashima, 1980), cooperative hunting is more typical, and before projectile weapons, cooperation may indeed have been important. The current situation suggests there may be other causes of abundant contact between men. Warfare has also been suggested but extensive male–male interactions persist even in the more peaceful populations. We may have under-rated the extent to which these contacts function to maintain alliances, and tolerance, as cost-reducing antidotes to the fundamentally competitive situation in which selection often forces men to exist. Rodseth (2012) gives a helpful overview of anthropology's long-standing interest in the complexity of relationships among males.

The data thus far do not contradict the common scenario. Among Hadza women, with unrestrained access to massive wild resources (Chapter 2) limited only by their own time and energy, the potential for competition is quite limited (perhaps limited to competition over helpers, examined in Chapter 19). Women's potential gains from increases in fertility are not dissimilar from their potential gains from improved offspring survival. Potential for competition between men is much greater and their potential gains from increased births appear to be greater than their potential gains from improved survival of offspring.

Some recent literature has attended to relationships between ASRs and mating systems (Brown *et al.*, 2009). In this connection, it is important to note that I have been discussing variance in RS, and not proposing a difference in mean RS between males and females. A difference in mean RS could only exist in an uneven ASR. The most carefully argued model (Kokko and Jennions, 2008) suggests that a high ASR favors more offspring care by males, because more males will be in competition for fewer females. However, that model is mainly aimed at understanding the evolutionary origins of anisogamy and sex reversals, and its application to humans may need extraordinary care. Our comparison of divorce rates in four hunter-gatherer societies (Blurton Jones *et al.*, 2000) shows something very similar to what Kokko and Jennions propose. Instead of ASR, we used a measure devised by Hurtado and Hill (1992), which takes account of population differences in fertility as well as of ASR. This measure, fertility units per male (the number of conceptions available per male), is the best predictor of divorce rates in this tiny sample. When there are more units per male (when there are fewer males than females), the prospects for a deserting male to find a new fertile mate are improved. Divorce rates are high. When there are many males, the contests for new mates would be tougher, and fewer men desert. Prediction from "father effect," the effect of the fathers' presence on survival of their children, was poor. This observation seems completely in line with those of Kokko and Jennions (2008) and others (Houston and McNamara, 2002), who had pointed out the constraints of the "Fisher condition" (every child has one mother and one father) that should have been incorporated in mate desertion models (SI 13.14). The ASR perspective accentuates issues about the forms of competition between males. When there are more males than females, desertion of a fertile female may be a poor strategy. A paired male may guard or sequester her more vigorously, and may need to conform to her requirements and preferences more readily. Unpaired males must either tolerate polyandry with or without "multiple paternity," seek surreptitious matings, or adopt opportunist or high-risk strategies, like raiding or emigrating. That all these strategies are known may suggest limitations to the application of the new models to some organisms or societies. We need additional factors to tell us which strategy males will adopt.

We now have some feel for the value of different targets to potential helpers. The sensitivity analysis showed that the impact on fitness would be greater for improvements in survival of children under age five years than of older children, and for increases in fertility of women aged 18–30 more than at a greater age. The present chapter suggests that there are limits to the amount by which a woman can increase her fertility, and she may have more effect on her RS by enhancing the survival of her children, for instance by enlisting helpers. The data on variance in men's reproduction emphasize men's competitive situation. A man might have more scope to increase his fitness by leaving for a younger, more fertile wife (if he can find one), even if there is some increase in risk to his deserted children. The standardized variances do not describe actual outcomes but a general bias in the opportunities, if they can be seized.

In Chapter 15, we will look at the relationship of RS to some features of marriage, and calculate a "Bateman's gradient" for the Hadza. The phrase is in quotation marks because we must consider carefully what the measures actually represent, and because I am not particularly interested in traditionally or popularly perceived implications about sex differences in sexual motivation (which are amply discussed by Brown *et al.*, 2009 and Schacht *et al.*, 2014). Before that, we will look at men's reputations among women as skilled hunters or traders, an arena in which Hadza men may compete.

# 14 Men's and women's reputations as hunters, traders, arrow makers, and diggers

Interviewer: "What is men's work?" Hadza woman: "To hunt, and to sit."

Early in our project, Hawkes found a puzzle in the apparent tendency for Hadza men to specialize in hunting large game. During O'Connell's follows of Hadza men, they seemed to pass by opportunities to take small game. When they took small game, they were as likely to eat it all themselves as to bring any of it back to camp, as described by Woodburn (1968a), Marlowe (2010, p. 120), and Wood and Marlowe (2013, p. 309). They ate honey or berries as they moved through the bush (Photographs 14.1 and 14.2). The men hunted nearly every day but most days came back to camp empty handed. They seemed to target their efforts toward the rare success at hitting something really large, like a zebra or buffalo. The resulting meat mountain was eaten by anyone who turned up, with (as Woodburn has described in 1982 and 1998) the hunter apparently playing little part in the distribution (sometimes he was not even present while others were dividing the carcass and loading up with meat to take back to camp). This seemed like a poor way to feed one's children but a good way to become popular.

To test our impression that men often ignored small game, Hawkes ran an experiment in which she paid men to hunt small game (Hawkes *et al.*, 1991). The men knew how to set traps; all had done this as children, like boys I had observed. As Obst (1912) and Woodburn (1968a, p. 51; 1970, p. 47) had commented, adult Hadza appear not to use traps in the normal course of events. Hawkes found that hunting and trapping small game could provide a daily supply of meat sufficient for a man's household. A daily supply of adequate protein would certainly be better for his children than the occasional feast. An even larger supply of food could be obtained if men brought home honey or plant foods in addition to small game more often. If hunting was a way to provide for wife and children, then big game hunting seemed to be not the best way to set about it. A simple payoff matrix (Hawkes *et al.*, 1991, table 4) showed that if men are foraging to feed their children, they should defect from any big game hunting collaboration and pursue small game. The argument was not that Hadza men were uninterested in their children but that big game hunting may relate more closely to the competitiveness of men's lives, which follows from their high variance in RS, and tends to become expressed in a variety of forms of "status seeking."

What is the selective advantage of "status"? Hawkes began to propose a series of ideas about big game hunting as a "show-off" strategy (Hawkes, 1993). She proposed that recipients of the large shares of meat might make concessions to the hunter to

**Photograph 14.1** While hunting, men snack on berries as they go. © James F. O'Connell, 2015. Reproduced with permission.

**Photograph 14.2** While out hunting, men and boys stop to feed themselves. Here they eat large chunks of honey comb. © James F. O'Connell, 2015. Reproduced with permission.

keep him in camp and continuing to bring in the occasional meat bonanza. However, others quickly pointed out the "secondary public goods problem": why be the one who makes a costly concession, why not sit back and rely on others to do so? Nowadays, a broader range of routes by which such phenomena can evolve have become better known (competitive altruism, Roberts, 1998; biomarket processes, Noe *et al.*, 2001, Barclay, 2013; partner choice, Martin Novak and colleagues, such as Lotem *et al.*, 2002; Barclay and Willer, 2007; Nesse, 2007; Fu *et al.*, 2008; and reputation, Milinski *et al.*, 2002, and others). At the time, the analogy with Zahavi and Zahavi (1997) was more evident. Selection would favor successful big game hunters if their bonanzas served as a visible and difficult-to-fake signal of qualities, which the audience benefits from by attending to them. Big game hunting was as much an advertisement of "quality" as an economic technique. Prospective wives might be drawn by the difficult-to-fake signal of genes for strength, intelligence, persistence, and courage, qualities that, if transmissible, might enhance the quality of their offspring. Prospective rivals might be deterred by the same qualities. "Show-off" as costly signaling was set out by Hawkes, Bird, and Smith (Smith and Bliege Bird, 2000; Bliege Bird *et al.*, 2001; Hawkes and Bird, 2002).

A chance conversation in an Mbulu hotel set the issue in a more direct light. Gudo and I had stopped for lunch on a refuelling trip (by 1995, one could buy diesel in Mbulu) and somehow met with the economist from a district resource survey. He eagerly set about asking Gudo questions, for he was expected to submit a report on each lifeway in the district and had not had the chance to visit any Hadza. The Tanzanian economist's questions were direct: What do you like to hunt? Buffalo. Can you keep all the meat? No. So why hunt them? Hunters get more women. We will see that there is some truth in Gudo's reply.

Some of our ideas about Hadza men's hunting had been well foreshadowed in Woodburn's account of Hadza life, especially in his (1998) extended discussion of sharing and hunting. Having pointed out that Hadza "believe that, to be successful at hunting, the hunter has to be modest and play down his success. Ostentation damages hunting. It enables animals to escape" (p. 59), and having also posed the question "why do Hadza men hunt large animals ... if they get so little material benefit from it ... ?", he remarked "you might think that the answer is prestige, that the status of successful hunters is in various ways enhanced. Well, is it? The answer is, not by much." But he closes this section by commenting "A successful Hadza hunter may have more chance of entering into a marriage with a woman who is seen as desirable and of maintaining his marriage than an unsuccessful hunter. But the solitary disposition of many hunters tends to conflict with success in marriage and I do not consider that enhancement of marital prospects is the central factor in motivating Hadza hunters to hunt." Woodburn (1968a,b) described some requirements of Hadza husbands, as quoted earlier, and commented that "Whether a man hunts is his own affair. Other men will not put pressure on him. He may, though, find it more difficult to marry a wife, or, once married, to keep a wife, if he is unsuccessful in hunting big game." (1968a, p. 54), and that "the husband should keep his wife and mother-in-law supplied with meat and with trade goods."

The "show off" idea elicited immediate opposition and has proved quite unpopular among some of our colleagues, even including Hadza researchers (Marlowe, 2003; Wood and Marlowe, 2013, 2014), while others have found it usefully provocative in leading to an account of, for instance, the circumstances under which women hunt, and the origins of a sharing ethic (Bird, 1999; Bliege Bird and Bird, 2008). The opposition may have arisen partly from the unintended baggage carried by the phrase "show off" (Hadza men are not boastful, nor flamboyant), but mainly from the contradiction of our long-held belief that paternal provisioning was a key piece of the evolutionary puzzle. Only very recently have there been papers that go beyond a simple dichotomy of theories to consider the complexities of male reproductive strategies long known in other species (Gurven and Hill, 2009; von Rueden *et al.*, 2011; Winking and Gurven, 2011).

Among the opposition arguments was the idea that a hunter of big game benefited by reciprocation (as meat or as other services) from those who received shares of the meat from his hunt. One result would be that a successful hunter received more frequent shares of meat. This reciprocity would be distinct from the classical anthropological view of hunters as showing "generalized reciprocity" in which gifts continued to be made even to those incapable of reciprocating. Men's neglect of plant foods was countered with evidence about the special nutrients in meat, which were claimed to be sufficient reason for men not to target more honey or plant foods (but see Speth, 2010). The theoretical details of reciprocation are restrictive and "reciprocal altruism" is too easily used as a blanket solution to almost any form of economic arrangement or social interaction.

Trying to find out which men Hadza women believed were good hunters might offer some clues. This could be combined easily with the reproductive history interviews. First, would they have any interest in the question? Second, were some men singled out, or were names given randomly? Third, were husbands, brothers, fathers preferentially nominated? Fourth, what could we find out about the lives of men who received many nominations and men who received few or none? Whether we think of big game hunting as advertising "good genes," or as evidence of the ability to provide for wife and children, we might expect women to be interested in the hunting, trading, and other abilities of the men around them.

Collecting a large body of data on the actual hunting effort and success of a good sample of Hadza men would have required a completely different field method (and may even be quite difficult, Hill and Kintigh, 2009). A reputation might be built on a closer and longer period of observation than we could make, given the scarcity of success at big game hunting. People's impressions of expertise or success might be an adequate proxy for success. Perceived success could even be more important to the Hadza than actual achieved success; a man's reputation might be as important as his observable, local performance. Reputations might differ from performance in important ways. For example, they may last longer. People in other camps hear about catches of large animals, and it would be easy for the people in a region to build a communal impression of who were exceptional hunters. People might base the reputation on too small a sample, but nonetheless, it may grow and endure by gossip, being reinforced by talk as much as by performance. This may give it a high "inertia"; it may take time to build a reputation (or perhaps not), and even more time to lose it. A man's reputation may be known to people who have never witnessed his performance firsthand. This seemed to be as true among the Hadza as elsewhere. A reputation may not be totally true (that is, not strongly correlated with success over some duration). It may span periods of ill-health or injury. It resides in the observations of people who live together every day, and mingle and converse often with people around their region, and sometimes with people from other regions.

In Blurton Jones *et al.* (1997, p. 300), I reported correlations between observed hunting success and later nominations. Hawkes *et al.* (2001, appendix A) show data on hunting by 18 married men in the camps where they stayed during their year in the field. The number of large animals killed by each was correlated with my 1992–1997 score of their number of nominations at $r=0.731$, $p<0.05$. From my stay in late 1986, I found a correlation of 0.682, $p=0.05$ between men's observed success and their 1992–1997 nominations. Marlowe (2010, p. 215) reported that men's hunting reputations correlated with their overall food returns with $r=0.319$, $p=0.014$, $df=56$; with hourly return rates $r=0.310$, $p=0.018$, $df=56$; and with the amount of meat they gave away. My intention is to use nominations of the larger population as a proxy for actual hunting success, or for some clear opinion of the population of women. If there are correlations with other measures, we cannot tell for certain whether these are due to the meat, or to the reputations. Nor can we tell whether the reputation is because of luck at hunting, skill at hunting, skill at choosing the best places to camp and to hunt, or to good salesmanship, least likely because in our observations, successful Hadza hunters displayed the modesty described by Woodburn.

## 14.1 The interviews and computations

I added the following questions to the reproductive history interviews: Can you tell me the names of some expert hunters, people who often shoot animals? Subsequent questions asked about experts at trading with non-Hadzas or white people; experts at collecting honey; experts at making arrows; and in 1997, I asked a few women about expert farmers. I also asked women to say who were some hard-working women. Women were encouraged to list up to three individuals in each category.

One hundred and twenty seven women were asked to nominate experts. In Mangola region, 47 women were asked to nominate experts; in Tliika, 57 women; and in Siponga, only 23 women. The median age of the women asked for nominations was 30, with a range of 14–67 years. Some women were asked again in another year. Nine of the women were interviewed in a different region on their second occasion. Similar numbers were asked about expert hunters, traders, and arrow point makers. Many fewer were asked about honey-gatherers (in 1992 only) and experts at farming (in 1997 only). These questions were not asked of every woman at every visit. They were among the "filler" questions, added if the essential questions about children had been covered quickly. Statistical considerations were far from my mind. The questions were asked during the 1992, 1995, and 1997 field visits.

Most women easily named two or three or more men (mean, 3.3 men). Some women failed to identify any experts. Eight responses (from eight different women) were generalizations that implied that most men could hunt, such as "there are no expert hunters," or "all the younger men." There were 25 other answers (from 23 different women) that I interpreted as "don't know." More interesting were apparent displays of reluctance to name the husband, although 28 husbands were named. Fifteen women, mostly very young, named only their father. Roughly equal numbers of men received nominations as expert hunter and expert trader, slightly fewer as expert arrow maker. The 436 scorable nominations include 397 nominations of an identifiable man; 104 different men were nominated as an expert hunter, 57 of them were named more than once.

Men's reputations were scored by the number of different women from whom they received a nomination. A second nomination by the same woman on another occasion was not counted. This was important; I did not want the women who were asked the questions on two occasions to weight the men's scores toward their preferences. Different numbers of women were interviewed in each region, and most men were nominated only in their own region. Men in a region where more women were interviewed are likely to receive more nominations. Men in the Siponga region would be especially handicapped, for only 23 women were interviewed about experts in that region. Hence, I pro-rated the men's scores for the number of women interviewed in their region. The final score for each man in each skill was 100 times the number of women nominating the man (he scored 1 for each woman, no matter how often she had nominated him), divided by the number of women interviewed in his region. In other words, the score represents the percentage of the interviewed women in his region who nominated this man. There may be real differences in hunting success between the regions, but my interest here is in success of individual men relative to other men, and the pro-rating should help with this goal by leveling out effects of opportunity and number of interviews.

## 14.2 Descriptive results

The frequency distribution of pro-rated hunting nominations is shown in Figure 14.1, and raw counts and trader nominations are given in SI 14.1. The frequency distribution of nominations is far from normal. For regressions, I have transformed the data by adding 0.5 to the zero scores, and computing the logarithm to base 10 of the zeros and the pro-rated nominations. Plotting the logged measure against the original measure suggested that a dichotomous measure might be just as useful. For most analyses, I dichotomized the scores, separating men with pro-rated scores greater than five from the remainder. There were only 37 of these "expert hunters." The analyses that follow were repeated for the untransformed score and the dichotomized score; results were similar. The real distinction seems to be between men who get some nominations and those who get none.

Split-half reliability, and a comparison of the first and second interviews (for the 50 women interviewed twice) gave encouragingly consistent results (SI 14.2). Women appeared to have some consensus on which men were good hunters, and which were not. Certain men were repeatedly named, but the majority were not named. The pro-rated scores showed no significant differences between regions.

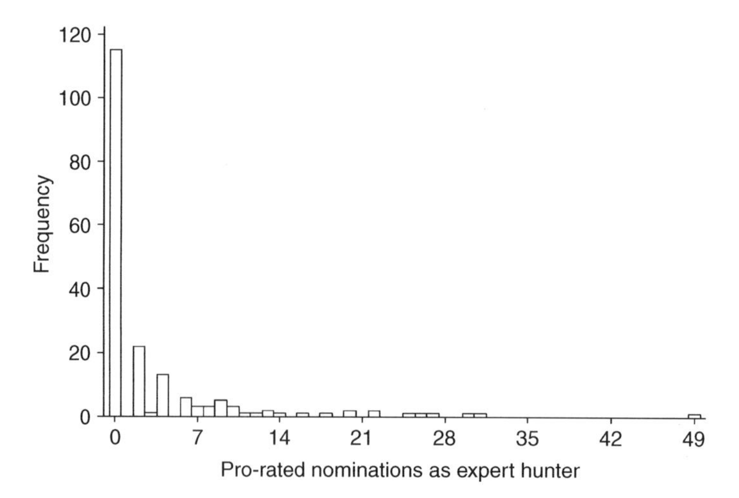


**Figure 14.1** Histogram of pro-rated nominations as expert hunter. (see table SI 14.1). Men with score zero were men seen in two or more censuses in the core study area who were not nominated; 104 men received one or more nominations, 37 men received a pro-rated score of five or more.

## 14.2.1 Experts at making arrow heads

Arrow making was included because it seemed that some men were distinguished craftsmen from whom others sought help and advice. When you sit at the men's place, there is always someone tinkering with arrows (Photographs 14.3, 14.4, 14.5, 14.6). O'Connell *et al.* (1991, table 2) show men occupied in weapons' maintenance in 129 of 200 scans of men in camp (a small sample of O'Connell *et al.*'s total record of scans). Hawkes *et al.* (1997, table 1) show adult men working on manufacture and repair for 7.99 hr/wk (1.14 hr/day). Marlowe (2010, table 4.6, fig. 5.7) implies a similar amount of time spent repairing arrows and bow. For much of the time, each man is working on his own arrows or arrow heads, but from time to time, arrow heads are passed to and fro and a recipient works on what is apparently another man's arrow. Sometimes it becomes clear that one of the men is the center of such exchange, help, and advice. Their expertise as arrow makers seemed little related to their own use of arrows. Some of these apparent experts were men who we had never seen catch anything.

## 14.2.2 Experts at getting honey

Questions about expert honey gatherers were dropped prematurely; it seemed as if women felt that all men could do this. The most interesting finding concerned men who were described as being afraid of bees. Asking about the whereabouts of one man, informants chuckled when asked if he had a wife. Their comment was "he is afraid of bees." At any time of year, men bring home a small amount of honey from time to time, which they usually keep well hidden until inside their house (see also

**Photograph 14.3** At "the men's place." Resting, watching, talking, listening, and working on arrows. © James F. O'Connell, 2015. Reproduced with permission.

**Photograph 14.4** Some men are regarded as experts at making arrow heads. © James F. O'Connell, 2015. Reproduced with permission.

Marlowe 2010, p. 234). Our questions to women turned up three more men who were described as being "afraid of bees." The attitude to these men seemed to imply this was a significant handicap in being taken seriously by women. I still wonder whether these are men who had the misfortune to have developed allergic reactions to bee

**Photograph 14.5** Fletching. Vulture feathers are preferred for some arrows, a tuft of guinea fowl feathers are used for others. Detailed illustrations and description in Woodburn (1970). © James F. O'Connell, 2015. Reproduced with permission.

**Photograph 14.6** Making a bow. Some men are widely recognized as expert hunters, and our records suggest their reputations are deserved. We do not know whether they succeed because they make themselves better bows, better arrows, can handle stronger bows, try harder, understand animals better, or have runs of good luck. © James F. O'Connell, 2015. Reproduced with permission.

stings. Recent researchers among the Hadza have paid closer attention to honey, perhaps because we reported little about it, or perhaps because it now forms a more significant part of the Hadza diet (Crittenden, 2011; Wood and Marlowe, 2013, 2014).

### 14.2.3 Experts at farming: constraints on the farmer's wife

In one field visit (1997), I also asked about expert farmers. Many had had the opportunity to farm during settlement attempts. A Hadza man who farms successfully for a period of years might seem to have an economic advantage over one who pursues the unpredictable career of hunter. From the point of view of a woman contemplating marriage (my working assumption that Hadza women have a large amount of choice in this matter is supported by the way the women talked about their marriages and divorces), a farmer, whether non-Hadza or Hadza, offers a different choice from a hunter. The farmer may supply food more constantly (on a day-to-day basis), and food over which s/he can perhaps exert more choice about who gets to eat it. However, the woman married to a farmer may not have the economic freedom and opportunities of a woman who lives in the bush. For the farmer's wife, distances to water may be greater (or less); distance to firewood will be greater (Biran *et al.*, 2004); distances to wild foods may be much greater. It seems likely that marriage to a farmer is a radically different situation from marriage to a hunter. If so, the implications for theoretical concepts of marriage could be profound.

I had hoped that some of the comments in the women's replies would help me understand why so few Hadza farm, a question that has long perplexed anthropologist and government development officers alike. A favorite reason given by Hadza is that they do not know how to! Enough of them have had experience at settlements or working for farmers to make this an unconvincing explanation. Another reason given is that when you grow some food everyone will come to eat some of it and you will not be left with enough to store until next harvest. Bob Ward (1999) reports of Danieli Tawashi after the first harvest at Munguli that "tribal members came in from all directions and they were taking the harvest he had laid up for the dry season. It's impossible for one Tindiga to deny another in need if he makes a request for help." Harvesting enough to last to the next harvest is a problem for farmers even in farmer society (Jenike, 1988, 1995, 1996; Bailey *et al.*, 1992; Wilkie *et al.*, 1999). Like Woodburn (1982, p. 447), I find this Hadza explanation quite persuasive; it seems more in tune with the realities of Hadza life (Blurton Jones, 2015). Howell (2010, p. 194) makes a similar comment with respect to the !Kung sharing and failure to farm consistently.

It may be that the effect on women's lives due to marriage to a farmer is another contributor to this issue. The comments of Hadza women who have returned from life married to a non-Hadza farmer displayed their deep objection to their relative confinement and servitude as a farmer's wife. The picture is complicated by the activity of some women as farmers. The 32 women asked to name good farmers nominated nine women (two of these nominated themselves) as well as 25 men. Five said there were no expert farmers among the Hadza. Ten said there were some who knew how to farm but mostly at Mono (the 1989–1991 settlement site). Even where women farm and control the farming productivity by their own effort, access to land (on a small and local scale) would be much more easily taken from them than access to bush food can be when they are forager women.

## 14.3 Nomination of husbands, fathers, and brothers

The nomination score could be biased away from representing the nominee's reputation if the women merely nominate the men best known to them, such as husbands, brothers, father, uncles. In this case, a man's nomination score might be merely a reflection of the number of kin he has.

Some women nominated their fathers; for example, among fifteen different women, two were sisters who both nominated their father, thus there were 14 men nominated by a daughter, and only eight of these men were in the reproductive success (RS) file. Women who named their fathers were significantly younger (mean age 24) than those who did not (mean age 31, $p=0.009$). However, they included a woman aged 40 and another aged 45. Five of the oldest women nominated men who had died before our study began, including the oldest to name her father. Expecting that hunting reputation would be a criterion by which women chose a husband, I was surprised during my fieldwork to find some of the youngest women apparently unable to name any good hunter, other than their fathers or current husbands. Nonetheless, on examining the data, I found that although on average significantly younger, most of these women did name at least one man although they tended to name fewer than did older women.

Thirty-one women nominated their husbands, some more than once (28 men, of whom 23 were in the RS file). Twenty-one women included a nomination of a brother among their nominations (20 men). Husbands comprised about 7% of the total identifiable nominations, and brothers about 5%.

A wife can add just one nomination to her husband's score. Thus, wife nominations make a small proportional difference to a man's score. Furthermore, men who are nominated by their wives are also nominated by several other women. In later analyses, I divided men into those with a pro-rated score of five or more, or less than five. A table in SI 14.3 shows that if a husband is nominated, he is far more likely to fall into the highly nominated category. The same was true for brothers. I also looked at how many men would change category if nominations by a wife were omitted. A mere two men would change category. Therefore, we can probably ignore the issue of wife nominations for the present.

## 14.4 Careers: age distribution of reputations

Figure 14.2, and SI 14.4 illustrate the distribution of nominations by men's age. Age was taken as the age in 1995, the approximate mid-point of the interviews about experts. The fitted lines are from regressions of nominations on age, $age^2$, $age^3$, $age^4$, and $age^5$. This was chosen to increase the sensitivity of the fitted lines to age changes. Stepwise

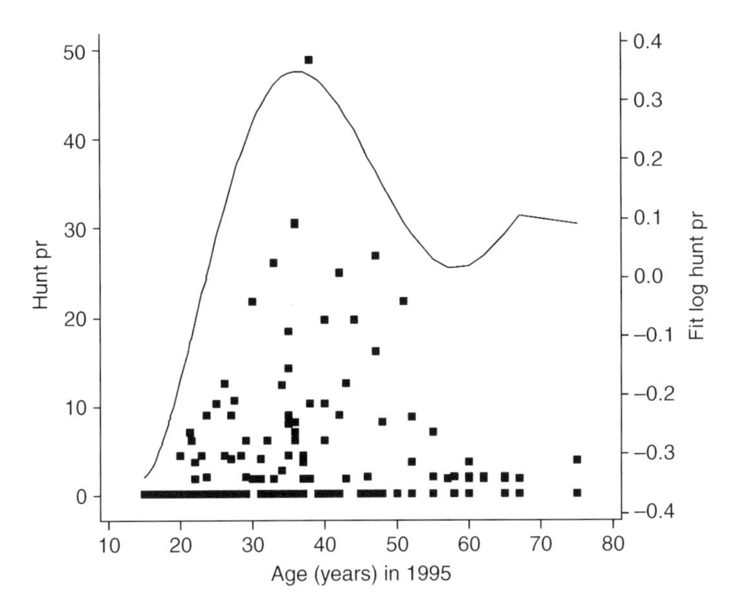


**Figure 14.2** Pro-rated nominations as expert hunter (hunt pr) by men's ages. Fit of log-transformed nominations by age. Fitted line is to age + age$^2$ + . . . + age$^5$. Plotted against men's ages in 1995, the mid-point of the interviews about experts.

regression shows the age pattern is far from linear. The contribution of age$^2$ is substantial and very significant. A plot of hunting nominations by age (Figure 14.2) shows that the highest scores are among men in young middle age, from 35 to 45. Some young men score high, but fewer over-60s do, although I have recorded kills by men as old as 70. The fitted data suggest that different skills may peak at slightly different ages; hunting around 40, trading around 50, and arrow making reputation continues high into old age.

## 14.5 Hard-working women

Because of our interest in grandmothers, I asked 72 women to nominate women who dug a lot or who were hard workers, making no mention of age. In the first year that I asked this, women named only older women. It seemed automatically assumed that young women did not work hard. Thus, on my next visit I asked for names of some hard-working women, and then followed this with asking for names of some young women (tlakweko) who work hard. I combined the results, and pro-rated individuals' scores by the number of women asked the questions in the subject's home region, in the same way as for the men. Results are shown in Figure 14.3. While I could elicit names of some hard-working young women, the Hadza bias toward thinking of older women as exemplifying hard work shows clearly on the scattergram. Cashdan *et al.* (2012) recently reported a similar observation among the Hadza. Women nominated as hard-working foragers were older than others.

My data also show a decline in women's nominations after the age of 70, perhaps an age at which older women's vigor begins to decline (strength tests were used in

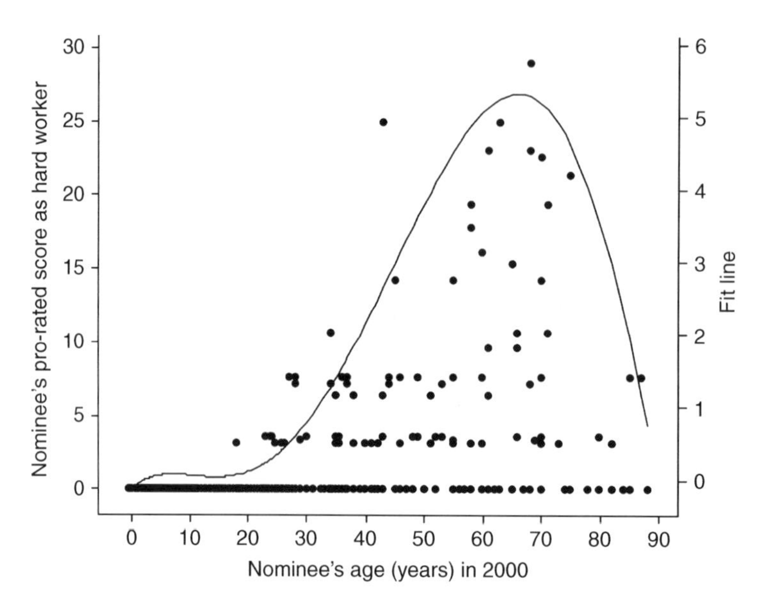


**Figure 14.3** Women nominated as hard workers by other women by age of nominee. Nominations pro-rated for the number of women interviewed in the nominee's home region. Women in >2 censuses, and >60% of observations in core area; 436 women never nominated; 85 women nominated more than three times. Note that peak age is many years later than for hunters.

Blurton Jones and Marlowe, 2002, and body weight discussed in Chapter 16). That the data support a rise and fall with age is confirmed by the greater $r^2$ of a model that includes age, and $age^2$ as compared to a simpler model. $Age^3$ also shows a significant relationship to nominations. Adding further powers of age does not improve $r^2$ but appears to give a more realistic fit at very low or very high ages.

## 14.6 Men with high reputations have greater reproductive success

When looking at a possible relationship between reputation and RS, we need to take into account two findings reported earlier: nominations vary slightly (but not significantly) with region, and with age. Some aspects of RS could also vary with region. We must control for region and for the curved relationship to age.

Reproductive success (sslive, which is independent of age) is significantly predicted by combined nominations, by hunting nominations, and by trading nominations (Table 14.1, top panel). The most nominated men, controlled for their age, had the most living children. The standard score of the number of children born (ssborn) to a man is even more strongly predicted by these variables (Table 14.1, center panel). Regressions in which the untransformed data are used give the same result but account for less variance. The categorization of men into those with more than five or less than five nominations also shows a difference in RS (SI 14.5). Stepwise regression shows that the contribution of nominations (entered after region, and age and $age^2$) to the prediction of the RS measures is significant but modest. This

**Table 14.1** Men's reputations and reproduction. Men with more nominations have more births, and more living children for their age. A smaller percentage of their children survive. Log transformed pro-rated nominations. From stepwise regression model controlling for region and the man's age: RS (sslive) = Mangola + Tliika + Siponga + end-age + end-age$^2$ (step 1). Step 2 adds a nomination score. NS: fails to meet alpha criterion 0.15 to enter. There are 188 men in the RS file and available to be nominated; 63 received at least one hunt nomination, 76 with trade nominations, 64 with arrows, and 111 with at least one nomination for any skill. Top panel: dependent variable is men's RS ("sslive") at end of observation. Center panel: dependent variable is men's births ("ssborn"). Lower panel: is for survival of children of the 125 men who had at least one birth. Survival is measured as standard score of ($N_{survive} \times 100/N_{born\ by\ end\text{-}age}$). ">" separates step 1 from step 2

| RS | Beta coefficient | p | Adjusted r$^2$ | Change in Mallow's Cp |
|---|---|---|---|---|
| Log hunt | 0.31 | 0.024 | 5.4>7.6 | 10.2>7.0 |
| Log trade | 0.44 | 0.005 | 5.4>8.9 | 13.1>7.0 |
| Log arrows | NS | NS | | |
| Log general | 0.40 | 0.001 | 5.4>10.89 | 17.4>7.0 |

| Births | Beta coefficient | p | Adjusted r$^2$ | Change in Mallow's Cp |
|---|---|---|---|---|
| Log hunt | 0.57 | <0.000 | 4.3>11.8 | 21.6>7.0 |
| Log trade | 0.55 | 0.001 | 4.3>9.4 | 16.3>7.0 |
| Log arrow | 0.36 | 0.014 | 4.3>6.9 | 11.1>7.0 |
| Log general | 0.60 | <0.000 | 4.3>15.9 | 31.0>7.0 |

| % survival | Beta coefficient | p | Adjusted r$^2$ | Change in Mallow's Cp |
|---|---|---|---|---|
| Log hunt | −0.41 | 0.018 | 4.7>8.35 | 10.7>7.0 |
| Log trade | NS | | | |
| Log arrow | −0.27 | 0.126 NS | 4.7>5.8 | 7.4>7.0 |
| Log general | −0.33 | 0.045 | 4.7>7.1 | 9.1>7.0 |

finding replicates, with a much larger sample, and better controls for age, the results that I reported previously (Blurton Jones *et al.*, 1997, 2000).

The results are different for child survivorship (Table 14.1 lower panel). Coefficients were negative – having a high reputation as a hunter predicted lower child survival. Both log hunt nominations, and the combined nominations (but not the log trade nominations) entered on step 2 with significant p-values and negative effects. We need to recognize that the measure of survivorship is missing for any man who fathered no children. Thus, two of the low fertility, high nomination outlier men visible in Figure 14.4 dropped out of analyses of child survival.

The four men reported as "scared of bees" had very low RS (sslive, b=−0.846) and fertility (ssborn, b=−0.903), but the two who had children were no less successful at keeping them alive (sssurv, b=0.292). Comparing only four men with the remaining 184 cannot possibly give a statistically significant result, but the low RS of these four men reminds us

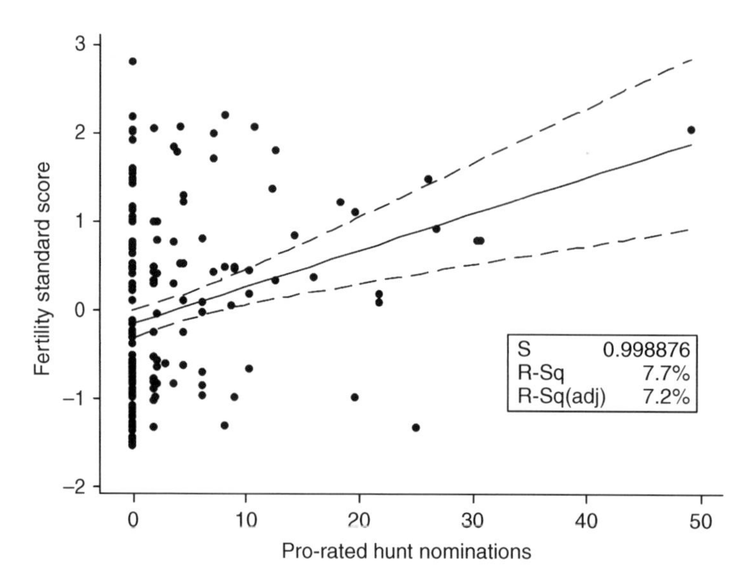


**Figure 14.4** Men's achieved fertility ("ssborn" at final observation) by nominations as an expert hunter. Fitted linear regression line with 95% CI. N=188 men.

of the role of honey in relations between the sexes. Married couples occasionally take a day off, apart from other people, to roam about in the bush "to eat honey." In some seasons, collecting honey as a nuclear family is a prominent activity for some weeks.

Women's nominations as a hard worker showed no relationship to their RS or child survival ("sssurv"). Because women's nominations increase with age, like the work hours reported by Hawkes *et al.* (1997), we may wonder whether they relate to women's success as grandmothers (Chapter 18).

## 14.7 Discussion: the inexpert majority, hunting careers, and competition

Men who are frequently nominated as expert hunters or traders father more children and are likely to end their reproductive careers with more live children. The few expert hunters, however, have lower child survivorship than others. We would expect higher survival if the benefit of big game hunting is better provisioning of their children. Even if better provisioning does not distinguish good hunters, we still need to account for their greater RS. They father more children. But why? What would be the advantage to prospective wives or mistresses? We will be attending to these questions in several of the following chapters as relevant data emerge. Meanwhile, there are some topics specific to reputations that need discussion.

### 14.7.1 The inexpert majority

Only a minority of men received many nominations in any category. What about everyone else? Do the poor hunters specialize in some other strategy? In his 1977 census, Lars Smith asked men to say what work they did. Almost every Hadza man described himself as a hunter. In my study, most men were not nominated by anyone. As far as we can see, they all try to hunt, and given the chancy nature of hunting, any of them may succeed once in a while. Because expert hunters have lower child survival than average, then "regular Joes" must have higher child survivorship than expert hunters. Do they make up for their lack of distinction at big game hunting by bringing back more small game, honey, and plant foods? They may, but then it is surprising that we saw so few men bring home these items during our 1980s behavioral ecology field seasons. It is our firm impression that we saw no men who brought back food every day. Nonetheless, impressions and memories can be incorrect. As Wood and Marlowe (2014) suggest, we should go back to our notebooks and test our impression of how rarely we saw men bring food home. However, our 45 naturalistic follows of men in 1985–1986 recorded men taking only 14 individual small animals, obtaining 0.062 kg/hunter-day (Hawkes *et al.*, 1991, p. 244). This suggests that bringing home a small animal cannot have been a frequent event during our fieldwork. On some days, men will have eaten while in the bush hunting, including eating a bird they may have shot. By eating in the bush, on returning to camp they may have eaten less of the plant food their wives brought home, leaving more for their children, although failing to give their children meat in exchange for plant food.

### 14.7.2 Is hunting expertise a mate choice criterion?

Do women seek out good hunters and traders? Marlowe (2010, p. 184, table 7.4) asked 36 women what characteristics they would prefer in a spouse. Foraging ability was the most frequent, followed by character, looks, and intelligence. If being married to a good hunter is important to a woman's life, would we not expect the replies of young women to reflect more knowledge and concern for the hunting abilities of the men around them than was displayed by several of my youngest informants? Perhaps young women select on the basis of some cue to their suitor's potential later success; or for some of them, a more important criterion may include the compatibility of his personality.

Very few Hadza men have much of a reputation (among interviewed women) in their young to mid-20s when men first get married. Perhaps women can recognize and are attracted to qualities that predict becoming a highly nominated man. Indications of energy, sound health (such as facial symmetry), intelligence, ambition may predict later achievement. They feature in the many studies of mate choice in industrial societies as much as qualities more suggestively predictive of paternal investment. In some ethnographies, success in a single hunt is described as a significant qualification for marriage (Lee, 1979, pp. 236–242; Marlowe 2010, pp. 57–58). It could be that women believe their husband is a good hunter, and that young women are especially inhibited from naming him in this context, or perhaps realize he has yet to prove himself in the eyes of a wider audience. It seems at least as plausible that we should take their replies at face value and they actually have not noticed many specially good hunters, it is just something men do, but their fathers did indeed bring home meat from time to time. If this is the case, then how do they choose a husband? The picture is further muddied by the observation that on the few occasions any of us asked Hadza women what attracted them to their husbands, the replies note seemingly uneconomic factors. "His words" or "he does not fight" were examples heard from time to time.

### 14.7.3 Men's hunting careers and competition

The age course of hunting success shows some consistency across populations. The Ache show "a strong convex age pattern . . . over the life of a single individual." (Hill and Kintigh, 2009; and see Walker *et al.*, 2002). Both Lee (1979, pp. 236–244) and Howell (2010, pp. 114–115, 116, table 5.2) describe a similar trajectory among the !Kung. Howell comments "Lee (1979) dates the start of a hunting career from the age of 25, and describes it as continuing for 50 years, or until the hunter dies, reaching a peak between the ages of 45 and 60. Men in their 50s or 60s may exceed younger hunters in knowledge and judgement but usually not in speed or stamina. The hunting success score starts to decline around age 55, but may continue substantial into the 70s." This suggests a slightly longer and later peak than the Hadza reputation data. Lee's data were observed hunting success, and interview counts of the number of large game that men had taken. He did not interview women about good hunters as Marlowe and I have done.

If reputations, or hunting success, are relevant to male–male competition, we may get clues about the nature of hunter-gatherer male competition by comparison with reports of other measures of male competition. Graves (2010) reports on the careers of boxers as a reflection of contests of physical strength and agility. The age course (mean age at wins, 26.44, with some variation by weight class) is very different from that of Hadza hunters, traders, and arrow makers. The boxers peak at an age when !Kung and Hadza men are just beginning to enter "adulthood" and gain reputations (Figure 14.2). The boxers' peak success lasts only about two to three years. Studies of the age course of success or eminence in other competitive pursuits might be informative, especially if they were conducted as carefully as Graves' study. The 50 players ranked highest at the Soccer World Cup in 2010 by an ESPN website had an average age of 26.8, similar to that of the boxers. A systematic assessment of professional tour golfers (Mills, 2011) suggested a peak in earnings from age 35–45, rather similar to the Hadza nominations of hunters. Wikipedia lists U.S. Presidents and gives the median age at accession as 55. Between 1980 and 1988, CEOs of large companies averaged age 59 in the United States and 66 in Japan (Kaplan, 1992).

Competition between men takes many forms, from killing each other to being influential about communal decisions (politics). The requisite abilities may have different age trajectories. In male–male competition, natural selection is ultimately concerned only with competition for conceptions. How this translates into behavior will depend on many factors including locally shaped costs and benefits. Where women are supported by their kin (and like Hadza women, can be economically self-sufficient), and can exert choice over prospective mates, men must compete partly over characteristics that appeal to women. Under some conditions, forms of patronage are effective. Under others, skill at making and maintaining alliances will predominate, or ability in warfare will be important. Killings are always a possibility, sometimes a successful tactic (Chagnon, 1968, 1979; MacFarlan *et al.*, 2014), but sometimes not, when retribution follows rapidly and effectively. Woodburn (1979, p. 252) has described the ease with which a Hadza man could assassinate another without trace "Conflict among the Hadza can easily be lethal. Every man has constantly at hand his bow and hunting arrows and everyone is aware of the danger of humiliating or antagonizing a man so that he may be tempted to retaliate with a poisoned arrow. The Hadza know that, because they sleep in open shelters and hunt individually in the bush, protection against a determined adversary who is willing to use a poisoned arrow is impossible."

Lee, Howell, and other researchers who have lived among the !Kung (Lee, 1979, pp. 244–246; Howell, 2010, p. 193) comment on the vigorous leveling processes that the !Kung direct toward successful hunters, who are expected to maintain a modest demeanour. The !Kung feel that "arrogance or bossiness could be a problem from a successful hunter and serve notice that there are limits on the power of a good hunter to control others." Boehm (1993) discusses more examples of societal restraints on dominance and "bossiness," contrasting these with the non-human primate context. I have previously quoted Woodburn (1998) reporting that Hadza believe that "Ostentation damages hunting."

Among both the !Kung and the Hadza, with every man lethally armed, and seldom more than an arms length from his weapons, unbridled male–male competition: (a) could easily result in large numbers of rapid deaths, and (b) need no longer follow the age course of physical strength and agility. The Hadza homicide rate is quite low (Chapter 8). Agility or strength could, only exceptionally, determine the outcome. We heard of one, and only one, incident in which it became necessary for a potential victim to dodge arrows. The elderly target succeeded.

While human contests may retain some aspects of contests among our primate forebears, they take a much wider array of forms. We should note especially that mammalian male reproductive careers differ greatly in duration. The meteoric rise and brief reign of a dominant chimpanzee or baboon (see, for example, Alberts *et al.*, 2006, fig. 2a) is very different from the lengthy reproductive careers of men in human societies such as the Hadza (first birth around age 25, last birth for some is in their 50s or 60s) and other hunter-gatherers.

### 14.7.4 Personalities of different kinds of expert

Previously, I cited Woodburn's comment on the "solitary" personality of good hunters. We may contrast them with the good traders, and although there was a significant positive correlation between scores as a hunter and as a trader, it was not strong (SI 14.6). Many men score as a good trader but not a good hunter, and vice versa. Good traders must overcome the fear and distrust of outsiders shown by most Hadza. Men noted as skilled arrow makers tended not to be nominated as good traders. Traders have to be more outgoing, and unafraid of negotiation with strangers. While today most "trading" involves interaction with people from other tribes, archaeology shows us that trade was extensive in a world of hunters among hunters. Even in the distant past, it may have been possible for men to follow their personality traits to achieve success either as a trader or as a hunter. It may be that the existence of these "niches" selected for the maintenance of suitable personality variants (Gangestad and Simpson, 1990; MacDonald, 1995; Nettle, 2006). I doubt that it accounts for the origin of these variants, given the existence of similar individual variation in other primates (Higley *et al.*, 1996; Schwandt *et al.*, 2010 from Steve Suomi's macaque research group; and vervets, McGuire *et al.*, 1994; Fairbanks, 1996). Theory papers appear to have neglected the effect of fortuitous individual variation in circumstances, size, or vigor, which may determine which of a suite of alternative strategies will maximize fitness of a given individual (Draper and Harpending, 1982).

### 14.7.5 Cautions

Hadza women seemed to agree about who were successful hunters, in the time frame they had in mind when they answered my questions. These time frames clearly vary, as witnessed by the few older women who named long-gone but well-remembered men. Kelly (2013, p. 223) suggests that reputation as a good hunter includes being a man who is less likely to come home empty handed. I do not think our questions were taken this way; I think they were asked, discussed, and responded to as concerning men who successfully hit large animals. A reputation for bringing home small prey and plant foods would less obviously spread outside the household than a reputation for providing conspicuous masses of meat. Our correlations of nominations with observed hunting success concerned the record of large animals (40+ kg, wart hog and heavier). However, Marlowe (2010, p. 216) reports that men most successful at shooting large animals were also most likely to bring home food of any type, much as Kelly suggested. It may be important to remember that our hunting data were collected in the area of richest large game; Marlowe's in a wider variety of locations, and apparently in an era of greatly depleted wildlife (Chapter 2). In trying to discern an enduring "correct description" of Hadza behavior, we may be in danger of failing to make use of the opportunities provided by changing circumstances.

The effect of reputations may be associated with social or long-evolved processes that affect mate choice, male competition, and mate retention. We need to remember, in thinking about the relationship between reputation and RS, and other such findings to be discussed in later chapters, that my reputation measure is not a direct calculation of meat consumption. Men often nominated may bring in more meat, and it may be the increased consumption by the wife and children who live with him more consistently than other individuals that accounts for the man's high RS. However, this suggestion is not compatible with the observation of lower survival of children of expert hunters.

The paper by Hill and Kintigh (2009) contains many important points about studying the relationships between hunting and other measures, too many to do them justice even in this lengthy book. The time scale upon which we are thinking is important. As they say, and as we have seen among Hadza, hunting success and reputation vary with age, and so does the number of children a man has fathered. Controlling for age is essential, and studies that do not control for age of men, nor age of children if the dependent variable is child growth or survival, are at risk of finding either erroneous confirmation, or erroneous contradiction of a hypothesis.

Despite their thoroughness, Hill and Kintigh have only scratched the surface of their topic. It is naturally concentrated on the justly celebrated long-term Ache data, and on the feasibility of the goal of distinguishing meaningfully between men who are all (by Hadza or !Kung standards) consistently very successful hunters. The habitat, and aspects of the hunting methods, may reduce differences between men. Much depends on which aspects of differences between men, which components of success, we are trying to assess. Several components might be interesting: average amount of food returned to camp, amount of time actually trying to hunt, kinds of prey targeted, return rate while hunting by various methods. This is presumably the reason for Hill and Kintigh (2009, p. 374, fig. 5) omitting the data on the 73% of hunter-days on which !Kung hunters caught nothing. Had we been interested in which men made most meat available to camp members, the failure to catch anything would be a very important piece of data. The distribution of skills might differ from one place to another and according to the aspect of "success" being measured. It may not be "unfortunate" (Hill and Kintigh, 2009, p. 375) that differences between successful Hadza hunters and others are extreme, and that there are only a few very successful hunters; it may be an important feature of reality. Everything we know about the Hadza suggests that this is exactly how life is for them. All men hunt for many days without catching anything. A few men catch big animals and some of them do so more often than most men. We can certainly make no claim to know why they are good hunters or whether experts differ in the component skills, some being better spotters (an ability !Kung informants emphasized to Blurton Jones and Konner, 1976), some better stalkers, or more patient and alert during a cold night in the hunting blind, others may be better shooters or build themselves better bows and better arrows.

# 15 Marriage

"Haine (the sun) was married to Seta (the moon). One day they argued and Haine sent Seta away. Later on they missed each other and have been trying to get together again. When it rains in the evening, you are seeing the tears of Seta trying to catch up with Haine. When it rains in the morning, these are the tears of Haine trying to catch up with Seta."

The tragic romance of this many-versioned Hadza story is a stark contrast to our early field experiences. Soon after we began our behavioral and ecological observations, staying in a camp and observing people's daily activities, we began to wonder why Hadza women kept husbands. Women toiled in the hot sun all day digging tubers, then came home and hammered baobab seeds until dark, while men sat in the shade talking, smoking, and fiddling with their arrows (Photograph 14.3). At about four o'clock, soon after the women had returned to camp, men would walk to their house, take a handful of tubers and march quickly back to the men's assembly place to eat the tubers. Seldom did men, avidly professed hunters, materialize with meat. Why did the women put up with this? Occasionally a man would arrive with a small bag of honey, or a game bird, or even a bundle of baobab pods and leave them at his wife's house for her to do the lengthy processing. Only rarely, a man would arrive in camp with meat over his shoulder, park it on the roof of his house, report that people could carry more meat if they cared to go and get it, and then retire to the men's place to smoke. Men, women, and children would rush off into the bush and come back an hour or so later, laden with meat. Every house, even the house of a single old woman, ended up with some meat. If waiting around for a hunter to succeed was a reason to keep a husband, then why not just wait around for someone else's kept man to announce a kill?

We still puzzle about this, and dispute with friends and colleagues over the fitness economics of hunting (Wood and Marlowe, 2013, 2014; Hawkes *et al.*, 2014). Nonetheless, Hadza women do keep husbands, although every Hadza man would claim he keeps a wife, and say that his job is to get food for her. The life of a Hadza man is not all lolling about in the men's place. Long cold nights in a salameda (hunting blind) are surprisingly debilitating and are sometimes immediately followed by tracking an animal throughout the early cool and into the midday heat.

Adult couples sleep together in a house, along with their children, or step-children. Sometimes they eat together in or by their house, especially after dark. At the door, they have a fire. Sometimes the couple go off coyly (occasionally one can see them leave separately and join up a little way out of camp) in the morning for a day

**Photograph 15.1** Collecting honey, something couples and nuclear families do together. Most years there is a period of perhaps two months late in the wet season when honey is abundant. People can eat enormous amounts but it is also saved up to trade, especially with Datoga who make mead for ceremonial use. © James F. O'Connell, 2015. Reproduced with permission.

together in the bush "to eat honey." At some times of year couples and their children go out all day collecting honey (Photographs 15.1, 15.2) but most of the daylight hours they spend apart (Photograph 15.3). Despite the separateness of their daytime lives one can now and again glimpse friendship, care, and intimacy between husband and wife. The Hadza clearly recognize couples, and anyone can tell you who is

**Phtograph 15.2** A couple collecting honey from a small nest. © James F. O'Connell, 2015. Reproduced with permission.

**Photograph 15.3** A Hadza family beside their house. The Hadza liked to have portrait photos taken. During the day, couples are seldom together, women gathering with other women, men hunting on their own or resting at "the men's place." © James F. O'Connell, 2015. Reproduced with permission.

married to whom. Some of these couples have been together at least since Lars Smith's census in 1977 (23 years before my last census in 2000), and to judge from the parentage given for their older children, quite a few years longer, well past 25 or even 30 years. Many couples do not last. As Woodburn described (1968b), divorce seems easy, even frequent, but it is not without bitterness. Liaisons with another woman (or man) are a common reason. Remarriage is common. "Serial monogamy" is a fair description of Hadza marriage patterns.

The literature, as far back as Obst (1912), speaks of marriages among the Hadza as comprising a man and a woman sharing a house. As remarked earlier, Hadza see themselves as married or not married, and see each other as married or not. Hadza stories seem to assume that marriage simply existed always (Berger, 1943; Bala, 1998). A current husband seems to be seen as responsible for all his wife's young children, whether from one of her previous marriages or not. Most houses are occupied by an adult man and an adult woman, along with their children if they have any. Some houses are occupied by a single old person, some briefly by teenage same-sex friends. Woodburn (1968b) described Hadza marriage in the 1960s in essentially the same form as we saw it in the 1980s–1990s. He also commented on the obligations of a married man to give meat and trade goods to his wife and mother-in-law, and on the dangers of collapse of a marriage if the man leaves for a stretch of time.

Sexual jealousy can be quite strong among Hadza. Kohl-Larsen (1958) discusses the inconsistently severe penalties for adultery. This was echoed by Gudo Mahiya during his impromptu questioning by the economist in Mbulu mentioned in the last chapter: what happens if you catch another man with your wife? "Arrows, that's all," accompanied by the gesture with which Hadza illustrate shooting. I had previously heard the complaints of another man, shortly before he was thrown out by his wife, that people were saying he should be shot. We had no record of such an event as a cause of death. Woodburn reports one likely case; Kohl-Larsen (1958) described the serious tensions of a case in which a Hadza man asked him to adjudicate. He describes being the potential target of the poised arrows of the offending woman's sons and he seemed to feel that the sentence was usually somehow avoided. Marlowe and I witnessed an earnest discussion in which concerned elders were urging a girl to choose between two suitors before trouble broke out.

The evolutionary minded have paid attention to social monogamy and the idea that human males differ strongly from other primate males in their investment of time and resources in their offspring. The exchange of care for paternity has long been held to be the core of human mating systems (Washburn and DeVore, 1961; Lovejoy, 1981; Hill, 1982). However, we have not always distinguished between two fitness payoffs that males may gain from providing resources or care. Males may gain fitness from increased access to mates, or from improved survival of offspring. For male care to increase access to mates, there must be the possibility for females to choose among suitors, and some advantage to females who choose care-giving or less costly males. The advantage to females could be that caring males signal a low probability of infanticide, or that she gains survival, fertility, or offspring survival from the male care or help. In the case of the Hadza, we have questioned established views about male strategies, especially whether the selective benefits for hunting or scavenging large animals have to do with provisioning offspring or with other outcomes such as status in male–male competition, including signaling of genetic quality. Our label for this male strategy, "show-off," may have led some readers to attach meanings to the term that we had not intended.

My aim in this chapter is to report quantitative aspects of Hadza marriage, and to show data on the relation of marriage to reputations and reproductive success (RS) and other analyses that may illuminate theories of marriage.

## 15.1 Marital status in the census household lists

The census household lists allow us a preliminary look at quantitative patterns of Hadza marriage. Figure 15.1 shows the proportion of men and women of each age who were in a marriage during a census. More women than men are in a marriage in their early 20s and more men than women in their 50s and 60s. Some old men are married to much younger women, but the main reason for the relative excess of elderly married men over elderly married women is probably the female-biased sex ratio among older people.

Woodburn (1968b, p. 108) reported a similar observation. Howell (1979, tables 12.1 and 12.2) reports a very similar age course for the !Kung. Slightly more !Kung women are married at any one time. The !Kung resemble the Hadza in showing a large proportion of old men married and more old women unmarried (SI 15.1). This is likely mainly because of the rather similar mortality pattern of these populations. A comparable observation in western populations, where the sex ratio does not change so markedly, has been attributed to older women choosing to remain single rather than support an aging and "costly" man (England and McClintock, 2009).

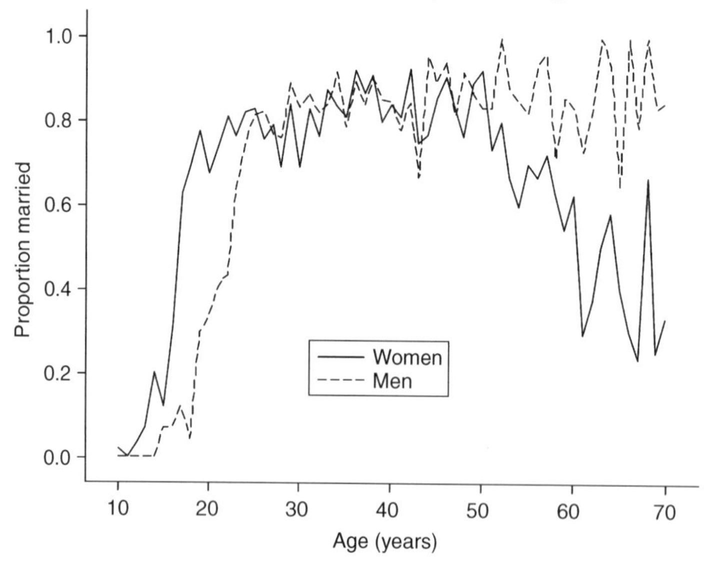


**Figure 15.1** Proportion married by age. Hadza in census lists. Table at SI 15.1.

Even at the ages of peak fertility, there appear to be a number of Hadza men and women who are single. Most are newly divorced and will remarry soon. It is interesting, and probably important, that there is always a "reservoir" of singles. Among people aged 20–40, there are between 10% and 20% of them unmarried most of the time. This must present both opportunity and competition for new partnerships, extra-marital affairs with unclaimed, unguarded partners, and divorces upon discovery followed by remarriages. We will see that divorces are quite numerous, and remarriage is rapid.

## 15.2 Marital histories of men and women

For further analysis, I constructed a set of marital histories for as many individuals as possible. The core of the annual marital history was a sequence derived from the household lists at each census. To this was added information on husbands from the interviews, anthropometry records, and from the "where are they now?" interviews, which included notes about divorces and new marriages. Men named as fathers of a woman's children were entered as husbands at the time the child was born (a procedure with disadvantages as well as advantages, discussed later). Especially for the period 1985 to 2000, a reasonable marital history could be built, including approximate years of marriage, divorce, and remarriage. A file was built for women and their marriages (if any) and another for men, and these were cross-referenced and edited (SI 15.2). This process did not resolve all contradictions. Some were clearly indications of an "affair" or temporary polygamy.

The marriage files initially included a record for each recorded year of the life of anyone in the population register from age 10 onward. The women's file included 12,630 woman-years of data. The men's file included 11,504 man-years of data. However, for many people the information is much too sparse. For most analyses, I have restricted the use of the files to people who have been encountered in the core study area on more than 60% of their records, and who have been recorded in three or more censuses (323 women and 306 men). Thus, the files did not include every Hadza who lived during the study period. Nor did every person's record include a marriage. For this and several other reasons, the men's file and the women's file are not two pictures of the same set of marriages but represent slightly different sets of marriages. When marriage data is combined with RS or nominations, for example, the number of individuals is reduced to the number for which RS or nomination data were available.

The marital histories were used to build three more files: marriage spans, divorce spans, and marriage summary:

1. Marriage spans comprised a record for each marriage of each subject (a nested structure) comprising the duration of the marriage, the year the marriage began, the year it ended, whether it outlasted the observation period (right censored), whether it ended by divorce, widowing, or death of the subject. Each record included the number of the marriage – was it the first, second, etc. for this subject. There were 595 marriages of 301 women, 418 marriages of 224 men.

2. Divorce spans included a record for each divorce and ensuing divorced status of each subject. Each record included the year of the divorce, the year of remarriage, if any, the duration of the divorced period, whether the divorce ended in remarriage, or death, or the end of the observations (right censored). Each record included the number of the divorce – was it the first, second, etc. for this subject. There were 279 divorces of 165 women, and 230 divorces of 138 men.
3. Marriage summary comprised a single record for the marital history of each man or woman seen in three or more censuses. The record showed: (i) the number of years of observation of this individual; (ii) the number of years of observation after the age of 15 for females and 20 for males; (iii) the proportion of this time in a marriage; (iv) the proportion of the time in a divorced state; (v) the number of marriages; (vi) the number of divorces; (vii) the number of marriages to a non-Hadza husband; (viii) the year of the first marriage, which could be determined approximately for young individuals who had appeared in the earlier household lists as single. There were marriage summaries for 323 women and 306 men.

Because, as described here, a record as a parent was used to contribute to building the detailed marital history, there was some danger of circular argument when relating marriage to RS. Consequently, I used the original census household lists to check some of the most important and potentially vulnerable findings. For both men and women, a summary measure of the household lists (percentage of each individual's observation, age >15 for women and age >20 for men, in a marriage) was correlated at 0.628 (p=0.000) with the analogous measure (percent of career in a marriage) from the full marital history files. It gave the same results with respect to the sex difference in relationship between marriage, fertility, and RS.

## 15.3 Polygamy

Only a few Hadza men have spent more than a year or so with two concurrent wives. Usually the wives live in different places and the man divides his time between these locations. More men have tried such an arrangement but few have made it last more than a few months. At first the women may not be aware of each other. When one of them finds out she quickly puts an end to the situation; the man is "thrown out." In one dramatic case, the wife set light to the house, which quickly burned down. Hadza women strongly object to sharing their husbands, even if we do not yet agree on an economic reason for their objection! The exceptions, women who have lived for more than a couple of years with a man who has another wife, or described such a man as her husband, are few. Marlowe (2010, p. 179) found that while 19% of women said it was acceptable for a woman to have two husbands, no men took this view. More people said it was acceptable for a man to have more than one wife (65% of men, 38% of women). Nonetheless, Marlowe describes similar outbursts of anger by women when they suspect their husband is pursuing another woman.

The Hadza men who have kept two houses going far beyond the initial discovery have been very few, but noteworthy. Woodburn (1982) discussed a case from before the time of our fieldwork. In our era, there were four. One lived in a Singida village and was evidently a successful farmer. After his death, his wives came to live in the bush in the same camp as each other and their grown children. One of his brothers, who lived full-time in the bush, also had two wives for an extended period. Another, an older man who also lived primarily in the bush and as a hunter, was living only with the younger of his wives during our time, but had children by both wives whose ages were closely intermingled. Thus, he had fathered numbers of children by each woman during the same period. During our study period, the women tended to live in different regions. The third bigamous arrangement lasted 11 years and both wives lived near each other in the same "settlement." The man earned a small government wage. There are other Hadza men who have lived as farmers, or earned a wage, and who have only been married to one woman at a time. If we count only these four long-lasting and widely recognized cases, we could try to derive a measure of the incidence of polygamy. If we use the average number of Hadza men's marriages recorded in a census (117 marriages), this gives 3.4% of marriages as polygamous. If we use the number of Hadza men's marriages in the marriage summary file, restricted to the years 1985 and 2000 (319 marriages of 168 men), we get a very low estimate, 1.25% of these marriages were polygamous, and 2.4% of men having been in a polygamous marriage. In a later section, I look at the incidence of briefer "affairs."

A few women have been involved in similar arrangements, being recorded with two husbands at the same time. These arrangements lasted less time than the men's second wife arrangements. One exception was quite memorable. The woman lived some of the time with a non-Hadza farmer close to some Hadza camps. She joined us as a guide to another quite distant camp and did not return to her non-Hadza husband for some years (our scoring system recorded this as a remarriage, not as a marriage to two men at the same time). She was often away and bore children, some of whom she attributed to the farmer and others to a particular Hadza man during the same period. Another time, we found her taking care of the young wife of a third man, an earlier ex-husband, after we had brought the couple to a village for medical attention. There are other surprising anecdotes about her. She was a notable character.

## 15.4 Nuptiality, and age at first marriage

The proportion of the interviewed women who had been married at any time up to age x years was calculated from the file of women's marital histories.

The results are displayed in Figure 15.2 (and in SI 15.3 along with numbers from England and Wales from Hinde, 1998, table 7.2). Every Hadza woman in the marital history file who reached the age of 30 had been married; no interviewed woman aged >30 had never been married. Marriages also occur quite early; half the women had been married by age 18–19 (median age for Africa is 18.2, and for Tanzania is 18.5, Garenne *et al.*, 2011). Several women remained unmarried in their early 20s, and we may wonder whether they differed in any interesting features.

Men appear to show a different picture, with median age at first marriage 25–26 (a little older than Africa as a whole, 24.1, and Tanzania, 24.0 years old, Garenne

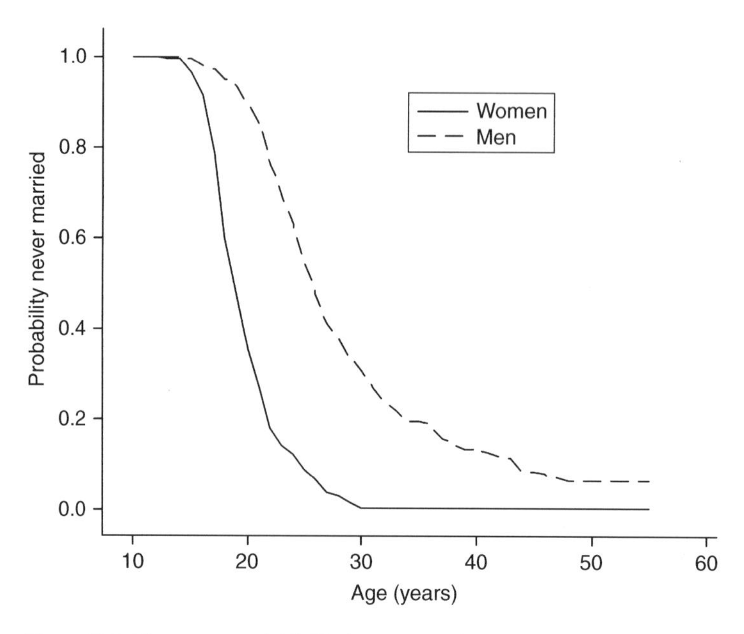


**Figure 15.2** The ages at which Hadza men and women first marry. Survival in the never-married state.

*et al.*, 2011), and with some men never marrying (Figure 15.2; SI 15.3). The data were derived from the men's marital history file by the procedure described for the women. The sample was restricted to men seen three or more times in censuses and in the core study area on more than 60% of their records, so as to limit the sample to men on whom we had better data. Comfortingly, if we use this criterion for women (instead of limiting the sample to interviewed women) we get results almost identical to the women's data reported here (SI 15.4).

The rate of arrival in the ever-married category is much greater for Hadza women than for men. The women's curve drops from close to 100% never married at age 15 to 20% at age 22, an interval of seven years. Men take from ages 18 to 34 to drop the same amount, a period of 16 years, just over twice as long. Thus, Hadza men's age at first marriage varies much more than women's. The result was the same (not surprisingly) for the ages at first birth (Chapter 13).

We may wonder why the sexes differ in their rapidity of marriage. Such a pattern could arise if there were strong societal or personal pressures on women to marry and bear children (and women did not resist the pressures) and the pressures were weak on men (or men were better at resisting the pressures). To be satisfied with such an explanation is not just to fail to seek evidence about whether there are such pressures, but it is to fail to ask why the pressures might be expected, complied with, or resisted. Likewise, men appear to mature physically more slowly than women (and perhaps mentally too, Tanner 1962, pp. 216–222), and this may affect their marriage prospects. But what could have led to such a sex difference in maturation patterns?

In a natural fertility population, women's RS is limited by time, the time needed to gestate and suckle a baby to weaning size, and perhaps further by the difficulty of caring for and feeding a succession of weaned juveniles. Given that she will cease childbearing sometime in her 40s, the sooner she begins, the more children she can pack in to the available time. Age at marriage has been widely reported in demographic literature as a predictor of completed family size. Men have a less distinct upper age limit to their reproductive careers. A man's RS is limited by the number of women he can get to give away their limited time. For example, some Hadza men, and a few American men (Lockhard and Adams, 1981; England and McClintock, 2009), divorce in middle age, marry a younger woman, and raise another family with her. In Chapter 7 we saw that Hadza men's fertility extended up into their 60s. In Chapter 13 we saw that the age at which an individual first reproduces has less influence on eventual RS for men than it does for women (the slope of the regression line is -0.08 for men and -0.14 for women). We saw from sensitivity tests on the demographic life table that fertility had its greatest influence on population increase early in a woman's reproductive career. The greater variance in men's age at first marriage may arise from the weaker time constraint on men, and the importance of competition between them. Nominations as an expert hunter are associated with an earlier age at first marriage (see later).

## 15.5 Number of marriages

Hadza often divorce, and divorcees and widows often remarry. The number of marriages was counted from the beginning of each person's annual marital history file at age 15. It is the closest I can get to a lifetime total of marriages. The final number was also entered in each person's marriage summary file. The highest number of marriages recorded from a woman was eight, from a man nine.

I plotted the number of marriages against age at the end of observation with a Lowess smoothing, using the marriage summary files (SI 15.5). For women, the smoothing suggested that the number of marriages appeared to flatten at a level just over two marriages, and at an age just over 40. This conceals much variation; the greatest number of marriages was found in a woman in her early 30s who had been married eight times. The two with seven marriages were aged 36 and 52. Limiting the sample to 104 women whose observations ended beyond age 45 (Table 15.1), the mean N of marriages was 2.202 and the median 2.0 (variance 2.124).

For men, the picture was very similar, leveling off at just below two at about age 45 (SI 15.5). The number of marriages is very variable, one man had had six marriages by the age of 40, others apparently only one in their entire lives. Among 70-year-olds, the number ranged from zero to six, and among 60-year-olds, from zero to nine. One 26-year-old had had four marriages. Limiting the sample to the 110 men whose observations ended at age 45 or older (Table 15.1) gave a mean at 1.973 and a median of two marriages (variance 2.229).

Figure 3 in Kreider and Fields (2002) shows the number of times married for Americans aged 45 or older. Twenty-six percent of "white non-Hispanic" Americans

**Table 15.1** Number of marriages recorded for 110 men and 104 women aged 45 or older. For example, six women had five reported marriages, six men had four marriages. One woman had seven marriages, one man had nine

| | 0 | 1 | 2 | 3 | 4 | 5 | 6 | More |
|---|---|---|---|---|---|---|---|---|
| Women aged >=45 | 4 | 42 | 19 | 18 | 13 | 6 | 1 | 1 at 7 |
| Men aged >=45 | 10 | 39 | 33 | 14 | 6 | 5 | 2 | 1 at 9 |

(the group with the highest percentage) had married more than once. The corresponding figure for Hadza women is 58 out of 104 women last seen at or after age 45, or 58%, which is more than twice the white American level. The figure for Hadza men was 61/110=55%. Hill and Hurtado (1996, figs. 7.8, 7.11) show much larger numbers of marriages among the Ache, with a maximum over 25 marriages, and many individuals reported having had 15 or so marriages and a mean of close to 10 spouses by the age of 30. Howell (1979, table 12.3) reports many fewer marriages among the !Kung, and an average of two marriages (range 0–5), much like the Hadza.

## 15.6 Durations of marriages

Many Hadza marriages are quite brief but others last several decades. If we want to estimate the normal length of marriages, we have to be careful about using these extreme cases. The data present serious problems of censoring (both left and right censoring) and over-sampling the longest marriages. For some of the longest marriages, we have little indication of exactly when they began. More important, for many of the older people, we have very poor information on their earliest, short-lived marriages. Most important, long-lasting marriages, for example, 35-year-long marriages, do not represent the typical marriage of 35 years ago. If the early years of these marriages were allowed to contribute to our record of survival of marriages during their first few years, we would severely bias the data toward high survival of marriages in the early years. The early years of the long-term marriages would be included while the early years of long-ago-dissolved marriages were omitted entirely. We would over-sample the early years of long marriages.

After lengthy experimentation with more orthodox methods, preparing data as input to Minitab's several survival analyses, I decided that two methods were the most transparent, honest, accurate, and of knowable accuracy. One was to follow the procedure I had used for mortality, using a purpose-built program in Visual Basic. The other method was to run a logistic regression on the marriage histories to predict the probability of occurrence of divorce, bereavement, or death for each year of a marriage in the marriage histories. We can find the median duration, the duration at which the probability of breakup reaches 0.5. I did this in Minitab, and in MLwiN where I could take account of the individual differences in contributions to the data. Some individuals contribute more than one marriage to the sample, some have long marital histories, others short ones. We might be concerned about over-representation of particular people who tend to have short marriages.

In the Visual Basic program, the fate of marriages was followed over the period of observation from 1985 to 2000. People who began the period in a marriage were allowed to represent the married people of only the 1985–2000 period. The early history of their marriage was not used, for we have no matching information on the brief marriages that occurred among their peers when their marriage was young. Their marriage entered the at-risk category only during the years observed and was taken as at risk of dissolution only during those years of observation (1985–2000, or whenever one partner died or was lost track of [censored]). However, the age of the marriage at which they began to be at risk was the estimated age of the marriage in 1985, or later if first observed later. This left a sample of 375 marriages of women and 206 of men.

The number of marriages that entered each year of marriage duration was noted, and the number of dissolutions by divorce, widowing, and deaths in each year of duration was collected. These numbers were used to calculate the age-specific death rates, $q_x$ (dissolutions/enter), and the $q_x$ numbers were used to calculate probability of survival from birth, $l_x$ for the survival of a hypothetical cohort of marriages. The program saved data on dissolution by divorce and/or death.

Figure 15.3 shows the survivorship of marriages of women, and of men against all forms of dissolution. This suggests that about half the marriages have ended by seven years. A minority, about 16%, continue to their "silver jubilee" at 25 years. Only 9% can match President and Mrs Mkapa's 34 years (President of Tanzania 1995–2005, married 1966, a marriage of 34 years by the end of my observations in 2000). A few

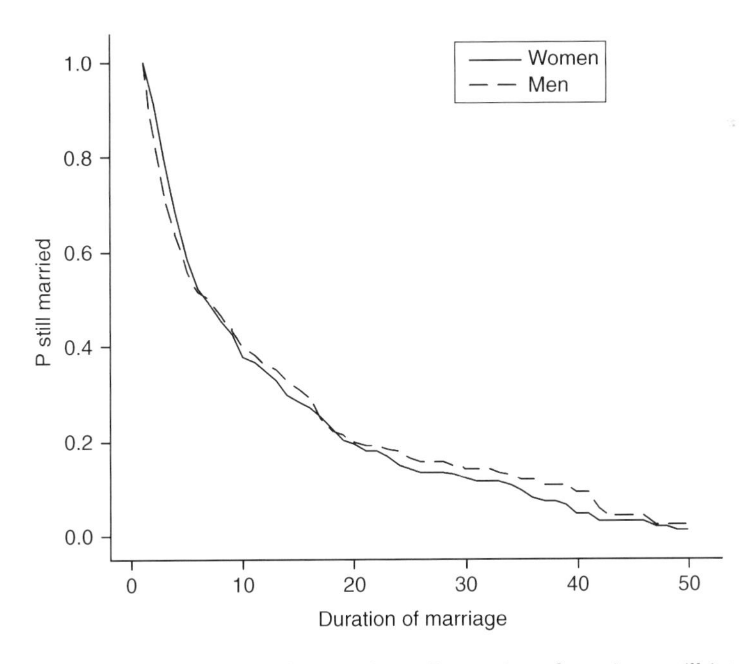


**Figure 15.3** Survival of Hadza marriages. Proportion of marriages still intact at x years from formation; 375 marriages of women, 206 marriages of men.

Hadza marriages last a very long time, just over one in 20 (6%) may last 40 years. Virtually none (2%) reach 50 years. Just one may perhaps have matched Queen Elizabeth II's 60 years (married 1947).

The curves in Figure 15.3 for men and women are comfortingly close to each other despite the fact that the marital history files comprise an only partially overlapping set of marriages. Comparison of Hadza marriages with United States marriages is complicated by the large secular changes in the American figures. While American marriages have been forming later and lasting less time, we have assumed Hadza marriages did not change. By logistic regression we can show that within the 1985–2000 data record, calendar year did not contribute to predicting divorce, b=0.0113, p=0.486 (OR=1.01, CI 0.98–1.04). If we go further back in the marital histories, we are vulnerable to the under-representation of the short marriages. Kreider and Fields (2002, table 2) show that in the 1980–1984 cohort of U.S. women's marriages, 86.4% reached their fifth anniversary and 73% reached their tenth. The corresponding figures for the Hadza are 60% and 40%. In the 1950–1954 U.S. cohort, 67.4% of the marriages were still intact at their fortieth anniversary, a striking contrast with the Hadza estimate of 6%.

Marriages to a non-Hadza husband did not last significantly longer than to a Hadza husband, nor did those to a wage-earning Hadza husband. Although, in both cases, the Kaplan-Meier survival graphs looked as if marriages to non-Hadza and wage earners lasted longer, the differences were far from significant.

## 15.7 Divorce rates

Woodburn (1968b) reported a high crude divorce rate and I compare his report with my data in SI 15.6. There I also show the difference between survival of marriages against divorce and against death. Here I describe the probability of divorce by duration of the marriage and by age of partners.

Dissolutions by divorce are abundant early in a marriage (half the marriages have divorced by seven to eight years), but divorce becomes a much weaker force if the marriage survives beyond 10 or so years. Very few really long marriages end in divorce. Dissolutions due to death of either partner occur slowly at first (most of the early years of most marriages are during young adulthood) and gather pace as the couple ages. In the 595 marriages of women in the span file, 338 had ended, 76 ended when their husbands died, and 16 when they died; 246 ended in divorce, and 257 were still intact when the study ended. In the 418 men's marriages, 257 had ended. Twenty-six of the marriages ended when their wives died and 29 when they died; 202 ended in divorce, and 161 had not ended at the end of the study.

It used to be said that divorce rates in industrial countries had risen because the oath to stay together "until death do us part" was easier to keep in the days of high mortality. The Hadza data, from a population with a mortality level comparable to that of medieval Europe, do not support this idea. This idea ignored the common observation that divorces are more frequent early in a marriage. However, the Hadza data could be seen as support for another commonly held view: that increasing "western" divorce rates have been because of changes in transaction costs. There is a massively lower "transaction cost" in a society like the Hadza than in a property-owning, lawyer-bound, industrialized society.

The instantaneous hazard of a marriage ending peaks very early in a marriage, steadily declining once the marriage has survived to about four years (SI 15.6). It is commonly found that marriages are most at risk in their early years. While colloquially we claim that in the early years of marriage the couple are finding out what each other is "really" like, economists suggest that this is the time when the "coordination problems" inherent in a marriage are being negotiated. A sociobiological interpetation of the hazard curve was offered by Fisher (1989). Having amassed an impressive set of figures about divorce in 45 countries showing the abundance of divorce in the early years of marriage, she noted that among foragers, at four years the first baby would be likely to have been safely weaned. She next suggested that effects of father's desertion would by then have dropped to a minimum, rendering his departure least costly to him (and indirectly of little cost to his wife). Other sociobiologists immediately wondered why a man would desert just as his wife was becoming fertile again. In light of recent findings about helpers, and the vulnerability of newly weaned children and concomitant sensitivity of fitness to improvements in their survival, we might also suspect that the desertion costs do not drop so soon. I will discuss the special place of weanlings in human life history again in the next chapter and in the chapter on grandmothers as helpers.

In logistic regressions on the marital history file, occurrence of a divorce was significantly but very weakly predicted by the woman's age ($b=-0.034$, $p<0.000$), and by duration of the marriage (longer into the marriage, lower probability of a divorce, $b=-0.0848$ $p<0.000$). The relationship with age became insignificant if we included duration of marriage as a predictor. It is not young women's marriages that are vulnerable. It is the early years of a marriage that are vulnerable. It might be that, in the early years of a marriage, crucial aspects of the "coordination problem" are being negotiated with greater or lesser success. Only later, or as the wife ages, might men's opportunities and costs of new marriages to younger women become a cause of divorces. Other factors, such as the wife's fertility, or ability to raise children, and the husband's success as hunter or trader, would also have to be considered.

To look at men's divorces, I used the subset of annual marital histories for men aged 20 years and older between 1985 and 2000. Probability of divorce declines as the marriage continues ($b=-0.1425$, $p<0.000$, OR=0.87, CI 0.83–0.90) just as for women. Older men were significantly less likely to divorce (logistic regression, $b=-0.0503$, $p<0.000$, OR=0.95, CI 0.94–0.97). In apparent contrast to women, when marriage duration was added to the model, both were significant predictors (age, $b=-0.0174$, $p=0.058$, OR=0.98, CI 0.97–1.00; duration, $b=-0.1284$, $p<0.000$, OR=0.88, CI 0.84–0.92), although the already weak effect of age was much weakened.

## 15.8 Remarriage and divorce spans

The special file "divorce spans" was extracted from the annual marital history. It included a record of each divorce of each subject (165 women, 138 men), and contained information on the ensuing marital status of the subject. This file could be appropriately analyzed with Minitab's Regression with Life Data, which allowed us to examine effects of independent variables on the speed of remarriage.

For women, the mean period of a divorce is 4.4 years, but many remarry very quickly and the median is 2.1 years; only a few remain single for 10 years or more. For men, the mean was 4.1 and the median 2.1 years, comfortingly close given that the sample of men's divorces is not the same as that of women's divorces. If we limit the analysis to divorces that end only in remarriage (not death), the mean is 4.8 and the median 2.1 years for women, and the mean is 4.7 and the median 2.1 years for men.

The amount of time someone remains unmarried after a divorce should be an interesting variable. It will have some effect on the amount of adult lifetime for which reproduction is most likely (assuming more frequent intercourse within marriage than outside). It may tell us about factors that affect RS by their effect on an individual's marriage prospects (marriagability). For example, later we will look at men's reputations and speed of remarriage, and women's "burden" of small children and their speed of remarriage.

## 15.9 Affairs, second wives, and second husbands

Previously, I discussed some of the more obvious instances where an individual had more than one spouse for more than a few years. The marital histories, built from records of households, and paternity claimed by women, allow me to make the computer look for shorter periods in which an individual appears to have a second spouse, or to have produced a child by someone other than the current spouse. These data cannot be error free, nor can they capture all extra-marital affairs, or would-be polygamous marriages, only the most overt of them.

Some men are listed as husband by two women in the same year. One of these women will be featured in the man's marriage history as his wife, the other woman will not appear in the man's record. I will call them first wife and second wife (for my convenience, I have no indication that Hadza see them this way). The first wife appears in the man's marriage history because she was the woman he was seen with in the census, or was the woman (out of the two) who appeared earliest in his record. The first wives are usually older than the second wives. During the editing process, I had tried to minimize the number of apparent second wives by checking each of their marriage histories, the list of children of each person, the "where are they now?" interviews, and any other information. I limited this analysis to the parts of the marital history file that covered the years 1985–2000.

These relationships were mostly very short. In the final edited files, the computer found the presence of a second wife in 311 man-years out of the total recorded 3811 man-years of marriage (8.16%). As I remarked earlier, men managed to keep two women for only a short time (SI 15.7), but 23% of men whose marital history between 1985 and 2000 was coded had at least one year (or part of one year) during which they were scored as having a second wife. The mean duration of the bigamous arrangements was 3.3 years, but with such a skewed distribution, the median of two years is a better representation. Men were more likely to have a second wife when aged 30–50 years. Most women who were second wives tended to be in their early 20s, although this status was almost as common in the 30s and 40s.

How many births did men gain from the second wives? Of 907 children of the men with marital histories, 75 children were with the "second wife." Thus, having a child by a concurrent "second wife" increased men's fertility on average by 9% (75/(907–75)=9%). This is rather less than the gain from starting a second family with a new young wife after leaving the first near the end of her childbearing years (Chapter 22).

If men's provisioning has a substantial effect on wives' fertility, then we might have expected its division between two women to have greatly decreased their fertility. The loss of paternal resources to another woman is often proposed as a reason for women's intolerance of spousal infidelity. We can look at the probability of concurrent second wives conceiving per year of marriage as the second wife. Men were fathers of 75 children whose mother was the second wife during 311 man-years with a second wife. The second wives were contributing children at 75/311= 0.2411 births/ yr, and they tended to be younger than "first wives." The first wives contributed 832/3811= 0.2183 births/yr. These are not very different from the figures for marital fertility of the interviewed women (average marital fertility for interviewed women aged 15–44 is 0.2487 births/yr., Chapter 8). We cannot take these figures as indicating a significant loss of fertility by either woman during the years that a man has a second wife. Logistic regression on the birth hazard file (controlling for mother's age, among married women only), with a new independent variable "is a co-wife" entered for each year, shows no significant effect of being a co-wife on probability of a birth. A similar analysis on the child annual survival file gives a suggestive but not significant effect on child survival from father's "divided loyalties" (probability of death is higher if the mother is a co-wife, b=0.4364, p=0.105, OR=1.55, CI 0.91–2.62).

Some women briefly had more than one husband (30 women out of 328 with a marriage recorded during 1985–2000), for a total of 54 woman-years. The mean duration of these arrangements was therefore 1.8 years and the median less than one year. In the entire marital history record, women's two-husband arrangements existed for 89 years out of 3911 marriage-years, 2.28%.

We might not expect such fertile women as the Hadza to be able to increase their fertility by the precarious means of keeping two husbands for a year or two. However, if this arrangement gives them greater access to resources, it might make a difference to their RS. Several of the first and second husbands were non-Hadza. Hadza women with a long-term marriage to a non-Hadza man may actually have liaisons (and children; we know such cases) with Hadza men. Hadza women with long-term marriages to Hadza men, may have brief liaisons with non-Hadza men. We think these are especially common around Mangola, when tourist money and alcohol are available. We can look at the births attributed to second husbands. There were 26 such children born to 23 women;

26/89 woman-years= 0.2921 births/yr. This rate is rather higher than the average marital fertility rate for age 15–44 (0.2487), but it is almost the same as the rate for age 15–29 (0.3180). A woman's chance of being a co-wife and of having a "second husband" is greatest in her early 20s when marital fertility is highest, so we cannot confidently claim to show that the second husband brings a substantial advantage. In the children's annual survival file, we find that having a second father has a highly non-significant and very noisy relationship to child survival, b=0.1259 (child of a married woman is more likely to die if it has a second father, p=0.798, OR=1.13, CI 0.43–2.98).

## 15.10 Women's marriage prospects track their age and fertility

The cross-sectional household census data suggested that the greatest proportion of women were in a marriage during their reproductive years. Few women in the early teenage years were married and fewer after menopause than before. The same thing can be seen in the marital history files. Logistic regression estimates the probability of being married predicted by age. To allow the regression to track changes in probability of being in a marriage most closely, the model includes age, $age^2$, $age^3$, $age^4$, and $age^5$. A clear difference between men and women can be seen (SI 15.8). For women, the probability of being married declines steadily from its peak in the early 30s, falling below 0.7 at age 50 and below 0.5 in the early 60s. The probability for men remains high from the early 30s until the late 60s, falling below 0.7 just after age 70.

Divorce spans are especially interesting because they show the propensity of single women of a wide range of ages to get married, and the readiness of men to marry women of a variety of ages. Divorce spans are longer for older women. Curves of age by probability of remarriage by one, two, and three years from divorce were generated from Cox regressions. Each run can be told to estimate percentiles for 50%, for example, and probability for times such as one year, two years, three years, and each run saves these probabilities. Hence, I ran analyses to get survival of the divorced state to one year, two years, three years. A plot of the results (Figure 15.4) suggested that women remarried quickly in their early 20s, but less quickly thereafter, and women in their 40s were very likely to remain unmarried at three years after divorce.

Thus, divorce spans are best predicted by marital fertility, which is itself best fit to age + $age^2$ + $age^3$. The higher the expected marital fertility, the shorter the divorce span ($b=-3.328$, $p<0.000$). Women remarry faster if they are of an age near to their maximum marital fertility. Marital fertility peaks at about age 20, being slightly lower in the late teenage years, and declines steadily into their mid-40s. Therefore, the marital fertility curve is very similar to the curves of age by probability of remarriage one, two, and three years from divorce. The relationship of remarriage to women's age probably tells us more about men than about women. The data reveal men's preferences for younger women, an issue on which neither research nor opinions are lacking.

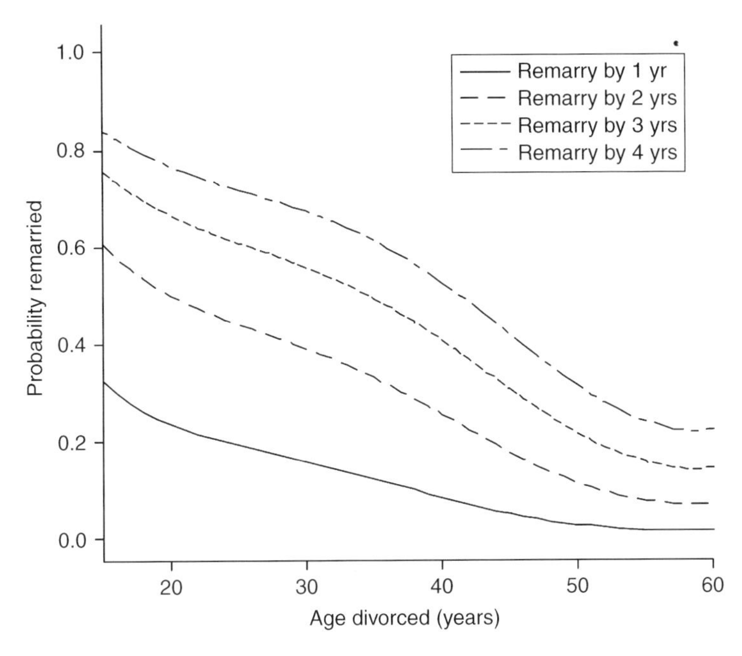


**Figure 15.4** Relationship between age at divorce and probability that a woman remarried by one, two, three, or four years later.

## 15.11 Do Hadza men avoid small step-children?

If men gain fitness by investing resources in their children's survival, they would avoid squandering their resources on another man's children and women with more small children (the most vulnerable and costly children) would be expected to take longer to remarry. Buckle *et al.* (1996) found that in western cultures they do. I used the divorce span file to see whether a woman's "burden" of small children affected the speed with which she gets remarried after a divorce. The burden measures are available only for interviewed women.

I scored the number of children aged under five years ("burden5") in the year of remarriage. If a child were born during that year, it was normally attributed to the new husband and was excluded, giving a modified burden score ("newburd5"). The aim was to reflect the number of small children that the woman was caring for immediately before the new husband married her. If a woman did not remarry, the divorce span was right censored at the end of observation, and the burden score at the end of observation was used.

Using Minitab's Regression with Life Data, we see that women with more small children remain divorced for a shorter time (Figure 15.5). Even if we control for the age at which the woman gets divorced, and her number of divorces, having more small children predicts faster remarriage. This was true of both versions of the burden measure and the full sample as well as the 1985–2000 sample (Table 15.2). Women with more small children remarry more quickly. This is the opposite of what I had predicted from the idea that men would not waste investment on other men's children.

The paternal investment idea could be defended by suggesting that these women were more eager for help (if help comes mainly from husbands). They might expend more effort or take more costs to get a new husband, such as putting up with a lower quality husband. Thus, I looked at the general nomination scores of the original and the new husband. In the 127 remarriages after 1985 (and the 214 in the entire file), there was no difference between the pro-rated general nominations of the old husband or the new husband (mean old husband, 10.92; mean new husband, 9.66, p=0.759). The result was the same for hunting, trading, and arrow making (SI 15.9). There were no significant relationships between the new husband's reputations and the woman's child burden. Women with more small children are not taking a husband with a lower reputation in order to remarry sooner.

Not only were the new husbands no less nominated but there was a significant positive correlation (r=0.236, p=0.007) between the old husband's combined nominations and the new husband's combined nominations. Old and new husband's

**Table 15.2** Women with small children remarry faster. Predictions of divorced women's times to remarry from "burden" of small children. Cox regression, divorce span (right censored) controlled for age divorced. Summarizing four different models. For example, line 1 reports model: woman's time to remarry = her age + N of live children aged under five years, for entire marriage history file. "1985+" refers to sub-sample comprising years 1985–2000

| | beta | probability | N divorces |
|---|---|---|---|
| All newburd5 | -0.4611 | <0.000 | 140 |
| All newburd7 | -0.3083 | <0.000 | 140 |
| 1985+ newburd5 | -0.3765 | 0.002 | 111 |
| 1985+ newburd7 | -0.2123 | 0.03 | 111 |

**Figure 15.5** Divorced women with small children remarry faster.

hunting nominations were also positively correlated, with r=0.309 and p=0.005. This remains true after controlling for the woman's age at divorce. My first reaction was that this might reflect some enduring quality of the women (men like some of them more than others). However, there was no association between these nomination scores and the length of the woman's divorce span.

It is possible that men take the existence of small children as a signal of potential fertility supplementary to the woman's age. In the chapter on measuring RS, we showed that women who start out reproducing faster, tend to continue having children. Current fertility, therefore, could be an accurate signal of future fertility. Marrying a woman with small children may increase the new husband's chances of being on the scene for more conceptions.

## 15.12 Men's reputations and marriage

Introducing the enquiry into Hadza men's reputations, I mentioned Gudo Mahiya's comment that good hunters "get more women," and the view expressed by Woodburn that to get a wife and remain married, a man should have some success at hunting and acquiring objects by trade. Here I report the search for links between reputations and marriage in our data. Statistically significant links were not hard to find, and we can try to see how well they account for the greater RS of much-nominated men.

**Age at first marriage** Men more nominated as expert hunters tend to have married a little younger than others (hunt pr, b=–0.1983, p=0.005). The relationship was also significant for log hunt pr and for the dichotomous measure, huntpr ≧5. In every analysis reported, these measures give similar results. Trading nominations were not associated with men's age at first marriage. (The sample was marriage summaries for men, restricted to those with the final observation age at 25 or older).

**Duration of marriages** I could find no reliable or significant relationship between nominations and duration of marriages.

**Number of marriages** Regression controlling for men's age at the end of observation show a small significant positive relation between hunt nominations and number of marriages. Expert hunters had slightly more marriages. Among 109 men aged 45+ at the end of the record, hunting reputation was slightly but significantly correlated with number of marriages (b=0.0549, p=0.024) and t-test using hunt >5, t=–2.33, p=0.033, and means of 1.85 and 2.93. Reputation as a trader did not have a significant relation to the number of marriages (b=0.0473, p=0.073).

Plotting the number of marriages against age at the end of observation separately for high-reputation men and for others shows the number of marriages reaching a higher level at advanced ages for the high-reputation men (SI 15.10). Their smoothed curve peaks at three marriages at age 50 while the other men's curve stabilize at about 1.5 marriages at age 40. Comparing means for the experts and the others among the over-45-year-olds gives a mean of 2.93 versus 1.85 (p=0.033). Stepwise regressions in which age is entered first show a significant contribution of reputation but not of age to predicting the number of marriages. The result is significant both for general nominations, and for nominations as expert hunter. Thus, reputations seem to enable men to enter more marriages and perhaps to do so to a greater age, presumably lengthening their reproductive careers.

A box-and-whiskers plot of the number of marriages among successful hunters and others in the over-45-year-old sub-sample, shows seven striking outliers (SI 15.10). These are seven men, rarely if ever nominated, who nonetheless had unusually many marriages. These men may exemplify a minority, "philandering" strategy that I discuss in Chapters 21 and 22.

**Men's affairs by reputation** There was a small but significant tendency for men nominated as good hunters to have a "second wife" more often (N co-wives 1985–2k = age-end + hunt pr, b=0.0524, p=0.003, 175 men). This was also true for men nominated as expert traders (b=0.1076, p=0.000).

**Men's marriage prospects by reputation** The length of men's divorce spans provide an informative contrast to those of women. The mean time divorced is 4.6 years with a median of two years. Among men, unlike women, age at divorce is not related to length of divorce spans, divorced men of any age get remarried equally quickly ($b_{age}$=−0.0013, p=0.895 in Cox regression). Men's speed of remarriage is very significantly influenced by their reputation (combined reputation score predicted by length of time divorced, b=-0.0240, p=0.000). Hunting reputation is also an important predictor (log huntpr b=−0.6882, p=0.000; also significant for huntpr [b=−0.036, p=0.000]). I also categorized hunting reputation as five or more pro-rated nominations (class 1), and fewer than five (class 0). The Kaplan–Meier survival plot for the divorces of such men is shown in Figure 15.6. The difference is quite striking, and is significant in both log-rank ($\chi^2$=11.3, p=0.001) and Wilcoxon

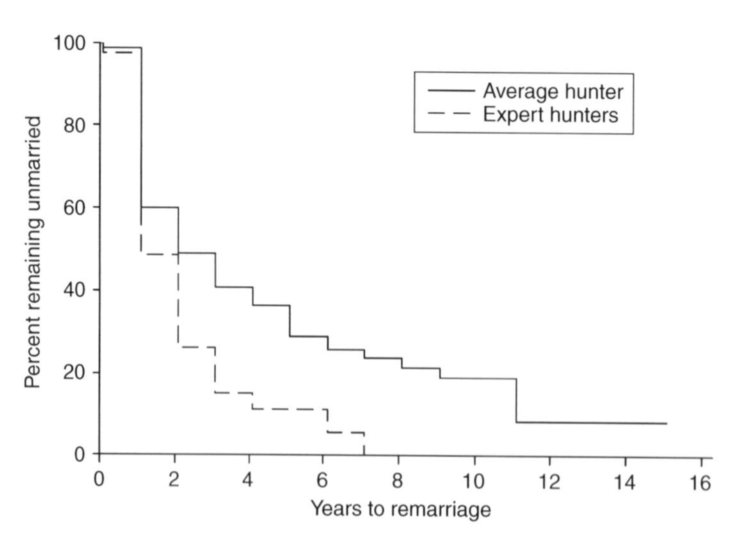


**Figure 15.6** Good hunters remarry faster. Kaplan–Meier survival plot for divorce periods of expert hunters (men with ≧5 pro-rated nominations (scores 1), or average hunters (<5 pro-rated nominations (scores 0). Log-rank test $\chi^2$=8.16, p=0.004. Wilcoxon test $\chi^2$=4.23, p=0.04.

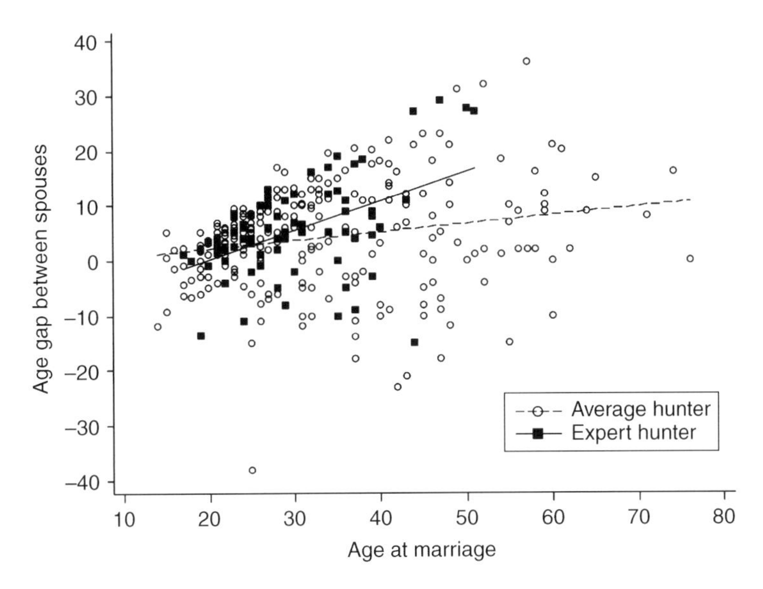


**Figure 15.7** Difference in age between husband and wife (positive when the husband is older than the wife) for Hadza men with five or more nominations as an expert hunter (solid line), or fewer than five (broken line). Regression: age gap = age at marriage + huntpr, b=0.1602, p=0.009.

tests ($\chi^2$=6.4, p=0.011). For men with five or more nominations, the mean time to remarry was 2.25 years, median 1.1. For the other men (<5 nominations), the mean was 5.41 years and median 3.1 years. Perhaps the remarriages are rapid because the man had been having an affair with the new wife. Either way, nominated men spend less time unmarried.

**Age gap between spouses by reputation** Because, on average, Hadza men first marry at age 25 and women at 18, people begin their marital careers with a substantial gap in age between spouses. For most men, the gap gets larger as they get older, because when they divorce and remarry, they tend to remarry a woman younger than the one they divorced. In the men's marital history file, we find the gap in age between spouses increasing with the sequential number of the marriage (r=0.358, p<0.000) and with the age at marriage (r=0.2317, p<0.000) (Figure 15.7). Women showed the opposite, the age gap decreased with the sequence of the marriage (r=–0.158, p=0.002) and with the age at marriage (r=–0.483, p<0.000). This comes about because some young women marry much older men, and a few older women married much younger men. The marriages of older women to younger men do not last as long as other marriages (SI 15.11). When they break up, the young man tends to marry a woman nearer his own age. Men who marry a woman five or more years older than themselves tend to marry late (two-sample t-test, 43 men with a much older wife get married at a mean age of 44, other men at a mean age of 24.9, t=9.54, p<0.000).

Among men, not only does the age gap increase with successive marriages, but it also tends to be greater among more-nominated men. Men nominated as expert hunters or traders tend to have wives who are younger than themselves by a greater amount than do other men. Regression analyses with the age gap as the dependent variable, and controlling for either marriage number or age at the beginning of this marriage, show that nominations predict greater age gaps. (General nominations, b=0.0943, p=0.001; hunt nominations, b=0.1376, p=0.009; trader nominations, b=0.2581, p=0.002). Stepwise regression confirms the importance of the nomination score in improving the prediction of the age gap. Marlowe (2010, p. 193) reports essentially the same results on reputations and the age gap between spouses.

## 15.13 Marriage and reproductive success

Sociobiological orthodoxy holds that couples stay together as a result of men exchanging resources for sexual access, conceptions, and paternity. Women gain increased access to resources, which enable them to bear and raise more children. If this is the case, then individuals who spend more of their reproductive years married, should have greater RS. This should apply to men, because they have more sexual access and more conceptions, and to women because they have a steadier supply of resources. Let us look at this among the Hadza. I will look at the relationship of the proportion of the reproductive years in a marriage to RS, measured as the residual of the number of live children multiplied by parent age.

The largest proportion of Hadza women are married from age 20 to 40, and the proportion is roughly constant between these ages. For men, the corresponding ages are 25 and 50. Thus, from the marital histories I extracted the years 20–40 for women, or 20–end-age (the final observation) if this were before the woman reached the age of 40. For men, I extracted the years 25–50 or to end-age if this were less than age 50. These were labeled "adult life" or "career." To discard records based on inadequate information, I limited the sample to individuals (born in or after 1950) with recorded careers of more than five years. These people had been seen in an average of six censuses. Then I counted the number of years during these periods in which each individual was married. I expressed this as a fraction of the observed career. This variable will be labeled "pctmarr" but used as a proportion, scored from 0 to 1.0.

There was a remote possibility that the detailed marital histories would show more time married if the individual appeared as the parent to more children. If this were the case, it would generate a positive association between fertility and proportion of time married. Therefore, I derived a second measure directly from the census household lists and no other source. The census household lists are independent of the number of children recorded. There is no danger that this measure is contaminated by the individuals appearing as parents in records of births. This census measure ignores much good information, such as that contained in the women's interviews, and not surprisingly, is much "noisier" than the full measure, as can be seen for the lower $r^2$ values. Consequently, I show the results for both measures in Table 15.3.

The 114 men with RS and marriage information were married for a mean of 0.6348 of their adult life, median 0.75 (inter-quartile range 0.34–0.96). The 130 women were married for 0.7359 of their adult life, median 0.89, inter-quartile

**Table 15.3** "Bateman gradients" for men and women (graphs in SI 15.12). The effect of "success in mating competition" (proportion of adult life married: "pctmarr") on reproductive success (RS) is greater for men than for women. Reproductive success is measured as the number of living children for age at the end of observation. "Births" is the number of children born for age at the end of observation. The Table shows b and p for the contribution of pctmarr. The sample was restricted to individuals whose record of adult life was ≧5 yr. Women's mean "pctmarr", proportion of adult life in a marriage, was 0.7359 with a variance of 0.1029. $\text{Mean}^2$ =0.5415, standardized variance =0.1029/0.5415=0.1900. Men's mean "pctmarr" was 0.6348 with a variance of 0.1274. $\text{Mean}^2$=0.4029, standardized variance=0.1274/0.4029=0.3162

Panel A. Proportion of adult life in a marriage from marriage history files

| Panel A. | b for pctmarr Marital histories | p | Adjusted $r^2$ | N individuals |
|---|---|---|---|---|
| Women live Cs | 2.97 | <0.000 | 14.7 | 130 |
| Women births | 4.29 | <0.000 | 19.5 | 130 |
| Men live Cs | 4.39 | <0.000 | 36.8 | 114 |
| Men births | 7.01 | <0.000 | 52.0 | 114 |

Panel B. Proportion of adult life in a marriage from census data only

| Panel B. | b for pctmarr Census data | p | Adjusted $r^2$ | N individuals |
|---|---|---|---|---|
| Women live Cs | 0.5019 | 0.233 | 0.3 | 155 |
| Women births | 1.2327 | 0.024 | 2.6 | 155 |
| Men live Cs | 2.1349 | <0.000 | 12.7 | 187 |
| Men births | 3.3926 | <0.000 | 17.9 | 187 |

range 0.60–1.0. Forty-four percent of women were married for their entire adult life (age 20–40 or end-age), and 24% of men were married 100% of the time. Twenty-two percent of women, and 33% of men, were married for less than half their adult lives. The standardized variance in "pctmarried" was greater among men (0.3162) than among women (0.1900), very close to the standardized variances in RS. Note: the sample sizes do not represent the adult sex ratio which is reported in Chapter 13.

Table 15.3 shows the relationship of RS and fertility to percentage of adult life in a marriage (pctmarr) for men and for women. The measures of RS here were the residuals of the number of live children for parent's age at the end of observation. There is a significant positive relationship between the proportion of adult life married and RS. Marriage accounted for more of the variance in RS for men than for women (adjusted $r^2$=36.8% for men, 14.7% for women). There is also a difference in the slope. The regression coefficient is larger for men than for women (4.39 for men versus 2.97 for women). Apparently, marriage makes more difference to the RS of Hadza men than it does to the RS of Hadza women (SI 15.12).

We dig a little deeper by looking at the relationship between marriage and fertility, and marriage and child survival. For women, marriage predicts 19.5% of the variance in the residuals of fertility (N born, b=4.29, $p<0.000$, N=130). For men, marriage

**Table 15.4** The greater RS of good hunters is accounted for by their proportion of adult life in a marriage. Stepwise regressions, step 1 predictors are age + $age^2$ + hunting nominations (hunt pr). Step 2 adds proportion of adult life in a marriage (pctmarr) and this removes the effect of hunting nominations. Upper panel: men's RS (standard score of live children). Lower panel: men's fertility (standard score of number of children born)

| sslive end | b | p | Adjusted $r^2$ | N cases |
|---|---|---|---|---|
| Step 1 hunt pr | 0.0293 | 0.002 | 5.96 | 133 |
| Step 2 hunt pr | 0.0029 | 0.730 | | |
| pctmarr | 1.76 | 0.000 | 37.0 | |
| **ssborn end** | **b** | **p** | **Adjusted $r^2$** | **N cases** |
| Step 1 hunt pr | 0.0420 | 0.000 | 9.23 | 133 |
| Step 2 hunt pr | 0.0082 | 0.331 | | |
| pctmarr | 2.26 | 0.000 | 49.9 | 133 |

predicts 52% of the variance in their births (b=7.01, p<.000, N=114). Being married accounts for more of the variance in fertility for men than for women.

We saw, in Chapter 13, an indication that the proportion of their children that people keep alive (sssurv) apparently plays a bigger role in predicting RS of women than men. If women's children benefit from the resources they receive from men, then "pctmarr" should predict sssurv. Although the beta coefficient is large and positive, it is far from significant (b=0.6025, p=0.236, N=130), and "pctmarr" accounts for only 0.4% of the variance in a woman's child-rearing success. The survivorship of a man's children appears to be worse if he is married for a larger proportion of his adult life (b=−1.205, p=0.022), even when we take account of his nominations as a good hunter ("pctmarr" b=−1.107, p=0.047, hunt pr b=−0.0069, p=0.579, N=87).

With the census measure (lower panel in Table 15.3), despite the lower $r^2$-values, the same pattern shows up, more of the variance in men's fertility and RS is accounted for by percentage of time married than is the case for women.

We saw (in Chapter 14) that men's reputations as expert hunters predicted greater RS. We also saw here that expert hunters remarried faster and spent more of their adult life married. Table 15.4 shows that the high RS of good hunters is accounted for by the greater proportion of their adult life that they spend in a marriage.

These results support only limited aspects of the classical sociobiological view. Hadza women gain some fertility but relatively little RS from marriage, and gain no advantage to their ability to keep children alive. Hadza men gain quite strikingly in fertility and RS. While in both genders, RS is highest among those married for their entire adult life, women seem to show little difference between being married 80% of the time and 30% of the time. The success and failure of the classical theory (exchange of provisioning for sexual access) in accounting for the Hadza data can

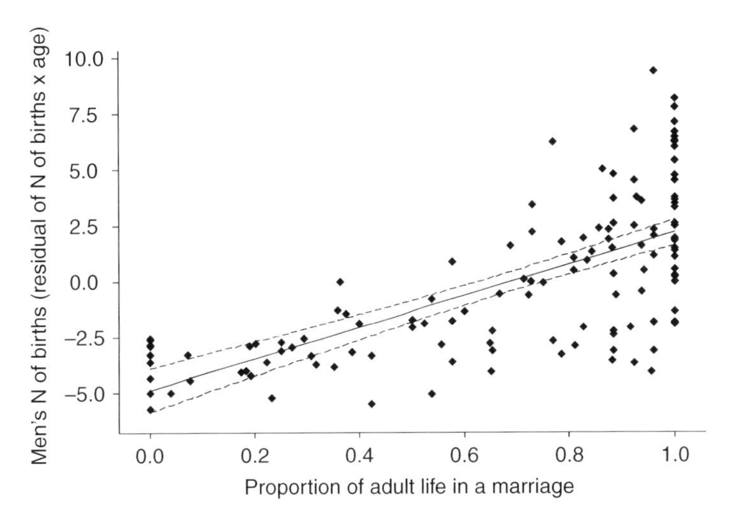


**Figure 15.8(a)** Marriage as an exchange of conceptions for care? Men gain paternity from marriage. "Fertility" of 133 men (N of births for age at the end of observation) by fraction of adult life in a marriage. "Fertility" = -4.826 + 7.030 pctmarr, $r^2$ adjusted=46.8%, p=0.000.

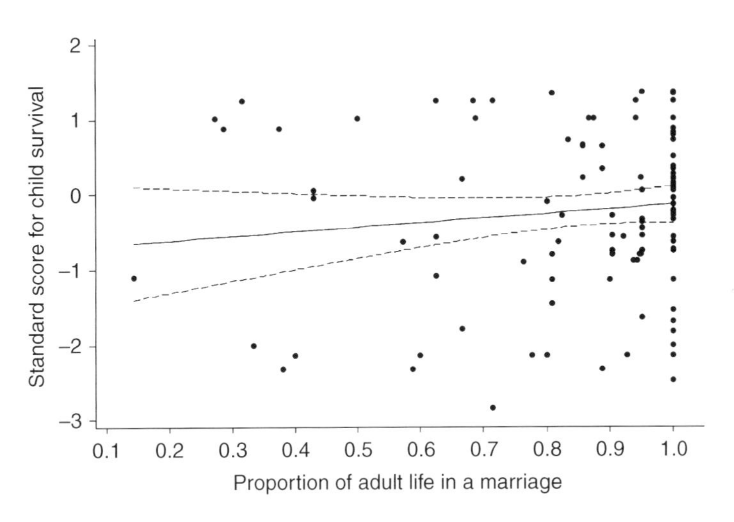


**Figure 15.8(b)** Marriage as an exchange of conceptions for care? Women gain little child survivorship from marriage. Regression sssurv = −0.7292 + 0.6025 pctmarr, p=0.236, $r^2$=0.4%. N=116 women with a measure of child survival.

be illustrated. Figure 15.8a displays the plot of births against "pctmarr" for men, and for women (Figure 15.8b), the plot of child survival against percentage of reproductive lifetime in a marriage ("pctmarr"). While men gain in births the longer they are married, Hadza women gain little or nothing in child survival from marriage.

The results are hard to reconcile with the idea that Hadza women's reproduction gains from the resources provided by a husband. Women gain little reproductive advantage from marriage. Without marriage, conceptions can be easily obtained, timed as she pleases, and husbands seem not to improve child survival at all. Men, in contrast, gain RS from marriage. The increased sexual access presumably increases chances of conceptions. Costs of competition are lowered, mate guarding (especially when marriage is socially recognized and accepted) may be less costly than competing for a new mate.

Apparently, being with a husband has no effect on a Hadza woman's ability to keep her children alive. I will be revisiting this issue several times in subsequent analyses where I look at other likely helpers. I do not intend to give up easily on the "exchange for care" theory or the "exchange for resources" version of it. If there are other forager populations in which child survival is affected by marital status, they may provide evidence that men can affect the survival of their children. This is apparently the case among the Ache. That much of the association between child deaths and father departure may have been because of children being killed, does not dismiss the Ache evidence. Somc of the societal support for the killings apparently derived from the difficulty of caring for fatherless children in an economy in which men acquired some 80% of the calories. The Hadza show us that marriage can exist without such an effect. Furthermore, Hadza marriage shows us marriage with little benefit to child survival but a lower divorce rate than reported among the Ache where there is a stronger effect of marriage on child survival (an issue previously discussed in Blurton Jones *et al.*, 2000).

## 15.14 If marriage is not an exchange of paternity for care, then what is it?

Hadza marriage shows several features found elsewhere in Africa and in the industrialized nations. Men marry later than women, and at more widely scattered ages. Men tend to marry younger women and seek to remarry much younger women. A few men remarry in middle age and raise a second family. Women's chances of remarriage decline with age and fewer old women are married than old men. Divorce is more frequent early in a marriage regardless of spousal age. Step-parenting is abundant. Extra-marital affairs are concealed, and are common grounds for divorce. Marriages to close kin are rare (SI 15.13), lower than the worldwide figure given by Bittles and Black (2010). Survival of older married men and women is better than it is for older single men and women (SI 15.14).

These similarities exist despite some obvious differences in the context of marriage. For Hadza, the "transaction costs" for marriage, divorce, and remarriage (ceremonial, legal, and property implications) are minimal. There seem to be no societal prohibitions on remarriage; ceremony and property barely exist. Hadza live in an information environment more like a small high school than a city. Hadza have better information on the most elementary issues: they know who is out there, available or not, who the competition is, and perhaps who their competitors are interested in. Marlowe (2010, pp. 181–184) describes the basic features of Hadza mate choice and courtship.

Some of the features of Hadza marriage are shared with socially monogamous birds. Extra-pair matings sometimes occur. Divorce is more common early in a marriage. Even in socially monogamous species, male care is not always essential (Bart and Tornes, 1989; Gowaty, 1996); a deserted female can raise her nestlings. Avoidance of a step-brood is not universal (Rohwer *et al.*, 1999). Additionally, as we will see in Chapter 21, among the Hadza as well as some birds, marriage breakup is more common after reproductive failure.

There are also some less familiar observations. Marriage to the brother of a dead husband is not unusual, its incidence resembles that among the !Kung (SI 15.15). Women with more small children remarry faster after divorce, even controlling for their age. They do not "settle for" a man with a lower reputation than their previous husband. Men do not seem to avoid step-parenting and the children are, statistically, a sign of their mother's future fertility and success at child-rearing. The most nominated hunters remarry faster than do others. The "Bateman gradient" (slope of RS against mating success) is steeper for men than for women (see further discussion later). It is important to note that, in our case, RS is the number of live children (controlled for parent age) and "mating success" is the percentage of the reproductive career in a marriage. Surprisingly, women who are married for a larger percentage of their reproductive career, have similar child survival to women who are more often single. This is not what I would expect if men provide effective resources or care for children, unless perhaps this care is offset by the dangers of shorter inter-birth intervals during marriage (Chapter 17). It is important to remember, first, that Hadza women have free access to abundant and nutritious plant foods, which yield many calories per hour of work, and second, that single women have access to meat from large game caught by any man in camp. The relationship between RS and marriage would probably be very different in a population in which plant foods were scarce or hard to acquire, in which meat was less widely shared, or in which access to resources was controlled by men.

In the past decade, there has been much discussion of the Bateman gradient and Bateman's experiment. Wade and Shuster (2005), in their "don't throw Bateman out with the bathwater," conclude that "we view his insight about the source of sexual selection as more fundamental than his speculation about the consequences . . . " Two issues are discussed: first, what causes the difference in slope of the regression lines? For example, is it differences in "handling time" (Hubbell and Johnson, 1987), similar to Kokko and Jennions' (2008) "time out"?; and second, whether the different slopes have implications for strength of sexual motivation in males and females (Borgerhoff Mulder, 2009; Brown *et al.*, 2009). When we think of application to people (or birds), we need many cautions. "Mating" can refer to copulation, or to initiation of a long-term relationship, and we often unthinkingly slide between the two meanings. In natural fertility human populations, as a marriage continues, more children arrive. Consequently, there is no inherent advantage in more marriages. Two marriages of five years each may yield the same number of children as one marriage of 10 years. There is no significant association between number of marriages and RS among Hadza men or women.

My use of percentage of adult career in a marriage is based on one primary expectation that had nothing to do with Bateman. If there is a benefit to women's reproduction from resources given to them by their husbands, time not married should lead to less reproduction than time in a marriage. This leads to a different prediction from Bateman's: women should show a steep slope of the graph of RS on percentage of time in a marriage, perhaps just as steep as the men's curve. This is not what we see. The women's slope is flatter than the men's slope.

If marriage is not about shared care of children, then what is it about? Although we see that it can be more advantageous to men than to women, this chapter gives no quick answers. The biological literature is full of alternative routes to monogamy, and I will discuss them in relation to the Hadza data in Chapter 22.

# 16 Another dependent variable: growth as a proxy for fitness

The clinical nutritional status of all the children was good by tropical standards . . .

Jelliffe *et al.*, 1962

Like other hunter-gatherers, Hadza are quite short. They are not particularly thin; men are quite muscular (Photograph 16.1), and most women have substantial subcutaneous fat. Adult Hadza women (aged 18–60) were 150.7 cm tall and weighed 46 kg on average. Men (aged 20–60) were 160.6 cm tall and weighed 51.3 kg. Body mass index (BMI) (kg/m$^2$) is widely used to assess obesity, and international reference standards have been publicized. Only three Hadza met the Centers for Disease Control and Prevention (CDC) criterion for "overweight," and Hadza are not excessively thin (Figure 16.1).

Hadza children are small but very active. Jelliffe *et al.* (1962) continued the earlier quote, ". . . in particular, the syndromes of kwashiorkor and nutritional marasmus, rickets, infantile scurvy, and vitamin B deficiency syndromes were not seen." Some might argue that this was because the sick could not endure the harsh forager life long enough to be observed in a brief visit. However, Jeliffe also noted the balanced diet on which Hadza children are raised. The broad patterns of development are similar to those anywhere. Chubby babies and toddlers (Photograph 16.2) grow into active and skinny "stick people" in middle childhood (Photographs 16.3, 16.4), and in the mid- to late teenage years are transformed into shapely young women and muscular young men. My data show general features of Hadza size and growth in close agreement with the descriptions of independently collected data by Marlowe (2010, pp. 141–149, table 6.1). Many aspects of Hadza growth could be explored with our data, perhaps best in combination with Lars Smith's data from 1977, and Frank Marlowe's from 1995 to 2010.

Here I primarily direct my attention to growth as another dependent variable, as a possible proxy for fitness. Height, weight, and upper arm circumference are commonly predictors of mortality and morbidity, and thus are possible proxies for fitness of the measured individual. In later chapters, I look for effects of helpers on growth as a reflection of the help children are receiving. Growth is likely to be a more sensitive measure than mortality. There are several reasons for treating the under-five-year-olds as special targets of helpers. I have shown in Chapter 12 that their survival is especially significant to population growth (and to natural selection). I review some of the other evidence later in discussing Bogin's argument that early childhood is a unique feature of human development.

**Photograph 16.1** Hadza men are short but quite muscular. This man is very close to Hadza average height and weight. He is holding two of the stone pipes that the Hadza make for themselves (the rock is only found in a few localities). Nowadays, they eagerly smoke tobacco. As archaeologists find such pipes from well before the arrival of tobacco, Fosbrooke (1956) and Phillips (1983) suggested they were used to smoke hemp. © James F. O'Connell, 2015. Reproduced with permission.

I also wanted to use anthropometry to obtain an indication of women's "condition" during their reproductive careers. Women lost half a kilogram of body weight as they aged from the beginning to the end of their childbearing years. Despite this, "maternal depletion," as loss of weight with parity, was not seen, nor did weight loss appear to depend on the supply of helpers (SI 16.1). We can also test whether adult size predicts reproductive success (RS), which would imply that selection might favor greater parental investment in growth than is sufficient to ensure survival.

These and other possible uses of anthropometric data encouraged us to measure people at nearly every camp that we visited. We measured height, weight, upper arm circumference, and triceps skinfold on people of all ages. While comparative biologists have looked at body mass mostly, obviously more comparable across samples of species of diverse shapes and sizes, researchers on human growth have focused on height. This is largely because weight is quite variable over short time spans, and the accumulated wisdom is that height is a reflection of long-term nutritional history. I report some data on each. Almost all my data were collected during the dry season

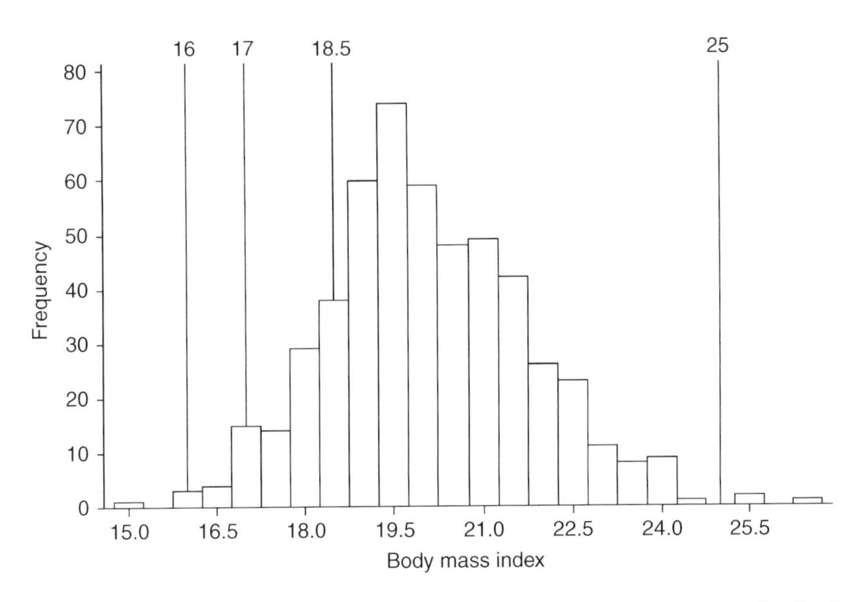


**Figure 16.1** Body mass index (BMI) of Hadza aged 20 and over. Frequency distribution of individual means of BMI for 285 women and 232 men aged over 20. Women have a very slightly higher mean than men (20.25 vs. 19.86, p=0.011, two-sample t-test). Women have slightly higher variance than men (SD, women=1.83, men =.49, Levene test=8.08, p=0.005). Reference values for the CDC standards, <18.5: underweight, 18.5–24.9: healthy, 25.0–29.9: overweight, >30: obese. (CDC "Overweight and obesity" at www.cdc.gov/obesity/defining.html).

**Photograph 16.2** Babies and toddlers can be quite fat. © James F. O'Connell, 2015. Reproduced with permission.

**Photograph 16.3** In middle childhood, Hadza are quite thin, although the boy shown here, having the use of his father's axe, hunted and collected honey for long hours almost every day. © James F. O'Connell, 2015. Reproduced with permission.

**Photograph 16.4** Some are very thin in middle childhood, although only in early childhood can we predict survivorship from body measurements. His necklace combines materials available for thousands of years (bones, and perhaps some grass sections), and several hundreds of years (glass beads). He is wielding a switch of the type sometimes used to discipline children, if the mother can catch up with them. © James F. O'Connell, 2015. Reproduced with permission.

and are not confounded with changes that might be expected when soft fruit first becomes available, or in a year when the honey yield is high toward the end of the rainy season. Contrary to such expectations, neither Sherry and Marlowe (2007) nor Marlowe and Berbesque (2009) found striking seasonal differences in Hadza BMI or percentage body fat.

## 16.1 Anthropometry methods and sample

Anticipating payment of a large cup of maize meal for each individual, the Hadza expected to be weighed and measured immediately on our arrival in camp. We cleared and leveled a 2 × 2 m area and set up our equipment. People arrived to take turns being measured. We encouraged parents to let their children wait until they had seen some adults being measured. Shoes and extra clothing such as a jacket or second shuka were removed. Each person was weighed on our Tanita load cell weigher, had their height measured with our portable stadiometer, upper arm circumference (UAC) measured with a steel tape, and triceps skinfold measured twice with our Harpenden skinfold caliper. Toddlers were sometimes too short to reach the lowest measure on the stadiometer and we made no attempt to measure height or length of infants. Infants were weighed on the mother, or a smaller helper, and their UAC and triceps measured while being held by the mother or helper. The few children who refused were nonetheless paid with their cup of maize "because s/he tried," a procedure everyone seemed to find fair. Details of the measurement equipment and techniques are described in SI 16.2, along with tests of our accuracy, for example on repeat measurements of the same people in the same field season. The mean absolute difference between pairs of such measurements of adult height was 7 mm. My brief survey of the literature suggests that this compares poorly with laboratory measurements but favorably with clinic settings.

One thousand individuals were measured and 754 of them were measured more than once; some were measured in as many as nine different field seasons. There was a total of 3195 person-measurement occasions. Our data are thus "mixed longitudinal"; each individual contributes a different number of observations at a different sample of ages. The right way to analyze such data is with multilevel analysis and I have used this to follow up particular questions at particular portions of the lifetime.

## 16.2 Early childhood: a special feature of human life history and an opportunity for helpers

Some significant divisions of the developmental period follow from studies of mammalian life history evolution, such as the idea that growth is driven by the mother's "biological productivity" until weaning, and after weaning the growth of juveniles is proportional to their own size and productivity (Charnov, 1993, p.92). Following weaning, juvenile mammal weight increase follows the equation $dW/dT = A\ W^{0.75}$ where W is current weight, and A is a taxon-specific coefficient. Hill and Hurtado (1996) found that Ache children's growth closely followed this pattern from the age of six years. As determinate growers, the age at which growth ceases, close to the age at first birth, also has some robust theory (Stearns and Koella, 1986; Kozlowski and Wiegert, 1987; Charnov, 1993), successfully applied by Walker *et al.* (2006), Migliano *et al.* (2007), and Stock and Migliano (2009).

Several authors have pointed out that humans tend to be weaned earlier than great ape infants, and upon weaning, continue to be highly dependent on food acquired and provided by older individuals. This dependence opens a novel opportunity for helpers. While other mammals, including primates, rapidly become self-sufficient after weaning, humans do not. Ecological reasons are proposed by Kaplan (1994) and Hawkes *et al.* (1995a). While in tropical forest, food is available and accessible to the newly weaned primate year round, in savanna there is at least a long dry season during which food is available in large quantities and large packages but obtained only with considerable strength and skill. While adults can acquire these foods in large quantities, the smallest children cannot. They are thus economically dependent on adults or older children. The significance of this for human life history evolution has been discussed by Kaplan (1994) and Hawkes *et al.* (1998). I will look at the Hadza data on age and size at weaning, and data relevant to Bogin's argument that the period from weaning to about age six years presents a condition novel to mammalian life histories.

### 16.2.1 Mother's weight and infant's weight

Growth slows down late in the first year of life and into the second in every population, whether breast fed, bottle fed, early weaned, or late weaned. The Hadza are no exception. A good fit of Hadza weight in the period 0–2.5 years, the natural logarithm of age, reflected the deceleration. After controlling for nlog age, I found a significant positive effect of mother's weight on child's weight, aged zero to 2.5 years ($b=0.04287$, $p=0.008$, $N=249$ measurements on 200 children), just as one would expect if growth before weaning is driven by the mother's "productivity." The mother's productivity is expected to vary with her weight to the power of 0.75, and infant weight predicted by nlog age and mother's $W^{0.75}$ was also significant with $b=0.154$ and $p=0.008$. There were no significant effects of birth order, father's weight, father's tribe, or infant gender, and only a marginally significant effect of mother's age at the birth of the child ($b=0.02263$, $p=0.051$, $N=243$ weighings).

### 16.2.2 Weaning weight

Weaning has interested human biologists and others in several ways. Public health researchers find that, under third-world conditions, weaning is a time of peril. In many such populations, "weanling mortality" and growth faltering are serious hazards. Under-nutrition is an important correlate of child mortality and even of cognitive and educational outcomes (Mendez and Adair, 1999; Scrimshaw, 2003; Glewwe and Miguel, 2008; Martorell *et al.*, 2010). These researchers draw attention to the consequences of age and size at weaning, and the process of weaning as factors contributing to the subsequent morbidity.

Definitions of weaning are often vague, and can encompass widely differing points during the transition from total nutritional dependence on the mother, and relative independence (Sellen and Smay, 2001; Sellen, 2006). There is almost always a time during which the infant suckles, and takes other food. This transitional time may be quite short, or quite long. Informants may also pay more attention to one feature or another. In the case of the Hadza, women seem to be most aware of the time when the baby ceases suckling, and they see this as a relatively sudden and complete change, sometimes because they enforce it. Hadza children will have taken a variety of foods before the final end of suckling. Baobab flour (which includes the shelled and pounded seeds, Chapter 2) makes an especially good toddler food.

Ideally, data on weaning would be gathered by observations repeated at quite short intervals. My data do not meet this ideal. At each interview, I asked the mother whether the child had ceased suckling, and made a note if I saw it suckling during the interview (Hadza infants suckle very frequently, if briefly, just like the !Kung infants described by Konner, 1972, and Konner and Worthman, 1980). The data available thus comprise two observations: the latest occasion on which the child was recorded as suckling (usually observed and noted on the interview sheets, as well as reported by the mother); and the first occasion on which the child was reported by the mother as having given up suckling. These occasions were often widely spaced (for instance between two or more censuses), and sometimes only one type of information was available on a child.

I used these data as arbitrarily censored, and providing start times and end times, start weights and end weights, to use as the dependent variable in Minitab's Reliability and Survival programs. Distribution analysis with arbitrary censoring showed that 50% of the children were weaned by 2.03 years of age (24.4 months) and between 10.3 and 10.4 kg in weight. The regression with life data (SI 16.3) showed that both infant sex, and the mother's body weight significantly predicted weaning weight but not weaning age. Boys were weaned at a greater weight than girls and children of heavier mothers were weaned at a greater weight than children of lighter mothers. Because children of heavier mothers were weaned at the same age as others, we must conclude that they have grown a little faster before weaning. We may regard this as a reflection of the larger mothers' greater biological productivity.

Comparative studies show that among mammals (a sample of 30 primates, 27 ungulates, and 20 pinnipeds, Charnov, 1993, fig. 5.4), weaning occurs when the offspring reaches 33% of the weight of the average female at the age of first reproduction. Among primates, the figure is a little lower, 28% (Lee *et al*, 1991, appendix 1, excluding *Homo sapiens*). At 10.3 kg, the Hadza are weaned at about 22% of the weight of women at the average age of first reproduction (46 kg). Hadza children reach the proportion of the mother's weight at which most mammals are weaned (33%) at about age six years (46 kg $\times$ 0.33=15.2 kg), and match the primate figure by age 4.5 years (46 kg $\times$ 0.28=12.88 kg). Thus, Hadza children, like other children summarized by Walker *et al.* (2006), are weaned at a much lower relative weight than other primates.

Among non-human primates, when offspring are weaned, they quickly become self-sufficient. Once weaned, they feed themselves (sometimes inefficiently, while the mother sometimes makes concessions to them by limiting herself to more juvenile-friendly foraging sites, Janson and van Schaik, 1993). Humans everywhere differ from this pattern. Children are weaned long before they are economically self-sufficient, even where the technology is the simplest.

### 16.2.3 What is so special about early childhood?

Bogin (1988, 1999, pp. 74–79, 173–178) described "childhood" lasting from weaning at about age three years to around age seven years, followed by the juvenile phase. He argued that the latter resembled the juvenile (weaning to maturity period) in other primates while childhood was a novel insertion into human development. I think our data can add to the support for "Bogin's insertion," and offer support for some of the suggestions about its place in human evolution (Hawkes *et al.*, 1998). Like Konner (2010), I will refer to this "novel" period, weaning to roughly six years old, as "early childhood," a term long used by developmental psychologists.

Bogin gathered several lines of evidence that distinguish early childhood. During childhood, "the rate of growth levels off at about five centimeters per year." Growth in brain weight is completed by seven years of age. The eruption of the first molar (M1) occurs between ages 5.5 and 6.5 years in most human populations, according to Bogin (1999, p. 76). In other primates, the eruption of M1 has been reported to coincide with weaning (Smith, 1992; Smith *et al.*, 1994; Bogin, 1999). However, the association is not close in wild chimpanzees, which continue to suckle for a year after eruption of M1 (Smith *et al.*, 2013). Bogin points out that the seven-year-old child has dentition more suited to processing less thoroughly prepared foods than the kinds often given to younger children. We noted earlier that the Hadza are weaned at a much younger age, and only by about age six years do they reach the relative weight at which other mammals are weaned. We see later that growth from six years old upward follows a different, gently increasing pace.

Bogin also cites evidence that the walking gait is not finally developed until age seven years, and that the energy consumed in walking declines between five and eight years old, when it reaches about 90% of adult efficiency. In our observations of foraging by Hadza children, the youngest child to make the two-hour trip from Ilomo to the distant berry patch was aged 6.5 years (Hawkes *et al.*, 1995a). Bogin also pointed out that the ages five to seven years encompasses the "5–7 shift," several cognitive changes described by developmental psychologists. This may relate to the greater foraging ability of Hadza children between ages five and seven years. Bogin argues further that third-world "street children," and so-called feral children, are of ages within his juvenile period, implying that early childhood does not equip them to survive without adult care. My data on Hadza orphans suggest few survive early childhood, while those who are orphaned later tend to survive but are smaller (SI 16.4).

To look at Hadza growth during early childhood, we have two difficulties in selecting a starting age. Weaning among the Hadza is earlier than the three-years-old mean of many populations. Weaning would be the logical starting point but we have no record of individual weaning ages. I showed that the mean age at weaning is two years old, but this implies that many are weaned before age two years. Infant deaths are defined as those up to the first birthday. Weanling deaths that occur late in the second year of life will appear in our record as deaths to one-year-olds. I am interested in relationships between growth, death, helpers, and other measures. To include the 1–1.99-year-olds in an early childhood sample seemed the best choice.

During early childhood, the relationship of Hadza weight and height to age is linear. For ages $\geqq 1.0$ and $<6.0$ years, the best fit of weight is a linear fit to age (wt = 6.85 + 1.43 × age, p=0.000, adjusted $r^2$ 56.4). Age$^2$ gives a non-significant beta coefficient and adding age$^2$ accounts for no more variance. In stepwise regression, age$^2$ does not meet the criteria for inclusion. This is also true if we restrict the period to ages 2.5 to $<6.0$ years. The 37 Hadza children with more than one height measurement in this period showed a mean rate of height increase at 6.2 cm/yr. The regression of height against age showed a gain of 5.34 cm/yr. In predicting the child's height in this age group, no significant result was found for gender, mother's age at the birth of the child, subject's birth order, or father's tribe. The mother's height (b=0.3012, p=0.001, N=127) added to the prediction of the child's height, but the father's measurements did not in this age group.

For the 187 pairs of weighings of children in this age group, the mean increase was 1.70 kg/yr. Linear regression gave 1.43 kg/yr (N=300 weighings), with a beginning weight of just over 8 kg at age one year and an ending weight of 14 kg at five years and 15.3 kg at six years. For the child's weight, the mother's weight was a significant predictor (b=0.07675, p=<0.000, N=299) as was the mother's age at the birth of the child (b=0.0311, p=0.011, N=300 weighings, perhaps some indication of terminal investment), but not the gender, father's tribe, birth order, or father's weight. The relationship of child's weight (and mortality) to mother's age at the birth of the child is not linear. The children of women aged <20 are about half a kilogram smaller, and the children of mothers aged 35+ are about half a kilogram larger. Firstborns are not at greater risk; the children of the youngest mothers are at the greatest risk.

Demographic data also suggest some special features of Bogin's distinction. Most child deaths occur before the end of the sixth year of life (that is, while aged five years). Because mortality decreases from birth to about six years old, this is a time during which life expectancy ($e_0$) increases, from 32.7 years (sexes combined) at birth to its peak (45.3 years) at age 6 years. Consequently, during this time the child's reproductive value increases faster than at any other time in its life, another indication that this age group may be a rewarding target for helpers. Thus, this would be a practicable period in which to look for effects of helpers on the nutritional status that may influence child mortality, giving helpers the greatest fitness gain. This observation, relying on a normal human demography, would apply to most human populations.

It may also be a period of wider vulnerabilities and importance for shaping subsequent development. Consequences have been suggested for several aspects of behavior (Draper and Harpending, 1982; Hrdy, 2009). Peculiarities of behavior before the five- to seven-year-old shift might be interpreted as vestiges of pre-weaning behavior, or more likely as special adaptations to the value of gaining and maintaining the attention of helpers, as suggested by Hrdy (2009) and Hawkes (2014).

When we call early childhood a novel insertion into development, we cannot, of course, have in mind the insertion of new time. What is new is the early weaning and its correlates. Being weaned before any approximation to self-sufficiency puts the youngster into new circumstances, being highly dependent economically while no longer nourished by the mother's lactation. At this point (and perhaps for a short while before final weaning), not only the mother but also others provide and process food for the youngster. This may be reflected in the linearity of their growth, and in the child's subsequent transition to an initially slower but gently accelerating growth as the focus of helpers becomes redirected to younger kin. The early weaning may have been made possible by the supply of opportunistic older female helpers (Hawkes *et al.*, 1998).

## 16.3 Diminishing fitness returns to growth in early childhood

There is plentiful evidence from other populations to show that slow growth or poor nutritional status predicts greater mortality risk for a child. There is also some evidence that childhood nutrition can influence adult health and RS.

We can examine the relationship of child survival to child growth with a file in which each entry represents a child who had been measured at least once, and lists whether it died or lived to the end of the study period. Of 46 dead children who had been measured, 39 were measured within the two years before their death. There were up to five observations on a child but each child has only one life. The file included measurements of children of all ages from birth to 17 (423 individuals, 1092 measurements of weight, 791 of height, 1005 of UAC). Whether the child had lived until the end of the study was treated as the dependent variable in multilevel logistic regression. The levels of analysis were: individual, and observation (measurement occasion). Height, weight, and UAC were significant predictors of death or survival after controlling for age (Table 16.1).

**Table 16.1** Children who are larger for their age are less likely to die. Growth and survival, 391 children aged zero to 17, total of 1005 measurement occasions. Results of multilevel logistic regressions predicting child mortality from anthropometry. Model: dead or alive = age + age$^2$ + age$^3$ + size measure, with random intercept. Body Mass Index (BMI) was a weak and non-significant predictor

| Dead/alive 2k | b | SE | Wald | p | Ss/obs |
|---|---|---|---|---|---|
| Height | -0.113 | 0.049 | 5.32 | <0.025 | 311/781 |
| Weight | -0.303 | 0.111 | 7.45 | <0.01 | 423/1092 |
| UAC | -0.441 | 0.170 | 6.73 | <0.01 | 391/1005 |

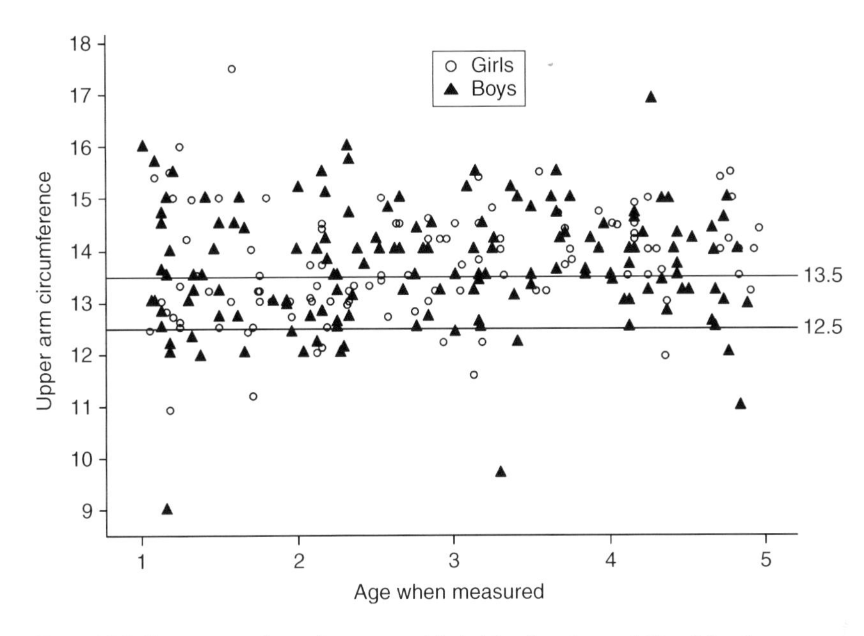


**Figure 16.2** Upper arm circumference and "at risk of under-nutrition" levels.

Upper arm circumference has been noted to change very little between ages one and five years (although Eveleth and Tanner, 1976, p. 14 note that during this period, fat decreases and muscle increases). A cheap and handy device to measure UAC during food emergencies, the "Shakir strip," has been widely used in the third world. Children with a UAC below 12.5 cm are regarded as in danger, those between 12.5 cm and 13.5 cm are a marginal risk group, and those above 13.5 cm are regarded as well nourished. Figure 16.2 shows that quite a number of Hadza children fell below these risk levels. Among the 269 measurements of UAC in this age group, 9.7% fell below 12.5 cm. (26 measurements) and 40.5% fell below 13.5 cm (109 measurements). By these criteria, Hadza children were at considerable nutritional risk during this age range.

The shape of the association of UAC and mortality among Hadza children shows the risks indeed to be high for measurements of below 12.5 cm (Figure 16.3). Upper arm circumference should be an important measure with which to look at effects of helpers. The mother's weight is a significant predictor of UAC; heavier mothers have children with larger UAC ($b=.00449$, $p<0.001$, $N=262$) but the father's weight is not.

Single level logistic regression in Minitab predicting survival to the end of the study, controlling for age, showed a substantial effect of UAC ($b=-0.6017$, $p=0.004$). Probability of survival, predicted from UAC, shows (Figure 16.3) something of the diminishing returns to size reported by Bairagi *et al.* (1985) and illustrated by Caulfield *et al.* (2004, fig. 1a). It gives some, but only some, support to the view that UAC below 13.5 cm and below 12.5 cm can be taken as indicators of severely

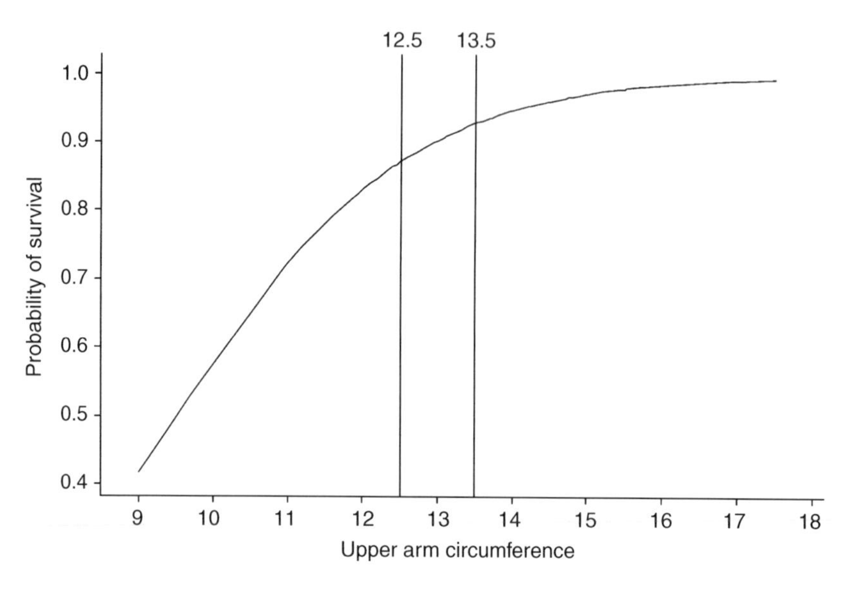


**Figure 16.3** Fitted probability of survival to the end of study by upper arm circumference (UAC). Logistic regression, alive/dead = upper arm circumference. Sample restricted to children aged one to five years old. 21 deaths, 255 survivors with UAC measured.

increased risk. Even when the fit is to UAC + UAC$^2$, which inevitably gives a steeper and more convex curve (but no better a fit), we do not see a step function. However, we do see almost equal survival among those with UAC at and above 13.5 cm. Below 13.5 cm, the child's prospects decline with increasing steepness, but a Hadza child with a UAC of 12.5 cm is by no means doomed. A mother whose child has an UAC of 13 cm or more gains little fitness by increasing UAC still further, and thus might enhance her fitness more by directing additional resources to another child or a new birth.

Some of the same diminishing returns can be seen in the weight of the one- to five-year-olds when we plot probability of survival against the residual of weight against age (SI 16.5). The gains in improved probability of survival are greater for growing from very light to normal (residual zero), than for growing from normal to heavy. Diminishing survival returns from increased birth weight can be seen in data cited by Guyatt and Snow (2004, p. 761) and diminishing returns to food supplementation during pregnancy are implied by Prentice *et al.* (1983, and elsewhere).

Diminishing returns are the key to solving many trade-offs and may have much to do with the trade-off between the number and size of offspring. This, in turn, relates to the association between population density and individual size discussed by Walker and Hamilton (2008). It is a key to doubting the "balloon theory" of growth (the more you pump in the bigger they get), and a reason to think twice about categorizing small-bodied populations like hunter-gatherers as "stunted." Each child might have grown bigger with more food, but descendants of mothers who arranged this would have quickly been outnumbered by those whose mothers' physiology followed the rule of diminishing returns to enhance their fitness. We will also see diminishing survival returns to time invested in the youngest offspring when we look at inter-birth intervals in Chapter 17.

Evolutionary ecologists emphasize the different evolutionary interests of offspring and parent, Triver's "parent–offspring conflict" (Trivers, 1974; Haig, 2014). If there are advantages in growing bigger, natural selection favors individuals who can grow big. Natural selection also favors adults who leave more descendants. Adults must trade-off the advantage of more offspring against the advantage from each of them being a little larger. Smith and Fretwell (1974) described the classic model of allocation of parental care. Walker and Hamilton. (2008) showed evidence for such a trade-off in a comparative study. Here I offered evidence about the fitness benefits of growth among the Hadza, as children facing severe challenges to survival, and as adults attempting to bear and raise children.

## 16.4 Describing growth after early childhood

Children continue to receive food from adults after early childhood. Even the eagerly foraging Hadza children can seldom provide their entire daily requirements until about age 15. We will need to look for effects of helpers on the growth of children aged 5–16 and thus include some description here. I report the fit of weight to two models: an empirical fit to age, and a fit to a widely used model of mammalian juvenile growth.

My endpoint for the "juvenile period" (age $<16.0$) was chosen as the age at which Hadza boys and girls rapidly diverge, girls height velocity decreasing while boys continue growing (see graph of height in SI 16.7). Differences between boys and girls become important toward the end of this period. Girls begin a rapid increase in weight and triceps skinfold around age 14 and stop growing taller at an earlier age than do boys. Hadza boys tend to follow a pattern reported for several other African samples in which growth continues into the early 20s (Figure 16.4 and SI 16.6). Contrary to my suggestion in Blurton Jones (2006), at least some Hadza boys appear to show an adolescent growth spurt in height (Figure 16.5). In this age group, children with a non-Hadza father were significantly taller than children with a Hadza father, and analyses are reported here on only the children with Hadza fathers. Fifty-three children attended school, few for more than a single year, and most belonged in the current age group.

I ran a three-level regression of 815 measurements of 335 children of 152 Hadza mothers and Hadza fathers. I tested for the best fit to age, for height, and for weight in multilevel regression, with the intercept random at j (child) and k (mother). Adding $age^2$ to the model improved the fit significantly, and adding $age^3$ did not. Thus, I used $age + age^2$ in subsequent models for this age section. The fit differentiates this period from early childhood when growth is linear.

1. Height: there was no significant gender difference in height. Children born to older mothers tended to be taller. Children born to taller mothers were

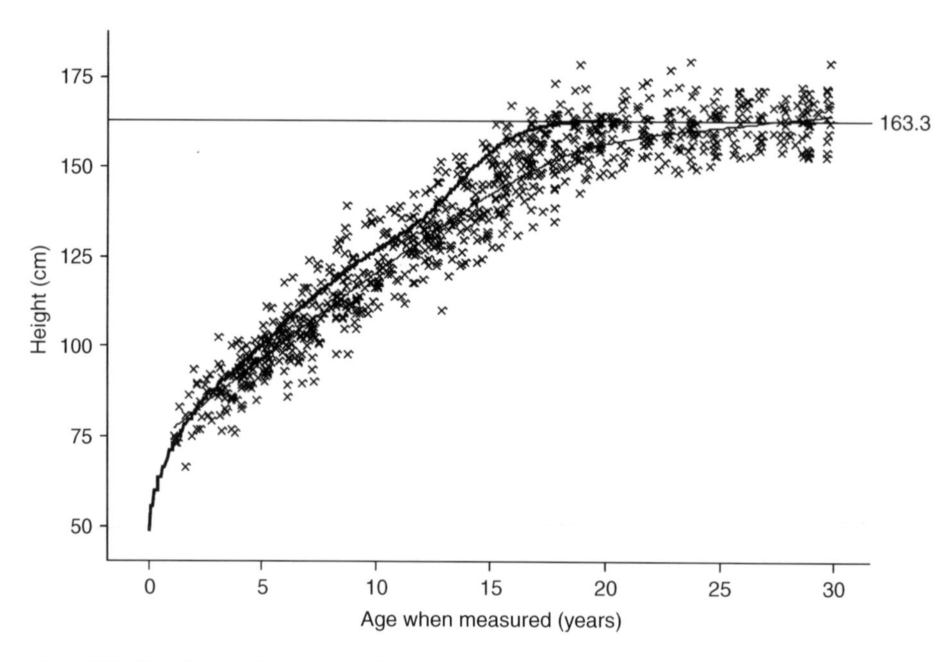


**Figure 16.4** The African boys pattern? Male height by age. Hadza plotted with CDC third percentile (heavy line), which levels off at 163.3 cm just before age 20. Hadza continue to grow taller after age 20. Lowess smooth of the Hadza records (light line) reach 163.3 between ages 25 and 30.

significantly taller. Children born to taller fathers were significantly taller. Children who attended school for more years were significantly taller.

2. Weight: the sexes appear to follow the same weight curve as each other until about age 14, when girls gain weight faster than boys. School attendance, and the mother's status (dead or alive) both have significant effects on weight. The mother's age at the birth of the child had a significant effect (b=0.176, SE=0.030, Wald $\chi^2$=0 34.4, p=<0.001). Children who were born to older mothers became heavier than children born to younger mothers. Neither the mother's nor father's weight was significantly related to the child's weight in this age group. The girls' weight gain slows between ages 15 and 20 (average age at first birth is 19). Boys continue to gain weight rapidly to age 20 but continue gaining weight slowly until the end of their 30s.
3. Female upper arm circumference: the UAC (SI 16.7) of girls begins to exceed that of boys as early as eight years of age, rising fastest at age 14 and 15, slowing sharply at age 20, but perhaps climbing throughout the childbearing years and into late middle age. If we link this to the triceps skinfold data, any growth in female UAC would appear to be a result of growth in both muscle and fat. Most of the teenage increase is fat, most of the adult increase is muscle.
4. Male upper arm circumference: the UAC of boys lags behind that of girls for most of the teenage years but exceeds them during the 20s, reaching a summit plateau that lasts from age 30 to 50, after which it declines slightly (SI 16.7). The triceps

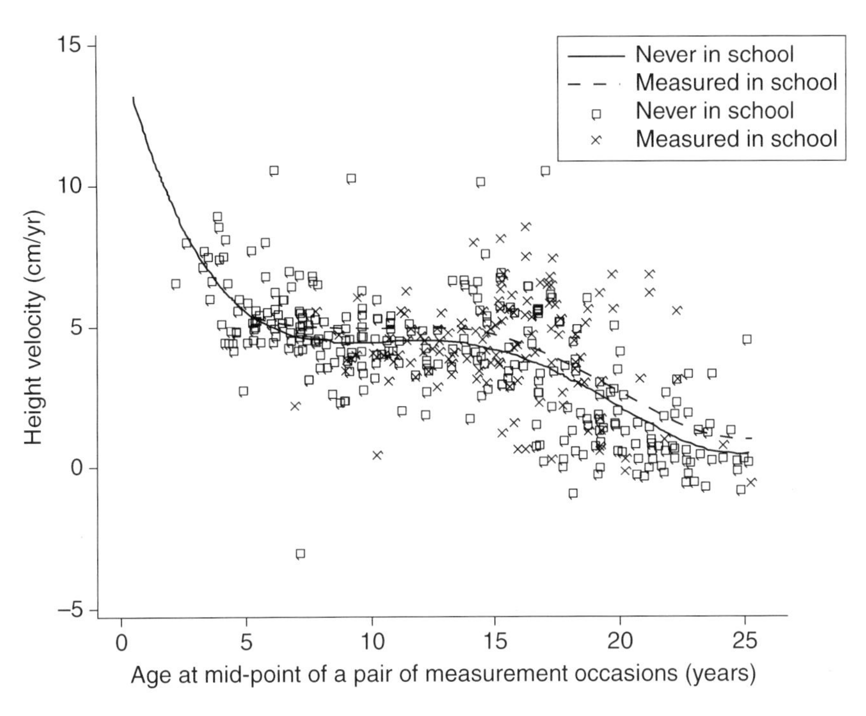


**Figure 16.5** Is there an adolescent spurt in height? Hadza boys height velocity by age. Lowess fits for boys who did or did not attend school.

skinfolds of boys tell us that none of this increase is in subcutaneous fat; the gain is mostly muscle. We can note that the age pattern of men's UAC parallels the age pattern of reputations as an expert hunter; we cannot be sure that this is more than a pattern of use, the expert hunter may use his bow (draw weight >>25 kg Blurton Jones and Marlowe, 2002) more than others, as well as more effectively. Blurton Jones and Marlowe (2002) found that accuracy at target archery was related to strength among Hadza men and boys. However, this age pattern may be followed by many aspects of male careers in various societies (as discussed in Chapter 14). Among the Hadza, it also parallels men's fertility career.

5. Triceps skinfold: this is generally assumed to be an indicator of subcutaneous fat, and help us understand the changes in weight and UAC. Among young children, boys and girls appear to have similar triceps skinfolds and these decline from birth to age nine or 10. At this point, girls begin an apparently accelerating increase in triceps skinfolds, while boys skinfold measurements continue to decline. Hadza boys do not show the increase in triceps skinfold reported in English boys (Tanner, 1962). The decline in male triceps skinfold continues into late middle age. In girls, the dramatic rise in triceps skinfold during the teenage years begins to slow down (in these cross-sectional data) around age 16 and its average value peaks at age 20. The late teenage accumulation of fat may be thought of as a preparation for reproduction.

6. Body mass index: figures in SI 16.7 show the age pattern of Hadza BMI. Not surprisingly, it seems to mostly reflect the pattern of change in weight and triceps skinfold. The most conspicuous part of the chart is the rapid increase among the teenage girls. BMI decreases during early childhood.

## 16.5 Fitting juvenile growth to the common mammalian growth equation

According to Charnov (1993, and others), during the mammalian juvenile period, growth rate varies with current size to the 0.75 power multiplied by a taxon-specific factor. If this pattern applies to children's growth, it implies that adult size can be predicted from the size at which the child is launched onto this trajectory, and the time at which growth ends.

Hill and Hurtado (1996) observed that Ache growth from age six years fitted Charnov's growth equation, $dW/dt = A \times W^{0.75}$ with A at 0.29 for females and 0.23 for males. The upward curving fit for weight of Hadza boys and girls aged 5–16 suggests that we try the same equation on Hadza growth. Several tests on the final dataset of individual measurements indicate that their growth follows the power function with values similar to those of the Ache. At first I looked at the sexes together and assumed A=0.25. A regression model with age + $age^2$ + $0.25W^{0.75}$ gave a significantly better fit than the model with age + $age^2$ alone ($0.25W^{0.75}$ b=1.06, p=0.000, $r^2$=27.8 versus 21.0%). The result was similar when the sexes were analyzed separately, with A=0.23 and 0.29.

In another test, I added $0.25W^{0.75}$ to a regression model predicting the second of a pair of weighings from the first weight and the age at second weighing. The power measure added significantly to the prediction. In stepwise regression, if the power measure was entered first, then age failed to enter. In the 6–16-year-old sample, the power function significantly improves the prediction of weight ($weight^2$ = age1 + age2 + $0.25W^{0.75}$, the power function beta coefficient was 13.07, p=0.000, $r^2$=70.6% versus 52.7%).

Figure 16.6 shows the observed, empirical fit of weight to age + $age^2$. If we apply the equation $dW/dt=0.25W^{0.75}$, beginning at the observed weight of six-year-olds (15 kg), we generate later weights that are within 0.3 kg, on average, of the observed, empirical fit. The predicted second weight of any pair of weighings was 0.3 kg higher, on average, than the observed weight (N=464 pairs of weighings on 242 children). When the sexes were examined separately, with their different values of A, the differences between predicted and observed second weighings were reduced to zero for males, and 0.18 kg for females, and the adjusted $r^2$ to 91.5% and 91.7%. Thus, the "juvenile mammal model" predicts that a 15 kg six-year-old will become a 37 kg 15-year-old and a 48 kg 18-year-old, close to the observed means. After age 15, the growth of many girls slows and then ceases. Thus, the scatter of weights becomes wide in the late teenage years. There is some support for Hill and Hurtado's observation that girls' growth tends to follow this trajectory until their first birth.

Figure 16.6 also shows how a 1.5-kg difference in weight at age six years would translate into a 3.5-kg difference at age 18. This is about the difference between an

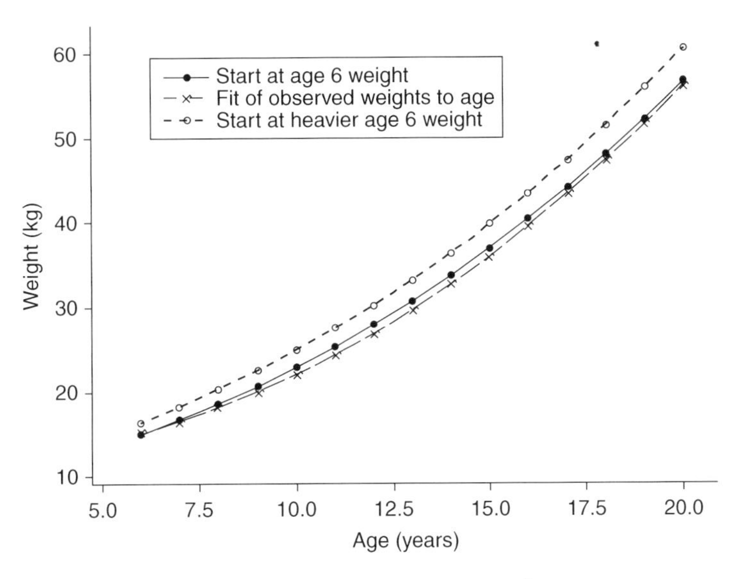


**Figure 16.6** Empirical fit of child weight to age + age$^2$ compared to weight predicted as W plus $dW/dt = A \times W^{0.75}$. The effect of the power equation on growth from two different starting weights are compared. A 1.5-kg difference at age six years becomes a 3.5-kg difference at age 18.

18-year-old Hadza woman at 48 kg and, at 51.5 kg, a Gambian woman as in Eveleth and Tanner (1976, fig. 75), close to the African median weight, or a Machiguenga or Mayan woman as in Walker *et al.* (2006, table 2). This calculation assumes no change in the "constant" A, and we do not know whether population differences in "growth rate" represent differences in A, or differences in the size at which growth begins to follow the power function. Thus, much of the variation in adult weight should be predictable from weight at the age when the individual begins to follow the trajectory given by the equation discussed earlier.

In contrast, among the under-five-year-olds, the power variable did not improve the prediction of weight gain or weight. Figure 16.6 shows the body weight predicted by $0.25W^{0.75}$ applied to the weight of Hadza children starting at six years old. If we apply $dW/dt=0.25W^{0.75}$ to the average age and weight at weaning (10 kg), we arrive at six-year-olds who weigh 16.4 kg instead of the observed 15.3 kg, and ten-year-olds and 15-year-olds who are several kilograms heavier than the observed weights at these ages. We may take this as another indication that Hadza children do not follow the power equation until near the end of Bogin's childhood (early childhood).

Thus, there is some evidence for the relevance of the widely used power model to growth during the "juvenile" period in two hunter-gatherer populations, the Ache and the Hadza. The fit of this equation suggests that growth in body mass during age five or six to 16 complies with the expectation for the juvenile phase of mammalian growth, after weaning, when growth is no longer directly dependent on the mother's productivity. But few if any, humans are economically independent of their parents before maturity. Even Hadza children, who forage as successfully as any recorded, seem unable to acquire all their own food until their mid- to late teenage years. We may wonder what the "productivity" of juveniles really means; perhaps it sets appetite more than it responds to foraging ability. Perhaps it describes allocation of resources between activity, maintenance, and growth more than it describes food intake. Possible implications for the stability and change of the mean body mass of a population should be deliberated.

## 16.6 Adult size and reproductive success

In most human populations, men are bigger than women and such sexual dimorphism is usually thought of as a result of sexual selection. Sociologists have observed in western societies that taller men have some social advantages. Hill and Hurtado (1996, p. 316) report a positive association between men's fertility and weight among the Ache.

In the Hadza data, I could find little relationship between RS and physique. Reproductive success (standard score of live children for parent's age) was not significantly related to height, weight, BMI, or triceps skinfold. Men's accumulated fertility (ssborn) is not predicted by their height, weight, skinfold, or BMI. However, there was a small positive effect of UAC ($b=0.0984$, $p=0.043$, $r^2=1.9\%$). By definition, the standard scores of numbers of births, or numbers of living children, do not correlate with the men's ages at measurement, thus no statistical control was needed in looking for a relationship of RS to adult physique.

Lee (1979, fig 10.1) reported that shorter !Kung men were more successful hunters than taller men. We may wonder whether, among the Hadza, hunting success or the reputation as a good hunter (Chapter 14) relates to height or strength. The Hadza may differ from the !Kung in this respect because their hunts cover much less distance in rather less extreme heat conditions. !Kung hunting involves covering great distances in a hot environment with no shade and some urgency. A large literature on heat stress implies that smaller men and women can work more successfully in extreme heat. Tanner (1964, and others subsequently) showed that Olympic endurance athletes tended to be both small and lean. While !Kung hunts often involve a daytime encounter with tracks, followed by a lengthy attempt to catch up with the animal, the Hadza rely much more on either waiting at a water hole, or a chance encounter with the animal itself in their more wooded countryside. Nonetheless, in both ecologies, tracking an animal to find its carcass is a demanding and sometimes lengthy task. While !Kung methods argue for an endurance physique, Hadza methods suggest no particular endurance demands. Hadza carry much heavier bows, which require greater strength to pull. In both populations, natural (or sexual) selection on hunting success exists in parallel with any sexual selection on men's physique.

I looked for relationships between adult Hadza physique and hunting reputations. Physique of each individual was represented by the averages of his/her measurements taken after the age of 25. The file of these averages was linked to the file of pro-rated nominations. Hadza men continue to gain weight, mostly muscle, until around 40 years old. Hunting nominations have a similar distribution with age

(Chapter 14). If we control for age (age + age$^2$, which can track the humped shape of hunt nominations multiplied by age), we find a significant positive contribution of weight (b=0.2217, p=0.015, N=210, and UAC b=0.5402, p=0.049). Height has a small positive but not significant effect (b=0.10006, p=0.176), and BMI is also not significant (b=0.3076, p=0.169). These regressions account for only very small amounts of the variance in hunting nominations (weight 9.0%, BMI 7.2%, UAC 8.1%). I tested the idea that extremes might be disadvantageous, by looking for an association between the absolute value of the standard scores of height or weight and hunting reputation. If average-height men were more often nominated, there should be a negative correlation, but there was not.

Thus, among the Hadza as well as the !Kung, expertise is mostly independent of size. At other times or other places, there may be clear advantages of greater size among men.

Do bigger women convert their greater productivity into more births? Because it is a basic expectation of some animal life history theory, and commonly found in a variety of animals (Roff, 1992), some anthropologista have expected greater fertility among larger women. I cannot show a relationship between size and fertility among the Hadza. I see no difference in the standard scores of the number of births, nor the annual probability of a birth in relation to the mother's weight. The picture can be complicated by variation in the role of helpers, the high mortality of children of young mothers, the relationship of infant and child mortality to inter-birth intervals, effects of disease on the reproductive system, and the free access that Hadza women have to resources.

There seems to be a common but not universal, worldwide observation of better survival by the children of taller women (Monden and Smits, 2009; Subramanian *et al.*, 2009). Previously, we saw that larger Hadza women tended to wean larger babies, and that heavier women had children who were heavier after weaning. Heavier Hadza children (for their age, and with diminishing returns) are more likely to survive. The Hadza data thus offer some support for the expectation that larger mothers may generate a greater mass of weaned offspring.

## 16.7 Summary

Hadza weight growth shares its general trajectory with other better known populations. A rapid deceleration during the first year of life is followed by a constant growth between weaning and about age seven years, then a gentle acceleration in weight from mid-childhood into the teenage years, and a decrease in height velocity until some point in the teenage years (Figure 16.5; and SI16.6). The sexes are the same height and weight until the early teenage years and girls cease growing earlier than boys. The few school attenders gained weight faster, but this may not influence adult height or weight. Other features are more notable.

1. Hadza children are weaned at about 22% of the weight of women at the onset of reproduction, contrasting with the mammalian norm of 33% and primate norm

of 28%. Larger mothers weaned heavier children. Several lines of evidence support Bogin's suggestion that the period from this very early weaning to about age six years is unique to humans. Survival of the under-five-year-olds has the largest effect on population growth rate and is expected to be a primary target for helpers.

2. As among the Ache, Hadza children's growth in body mass from age six years to about 16 followed the mammalian juvenile growth equation $dW/dt=AW^{75}$, in which W is weight and A is a taxon-specific constant. This should mean that individual differences at age six years would be reflected in differences in adult size.
3. Although larger women weaned larger babies and maternal weight was a significant correlate of adult height and weight, I found no indication of greater fertility among larger women. Good hunters were no bigger than other men, nor did larger men have greater RS.
4. In SI 16.1, I found little indication of maternal depletion (weight loss associated with parity) and could not link presence or absence of mother or husband to a woman's adult weight changes. Hiernaux and Hartono (1980) suggested that, compared to other African women, Hadza women were relatively well nourished. Hadza women's freedom of access to resources and control over weaning age, allows them to avoid the worst consequences of male control of resources and fertility.
5. Orphans were lighter than others, even as adults (SI 16.4).
6. Both child weight and UAC predicted child survival. However, there were diminishing survival returns to increased size. Above a certain size, it would clearly pay the mother to switch her resources to a new birth. If there are advantages to being a larger adult, the critical offspring size at which the mother diverts resources may be a little higher than indicated from child survival alone. Most studies of change in size with migration or of secular trends have neglected the often concomitant changes in fertility. If fitness consequences of fertility decrease, the point at which it pays to divert resources from the current offspring to a future birth may be delayed. Diminishing returns may make it more difficult to see effects. We should test for the curved relationships that we would expect from diminishing returns, even if we risk low power from adding too many variables.

# 17 Inter-birth intervals: a trade-off between fertility and offspring survival?

A woman who gives birth like an animal to one offspring after another has a permanent backache.

!Kung saying, Lee, 1972, p. 332

"You won't have enough milk for the older one," she was told. "You must feed the older one."

Ache women cited by Hill and Hurtado, 1996, p. 375

By one means or another, Ache women successfully follow this advice and show no significant effect of the following inter-birth interval (IBI) on child mortality (Hill and Hurtado, 1996, pp. 220–221, 380–385; table 12.5), and only a small effect of the preceding interval. Many other populations show significant effects of the preceding interval on mortality and morbidity of the "index child" (Rutstein, 2005). Effects of the following interval are more difficult to study but effects on mortality and morbidity have been claimed by many. Numerous studies across the decades and continents have linked mortality of each child to IBIs, beginning with Yerusalmy (1945) (Carlaw and Vaidya, 1983; Hobcraft *et al.*, 1985; Thapa *et al.*, 1988). Early doubts, caused by distortions that could arise in retrospective family histories (Potter, 1977, 1988) are overcome in more recent studies that use longitudinal data, such as by Ronsmans (1996).

All over the developing world, healthcare workers and mothers are exhorted to lengthen the intervals between births in order to lower child mortality and enhance family health. A rapid series of births is suspected as a cause of "maternal depletion," a syndrome that seems to endanger both mother and child in some populations (but not in all, Dewey and Cohen, 2007). "Weanling mortality" and morbidity, close to the time when the index child is supplanted by a new pregnancy and infant, is all too familiar in many countries (Jones *et al.*, 2003; Lamberti *et al.*, 2011). Reviews of recent, more statistically sophisticated studies, continue to argue for the massive worldwide effects of birth spacing on infant and child mortality and mother and child morbidity (Norton, 2005, and references therein). Effects are not restricted to the developing countries but have now been found in industrialized countries, and the type of effects expanded even into parental behavior and educational outcomes (Crowne *et al.*, 2011).

A series of papers by Mturi and colleagues report analyses of birth spacing, fertility, and child mortality in the Tanzanian national population, using the 1991/1992 Tanzania Demographic and Health survey. The median duration of breast feeding was 22 months with >90% of children breastfed for at least 12 months (Mturi, 1997). Intervals are longer among women aged over 30, in urban areas, and working in "the modern sector." Women in polygamous marriages had longer intervals (Mturi, 1997). Children of shorter intervals are at greater risk than those of long intervals (Mturi and Curtis, 1995).

Are there such effects among the Hadza? If so, do IBIs demonstrate a trade-off between number and care of offspring, between fertility and child survivorship? For the evolution minded, IBIs touch on several theoretical threads and offer to illustrate some theoretical points. Are there diminishing returns to care, as we saw in Chapter 16 for weight and upper arm circumference (UAC)? Is there an indication of parent–offspring conflict (Trivers, 1974; Haig, 2014)? Do the selective forces on the father and mother push in different directions? If larger mothers have greater productivity, we may expect larger mothers to have shorter intervals, or larger weanlings. In the next chapter, I look at whether the grandmother's presence is related to the length of the interval. Each of these is challenged by the importance of extrinsic mortality, mortality from causes to which neither parent nor child can evolve methods to avoid. These are, of course, hard to measure. Given the known interactions of nutritional status and susceptibility to disease, we should not assume extrinsic mortality to equal abundance of disease directly. Finally, I have made no attempt to use data on IBI to conclude anything about the proximate (physiological) mechanisms controlling fertility.

Physiological processes that could allow women or their reproductive systems to adjust IBIs have been quite well described (van Ginneken, 1974; McNeilly and McNeilly, 1979; Konner and Worthman,1980; Wood *et al.*, 1985; McNeilly, 2001) on frequency of suckling. The influence of suckling can allow fertility to respond to the condition and growth of the child, and perhaps to the mother's condition, and to the availability of weaning food and its effect on the infant's hunger and thirst. Milk production, and hence the growth rate and demand of the infant, could respond to the energy available to the mother. While these preferences indicate the effects of frequency of suckling, especially at night, Valeggia and Ellison (2009) offer strong evidence for an interacting effect of the mother's energy budget. They suggest that the mother's ability to expend energy on her reproduction, and the baby's demand, both influence the mother's return to ovulatory cycles.

I became interested in IBIs long ago, but mainly by coincidence. Richard Lee stopped by my office in London and dropped off some of his papers, including his well-known 1972 paper on !Kung women's work and birth intervals. The same evening, I joined up with old friends and colleagues from Oxford for a rail journey to the 1975 International Ethological Conference in Parma, Italy. Among them was Richard Sibly, who quickly saw the possibility for quantifying Lee's work in an effort to model the optimal trade-off between fertility and child survival, or at least between the number of children born and the work required to raise them. The exercise appealed to me for its possible contribution to the fledgling efforts to get empirical human sociobiology (human behavioral ecology) started. A common objection to this emerging paradigm was that people could not be maximizing fitness because they did not reproduce at physiologically maximal rates. Lee's paper seemed to us to offer a way to counter this objection, at least within the context of a well-studied hunting and gathering society. If very short inter-birth intervals lead to greatly increased child mortality, then less than maximal fertility might maximize the production of surviving descendants in the ecology of Dobe !Kung women. The analogy to David Lack's ideas of optimal clutch size in birds is obvious (Lack, 1966; Stearns, 1992). We did not consciously think about bird clutches, but the concept of optimization was at the heart of Oxford animal behavior at the time, and most whole animal biologists at Oxford had been influenced by Lack.

Later, Nancy Howell agreed to let me work with her reproductive histories of !Kung women, while at the same time warning me that data of this kind, on a population as small as the !Kung (or the Hadza), might be stretched too far by what I had in mind; a direct test of predictions that followed from the Blurton Jones and Sibly (1978) backload model. Central was the idea of optimal birth spacing, that !Kung women, less than maximally fertile, might nonetheless be maximizing their number of descendants (surviving offspring). In my 1986 paper, I reported what I thought was a careful test of this proposition. I felt that I had shown that under the ecological-economic circumstances of !Kung forager women around Dobe, they were reproducing at the fastest possible rate. An implication was that !Kung women, or their reproductive systems, were involved in a trade-off between number and survival of offspring.

Over the years, a variety of criticisms and alternatives have come to light. Some of them should have immediately been obvious to me.

1. The backload model was too specific to the ecology of the !Kung living in Ngamiland where dry season water sources and the very abundant, staple food are far apart. Any real life model of allocation of resources, energy, time, would have to attend to the "on the ground" details. Given the detailed ecological information, it should be easy to translate such models to energy or time budgets (SI 17.1).
2. It would be equally logical to expect, not a trade-off (negative relationship) between fertility and survival, but a positive relationship. Inequalities in access to resources, or intrinsic vigor of the women, might allow some women both to bear more and to raise more children while others could bear fewer and keep fewer alive. Given the extensive food sharing, and free access to resources, we might expect such effects to be quite weak among the Hadza or the !Kung women.
3. Because each woman contributed a varying number of IBIs to the data, it was impossible to tell whether the results were because of different intervals, or a result of differences between the women. Multilevel statistical methods to deal with nested data and investigate just such problems were only beginning to be developed at that time by colleagues old (Goldstein, 1984, 1986) and new (Burstein, 1980).
4. While anthropologists have found (Strassmann and Gillespie, 2002, among Dogon; Borgerhoff Mulder, 1993, among Datoga) and not found (Hill and

Hurtado, 1996, among Ache) relationships between mortality and IBI or fertility, my findings remain not generally replicated among studies of hunter-gatherers. Meanwhile, in public health research and practice, the importance of interval length has long been accepted as an urgent priority for intervention.

5. Censoring was ignored. Observations of IBIs always end sometime and leave some intervals unclosed that may have been closed soon after observations ended. Including or excluding the unclosed "intervals" may distort the data; methods of survival analysis should be used, as in this chapter.
6. A criticism that requires careful attention is that by Pennington (2001). The association between interval length and mortality may be spurious; it could arise merely because most child deaths are early in life (owing, at least in part, to threats unresponsive to parental care). Short intervals concern younger children than do long intervals. The association is not evidence that mortality would be lowered by delaying a birth. This argument is addressed later in this chapter.
7. Long IBIs can result from primary and secondary infertility caused by sexually transmitted diseases (STDs) and ensuing fetal wastage. Hadza have low levels of each (Chapter 7).
8. Optimal intervals, even the long four-year IBI of the !Kung, could not lower fertility enough to stabilize hunter-gatherer populations. I do not recall suggesting that it would, although Skolnick and Cannings (1972) reported that their (seriously opaque) simulation suggested that it could, and one may wonder how the total fertility rate (TFR) about four births reported for central Kalhari San are attained in the absence of STDs (Chapter 11).

I discuss most of these criticisms more extensively at appropriate places in this chapter. We could add another. Selection may work not only differently on mothers and children but also differently on children of each sex, and on fathers. I take a quick look at the different interests of fathers, and at the relation between interval length and sex of the children in later sections. A better account of the fundamental issue can be found in Haig (2014).

## 17.1 Categories and lengths of Hadza inter-birth intervals

My data on Hadza IBIs are drawn from the estimated birth dates of the 695 children of the interviewed women. Three files were built, the first mimicked the way I treated Howell's data for Blurton Jones (1986, 1987). In this file, each record comprised an interval, including unclosed terminal "intervals," with information about the fate of the child who opened the interval, and the child who closed the interval (if any). The "success" or "failure" of the interval at adding or losing children was entered according to their survival. In the second file, each record was a child, and its observed lifespan was entered, recorded as censored if it were still alive when the study ended. Additionally, in the file was the length of the interval that preceded the child's birth and the interval that followed its birth. In the third file, the "annual hazard" file of each child, each record is a year in the life of an individual child.

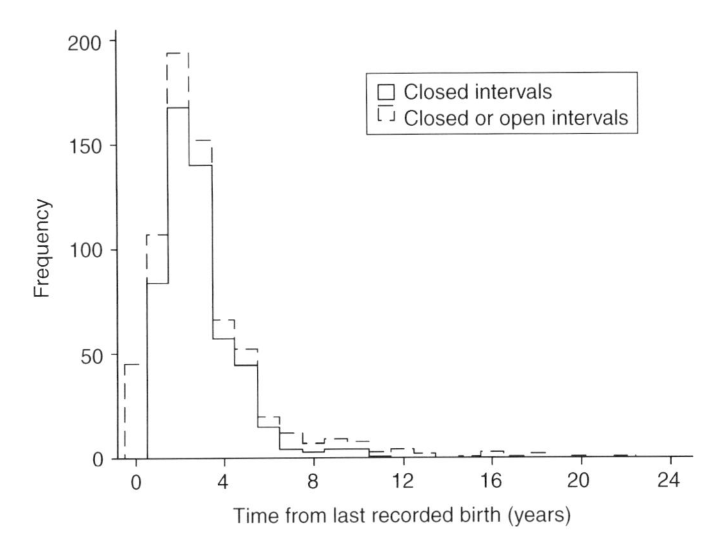


**Figure 17.1** Inter-birth intervals cluster between two years and 3.5 years (closed "non-replacement" intervals), mean 2.8 years, median 2.45 years. Histogram shows the number of intervals scored as centered on one year, two years, three years, and so on. The solid line shows closed intervals. The broken line shows the total of closed and "unclosed intervals."

The arrival of the next child is marked, denoting the subject as "supplanted" or not. The length of the preceding and following intervals is entered for every year.

Three classes of interval were distinguished (SI 17.2).

1. An interval was entered as "open" (and censored) if it were not closed by a subsequent birth before the observation ended. There were 154 open intervals. The time before a woman's first birth was ignored.
2. Among the 523 closed intervals, there were 83 "replacement" intervals. Intervals were classified as "replacement intervals" if the first child died before the next could have been conceived.
3. The remainder, and most interesting, were the 440 "closed non-replacement" intervals. The mean length of these was 2.81 years, and the median 2.45 years (Figure 17.1), perhaps surprisingly short.

### 17.1.1 Replacement intervals

The literature has long drawn a distinction between "replacement intervals" and the remainder. In freely breast-feeding, natural fertility populations, if a child dies very young, before the next pregnancy, the mother often speedily returns to ovulatory cycles and a new pregnancy. The rapid closure of the interval is thought to be a result of the sudden cessation of suckling at the time of the early death. A similar phenomenon has been reported for other primates (Stewart, 1988; Galdikas and Wood, 1990; Jones *et al.*, 2010). I marked each record in the file as replacement or not replacement. An interval was classed as a replacement interval if the child who opened the interval had died before its mother could have been pregnant with the next child. Because the estimates of birth date are only approximate, an approximate criterion was used: an interval was classed as "replacement" if the first child had died a year or more before the birth of the next child.

The aim was to distinguish the effects of death on interval length from the effects of interval length on death. I am not aiming to add anything to the knowledge of proximate causation of fecundability. I am interested in the payoff of different allocations of the mother's resources between the existing child and the next. If the youngest, still suckling child dies (and if it dies, it is most often at a very young age), its mother is likely to return to ovulatory cycles quite quickly. She may then become pregnant very soon after the last birth. The loss of the first child led to its rapid "replacement."

Replacement intervals, as well as open intervals, can be included in computing the average length of intervals. They are excluded whenever I look at relationships between interval length and mortality.

### 17.1.2 Open, unclosed "intervals"

Open intervals will be a mixture of three biologically distinct types. In our sample, there are many short, open intervals between the last recorded birth to a young woman in 1999 or 2000 and the end of my data collection. The shortest unclosed interval was 0.0153 years – a woman last observed with a newborn baby. There will be some very long open intervals, that cover the time from the final birth of an older woman to our last interview with her.

The third category are those that are classified as signifying "secondary sterility," usually assumed to result from disease of the reproductive tract. In some populations, these are quite numerous. Among the Hadza, there were a few striking cases, discussed in Chapter 7. The longest was 31 years, from an older woman whose last birth was early in her career. Especially in a population that has high rates of primary sterility, we might suspect that a long "unclosed interval" in a young woman indicates that this woman has "secondary sterility." However, in the absence of an endocrine and gynecological examination, we have no way to be certain. Recall that when trying to assess secondary sterility and age at the last birth in the Hadza (Chapter 7), we noted that very few women who lived a childless five years after a birth bore another child. If a woman was young at her last recorded birth, there would presumably be a substantial segment of her life during which there was still some small chance of another birth (SI 17.3).

### 17.1.3 Characterizing the length of IBI of a population

An "average" value for the length of interval of a population is often summarized as a mean derived simply by adding the lengths of closed intervals and dividing by the number of such intervals. For some purposes this may be the best measurement. However, this measure ignores the potentially important "censoring problem," the fact that we have also observed intervals that were not closed during our period of observation. Expression of the central tendency as a mean ignores the actual distribution of intervals, and the shape of the probability that they become closed as time elapses from the previous birth. Simply adding the open "intervals" to the sample and recalculating the mean leaves the reason why open intervals made a difference (such as the age of the sample, the frequency of STDs) unspecified. The open intervals that signify secondary sterility were probably rather important in the comparisons of the !Kung women living at cattleposts with those living a nomadic foraging life (Lee, 1972, p. 338; Blurton Jones, 1987; Pennington, 2001) (SI 17.4).

Deriving a mean IBI by dividing TFR or completed family size (CFS) by the observed mean length of the reproductive career should give a mean close to the mean of observed closed intervals. However, dividing TFR or CFS by a standard or assumed length of reproductive career gives a figure that is easily influenced by an unusually late or early age at first reproduction. It is even more likely to be influenced by variation between populations in age at the last birth. This can vary by many years and exert a large influence on so-called mean IBI.

I examined the distribution of Hadza IBI with Minitab's Reliability/Survival Distribution analysis (assuming right-censoring), including the "open intervals" and the replacement intervals in the sample. This examines the probability of an interval being closed as time progresses from the previous birth. The distribution is not a normal distribution. The closest fit to the distribution is the log-normal distribution, with a mean of 3.51 years and median of 2.78 years (loglogistic distribution, an equally good fit, gives a mean of 3.63 and median of 2.80). The hazard plot of the log-normal fit to the open and closed IBI data is shown in Figure 17.2. Births become more probable as time from the previous birth elapses. The instantaneous hazard of closure (the next birth) is very low a year after the birth,

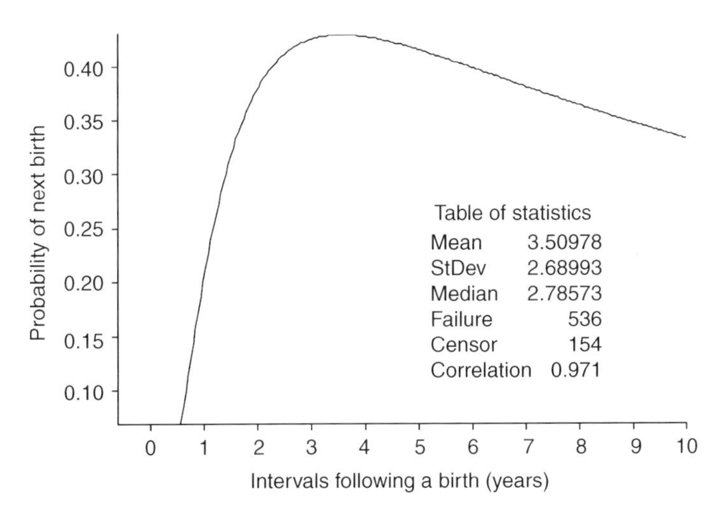


**Figure 17.2** Probability of an interval becoming closed by time from the previous birth ("hazard rate"). Combined closed and open intervals. Lognormal distribution.

but climbs rapidly during the second and third years to a peak soon after three years, it then slowly declines from five years onward.

The log-normal distribution is compatible with an underlying process in which the reproductive system gives a probability of conception that is very low immediately after a birth, then climbs quite rapidly to a peak at 2.5–5 years after the previous birth. The reproductive system is somehow responding to time since the last birth (for instance, by responding to the infant's suckling frequency and the mother's energy input, output, and reserves). By "reproductive system," I mean the broad process, including the physiology of the organs and endocrine system, and the neurophysiological, behavioral, and even social components that determine that a birth occurs.

This section has been concerned with estimating central tendencies for IBI lengths which is important for comparing populations. Most of the following analyses are concerned with what happens to the children when an interval is closed. For this purpose, the correct sample is the closed, non-replacement intervals.

## 17.2 Mother's age, parity, family size, and gender of the index child

The literature seems to contain varied results on the relationship of IBI to birth order, the mother's age, and various measures of the father's status. In the Hadza data, I found no consistent relationships of closed IBI to birth order, or mother's age, or father's tribe (Hadza or non-Hadza). There was a small positive effect on IBI from the number of living children the mother had at the time the interval began (longer intervals with more children), controlling for the mother's age, $age^2$, and the father's tribe (all non-significant) in the sample of closed, non-replacement intervals.

First intervals are no shorter and there was no relationship of IBI to the number of previous intervals in the mother's career. Nor was there any relationship of the length of closed IBI with the mother's age ($mom_{age} + mom_{age}^2$). The significant relationship to the number of living children is consonant with the idea that children may continue to be a costly burden on the mother, even among the Hadza where children forage quite successfully and helpers seem plentiful.

Some authors have reported longer intervals following the birth of boys and interpreted this as an indication of greater parental investment in males. This finding seems common but not universal in non-human primates (Simpson *et al.*, 1981; Takahata *et al.*, 1995; Bercovitch *et al.*, 2000; and see Silk, 1990; and review in Hinde, 2009). It is less often remarked upon in the literature on people, but Mturi (1997) reports longer intervals following boys in Tanzania, and Margulis *et al.* (1993) report later weaning and longer intervals following the birth of girls among Hutterite women. The primate researchers hypothesized that the condition of males would make more difference to their later reproductive success (RS) than differences in the condition of females, thus repaying the extra investment in males (employing a rough version of the Trivers–Willard hypothesis). The standardized variance of RS is lower in human males than among other primates, which may imply different payoffs for greater investment in male or female children. Margulis *et al.* (1993)

discuss the possibility that mothers invest more in the sex that is most likely to help them in the future. Daughters are sometimes observed to help more than sons. Hinde (2009) shows differences in the composition of milk taken by male and female offspring. Males and females may also grow at different rates, reaching different weaning weights (as we saw among Hadza children in Chapter 16).

There was no difference in closed IBI length following the birth of Hadza boys or girls (2.76 vs. 2.89 years, 251 boys, 215 girls, t=0.86 p=0.389), nor in the sample of closed and open intervals combined, nor in the tendency for an interval to remain open. Gender made no significant difference to the relationship between survival and IBI when examined in regression models, reported later.

I looked separately at the lengths of Hadza closed intervals, and at the probability of an "interval" being closed (the probability of a birth being followed by another birth). Very little of the decline in fertility seen among women in their late 30s and early 40s is due to lengthening of closed IBIs. It is mostly because of an increase in the number of intervals that are never closed (Figure 7.9). A similar result was reported by Hawkes and Smith (2010) for Utah women in the nineteenth century.

## 17.3 Inter-birth intervals and growth

I could find no consistent relationships between growth and IBI. There may be effects (they show in some samples and models) but the conditions must be too subtle to allow a consistent set of results. It may be impossible to separate actual effects on growth due to being supplanted at different ages from two selective effects (frailty effects). Children who lost weight at this time would be among the most likely to have died, and thus not had a second weighing. Mothers in good condition may grow larger children before and after weaning, and may have shorter IBI than women in poor condition. Only with many more closely spaced pairs of measurements of growth rate could we hope to separate these possibilities.

## 17.4 Is there an effect of interval length on Hadza child survival?

The backload model of the !Kung implied that as IBIs shorten, the amount of work to acquire the food needed by the resulting children would escalate. The relationship is not linear; length of intervals above four years made little difference, reductions in intervals below four years made increasingly sharp differences, reaching unsustainable workloads at repeated two-year intervals. The core of the backload model is much more general than is apparent from its concentration on measurable specifics of the "mongongo nut economy." Think of any mother caring for a series of children, and especially a mother in a subsistence economy in which women care for children, including frequent breast feeding into at least the second year, and conduct significant amounts of subsistence work. It is not hard to imagine something of a threshold effect – an interval length below which raising two children is very difficult, and an interval above which further elongation makes almost no difference. For example, care of a five-year-old may not impede care of

**Photograph 17.1** A Hadza toddler wearing ankle bells. It would be very difficult to raise a still-suckling baby and a weaned but still helpless toddler at the same time, the result of a very short inter-birth interval. © James F. O'Connell, 2015. Reproduced with permission.

an infant at all. Being a mother to a newborn and a one-year-old would seem to be far more challenging than coping with a newborn and a five-year-old in almost any setting (Photograph 17.1). For a woman who exerts substantial energy to collect a significant portion of the day's food, as well as suckling the youngest and therefore taking it everywhere she goes, the challenge would seem to be even greater. If she does not have to go about collecting food, the challenge is presumably less. A major transition in her effort would be expected around weaning, especially if this is a relatively rapid process. Not only does the mother no longer have to convert resources to milk but she also no longer has to carry the nursling on her foraging trips (she can use a baby-sitter if she can find one). She no longer has to expend the energy of carrying the nursling, nor is it riding on her and impeding the effectiveness of some kinds of foraging. Marlowe (2003) finds Hadza women with suckling children were less efficient foragers; I have data from 1986 and 1989 that says the same thing; Hawkes has data from 1985 to 1986 that shows no difference. At extreme short intervals, suckling two children at once, which I have seen attempted with Hadza twins, could only succeed if there are ample helpers, such as grandmother and sister carrying the children and getting the food. Weaning the first child very early, so as to suckle a new child leaves the first at risk of high weanling mortality (references in Ronsmans, 1996, and others). It is not difficult to believe that the risks associated with weaning also decline rapidly with age after a year or so. Thus, I do not expect a linear relationship of rearing success to IBI; I expect very low success at short intervals, high success at long intervals, and a rapid transition between these periods.

This common-sense child-rearing argument affects the kind of curve we should attempt to fit to the data on IBI and child survival. Some sort of S-shaped curve, or if very short intervals are scarce, an initial steep climb followed by a leveling might be expected. A Lowess fit of the raw Hadza data (SI 17.5) supported the idea that the relationship between success and IBI was not linear. Success (whichever way it was defined) increased steeply between short and average interval lengths, and leveled off at intervals of nearly four years. In regressions, we should test for non-linearity, expecting better fits from, for instance, interval length + interval length$^2$.

We may expect threats to the survival of the child from a short preceding interval to be different from threats due to a short following interval. The short following interval is likely to include early weaning, an often described threat to children under third-world conditions (Popkin *et al.*, 1990; Lamberti *et al.*, 2011). The supplanted child is no longer the mother's first priority. It no longer is given her milk, may be carried much less often, and may rely more on other adults or older children to give it food. In many populations, this transition is quite sudden. The supplanted child may be at quite some risk. The new infant will also be at risk because of its young age, but added risks have been suggested not only from the division of the mother's time and effort between the new infant and the previous child, but also from the rather elusive process called "maternal depletion." A series of closely spaced births are suggested to "deplete" the mother's capacity to grow the fetus and lactate adequately for the new nursling. Data sometimes support this proposition and sometimes do not (Dewey and Cohen, 2007). We may not be surprised by the varied results if we remember the variety of time or energy budgets under which women live, and the variation in the help they have available from others.

## 17.4.1 Child survival and the preceding interval

The most recent studies of birth interval look at the survival of children in relation to the interval that preceded them. This is relatively simple; interviewers can concentrate on the latest-born child as the "index child," and on the survival status and age of the preceding child when the index child was born. I looked at this in two compilations of the Hadza data.

**Child lifespan file and survival analysis** In the "span" file, each record is a child, and the interval preceding its birth is one variable, and the interval following its birth is another variable. The child's fate is recorded as lifespan, and noted as censored if it is still alive at the end of observation. Each of the intervals was entered as closed or open, and as "replacement" or not replacement intervals. These data were analyzed with Kaplan–Meier survival analysis, and with Cox regression, which can present predicted survivorship to any requested age. I used this to give the expected survivorship to age five years.

For the preceding intervals, I excluded replacement intervals, cases where the preceding child had died before the subject child could have been conceived. In this restricted sample, lifespan was longer if the preceding interval were longer (span = preceding IBI, b=0.1705, p=0.005, CI 0.0522–0.2888). Figure 17.3 panel (a) is a

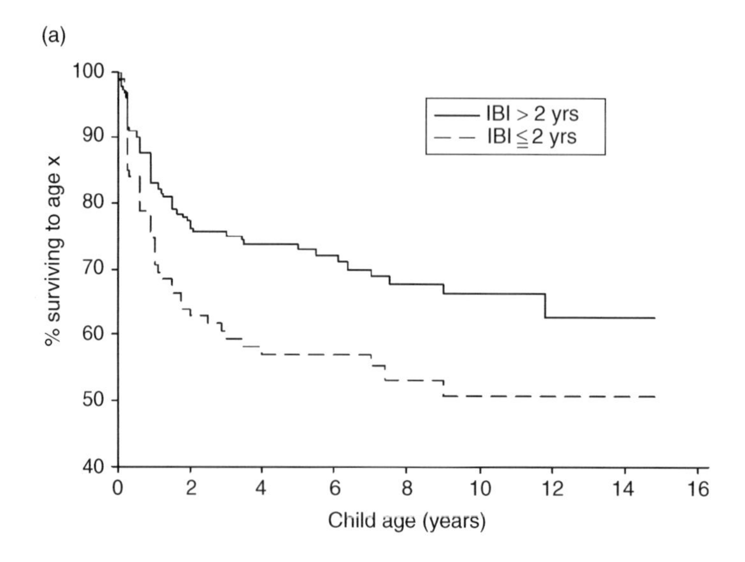


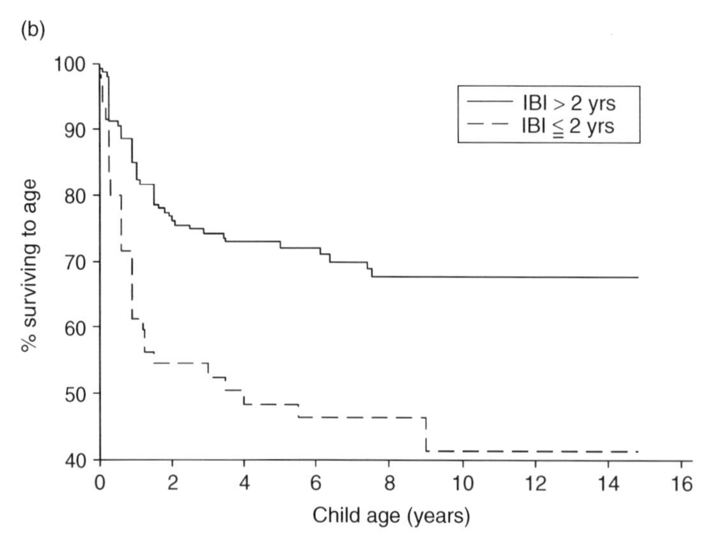


**Figure 17.3** Length of inter-birth interval and probability of survival to age x years (as a percentage). Kaplan–Meier survival plots. (a) Survival of children preceded by intervals $<2$ yr or $>2$ yr. Log-rank test=7.73, p=0.005, Wilcoxon test=7.53, p=0.006. (b) Survival of children followed by an inter-birth interval of $<2$ yr or more. Log-rank test=13.9, $p<0.0005$, Wilcoxon test=13.9, $p<0.0005$).

Kaplan–Meier survival plot that compares children who were preceded by an interval of two or more years with children who were preceded by shorter intervals. The children with a longer preceding interval lived longer than children with a short preceding interval. The difference is significant, with log-rank $\chi^2$=7.73, p=0.005, Wilcoxon $\chi^2$=7.53, p=0.006.

**Annual hazard file and logistic regression** I used the file of each child's "annual hazard of death." In this file, each year of a child's life comprises a record, and each

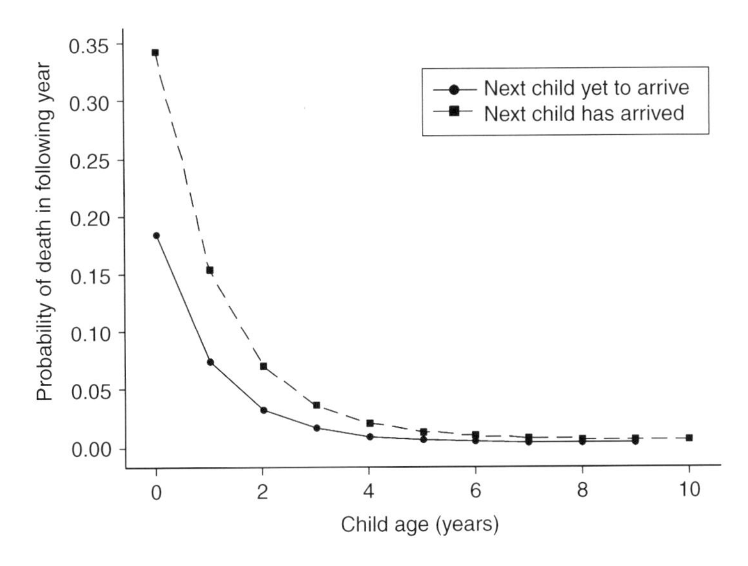


**Figure 17.4** Inter-birth intervals analyzed as age when supplanted by the next child. Probability of death at ages zero to nine, when a following birth has occurred, or has not occurred. Sample: closed inter-birth interval, replacement intervals excluded. Logistic regression model: dead/alive = age + age$^2$ + age$^3$ + has been supplanted (b=1.457, p$<$0.0005, OR=4.29, CI 2.71–6.80). A child of given age is more likely to survive if its following sibling has not yet been born.

record includes whether the child survived the year or died. The record of most children's lives are censored; observations end while the child is still alive. The record also includes whether, and when, another birth precedes or follows the birth of the subject ("index child"). The length of the interval between births was entered for each year of its record, and whether the interval was classified as a replacement interval or not, and as an open or closed interval.

The probability of death in any year of a child's record (controlling for its age) was lower if the interval between its birth and the birth of the previous child were longer (b=–0.1892, p=0.009, OR=0.83, CI 0.72–0.95). Replacement intervals were excluded; the previous child was alive until at least a year before the birth of the index child.

Intervals preceding the index child give a different picture from intervals following the child. The straight line model gives as good a fit to the preceding interval as do the curved models. The probability of death associated with short preceding intervals is much lower than for short following intervals. A preceding interval of a year is nearly as good as a preceding interval of two years. In contrast, a following interval of two years is much better than a following interval of one year (Figure 17.4 ).

### 17.4.2 Child survival and the following interval

From the child lifespan file, Figure 17.3(b) shows a Kaplan–Meier survival plot comparing the survival of children who were followed by intervals of less than two years, or more. The differences are quite striking, and highly significant.

But looking at the relationship between a child's survival and the length of the following interval is not so simple. Renee Pennington pointed out one problem in discussing my !Kung study (Pennington, 2001). An association between survival and the length of the following interval can arise simply because younger children are more likely to die than older children (SI 17.6).

Pennington's point shows that we must control for the age of the first child (the "index child"). We need to know whether the time of being supplanted, or the event or status of being "supplanted" by the next child changes the mortality level expected for the age of the index child. In other words, we need to control for the child's age, before looking for an effect of interval length on its survival. I did this in two ways; both include adding age to the regression model predicting survival (age + $age^2$; adding $age^3$ gives the same results but a slightly better fit). In one procedure I then added the IBI to the model. The effect of this variable should represent the effect of the time of closing the interval beyond the effects of child age (its effect on the residuals after control for age). Longer IBI ("ibilen") were strongly and very significantly associated with a lower probability of death ($b=-0.5921$, $p<0.000$, $OR=0.55$, CI 0.46–0.66). The sample was limited to closed intervals, excluding records of years when the child was aged 10 or more, and excluding replacement intervals. If the following interval will be long, or has been long, the probability of death in any year of life is lower than if the interval will be or has been short. Multilevel regression gave the same result (ibilen $b=-0.363$, $SE=0.065$, Wald $\chi^2=31.2$, $p<0.001$). The effect of the following interval was stronger than the effect of the preceding interval, as judged from the odds ratios and their confidence intervals. When both intervals were included in a model predicting probability of death in any year of a child's life, for the sub-sample with not replacement intervals, both had significant effects but the following interval had a larger effect ($b=-0.4229$, $p=0.000$, $OR=0.66$ [0.54–0.79] versus preceding interval, $b=0.1652$, $p=0.033$, $OR=0.85$ [0.73–0.99]).

### 17.4.3 The supplanted child

The other test is easier to describe and broader in scope. The year in which the next child was born was scored as the year in which the index child was "supplanted." In another variable, the child was scored in this year and all subsequent years of its record as "having been supplanted or not." Regressions were run with age (plus $age^2$ and $age^3$) and supplanted status. Being supplanted has a clear relationship to probability of death ($b=1.4570$, $p=0.000$, $OR=4.29$, CI 2.71–6.80) and so does the age at which supplanting occurs. Young supplanted children were much more likely to die than not yet supplanted children of the same age. (An equally striking result is obtained if open intervals are included, SI 17.7.) It can be seen from Figure 17.4 that the effect of being supplanted is greater for younger children than for older ones (the interaction between being supplanted and age is significant at $p=0.000$).

This seems to provide fairly robust support for the widely held view that, under third-world conditions, the arrival of the next child can endanger the existing child. The analysis avoids problems attached to the way I tested for effects of interval length in my 1986 paper. The analysis was repeated in Mlwin multilevel regression to control for nested observations and mother effects, and the effect of supplanting remained significant (b=0.835, SE=0.205, Wald $\chi^2$=16.6, $p<0.001$).

I suggested earlier that we should not expect a linear but a curved or step-like relationship between IBI and child survival. At very short intervals very few survive. At long intervals many survive. Somewhere in between, survival increases rapidly with IBI length. Figure 17.4 shows how the effects of being supplanted are large on the youngest children but rapidly get smaller (the two lines on the graph converge), and after about four years, are almost indistinguishable. We can see that the relationship is far from linear. We need to test for the shape of the effects of interval length. When we model effects of IBI ("ibilen"), controlling for the child's age, we find that including ibilen$^2$ makes a significant contribution ($p<0.000$). In the child hazard file (excluding replacement intervals), looking at closed intervals following the index child, death is significantly better predicted by IBI length + IBI length$^2$ + IBI length$^3$ than by IBI length alone (LL 607 vs. 612, $\chi 2$=10.0 with 1 d.f., $p<0.005$). Adding IBI length$^3$ improves the fit over IBI$^2$ by a just significant amount (LL –607.02 vs. ibilen$^2$ 609.27 vs. ibilen –612). These results support the view that the relationship of offspring survival to interval length is not linear.

The Hadza resemble numerous populations in the developing world in showing a relationship between IBI and child survival. As in other populations, the effect is limited to the shorter intervals. Extending intervals beyond four or five years gives no further increase in child survival. The effect of IBI is not linear; being supplanted when very young is much more dangerous than being supplanted at age three or four years (Figure 17.4). The results thus far support the idea that there is a trade-off between rate of bearing children and their survivorship. There is a payoff in child survival from spacing births. By spacing births, a woman lowers her eventual total number of births, but would it pay her in the currency of natural selection? Would it increase her number of surviving descendants? Can we assume that selection on the mother's reproductive system prevails over selection on the child, or its father (see Haig, 2014)?

## 17.5 Measuring the yield of an inter-birth interval

As the current child grows, its probability of survival increases, the effect of adverse events decreases, the risks that arise from weaning decrease, and the dangers due to being supplanted by the next baby diminish. The risks to the next baby, which seem to be smaller in the Hadza case, also decrease. As time goes by, another fraction of the mother's reproductive career is used up. Selection on the reproductive system would be expected to balance these, finding the right time for the "decision" to become pregnant again. To test this, we need to measure the mother's gains and losses from the interval. We need to score the outcomes for both children. Ideally but impracticably, we should look at all the other offspring a woman may be caring for (closed birth intervals are slightly longer when the mother has more live children, regardless of her age), and at all the adjacent birth intervals. By bearing a new infant, a woman increases the risks to the previous infant if it is still alive. She increases the risk more, the sooner she has the next infant. By a rapid next birth she also slightly increases the risks for the new infant. In the previous analyses, I looked at one child at a time, using familiar outcome measures: lifespan in survival analyses, death or survival in logistic regression analyses. These fairly rigorously measure the prospects for an individual child but they do not show us the adaptive value or success of an interval.

In my 1986 paper, I scored the success of an interval by whether it added a surviving child to the woman's family. Sometimes she loses both children, sometimes one dies and the other lives, sometimes both live. My criterion for adding a child was its survival to five years old. I scored these as −1, 0, + 1, but reported the relatively conservative results, testing whether the mother adds a child or does not add a child (contrasting score +1 with either −1 or 0). Here I use this measure again and report both versions of this "net success" (SI 17.8).

## 17.6 Is there an optimal interval?

Let us not forget the crucial first question: optimal for whom? Clearly, the analyses of the relationship between IBI and survival show that the longer the interval, the better the child fares. In this section, I am looking primarily at the mother and trying to see whether there is an interval length that combines child survival and maternal fertility in a way that maximizes the mother's production of living descendants. Child survival and their later RS will contribute, but so will their number.

There seems to be evidence that women can influence IBIs; for instance, by prolonged frequent suckling of the current infant (see review by Vitzthum, 2009). The energy costs of lactation are quite substantial and women's energy budget and fat reserves may also influence the time to the next conception (Valeggia and Ellison, 2009). In some societies, the endocrine mechanisms are supplemented by societal prohibitions on intercourse while lactating. Among the Hadza, male informants have indicated that intercourse is more difficult, less easily arranged, while a baby is still unweaned, it is always "in the way." There are indications in the literature that mothers sometimes adjust final weaning to the condition of the infant, a small and sickly infant being suckled to a greater age than a large and robust one (Simondon *et al.*, 2001, and additional references in Martin, 2001).

Closed IBIs follow a log-normal distribution. In such a distribution, the hazard of the event, in this case the probability of a pregnancy, increases gradually and then levels off. If changing IBI can produce a change in survivorship, then selection could have acted on the mechanisms that influence speed of return to fecundability and thereby influence IBI.

The next question is whether lowering mortality can ever be worthwhile (for the mother's fitness) when it entails also lowering fertility. Would the gains from lower mortality ever outweigh the losses in fertility? Is there an intermediate interval length at which a mother maximizes her rate of production of live descendants? Note that here the difference between the evolutionary "optimum" and the humanitarian "optimum" discussed in the public health literature is starkly obvious. Organizations and authors discussing public health are now arguing that intervals of three to five years are optimal. These are the intervals that accompany lowest maternal, and offspring perinatal, infant, and child mortality and morbidity, and thus mostly reflect the child's fitness. Intervals of this length will seldom have been favored by natural selection.

In my 1986 paper, I calculated the number of births that would be produced in a 20-year career of different IBIs if all intervals were the same. We could work as easily with the rate of producing survivors, the number of survivors per year. We want to know if there is some intermediate interval that maximizes the rate of production of survivors. Very long IBIs imply very low fertility, and even if all the children survive, are likely to always yield rather few surviving offspring. What about the very short intervals? As IBIs decrease, the rate of births increases dramatically. To overcome this, the mortality associated with short intervals would also have to increase markedly. Can the effect ever be strong enough among the Hadza to make the shortest intervals yield fewer live children than slightly longer intervals?

In another paper (Blurton Jones, 1989), I tried to set out some of the general circumstances in which one might find an optimum IBI for the mother's fitness. I calculated the combinations of mortality and birth interval that would lead in a fixed-length reproductive career, for instance, to two surviving teenagers, three surviving teenagers, four, and so on. I plotted these "isoteens" on a graph along with some hypothetical curves of relationships between IBI and mortality. In Figure 17.5, I have inverted the mortality axis of my 1989 diagram, which makes it easier to

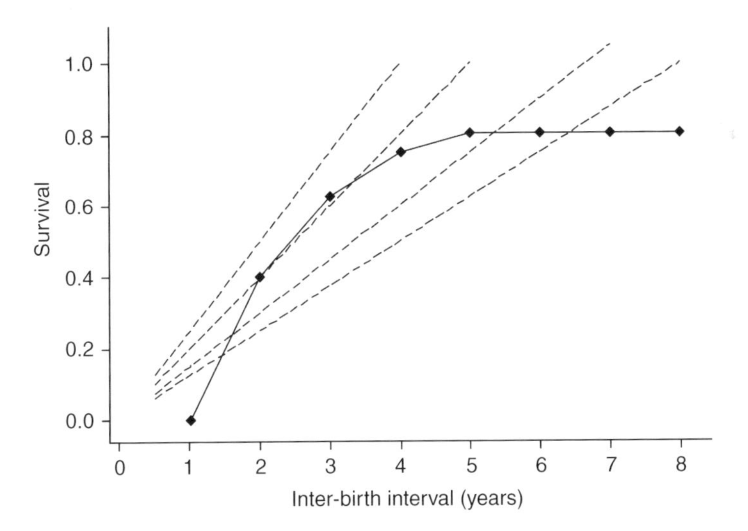


**Figure 17.5** "Isoteens" (broken lines) and the fitted line for observed success of inter-birth intervals. Each "isoteen", a product of survival and fertility, yields survivors at the same rate anywhere along its line. The uppermost "isoteen" yields 0.25 survivors per year. The next yields 0.2/yr, next 0.15/yr, and the lowest yields 0.125 survivors per year. The curved line shows the fit of the success of the interval at adding a surviving child to IBI + $IBI^2$ + $IBI^3$.

describe and follow (and makes it similar to that in Smith and Fretwell, 1974, fig. 2). Each "isoteen" connects points that yield the same number of survivors per year (surviving "teenagers"). The uppermost "isoteen" yields 0.25 survivors per year. This yield can be gained by the combinations of survival and fertility tracked by the isoteen line. If IBIs are four years, then if every child of such an interval survived, survivors would be being produced at 0.25 per year. At a three-year IBI, survival would only have to be at about 0.75 to produce 0.25 survivors per year. The other isoteen lines represent slower rates of annual production of survivors. Attending to annual production of survivors removes the need to worry about the whole career, and the unrealistic idea that any woman keeps to a constant IBI throughout her career. This should have been obvious to me long ago, although we will see later that some women tend to have short intervals, and others tend to have long intervals.

As discussed in the previous section, most curves of survival against IBI are likely to be either sigmoid, or "convex" (or "concave downward") like the curve shown in Figure 17.5, which results from fitting net success to IBI length + IBI length$^2$ + IBI length$^3$). An increase, for instance, from a five-year interval to a seven-year interval will produce smaller benefits to survival than an increase from a two-year interval to a three-year interval. Such a curve will produce an optimal IBI if it crosses any one of the infinite possible isoteens. The optimum will be where the highest isoteen meets the survival by IBI curve. Figure 17.5 suggests another requirement for the existence of an optimum – the survival by IBI curve must have a threshold level below which it is impossible to keep both offspring alive. In practice, there will always be a lower limit at about one year as the minimal practical time it takes to move from the birth of one baby to the birth of the next. If there is not, or if one fits to the data a curve that can show some probability of survival at an interval of zero, the predicted optimum will be at the smallest possible interval.

We can also see from Figure 17.5 that a straight-line relationship between IBI and survival cannot give an intermediate optimal interval. In the unlikely event that the straight line is an "isoteen," then any point on the line yields survivors at the same rate. If the slope of the straight line is relatively flat, then it meets its highest "isoteen" at an IBI of zero. The shortest physically possible IBI will yield survivors at the greatest rate. Such a line would mimic a situation with high "extrinsic" mortality, mortality that the mother's care or continued allocation of resources cannot affect. Alternatively, we might expect such a flat straight line when conditions are good, food for the mother and infants is plentiful, and the mother's work quite limited. All infants can receive good care even at short IBI. This may be the situation in some agricultural and industrial societies. If the straight line is relatively steep, then it meets its highest "isoteen" when it reaches maximum survivorship, and all the children survive. This condition is difficult to envisage under forager conditions.

It follows from this discussion that an S-shaped curve could give an optimum if its level of survival at very short intervals is low enough. A concave curve ("concave upward") could give two optima, one very short, the other very long. If the mother's survival prospects are less than 100%, we might expect the short intervals to be favored by selection.

As the isoteens all have their origin at IBI of zero length and zero survivorship, we can determine the optimum by a straight line from this origin, tangential to the observed (or best-fitting) curve of survival against interval length. The steepest straight line that touches the observed curve of survivorship by IBI gives the maximum rate of production of survivors. In Figure 17.5, we can see that for the graph of net success among the Hadza, this point lies at an IBI close to 2.5 years. We can also see that the curve runs just above the 0.2 isoteen for IBIs from about 2.25 to just over three years. There will be a range of sub-optimal intervals that are scarcely distinguishable from the optimum. The 95% CI (not shown) suggest a similar range of possible optima – from roughly two to a little under three.

## 17.7 Do Hadza women space births optimally? If not, why not?

There appears to be an intermediate band of IBIs (the optimal length of IBIs) that add surviving offspring more rapidly than either shorter or longer intervals. The next question is: does this optimal band coincide with the central tendency of the population? Do most intervals follow the optimum? If not, why not? If there is an optimum, and the central tendency matches it, then why do IBIs vary so much? Why is there not a rigid clustering around the optimum? Why do so many women so often have "sub-optimal" intervals? If very short or very long intervals produce low fitness, why do they occur? Why are these "mistakes" observed at all? Could they reflect the conflicts of interest discussed by theoreticians?

Figure 17.6(a) shows the frequency distribution of IBIs plotted with the predicted yield of added survivors at different IBIs. There are many intervals between two and

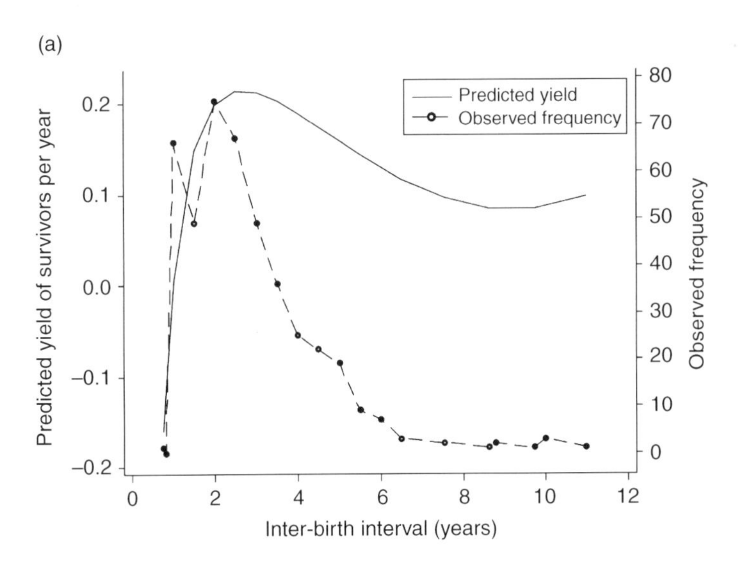


**Figure 17.6(a)** Too many short intervals. Comparing yield of living offspring and the frequency of inter-birth interval.

three years (the optimal range) but also two striking differences. Longer than optimal intervals are quite rare, shorter than optimal intervals are very numerous. Let us look at IBIs that yield around 0.15 survivors per year: intervals of 5.5 years, and intervals of close to 1.5 years. The long intervals (at five years that give a yield of 0.15/yr) are very scarce (nine of them), whereas shorter than optimal IBIs are common (at 1.5 years, which also yields 0.15/yr, there are 49 intervals).

The scarcity of long intervals is easy to understand. Let us refer back to the description of the frequency distribution of IBIs, and note that the course of the hazard underlying the log-normal distribution shows an increase followed by a plateau gently sloping downward. If we regard the hazard of pregnancy as reflecting the condition of the reproductive system, it is easy to account for the scarcity of the longer than optimum intervals. Once the hazard has reached its plateau, the distribution will resemble the exponential distribution; a constant proportion of women yet to close the interval are removed from the risk group each year and the number remaining in the risk group declines rapidly. The scarcity of long intervals is largely a result of the arithmetic of the recovered reproductive system. It is more difficult to think of and test explanations for the quite large number of IBIs that are shorter than optimal. Why are there "too many" short intervals? We can look at some of the many possibilities (others are discussed in SI 17.9).

### 17.7.1 Too many short intervals: using the quality rating of birth dates

The quality of the estimates of children's birth dates varies, and IBIs are merely the difference between two such estimates. I have suggested we had an erroneous IBI of 0.5 years. How many more are there? How likely are the very short intervals to be erroneous? Estimating ages of individuals was the first task in this study before any analyses could be attempted. When I estimated the birth dates, I also recorded a rough estimate of the quality of these estimates, rated poorest to best as one to four. Now, some years later, I look at the relationship between the quality estimates and IBI. There is a small positive correlation between quality and interval length. Among the 366 intervals for which both children had the highest quality age estimates (rating three or four), there are many fewer one-year intervals (7% of the intervals were at one year, whereas for the whole sample, 15% were at one year). The match of frequency of IBIs to yield shown in Figure 17.6(b) appears quite close. This suggests that perhaps as much as half the excess of very short IBIs may be a result of errors in age estimation. I recalculated the relationships between child survival and interval length for the high quality sub-samples. The results were the same as for the entire sample; even the optimal interval length remains at about 2.25 years (SI 17.10).

Age at death estimates could also be wrong and I had recorded no quality estimates of them. I therefore cannot test the possibility that some of the shorter intervals should have been classed as replacement intervals, nor how many of those classed as replacement should have been classed as non-replacement intervals.

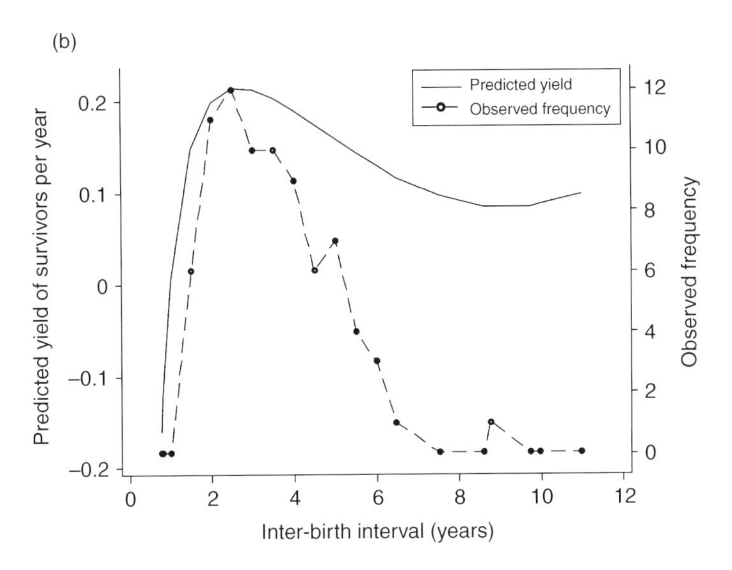


**Figure 17.6(b)** Comparing yield of living offspring and the frequency of inter-birth intervals. Restricted to intervals in which the birth date estimates for both children had been originally rated as highest quality (4 on a scale of 1–4). There is still a slight bias toward intervals shorter than the mothers' optimum.

## 17.7.2 Effects of extrinsic mortality

Where extrinsic mortality, mortality that cannot be avoided by parental care, is high, parents would be expected to invest less in each offspring (Harpending *et al.*, 1990; Lycett *et al.*, 1998; Quinlan, 2007). Thus, as Quinlan found in the standard cross-cultural sample, weaning may be hastened or delayed in accord with factors such as disease incidence. We could interpret the excess of short IBIs as reflecting the abundance of infectious diseases such as malaria. Previously, I suggested that such extrinsic mortality could be modeled as a near horizontal line on Figure 17.5. In such a case, very short IBIs should be favored.

The interaction between nutritional status, weaning, and susceptibility to infection (see Chapter 16 for references) suggests that theory may overdraw the distinction between care effects and non-care effects, especially in the case of disease. In each of the diminishing returns curves that I have shown, for UAC, weight, and IBI, the level area arrives at surprisingly high survivorships. Of course, unexpected, random, and undeserved deaths occur to Hadza children, but the amount of room left for mortality that is not affected by parental care seems quite limited.

## 17.7.3 Men have a different optimum and encourage earlier weaning

A dead baby contributes nothing to its father's RS, just as to its mother's. Therefore, in principle (at least in absolute monogamy), we might expect men to have been selected to favor the same IBIs as women. However, they are commonly found to prefer shorter intervals. In response to this observation, our obvious ruse is to switch explanatory levels and point to the short-term motivation of men and males of many species. The observation that women in polygamous marriages have longer IBIs (Mturi, 1997) can be interpreted as resulting from the wives sharing the husband's sexual attentions. In a survey of data from 10 sub-Saharan African countries, Gebreselassie and Mishra (2011) found that when couples disagreed about their preferred waiting time to the next birth, in all 10 countries, men preferred shorter times than women. Thus (if we understand how natural selection might favor this, or not), there is some tendency for men to want the next child to arrive sooner than does the woman.

Perhaps we can do better than rely on what we know about men's motivation, stay within our adaptationist paradigm, and see if there might be selective benefits to men that arise from maintaining this motivational system in this specific context. We have seen in Chapter 13 that the number of births makes a larger contribution to men's RS than to women's, and survival of the children relatively less. We have also seen in Chapter 14 that men nominated as expert hunters have greater RS, by virtue of achieving a greater number of births, despite lower child survival.

Given the high rate of divorce and remarriage, the current youngest child will quite often be the child of a previous husband. The child's direct contribution to the new husband's fitness is zero; the next child is an urgent priority, just as in infanticidal mammals. Among the 523 closed intervals, 88 (16.8%) involved children of different fathers. The IBIs between children of different fathers were longer than IBIs between children of one father (3.22 years vs. 2.75 years, $t=2.03$, $p=0.045$). Perhaps, even though men did not avoid under-five-year-olds (the number of under-five-year-olds cared for by a woman did not lengthen her time to remarry; Chapter 15), they did avoid infants from a previous marriage long enough to delay remarriage and slightly lengthen the mother's IBI.

Are there other reasons to expect men to have interests that differ sufficiently from their wives' interests to lead us to expect men to be so hasty? Could men be motivated by the chances of the current wife leaving? How much difference would this make to their optimal IBI? The high probability of a marriage ending during its early years (Figure 15.3) means that a Hadza man who (early in the marriage) waits four years before his wife is ready to conceive the next child has a 23% chance that they are no longer together. The prospects are better if he waits less time, only 5% at two years, and 16% by three years. Once separated, he faces the task of finding a new wife and competing with other men in this search (assuming costs of finding a wife are higher than those of guarding a current wife). Costs of mate guarding, even when marriage is socially sanctioned, may not be negligible, and could also be a significant component of selection pressure on a man to make his wife pregnant sooner rather than later.

If women gain from the presence of a husband (I have seen little evidence for this in these Hadza data), they may be subject to the same increased cost of long IBIs and benefit from a shift toward shorter IBIs. They are more likely to gain from complying with the husband's wish for an earlier weaning and more rapid resumption of ovulation and intercourse.

If we multiply the predicted yield of an interval by the probability the couple will still be together at the likely time of the next conception, we get an indication of the man's chance of fathering another surviving child with this woman. In all cases, 2.5 years is the optimal interval but there is a small shift in favor of shorter intervals. For example, in the baseline condition, three-year intervals yield more surviving offspring than two-year intervals. Taking the probability of divorce into account, two-year intervals yield the father more survivors than do three-year intervals. The longer IBIs are devalued most and this makes short intervals relatively more valuable (SI 17.11). The predicted effect is greatest early in the marriage when probability of divorce is greatest.

In conclusion, between accuracy of birth date estimates and husbands keen to hasten weaning, we seem to have promising explanations for the apparent excess of IBIs shorter than the woman's optimum.

## 17.8 Differences between women or differences between intervals?

Have I described an effect of intervals on mortality, or a pattern of differences between women? In particular, have I any evidence that differences between women, such as might be a result of differential access to resources, or differences in power to delay weaning, either generates or obscures the proposed trade-off. Are inequalities among Hadza so extreme as to generate an association in which short intervals are accompanied by higher survival? Are some so rich as to be able to shorten intervals, and keep more children alive?

Multilevel regression can help answer the question. Earlier I reported that in the logistic regression of the annual hazard of mortality, taking account of the mother's identity did not remove the effects of interval length or supplanting. I have also run multilevel analysis of variation in net success, of variation in IBI length, and of the relationship between them, using the data file in which each record is an interval. There are clear mother differences. "Caterpillar plots" of the mother residuals show numbers of women scoring significantly above and below the sample mean (SI 17.12). They also show that some mothers can get more success out of the same interval than others. The "within mothers" effect of interval length on success remained significant and positive despite the variation between mothers.

Thus, mothers vary significantly but there is no support in the Hadza data for the expectation that greater individual access to resources, or greater size or underlying vigor, results in shorter intervals accompanying greater success at keeping children alive. Using multilevel regression, I also looked for differences between individual maternal grandmothers related to their daughters' IBIs and net success in the same way. There was no tendency for the IBIs of sisters to be clustered according to their mothers' identities. We will see later (Chapter 18) that the grandmother's status (alive or dead) is associated with the length of successful IBIs, and this may account for some of the differences between individual women's IBIs.

## 17.9 Do inter-birth intervals affect women's reproductive success?

Does keeping to intermediate IBIs make a significant contribution to Hadza RS? I have shown that the rate of producing survivors can be maximized by intervals of two to three years between births. Longer or shorter intervals produce survivors at a slower rate. However, on a larger scale, is there any other indication that spacing births contributes to a Hadza woman's RS?

In the multilevel analysis, the mother residuals for net success are also positively correlated with the mother residuals for interval length; women who tend to have longer intervals also tend to have more successful intervals ($r=0.474$, $p<0.000$, $N=123$ women with one or more closed, non-replacement intervals). This suggests that the interval length helps account for some of the individual differences in success at keeping children alive.

In addition, I looked at the relationship between each woman's mean IBI and her standard score on number of live children, number born, proportion kept alive. The mean length of a Hadza woman's IBIs is positively correlated with the proportion of her children that she keeps alive (sssurv $r=0.484$, $p<0.000$). Not surprisingly, the mean IBI is negatively correlated with the number of births (ssnborn $r=-0.387$, $p<0.000$). We have seen that the survival of a woman's children (sssurv) is quite a strong predictor of RS, accounting for some 36% of the variance as assessed by adjusted $r^2$. In regression of sssurv on meanibi + meanibi$^2$, we account for 25% of the variance in sssurv. However, there is no correlation between women's mean IBI length and their standard score of live children (our measure of RS) ($r=0.011$, $p=0.907$).

Plots of these relationships show that the relationship is not linear. Lowess smoothings are shown in Figure 17.7. Not surprisingly, the negative association of the mean IBI with number of births ("ssborn") applies over most of the range of interval lengths, intervals from two years upward (Figure 17.7, upper left panel). The standard score of child survivorship increased with the mean IBI up to just below four years, after which it leveled off (Figure 17.7, lower left panel).

Most interesting was the plot of RS (sslive) against mean IBI (Figure 17.7 upper right panel). This showed the peak RS between IBIs two and four years. Women with longer intervals did worse, not surprisingly. Women with shorter intervals also had lower RS. The near zero correlation of RS and mean IBI merely shows that the relationship is not linear. The regression of RS (sslive) on mean IBI and mean IBI$^2$ give significant contributions of IBI (meanibi $b=0.8966$, $p=0.007$; meanibi$^2$ $b=-0.1289$, $p=0.005$). Stepwise regression that includes ssborn in the model gives $r^2=59.9\%$, and increases the significance of IBI and IBI$^2$. IBI contributes 10% to the adjusted $r^2$ and a major reduction in the Mallows Cp from 32.5 to 4.0.

Pennington and Harpending (1988, see especially fig. 3) showed that the RS of !Kung women increased linearly with CFS. A plot of our Hadza standard scores of RS against fertility (ssborn) also shows how powerfully fertility can determine RS. However, the fitted line suggests an inflection in the line among the most fertile Hadza women (Figure 13.3). There is no decline in RS at high fertility but there is a leveling. Hadza women seem to gain no RS from fertility more than one SD above

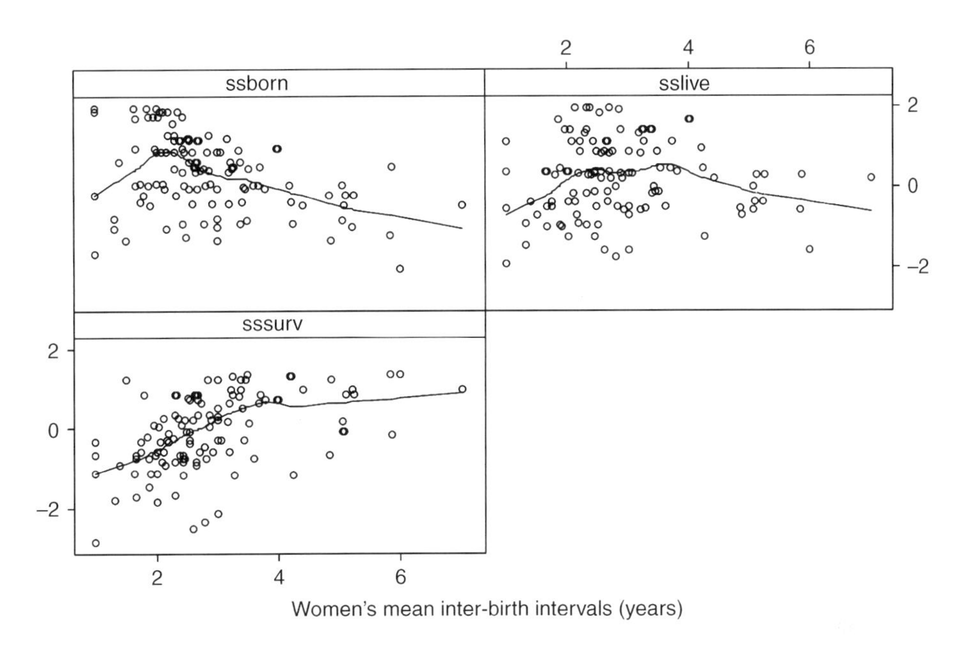


**Figure 17.7** Individual women's mean closed inter-birth intervals (X-axis) by their standard scores of RS ("sslive"), fertility ("ssborn"), and success at keeping children alive ("sssurv"). Lowess fits to women's individual means.

the mean for their age. This argument is similar to that of Strassmann and Gillespie (2002), which found that the greatest number of surviving offspring are produced by the Dogon women with a slightly lower than maximal number of births. It is not hard to imagine that in a population with a high incidence of infertility caused by disease, such as shown in the Pennington and Harpending report, individual differences in fertility would make a larger contribution to RS than child survivorship. Pennington and Harpending (1993) suggest that differences between their results and Howell's (1979, pp. 125–128) probably arise because their sample included infertile women with zero births whereas Howell's women of completed careers happened to include none.

## 17.10 Summary and discussion: mother quality, diminishing returns, and the trade-offs between child quantity and quality

The new Hadza infant is at some increased risk if it arrives while its older sibling is still young. However, the arrival of the next child is a greater risk to the supplanted Hadza child, even when the age of the supplanted child is taken into account. The relationship is curved. Risks are highest at short intervals between births, and decrease as the interval increases, up to intervals of about five years, after which no advantage accrues to either child from wider spacing. Child mortality would be minimized by intervals of five years or more. However, fertility would be very low. Short intervals imply greatly increased fertility but with increased mortality. The rate of production of surviving offspring is maximized by an intermediate rate of births. Most Hadza women give birth close to or a little below this rate.

Multilevel regression, and women's mean IBIs, standard scores of fertility and RS show differences between women. Mothers differ in fertility, child survivorship, and RS. Differences between mothers in child survival and RS are partly accounted for by their birth spacing. Women who show very high birth rates were not always the women with the highest RS. Some large-scale studies in the literature show poorer survival of children of the longest intervals, perhaps because their mothers are living a much harsher life than the majority who have shorter intervals.

The data used for this study are far from ideal. The sample size may be adequate but the dating of births and deaths is less accurate than one would wish. In comparing studies, the following are important to note. First, was there a test for a non-linear relationship between mortality and interval length? Were open intervals included, and if so, how many of these were the terminal open intervals at the end of the reproductive career? Were replacement intervals considered, and were they included or excluded? Was the effect of infant mortality on fertility (via replacement intervals) taken into account? Completed family size is influenced by age at first birth and last birth as well as by the length of closed IBIs. Age-specific fertility includes unmarried or otherwise non-reproducing women and is not a direct reflection of IBIs. Thus, indirect measures of IBIs, when CFS is divided by a supposed standard length of career may be misleading.

The length of closed intervals does not change very much with the Hadza woman's age. Reproduction ends, not with gradually elongating intervals but with a never-to-be-closed interval. The decline in age-specific fertility is not a result of elongating intervals but of an increasing number of women who simply have no more births. A similar result was found by Hawkes and Smith (2010) in the Mormon (LDS) database. We will see in Chapter 19 that the Hadza women who continue childbearing for longest are those with the most births in their early career and with the most helpers late in their career. The !Kung women were predicted to have to work harder to feed their family in mid-career. This is probably because the model allows for the long dependency of !Kung children, omitting any of the foraging that Hadza children do for themselves.

I listed and attended to some of these criticisms of my 1986 paper. Others are attended to in SI. Briefly, the backload model is more general than it seems at first sight (SI 17.1). In SI 17.13, I discuss the idea that forgotten births of dead babies would give an illusion of an association of mortality and interval length. For this to happen, there would have to be 5% more fertility and 20% more mortality than observed, near the limits of our 95% CI shown in the tests in Chapters 7, 8, and 9.

Especially in the large-scale studies in developing countries, the nature of the samples is changing. More women have access to contraceptives, more have access to healthcare, more are involved in a cash economy, wage labor has increased particularly among women, there is more education, more treatment of STDs, and more wealth differentials and social stratification. It becomes more important to identify the exact indicators of IBIs being used, and remember that fertility differences include later onset, and earlier last birth.

My use of the word "optimal" is different from its meaning in the international public health literature. The latter abundantly discusses the recommendation that spacing of two to four or three to five years is optimal (and a separate literature discusses whether exclusive breast feeding during the first six months is optimal). In this literature, "optimal" tends to mean maximizing the chance that an individual child survives. This is a good humanitarian goal, and a good adaptationist usage if we are thinking only of selection on the individual child. Selection is likely to have favored individual children who could maximize their chance of survival. Many have discussed the ways children do this, from Bowlby (1969) to Hrdy (2009) and Hawkes (2014). My usage has been from the viewpoint of natural selection on the mother's reproductive system in the wide sense, encompassing physical, endocrine, behavioral, and social components. These usages should not be confused.

A logical adaptationist prediction could have been that differences in women's access to resources would lead to a positive association of short intervals with high offspring survival. This could be expected where there are clear differences between women in their access to resources. Ford *et al.* (1989), in a study in Bangladesh, found that every 20 lbs increase in the mother's weight was associated with a three-month shortening of postpartum amenorrhea. The families varied in land holdings, available labor, and money for fertilizer, among other opportunities. Among the Hadza, women differ, but multilevel regression allows us to see that the differences are partly explained by the IBI that each mother maintains. I found no tendency for short intervals to characterize women who are better able to keep their children alive. Given the egalitarian ideology and habits of the Hadza, very wide food sharing, and apparent ideal free distribution across their land (Chapters 6 and 10), we might not expect wealth differences sufficient to generate a positive association between fertility and child survival. A rough proxy for wealth differences might be height or weight. Woman residuals for "ibilen," "net success," and net success controlled for "ibilen," were stored from multilevel regressions but show no significant relationship to the women's height or weight. While Hadza women's physique appears to have little relationship to their IBIs or the success of their IBIs, Hadza women do differ in their access to helpers, the subject of the next four chapters.

The worldwide data on survival of children at different IBIs implies a trade-off between fertility and offspring fitness. Smith and Fretwell (1974) outlined the classical model of a trade-off between number and size of offspring. Its core is diminishing returns in offspring fitness to increased parenting effort per offspring (Smith and Fretwell, 1974, fig. 2). They show that there will be an optimal amount of investment per offspring for the parent, and that increased access to resources should lead to an increased number of offspring, not more investment per offspring. We have seen two instances in which there are diminishing returns to Hadza childcare. In the chapter on growth, we saw that survival increased with size (UAC, and body weight), "up to a point," the advantages gradually diminishing and leveling off. Here we see the time between births, the time of "exclusive care" of one small child, producing advantages in survival up to around four to five years, after which there is little further advantage. Selection on mothers would favor those who redirected resources that might be available for extended care or a larger child toward some other fitness-enhancing target, such as producing more children. Natural selection on the mother's reproduction has resisted the advantages that may follow from raising the perfect child.

Smith and Fretwell (1974, p. 504) discuss some limitations of the applicability of their model, including cases in which "young from successive litters depend on a parent at the same time, as in beaver [. . .], chimpanzees [. . .], and humans." [their references omitted]. They regarded their model as best applied to litter size. However, aspects of their model have been more widely applied (for example by Walker *et al.*, 2008). Trade-offs between number and fitness of offspring have been described in many species belonging to diverse major taxonomic groups (Clutton Brock, 1991; Roff, 1992). So have "phenotypic correlations," where mothers in good condition or a better environment have more offspring and the fitness of the offspring is greater than average. Other factors have been found to influence the outcomes, including maternal age and early history, and seasonal and long-term environmental factors. Trade-offs are most often revealed by experimental interventions, least often seen in natural variation where differences in individual quality or access to resources often obscure underlying trade-offs.

I have made no attempt to conclude anything about proximate mechanisms from these data on IBIs. Those who are interested in the physiology of fertility have seemed not to need to test the intricacies of evolutionary theory. Overviews such as by Valeggia and Ellison (2009) seem readily able to account for many aspects of the data. Responses to suckling seem likely to allow some response to the infant's interests. The mother's energy budget seems to reflect her prospects for another viable birth excellently, and perhaps to relate her interests to the intensity of the infant's signals successfully.

# 18 Grandmothers as helpers

"What was you mother's mother's name?" "I don't know, I just called her Grandma [Amama]."
Young Hadza mother

Thanks largely to the Hadza (Photograph 18.1), grandmothers have drawn attention from many researchers in recent years (Hawkes *et al.*, 1989, 1997, 1998, 2011; Hawkes, 2003; Lahdenpera *et al.*, 2004; Voland *et al.*, 2005; Sear and Mace, 2008). The attention given to grandmothers illustrates an important scientific tradition: give maximal attention to things that appear not to fit the dominant theory. Grandmothers are a major evolutionary conundrum. Natural selection is generally weaker on the old than the young, and cannot act on characteristics of individuals who no longer reproduce. Hence, why do hunter-gatherer women live on average 23 years after their final birth (Blurton Jones *et al.*, 2002)? Many, since Hamilton (1966), have suggested that our species breaks the rule by virtue of the help that older women can give to their children and grandchildren, thus influencing their own inclusive fitness long after they have ceased childbearing. Some have argued against this idea, by suggesting that grandmothers would be few and ineffective in earlier times.

In this chapter, I address the availability and effectiveness of Hadza grandmothers' help. First, I show how many women and children have an older helper, and how many older women have small children they could help. Then I give special attention to whether grandmothers affect children's growth and survival, examine some alternative explanations for the results, and look at some of the details. I go on to look at whether grandmothers affect their daughters' fertility, inter-birth intervals (IBIs), and reproductive success. In the next chapter, I discuss models, theory, and debates in the field. While Hawkes (2003), Hawkes and Smith (2010), and Jones *et al.* (2007) have set the primary problem as accounting for evolution of post-reproductive life, others take lifespan as a given, and aim to account for shortened reproductive careers. Some have examined a trade-off between continued reproduction, and care for descendants (Hill and Hurtado, 1991), and others have looked at competition between the generations (Cant and Johnstone, 2008; Johnstone and Cant, 2010).

Hawkes *et al.* (1989, 1997, p. 552) and Kaplan (1997, fig. 10–4) suggested that in habitats in which weaned offspring could not feed themselves, and where food came in large packages that were difficult to acquire, selection would favor kin who provided food for the offspring (Hawkes *et al.*, 1989). This could select for continued vigor in women nearing the end of the childbearing years via the help they can give to grandchildren (Kim *et al.*, 2012). This selection would lead to the greater lifespan of

**Photograph 18.1** A hard-working Hadza grandmother. When was the last time you walked out into the mountains and dug fence post holes by hand in rocky ground for four hours in full sunlight? © James F. O'Connell, 2015. Reproduced with permission.

humans compared to other Great Apes. Hawkes *et al.* (1998) pointed to expectable consequences for human life histories: later maturity, larger size, and when the productivity of daughter and her post-childbearing mother are pooled, greater fertility. Hawkes *et al.* (1997) showed that among the Hadza, the growth of weaned children was positively correlated with their grandmother's foraging hours. However, we have yet to show any demographic effect of Hadza grandmothers. What will the current much larger sample show about growth and survival?

Hill and Hurtado (1996, p. 419) made a very important point about studying helpers. By directing their help efficiently, helpers could obscure their own effects. Grandmothers may choose to live where their help is most effective, for example with the more vulnerable of their daughters. This daughter's children may then survive as well as the children of the grandmother's other, more robust and fortunate daughter. The effect of the grandmother's presence would then be obscured by her own strategy. I reported (Blurton Jones *et al.*, 2005a,b; Blurton Jones, 2006) that Hadza women over age 40 tended to live where one would expect them to live if their help were effective, and was distributed in a way that enhanced their own fitness. Men aged over 40 did not conform to these predictions. We have yet to offer any data about effects of the grandmother on children's survival. Even if living grandmothers allocate their help strategically, dead grandmothers cannot help at all. Consequently, my analyses compare children who have living grandmothers with those whose grandmothers have died.

## 18.1 Measures and expectations: grandmother status

In the following discussion, I compare children whose grandmothers were dead or alive at the time of observation. The "time of observation" was different in different analyses. It was the time of a measurement of the child's weight, the time of the child's birth, or each year of a child's life. Grandmother status (GM status) was scored as follows. If the mother's mother were alive, grandmother status was coded as 1 (an alive grandmother), if the mother's mother were dead, coding depended on the father's mother. If she were alive, grandmother status was coded as 1. If she were dead, the coding was 2 – both grandmothers dead. I also included a separate coding for the mother's mother (MM status), and for the father's mother (FM status). In our papers on Hadza grandmothers, we have generally taken most interest in the mother's mother, while here the combined measure is given most attention for two reasons. First, during our behavioral observations, we came across very few women who had no older female helper. Some impressive helpers were the husband's mother. Second, I had shown that older Hadza women tended to live with a son and daughter-in-law if the daughter-in-law had no living mother (Blurton Jones, 2006). I took this to indicate that the mother-in-law could often be an important helper.

### 18.1.1 What Hadza grandmothers can do to help

While in the western world we tend to think of grandmothers as occasional babysitters, among the Hadza, grandmothers are most noteworthy as breadwinners. Older Hadza women are very productive foragers (Hawkes *et al.*, 1989, 1997). The !Kung seem to resemble the Hadza in the ability and willingness of older women to forage productively, finally "retiring" when they reach their 70s (Howell, 2010, pp. 40–43). Hill and Hurtado (1996, pp. 235–236) describe Ache grandmothers aged 45–60 "collecting food and doing the housework in order to ease the burden of their daughters or their sons' wives." Later, they write "when they are too old to engage in physically taxing activities, they baby-sit grand-children and enable their daughters or daughters-in-law to work unencumbered." Hill and Hurtado report small effects of older kin on the survival of small children. Among the Ache in the forest, before the 1970s, the extremely old and feeble were left to die or were killed. Sear and Mace (2008) reviewed studies of kin helpers including grandparents and suggest that maternal grandmothers show the most widespread tendency to improve child survivorship (nine of 13 studies). Results for paternal grandmothers were more variable. The percentage of studies showing positive effects ranks maternal grandmothers third, at 69%, behind mothers (100%) and older siblings (83%), but well ahead of fathers (32% of studies). I suspect that the importance of older women has been widely underestimated in anthropology. While the ethnographer interviews old men (who display their erudition in cultural lore), and the demographer interviews young women, the grandmothers are quietly keeping daily life running.

The opportunity for women to acquire food directly varies between one economy and another, and this may influence the tasks grandmothers can do and the ways they can help raise grandchildren. In some circumstances, opportunities to help may be limited and the grandmothers' help may be relatively ineffective. In some societies, men may be able to control distribution of resources and grandfathers may be important in those contexts. In the case of the Hadza, we have measured older women's foraging and consider this most likely to show an impact. Hadza grandmothers forage very effectively. Elsewhere, baby-sitting, cooking, or maintaining clothing and household equipment may be all that grandmothers can do. In Chapter 14, I reported the results of asking Hadza women to name some hard-working women. There was a great bias toward naming older women (Figure 14.3). Hadza seem to recognize that older women work as hard as we have observed them to.

### 18.1.2 Who should we expect grandmothers to help?

The sensitivity analysis reported in Chapter 12 showed that population growth is most responsive to the survival of children under age five years, and to fertility of young women. Increases in fertility of adult women have a lesser effect, a pattern apparently commonly found in long-lived mammals (Stearns, 1992) and humans (Jones, 2009). Thus, the grandmother's help would yield the greatest fitness (for the grandmother, her genes would spread faster) if it were directed at these targets. Hawkes *et al.* (1997) reported that weight of weaned children was most responsive to the grandmother's work effort. Weaning is widely found to be a time of heightened risk under third-world conditions. In each analysis, I have also looked for differences between younger and older women by controlling for the mother's age at the birth of the child. Child survival appeared to be especially poor if the mother was less than 20 years old when she bore the child, and especially good when she was aged 35 years or older. Later, I show analyses of grandmother effects for the whole sample, and for women aged less than 35 years.

I present results for grandmother effects on all children aged up to five years. Because we think the grandmother's main contribution is to the nutrition of the weaned children, weaning would be the logical starting point if we had records of individual weaning ages. I showed, in Chapter 16, that the mean age at weaning is two years, but this means many children are weaned before two years old. Infant deaths are defined as those up to the first birthday. Weanling deaths that occur late in the second year will appear in our record as deaths to one-year-olds. To include the one- to two-year-olds seemed the best choice for a sample of weaned children.

An appropriate endpoint could be anywhere between five and seven years. Most child deaths occur before the end of the sixth year (up to age 5.99 years). This would also be a practicable period in which to look for effects of helpers on the nutritional status that may influence child mortality. By age five years, a Hadza child can obtain some food by its own efforts. By seven years, the child can obtain about 800 kcal/day

(Blurton Jones *et al.*, 1997). Between five and seven years, the child can begin to buffer itself from inadequate provisioning by mother or her helpers.

Recent literature has attended to details of kinship in predicting whose grandmothers would be expected to help (Chrastil *et al.*, 2006; Rice *et al.*, 2010; Fox *et al.*, 2009, 2011). Targets of help might depend on paternity uncertainty (grandmothers should prefer their daughters' children to their sons' children), and patterns of inheritance of the X chromosome. The latter would predict that sons' daughters should be favored over sons' sons, with daughters' children of either sex favored at an intermediate level (Fox *et al.*, 2009). Some studies on large samples of farming peoples appear to support these predictions, with respect to the survival of children. I report Hadza data on these issues in SI to the next chapter (SI 19.1). In the relatively small Hadza sample, the first task is to see if there are indications of *any* effects of grandmothers.

## 18.2 How many grandmothers?

We may wonder about availability of grandmothers in high mortality populations, and especially at earlier times during our evolution. In an SI appendix to Chapter 8, I discussed the apparent contradictions between populations described by archaeologists from bone assemblages and populations with written records or direct observation described by historians, anthropologists, and demographers. Here we attend to how many Hadza women have a living mother to help them. How many Hadza women at the end of their childbearing years have live descendants they might help?

### 18.2.1 How many women have a living mother or father?

Figure 18.1 (data table at SI 18.1) shows the proportion of women in the population register between 1985 and 2000 who had a living mother or father at each age. Around 70% of women had a living mother during the first 10 years of their childbearing lives. A majority of women still had a living mother until the end of their 40s. Fathers disappear 10 years earlier, partly because an individual's father is often substantially older than her mother, and partly because mortality rates appear to climb earlier among Hadza men than women. By world standards, the Hadza are a high mortality population, yet most women have a living mother during their childbearing years. Howell shows similar data for the !Kung, whose mortality is a little higher than that among the Hadza (Howell, 1979, fig. 15.1). The time course is similar to that of the Hadza but from a very similar 70% or so of 20-year-olds with a living mother, the proportion of !Kung women with a living mother declines faster with age than among the Hadza, falling to 50% by age 35. If these high numbers are surprising, remember that for a woman to have been born, her mother must have lived at least to childbearing age.

### 18.2.2 How many children have a living grandmother?

Figure 18.2 (data table at SI 18.2) shows that most Hadza children had at least one living grandmother during their childhood years. However, the availability of living

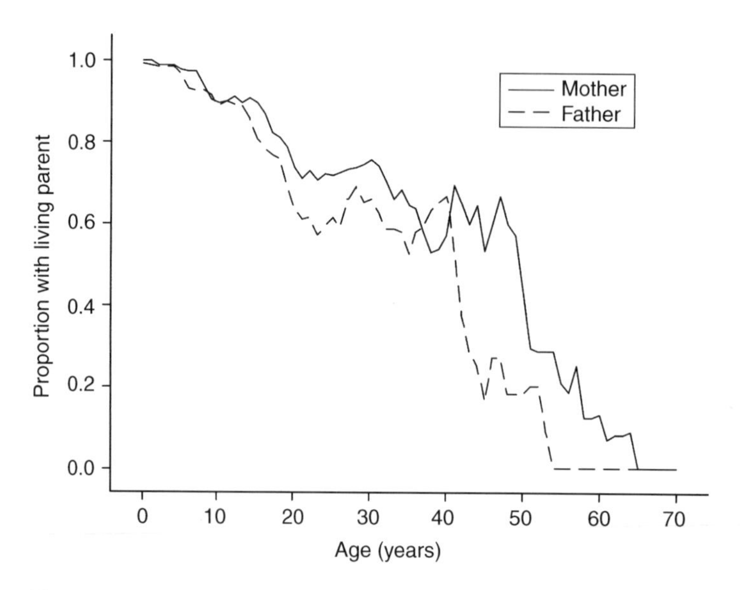


**Figure 18.1** How many women have a living parent? Proportion of women in each year of life who have a living mother or father (SI 18.1).

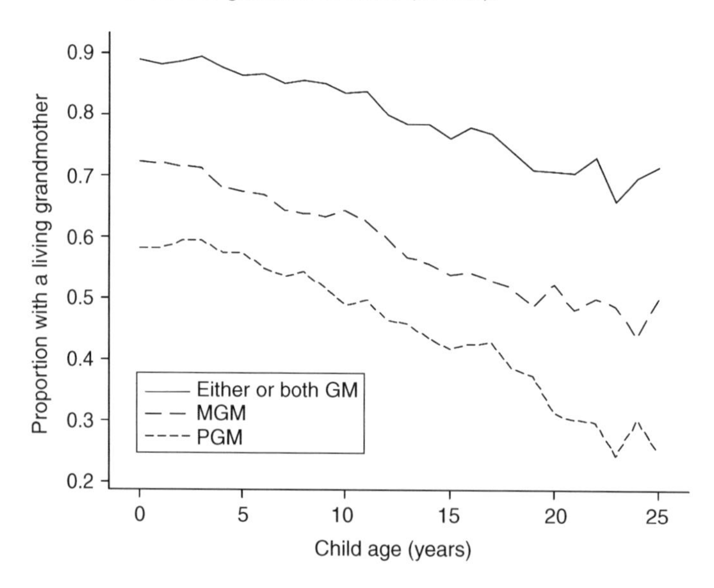


**Figure 18.2** How many children have a living grandmother? Probability that a Hadza at age x years has a living maternal (broken line) or paternal grandmother (dotted line), or either grandmother (GM status = 1) (solid line). Scored from the annual hazard file of the 695 children of interviewed women (SI 18.2).

grandmothers varies with the mother's age and with the child's age, which together with maternal age differences in child survival means that we must control for the mother's age in all the analyses. Later-born children are likely to lose their grandmothers at an earlier age than early born children. Nonetheless, the effect of a child's birth order on the availability of a grandmother is quite small; 90% of newborn to

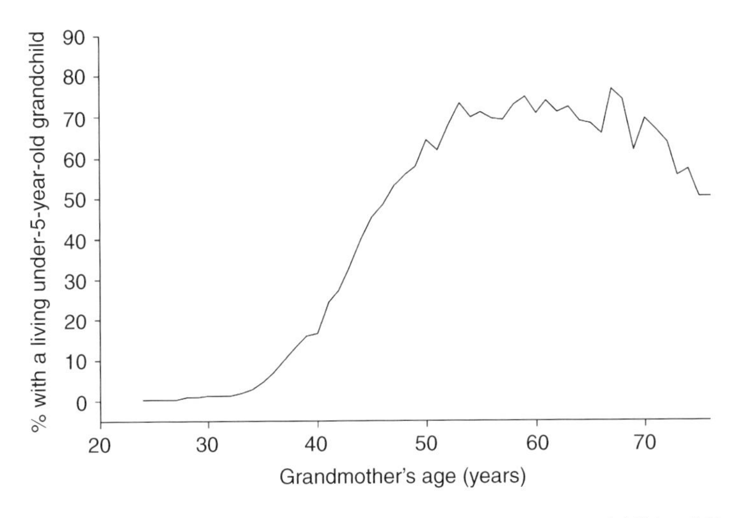


**Figure 18.3** At what age do Hadza women have young grandchildren? Percentage of women alive at age x years who have a living grandchild under five years old (SI 18.3).

five-year-olds from birth order one to three have at least one living grandmother. The figure for newborn to five-year-olds of birth order four and greater is 81%. Hadza grandmothers were available to the great majority of small children. Birth order, controlled for the mother's age at the birth, is not associated with the survival of Hadza children. First-borns tend to be at greater risk (b=–0.8204, p=0.094 in Cox regression).

### 18.2.3 How many older women have grandchildren they could help?

If we want to think about the evolution of longevity, it may be important to know whether older women (around the end of their childbearing years) have adult children, or small grandchildren they could help. Figure 18.3 (data table at SI 18.3) shows the percentage of living Hadza women of each age who have at least one living grandchild aged under five years. The percentage of women with small grandchildren increases rapidly during their 40s, passing 50% between ages 46 and 47. The percentage remains high during their 50s and 60s, and begins to decline during their 70s. Figure 19.1, in the next chapter, shows that by the time a woman is or would have been aged 80, her sons and daughters have finished childbearing. If grandmothering selected for longevity, it would not have selected for continued survival after about age 80.

### 18.2.4 Which grandmother lives with which grandchildren?

Census camp lists show how often young children are living with each grandparent (Blurton Jones *et al.*, 2005a,b; Blurton Jones, 2006). Grandmothers tended to live with a daughter more than a son, and more with a daughter who had children under age seven years, more likely with a daughter suckling a baby than not suckling, less likely with a daughter who has a teenage daughter (a significant food producer). An important observation was that if the daughter-in-law had lost her mother, her husband's mother was more likely to be living with her than otherwise. Looked at from the child's perspective, children under age 10 (boys and girls equally) are much more often with their maternal grandmother (boys 64% of their records, and girls 65%) than their paternal grandmother (boys 21%, girls 30%). This may be partly an outcome of a difference in timing. The mother's mother is more likely to be alive than the father's mother. Much of the time, daughter-in-laws have their own mother to be with if they choose; but if their mother is dead they are likely to be with their mother-in-law, if they still have one. A further complication is that because divorce and remarriage is common, a son may leave a series of children by different women in different camps. His mother has to choose which daughter-in-law's children to live with and help.

## 18.3 Grandmothers and children's growth

Hawkes' original evidence for the importance of Hadza grandmothers was that the weaned grandchildren of grandmothers who foraged longer hours gained more weight (Hawkes *et al.*, 1997). Do my data on many more individuals replicate anything like Hawkes' finding? I do not have data on grandmothers' work for my whole population sample but I can compare growth in children who have a living grandmother and those who have none.

Hawkes' sample was from Tliika where no Hadza women were married to non-Hadza husbands at that time. In Chapter 16, we saw that the size of children is quite strongly related to the size of their parents, and children of non-Hadza fathers and Hadza mothers tended to be bigger than children of Hadza fathers and Hadza mothers. Therefore, I looked for "grandmother effects" on growth only in the children with a Hadza father and mother.

There were few measures of growth rate (pairs of measurements for the under-five-year-olds) in which both grandmothers were dead. Consequently, I looked at single point measurements, which give a much larger sample by including those who were only measured once. The single measures assess size at any point in time. I expected living grandmothers to be associated with larger grandchildren (controlled for their age). This view is well supported in single-level regression even when the mother's average adult weight is included in the model.

The data were then examined with multilevel regression to check for effects of the nested structure of the data. Observations were nested within individuals; individuals are nested within mothers, and within maternal grandmothers. Apparent effects of grandmother status could indeed reflect differential access to resources (unlikely in a society such as the Hadza) or some heritable vigor, and not grandmother presence or absence.

The sample comprised all measurements of children aged one to five years ($\geqq 1.0$ and $<5.0$ years) with Hadza fathers, and mothers of any age. The results of multilevel regression are shown in Table 18.1 for weight, upper arm circumference (UAC), and

**Table 18.1** Children (aged 1–5 yr) whose grandmothers were both dead tended to be lighter and slighter than children with a living grandmother. Multilevel regressions for child weight, upper arm circumference (UAC), and triceps skinfold by grandmother status (1=either grandmother alive [274 cases], 2=both dead [33 cases]). Model: wt = age + mother's age at the birth + mother's adult weight + grandmother status. Child's age, mother's age at birth of the measured child, and mother's average adult weight were positively and significantly associated with child's weight and UAC

| | b | SE | Wald $\chi^2$ | p | N children/measures |
|---|---|---|---|---|---|
| Weight | −0.836 | 0.307 | 7.41 | $<0.01$ | 218/300 |
| UAC | −0.588 | 0.237 | 6.15 | $<0.025$ | 197/263 |
| Triceps | −1.053 | 0.501 | 4.42 | $<0.05$ | 143/189 |

triceps skinfold. There were significant effects for these measures but not for height or body mass index (BMI), and the result for triceps skinfold had borderline significance. Children with a grandmother were larger for their age than those without a grandmother.

Thus, there is a period, up to about age five years, when grandmother status makes a significant difference to child weight. However, there are later stretches of childhood in which the grandmother seems to make no difference. These results add to the picture of Hadza grandmothers as a significant influence on their grandchildren's lives. Child growth is, in many circumstances, a predictor of child health and survival. Therefore, these results support the idea that surviving beyond the childbearing years can enhance the older female's fitness by enhancing the fitness of her descendants.

## 18.4 Grandmother status and child survival

Although living grandmothers seem to distribute their presence in ways that might enhance their own fitness by help to grandchildren, dead grandmothers cannot help any of their children. Here I report differences in survivorship of children in relation to the survival status of their grandmothers. I looked at survivorship of the 695 children of the interviewed women; 181 of the children had lost their maternal grandmother by the time they were born and only 74 had lost both grandmothers while 147 had lost both grandfathers. There were 242 child deaths. The first aim is to see if there is "a case to answer." Given basic controls for the child's and the mother's age, does grandmother status appear to affect child survival? In subsequent sections, I will examine some other factors that might account for the result.

### 18.4.1 Grandmother status and child survival: survival analyses

The child "survival span" file was designed for use with Kaplan–Meier survival analyses and proportional hazards regression (Minitab 15's Regression with Life Data). There was one record for each child. The child's survival data,

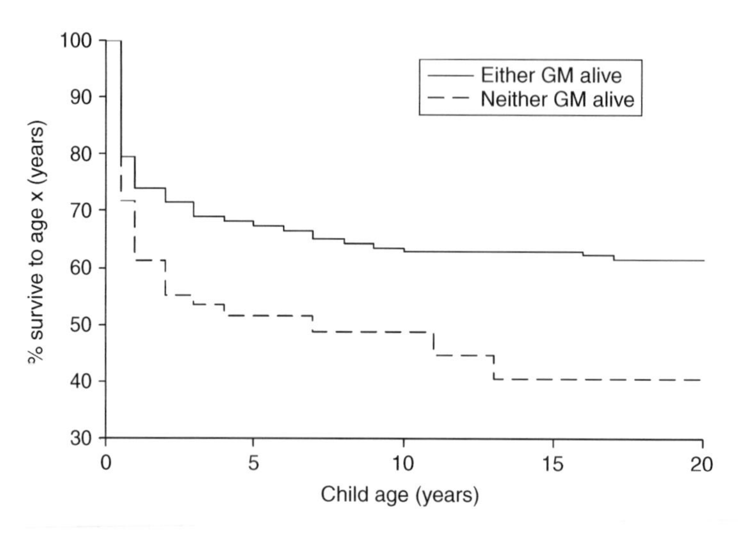


**Figure 18.4** If a grandmother is alive, grandchildren are more likely to survive. Kaplan–Meier survival to the end of the observation of children of women who were under age 35 at the birth of the child. Log-rank $\chi^2$=8.52, p=0.004, Wilcoxon $\chi^2$=6.42, p=0.011. At the time of their birth, 59 of these children had lost both grandmothers. At the time of their birth, 526 children had a living grandmother. In addition, 229 children died, 346 lived to the end of fieldwork, therefore were censored. Total: 585 children (10 of unknown outcome).

comprised: (1) "span," named for lifespan but actually the years from birth to either death or the last observation; (2) "censored," when a "1" was entered if the child lived to the end of observation, a zero if not. Most cases survived and were right censored. This file was most useful to obtain descriptive charts such as Figure 18.4. In this file, grandmother status was her status at the birth of the child.

Kaplan–Meier survival estimates for the children show a difference between those who had a living grandmother at the time of their birth and those who did not. With no grandmother, $l_{15}$ was 0.45, with a grandmother at the time of the child's birth, its $l_{15}$ was 0.63 (log-rank p=0.034, and Wilcoxon p=0.093). For the children born when their mother was aged <35, the Kaplan–Meier survival estimates show a striking and significant difference (log-rank p=0.004, Wilcoxon p=0.011) between children with a living grandmother (whose observed $l_{15}$ was. 063) and children with neither grandmother alive (whose $l_{15}$ was 0.41) (Figure 18.4). The population life table with the sexes combined gives $l_{15}$ at 0.556.

Cox regression allows us to control for the mother's age at the birth of the child. Older mothers have children who survive significantly longer. When the mother's age is used in the model, grandmother status has a strongly significant effect. Children with no grandmother live shorter lives. For the whole sample: span = $mom_{age}$ + GM status, b=–1.0645, p<0.004, with a Weibull distribution (CI –1.78 to –0.35). For the children who survived the first year, span = $mom_{age}$ + GM status, b=–1.084, z=–2.32, p=0.021 (CI –2.00 to –0.16).

### 18.4.2 Grandmother status and child survival: logistic regression in the child annual hazard file

In the child annual hazard file, each year of a child's life was a record, in which its death or survival was recorded. Each child had a record for every year of observation, from birth to death or the end of observation (the mother's final interview). The key dependent variable was "live or die" in the year. Most years, the child lived (score=0). For 242 children out of the 695, there was a year in which the child died (score=1). Logistic regression could be used in both single-level (Minitab logistic regression) and multilevel (MlWin) analyses. Logistic regression shows the odds ratios (OR) generated by each independent variable. This gives a useful way of summarizing the importance of an independent variable. I used the logistic regressions (and multilevel logistic regression) to look at the association between independent variables and the probability of a child's life or death.

Grandmother status was measured for every year and changed during the lifetime of a child if the remaining grandmother died. Because we know that child survival varies with the child's age and the mother's age, the starting model includes child age, child age$^2$ and child age$^3$ (which provides a quite good fit of mortality to age for people under age 20), and mother's age at the birth of the child.

If both grandmothers were dead, the child was much more likely to die. The effect was large, with an OR of 1.65 with 95% CI from 1.13 to 2.41 (Table 18.2, line 1). The effect was stronger on children of women who were aged <35 when the child was born (Table 18.2, line 2).

**Table 18.2** Grandmothers and child survival. Logistic regressions: child lives (scores 0) or dies (scores 1) = child age + age$^2$ + age$^3$ + mother's age at birth of child + other predictors listed in the table cells. Sample: 695 children, 5298 child–years, 231 child deaths; 181 children had lost their maternal grandmother by the time they were born, 74 had lost both grandmothers. Line 11: sample restricted to children aged 1–5 yr and mothers of all ages. "Supplanted" scores 1 if next child had been born, 0 if not yet born

| Line | Sample of mothers | Predictors | b | p | OR | 95% CI |
|---|---|---|---|---|---|---|
| 1 | All ages | GM status | 0.5017 | 0.010 | 1.65 | 1.13–2.41 |
| 2 | <35 yr | GM status | 0.6498 | 0.001 | 1.92 | 1.3–2.83 |
| 3 | All ages | MM status | 0.2587 | 0.087 | 1.3 | 0.96–1.74 |
| 4 | <35 yr | MM status | 0.3022 | 0.053 | 1.35 | 1–1.84 |
| 5 | All ages | Supplanted + GM status | 0.6614 | 0.002 | 1.94 | 1.26–2.97 |
| 6 | <35 yr | Supplanted + GM status | 0.7039 | 0.004 | 2.02 | 1.26–3.25 |
| 7 | All ages | Supplanted + MM status | 0.3131 | 0.059 | 1.37 | 0.99–1.89 |
| 8 | <35 yr | Supplanted + MM status | 0.3249 | 0.075 | 1.38 | 0.97–1.98 |
| 9 | All ages | Mom wt + GM status | 0.3969 | 0.047 | 1.49 | 1.01–2.20 |
| 10 | <35 yr | Mom wt + GM status | 0.5484 | 0.008 | 1.73 | 1.15–2.6 |
| 11 | Children aged 1–5 yr | GM status | 0.7013 | 0.006 | 2.02 | 1.22–3.34 |

## 18.5 Further questions about grandmothers and child survival

Eight issues arise about this preliminary result. The first three concern alternative explanations. The remaining add further details to the analysis.

1. Are some families more vigorous? Long-lived women have long-lived grandchildren.
2. Women tend to be smaller if they had lost their mother while they were children (Chapter 16). Some studies have found children of small mothers to be at a disadvantage. Does this account for the association of grandmother absence and child deaths?
3. What about a confounding effect of IBIs? In Hadza, child mortality is quite strongly related to the IBI following a child's birth (Chapter 17). Could this account for the apparent grandmother effect?
4. Is the effect really due to both grandmothers or mainly due to the maternal grandmother?
5. Are maternal grandmothers more important for young mothers?
6. Hawkes' original report concerned weaned children aged under five years, not all children under five years. Does it change the result if we exclude infants?
7. Hadza grandfathers appear to be ineffective (SI 18.4), but in farming populations the mother's father has been found to be important (Strassmann and Garrard, 2011).
8. Disease epidemics and famines tend to kill the very young and the old simultaneously. Of the 92 cases where both MGM and grandchild died, in only four were the deaths in the same year, and in only 16 were they in the same three-year period. Results were similar for the 88 PGM deaths (nine in the same three-year period) (SI 18.5).

### 18.5.1 Are some families more vigorous? Multilevel regression

The single-level analysis of grandmother status with Minitab's logistic regression leaves an important alternative to the claim that the results arise from grandmother help. The data have a nested structure. Child observations are clustered within 695 individual children, children are nested within 166 individual mothers, and mothers are nested within a mere 96 maternal grandmothers. Four old ladies in the sample each had 20 or more grandchildren. Another five had 15 or more.

The grandmother hypothesis claims to explain the evolution of a vigorous old age, so it must assume some form of heritable vigor. If the old ladies varied in heritable vigor, or some had simply run into better or worse economic circumstances, or more or fewer infectious diseases, the vigorous and fortunate might have better-fed, healthier, longer-lived grandchildren than the frail and unfortunate. The vigorous might live longer than the frail, and their grandchildren inherit the same degree of vigor or frailty. This would give an association between grandmother status (alive or dead) and grandchild survival. This association could have nothing to do with the help that we observe grandmothers giving to grandchildren day to day. The help may be ineffective, or much less effective than estimated by single-level regressions. An

**Table 18.3** Taking account of individual variation between maternal grandmothers does not remove the effect of grandmother presence. If both grandmothers are dead, a child is more likely to die. Multilevel logistic regressions predicting a child dead or alive in each year of its record. Nested by MGM (mother's mother), except in line 9 where nested by MGM, mother, and child. Model: dead/alive = child age + child age$^2$ + mother's age at child's birth. Lines 3, 4, 7, and 8: grandmother effects after "supplanted" (next birth has occurred or not) was added to the model. MGM status is significant for age groups younger than <35 (age <30, <25, same models as lines 3, 4, 7, 8). P-values are lower than in the single-level analyses of Table 18.2, as is common for multilevel analyses. Other analyses include nesting by MGM > mother > observation, and analyses for younger mothers as discussed in the text. Effects are stronger and more significant for younger mothers

| Line | Sample of mothers | Predictors | b for GM | SE | Wald | p | N GM/N child-years |
|---|---|---|---|---|---|---|---|
| 1 | All ages | GM status | 0.427 | 0.221 | 3.73 | <0.1>0.05 | 96/4943 |
| 2 | | MM status | 0.230 | 0.182 | 1.59 | >0.10 | 96/4943 |
| 3 | | Supplanted + GM status | 0.556 | 0.252 | 4.86 | <0.05 | 84/4346 |
| 4 | | Supplanted + MM status | 0.317 | 0.201 | 2.48 | >0.10 | 84/4346 |
| 5 | Age <35 yr | GM status | 0.587 | 0.231 | 6.45 | <0.025 | 96/4577 |
| 6 | | MM status | 0.269 | 0.188 | 2.05 | >0.10 | 96/4577 |
| 7 | | Supplanted + GM status | 0.590 | 0.260 | 5.15 | <0.025 | 84/4135 |
| 8 | | Supplanted + MM status | 0.320 | 0.210 | 2.32 | >0.10 | 84/4135 |
| 9 | <35 yr nested by MGM and Mom | GM status | 0.556 | 0.252 | 4.86 | <0.05 | 96/166/4577 |

association due to "vigor" should tend to apply to each grandchild, even those born after the grandmother had died. The best method for dealing with this problem is to use multilevel modeling.

Multilevel logistic regressions show that there are individual differences between mothers and grandmothers in the probability that children die, compatible with the idea of heritable vigor. However, these effects do not remove the effects of grandmother presence or absence. The effect of grandmother status endures this test, most robustly when IBIs are included in the regression models (Table 18.3, lines 3, 4, 7, 8). The significance (p-values) of the grandmother regressions are reduced, as often observed when multilevel analyses are compared with single-level regression. The "grandmother effect" is not explained away by individual variation between mothers or grandmothers.

### 18.5.2 Orphaned mothers may be smaller, children of smaller mothers are sometimes at a disadvantage

The sample of women includes some who had lost their mothers during childhood. In Chapter 16, I showed that children growing up without a mother tend to be smaller than others. Many studies and data sets (Monden and Smits 2009) have shown higher mortality among children of smaller mothers. Is this why more children without a grandmother die? Some analyses reported earlier argue against this. For example, I reported that taking account of the mother's weight did not remove the effect of the grandmother on the grandchild's weight. In Chapter 16, I found no relationship between a woman's average adult height or weight and the percentage of her children that she kept alive. Nonetheless, it seemed necessary to test for an effect of the mother's weight on grandmother effects.

Both in logistic regression (Table 18.2, lines 9 and 10) and multilevel logistic regression, the mother's mean adult weight failed to remove the grandmother effect. We may note that Hadza women have free access to abundant resources. Results may differ for women under other circumstances where they have less opportunity to make up for early stunting. Monden and Smits (2009) discuss this possibility and the several studies that appear to find effects of differing local circumstances (such as by Pollet and Nettle, 2008). Their multilevel analysis of the multi-nation data shows that few of these factors are significant in that sample, although maternal education did significantly effect the strength of the effect of maternal height.

### 18.5.3 Are there confounding effects of inter-birth interval?

Hawkes *et al.* (1997) showed that the grandmothers' work influenced the growth of the newly weaned. The newly weaned are vulnerable, and especially vulnerable to short IBIs (Chapter 17), which entail early weaning and competition with the new baby. I looked at the role of the variable "supplanted" (the next baby has arrived) alongside grandmother status (neither grandmother alive, mother's mother alive or dead) in each year of the child's record. Some children were supplanted at an early age, some later. Some had grandmothers, some did not. Adding the child's status as "supplanted" or not to the regression models strengthened the beta coefficients and significance, and raised the ORs for grandmother status (Table 18.2, lines 5–8, and SI 18.6).

I also looked at this issue in multilevel regression. Comparing the ORs, we see that the effect of grandmother status is slightly strengthened by including "supplanted" (compare lines 1 and 3, 5 and 7 in Table 18.3). Both supplanting and grandmother status appear to have equally important effects.

### 18.5.4 Are paternal grandmothers important?

We can show that the paternal grandmother makes a difference by separating a sub-sample in which all maternal grandmothers were dead. The status of the paternal grandmother made a significant difference to the survival of children (Table 18.4). If their paternal grandmother (FM, PGM) were alive, there were fewer deaths to children whose maternal grandmother (MM) was dead.

**Table 18.4** Testing for an effect of paternal grandmothers (PGM). The sub-sample comprises all those records in which the maternal grandmother is dead. Model: child dead or alive = child age + age$^2$ + mother's age at birth + PGM status. Sample: 1814 child–years, 71 child deaths; 33 deaths when father's mother is alive, 38 when father's mother is dead

| $Mom_{age}$ | FM status b | p | OR | 95% CI |
|---|---|---|---|---|
| All | 0.4921 | 0.053 | 1.64 | 0.99–2.7 |
| <35 yr | 0.6798 | 0.011 | 1.97 | 1.17–3.34 |

### 18.5.5 Are maternal grandmothers more important for young mothers?

There are indications that the maternal grandmother is more important and the paternal grandmother less effective among the youngest women. In every regression model, the maternal grandmother shows a significant effect among women under 25 years old. In multilevel regression on the women aged under age 25, PGM status gives a non-significant result while the maternal grandmother shows a strong and significant effect (SI 18.7).

### 18.5.6 Is the effect mainly on weaned children?

Hawkes' original observations and theorizing about grandmothers emphasized the care of weaned offspring. The preliminary result given in Table 18.2 (lines 1 and 2) concerned all under-five-year-olds. We can best approximate weaned offspring by looking at children aged one to five years. Age 1–1.99 years was included because it would cover many of the newly weaned children (discussed previously). When the data from the first year of life were removed, the grandmother effect remained strong and significant. In the Kaplan–Meier analysis, when we limit the sample to children who survived the first year of life (Figure 18.5), we again find a striking and significant effect of grandmother status. With no grandmother, survival from age one to age 15, "$l_{15}$"=0.5657, with a grandmother, "$l_{15}$"=0.7918. Grandmother effects appear to be greatest for year two and year three of life. In the logistic regression analysis, children aged one to five years with no living grandmother had more than twice the odds of dying compared to those with a grandmother (Table 18.2, line 11).

## 18.6 Grandmothers and length of successful inter-birth intervals

In Hawkes *et al.* (1998), we argued that grandmothering may be ultimately responsible for the much shorter IBIs of humans contrasted with our nearest relatives. Wondering whether the reproductive system retains the flexibility to adjust fertility to availability of helpers, I looked at the length of closed IBIs as predicted by grandmother status. Survival analysis of closed intervals (no censoring but intervals closed at any time between one and 10 years) showed no significant difference between interval lengths in relation to grandmother status.

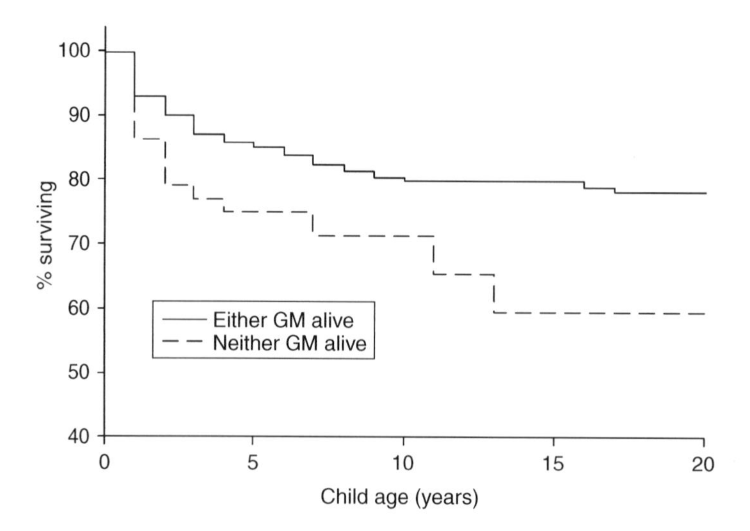


**Figure 18.5** Grandmother effect on the 1–5 year olds is significant. Survival of children who were alive at age one year, and born before the mother was aged 35. Log-rank $\chi^2$=6.1, p=0.013; Wilcoxon $\chi^2$=5.4, p=0.02. (47 children with both grandmothers dead, 385 with grandmothers alive, 20 unknown grandmother status; 97 died, 355 survived until the end of observation).

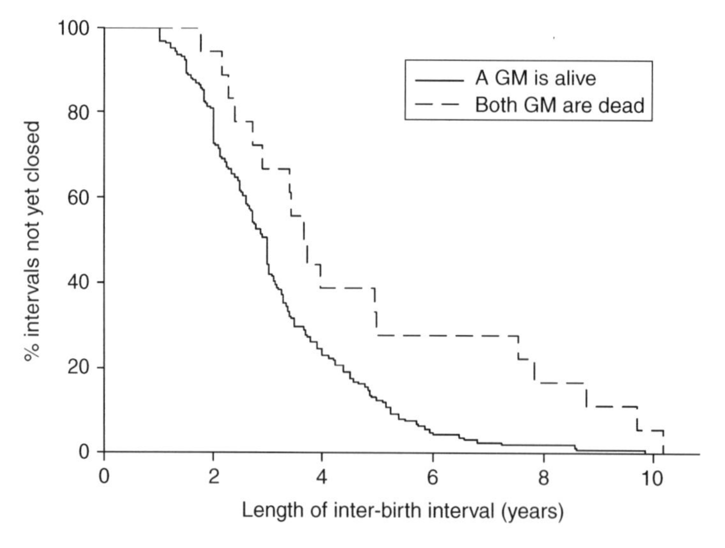


**Figure 18.6** Successful inter-birth intervals are shorter when a grandmother is available. Births to women aged under 35, 180 closed intervals. Mean IBI when at least one grandmother (GM) was alive was 3.2 years, when both GM had died (by the time the interval was closed) the mean IBI was 4.8 years. Medians were 3.00 and 3.67 years. Log-rank test=8.87, p=0.003, Wilcoxon test=5.44, p=0.020.

If we look at the length of successful intervals, we get a much more striking picture (Figure 18.6). Successful intervals are those which add a surviving child to the family. The child who opens the interval survives, and so does the child who closes the interval. Looking only at successful intervals, we see significantly shorter mean and

**Photograph 18.2** A successful but quite closely spaced family. This woman's mother died shortly before our study began, but as commonly happens, her husband's mother lived with them and helped keep them supplied with tubers, baobab, and berries for several years until her own daughter began to have children. © James F. O'Connell, 2015. Reproduced with permission.

median intervals for women of any age who have a living mother or mother-in-law (198 intervals). Photograph 18.2 shows such a closely spaced but healthy family. Among women aged less than 35 (180 intervals), the means were also significantly different (log-rank test $\chi^2=8.20$, $p=0.004$, Wilcoxon $\chi^2=5.44$, $p=0.020$). If the grandmother was dead (by the time the interval was closed), the mean successful interval was 4.8 years (median 3.67 years). If at least one grandmother was alive, the mean was 3.2 years and median 3.00 years. The medians differ by 22%. A young woman with a mother or mother-in-law added living children to her family 22% faster than a woman with no such older helper.

This finding could arise simply from young women being more fertile and more likely to have a living mother or mother-in-law. Cox regressions in which the woman's age can be controlled (age + $age^2$ + $age^3$, a good fit to age-specific fertility) support the strongly significant difference in length of successful IBIs with respect to grandmother status. Women with senior help can add a surviving child to the family faster than women with no such help. This seems like strong support for Hawkes *et al.*'s (1998) contention that grandmothers account for high human fertility.

## 18.7 Grandmothers and fertility

There were consistent and quite large effects of the grandmother on the proportion of children surviving each IBI, and on the length of successful intervals. If IBIs are truly optimized over all conditions, and women with living mothers (or mother-in-laws) are able to produce survivors at a faster rate, we would expect women with helpers to have higher fertility. Looking at the women's annual reproductive history file suggests that they do not.

Instead of increasing fertility, grandmothers are associated with lower fertility. The analysis, using logistic regression, controlled for woman's age, age$^2$ and age$^3$ (which gives a good fit to fertility). The effect of grandmothers' status on probability of a birth was positive, moderately strong, close to significant (Table 18.5 line 1, b=0.361, p=0.056), and in the opposite direction to my initial expectation. If both grandmothers are dead, a birth is more likely. Why? Are there circumstances under which grandmothers gain from restraining their daughter's fertility? Behavioral ecologists familiar with the literature on communally breeding mammals would immediately ask if this were a sign of competition between older and younger women. I will report on this in the next chapter, but there are several possible reasons for the failure of grandmothers to increase the fertility of their daughters.

1. Perhaps the reproductive system cannot facultatively adjust to grandmother presence or absence and the success with which she is supplementing the diet of the weaned children. The grandmother's influence on frequency of suckling, especially at night when its influence on fecundability is greatest (McNeilly and McNeilly, 1979), is likely to be weak. In so far as she improves the child's and mother's nutritional level, we might expect faster infant growth and earlier

**Table 18.5** Grandmothers do not increase the fertility of their daughters. Births are more likely if grandmothers are dead. Logistic regression: anybirth (birth/not) = age + age$^2$ + GM status or MM (young woman's mother) status. Number of non-Hadza husbands is a count, analyzed by ordinary least squares regression. Lines 1–5, women of all ages; lines 6–8, based on 43 births in 1577 woman–years of women aged <18

| Daughter: dependent variable | Grandmother: independent variable | b | p | OR | 95% CI |
|---|---|---|---|---|---|
| 1. Anybirth | GM status | 0.361 | 0.056 | 1.43 | 0.99–2.08 |
| 2. Anybirth | MM status | 0.190 | 0.070 | 1.21 | 0.98–1.49 |
| 3. Anybirth | GM status + %children dead | 0.042 | 0.822 | 1.04 | 0.72–1.50 |
| 4. Anybirth | MM status + %children dead | 0.057 | 0.597 | 1.06 | 0.86–1.31 |
| 5. N of non-Hadza husbands | MM status | 0.038 | 0.020 | | |
| Under-18-year-olds | | | | | |
| 6. Any birth | MM status | 1.016 | 0.009 | 2.76 | 1.29–5.94 |
| 7. Married/not | MM status | 0.746 | 0.009 | 2.11 | 1.2–3.69 |
| 8. N of non-Hadza husbands | MM status | 0.016 | 0.021 | – | – |

weaning, and an earlier return to the ability to support a pregnancy. Apparently, the grandmother's presence does not encourage the woman to wean a little earlier and move on to the next baby more quickly.

2. We might wonder whether, as more children die in the absence of grandmothers, there were more "replacement" intervals, which slightly increase fertility (the rate of births). If this effect was accounted for, we might find that grandmothers actually do slightly increase fertility. I added child mortality (% dead/born during the mother's record) to the model (Table 18.5, lines 3 and 4). The apparent fertility reducing effect of the grandmother was abolished but not reversed. On this evidence, we can say that the effect of Hadza grandmothers on their daughters' fertility is neutral.
3. In some societies, parents attempt to exert control over their children's reproduction. Are Hadza grandmother's trying to restrain the reproduction of their daughters or daughters-in-law? Although no formal socially recognized process of this type has been described among the Hadza, Hadza women do oppose sending their daughters to school for fear of an early, unwanted pregnancy, especially by someone of another tribe. Some Hadza comment adversely on the marriage of young Hadza women to non-Hadza men. In the total sample, there was a weak indication that the absence of a woman's mother was accompanied by a greater incidence of marriage to a non-Hadza man. (Table 18.5, line 5). My field impressions were that the resistance applied, especially to the youngest women. I looked at the under-18-year-olds, rather below the average age at the first birth, which is 19. If the young woman's mother were dead, she was much more likely to have a birth during the years up to age 18 (line 6), much more likely to be married (line 7), and significantly more likely to be married to a non-Hadza man (line 8). These results are compatible with the idea that mothers restrain the reproduction of their young adult daughters. Would the mother benefit by making her daughter wait for a better prospect, one more likely to keep his mother-in-law supplied with meat and trade goods?
4. We might wonder whether, under Hadza circumstances, the mother gains fitness by retaining an especially useful helper toward the end of her own childbearing career. Perhaps this is less costly to her daughter than we might expect. A percentage increase in births to women aged 15–20 contributes less to population growth than among women in their 20s (Chapter 12). Children born to the youngest mothers suffer the highest mortality, thus by preventing them, the grandmother may lose less than one might suppose. The number of births between ages 15 and 19 is a good predictor of RS at the final observation (brt15 × sslive, b=0.428, p=0.000, N=172), as is the age at the first birth (Chapter 13). Therefore, we may expect daughters to benefit from resisting their mother's restraining influence. I can offer only weak hints that beginning the childbearing career very early brings a disadvantage. In the principal components analysis in Chapter 13 SI, there was a dimension with a high positive loading for early fertility but negative loadings for later fertility. In a plot of RS by age at first birth, there are a handful of women among those whose first birth was before age 18, who unlike

the majority had very low eventual RS. Very early pregnancies and births have poor outcomes for the child but apparently do not entail greatly increased mortality for the women, especially when compared with the increased risks to older mothers (Makinson, 1985, for U.S. and northern European data; Stover and Ross, 2010, for worldwide data). As maternal deaths are always rather rare, as are very early births, it may be that the statistics do not accurately portray the increase in risk to the very youngest mothers. I cannot measure these risks among the Hadza; the sample is much too small. The mother may lose contact with her daughter and grandchildren if the daughter and her non-Hadza husband move away to a village. The move may, in actuality, give the mother greater fitness via the daughter married to a farmer. However, in a world of hunters among hunters, when the daughter moved to a neighboring tribe, the grandmother might lose control and the opportunity to promote her daughter's fitness, in the absence of an expectable greater fitness among the "foreigners."

5. Can grandmothers influence the disagreement between husband and wife over IBI discussed in Chapter 17? Noting references in the literature to men's preference for shorter IBIs, I suggested that with frequent divorce and presumed costs and delays of competing for a new wife, men might gain fitness from IBIs shorter than intervals that were optimal for women. Perhaps grandmothers favor late weaning while husbands favor earlier weaning and an earlier subsequent pregnancy. The mother's mother might side with her daughter, and the father's mother side with her son, each expressing opinions about the current infant's readiness for weaning. Hence, closed IBIs should be longer when MM is alive and shorter when she is not, especially if FM is alive. However, there are no indications of opposite effects of FM and MM on the probability of a birth, nor of either on its own. MM and FM's interests might diverge most where divorce is frequent and remarriage difficult, but the limited Hadza data is not the place to pursue such possibilities.

## 18.8 Grandmothers and young women's reproductive success

Grandmothers promote survival of the under-five-year-olds; but I could see no statistically significant effects on fertility in the young women's annual fertility hazard file, nor in the average IBI. We should try to look at the net fitness outcome.

I looked at the RS measures at the last time each woman was interviewed. I made an approximation of the grandmother status measure. The mother's mother was easy to assess; I noted whether she was dead or alive at the mother's final interview. The father's mother was less easy to assess because of the high divorce rate, and different children had different fathers, different paternal grandmothers. I made an approximate measure by identifying the man who had fathered most of each woman's younger children, then looking to see whether his mother was alive or dead at the time of the mother's final interview. From these results, I made a combined measure analogous to GM status, in which a grandmother was scored as 1, alive, if either MM or FM were alive at the time of final interview, and 2 if both were dead.

Controlling for age at the last interview, there is a significant contribution of GM status to predicting the standard score of RS (sslive) and a significant contribution to predicting the number of living children (controlled for the mother's age) at the final interview (b=−0.6399, p=0.044, N=171 women). Reproductive success is lower if both grandmothers are dead. The fitted lines show a slight excess of live children when a woman is aged below 20 and has no grandmother for her children, but thereafter they produce surviving children more slowly than women who have a live grandmother for their children. The fitted lines suggest that by age 45 women tend to have about 1.5 more children if the children have a living grandmother than if they do not. These results seem consistent with the observations of better survival of children with a grandmother, and greater fertility of the youngest women without a mother.

## 18.9 Discussion: circumstances that may shape effects of grandmothers

The data suggest quite strongly that Hadza grandmothers enhance the growth and survival of their grandchildren, especially those born to younger mothers. The combined measure (GM status, either grandmother alive, versus neither alive) had a strong effect that resisted attempts to remove it by entering other relevant independent variables. The size of the effect was not influenced by including or excluding grandfather status, father presence, or presence of older siblings (Chapters 20 and 21). It is quite surprising that the results are as robust and consistent as they are. There were only 86 children with both grandmothers dead, and only 242 child deaths. Subdividing the sample becomes quite risky; the numbers of absent grandmothers or child deaths become even smaller. Divisions by the mother's age nonetheless showed a consistent pattern, suggesting that grandmothers were more important for children of younger mothers, and that the mother's mother was relatively more important than the father's mother for the youngest women.

One important question concerns the children of non-Hadza fathers. The non-Hadza fathers were either absent, or living among the Hadza. The 14 non-Hadza grandmothers were not co-resident with the children, and most lived far away and we had no way to tell whether they were alive or dead. They were coded as dead. Bear in mind that a coding of gnmostat=2 also requires the mother's mother to be dead. It is conceivable that some of the non-Hadza grandmothers were able to send food back after visits by their son. Such visiting would not be frequent and should work against our seeing differences by grandmother status. The father's tribe has a small non-significant effect on child mortality (b=-0.1683, p=0.378, OR=0.85, CI 0.58–1.23); children of non-Hadza fathers may be slightly less likely to die. This would also work against our result; these children add to the count of living children with dead grandmothers. Including the father's tribe in the regressions does not change the pattern of results for grandmother status. If children of non-Hadza fathers are removed from the sample, the pattern of regressions does not change but the number of child deaths is reduced and fewer of the results are statistically significant.

There is support for the view that grandmothers primarily affect survival of the weaned under-five-year-olds. For example, the grandmother effect was as strong for children who survived to one year as for all under-five-year-olds (Figure 18.5). Effects on growth were seen in the age group one to five years. Many infant deaths are probably out of the control of mother, grandmother, or anyone. Demonstrating an effect of grandmothers shows us that helping could enhance the fitness of post-reproductive women. To fully make use of these data, however, they should be incorporated in a model that tests whether they are sufficient to evolve a vigorous post-childbearing life. Several such models have been built and aspects of them are discussed in the next chapter.

As Sear *et al.* (2003a) found for the rural Gambia sample, Hadza grandmothers seemed to have no positive effect on the fertility of their daughters. Since we suggested in Hawkes *et al.* (1998) that grandmother help allowed humans to have much higher fertility (and shorter IBIs) than other Great Apes, this result received some attention in this chapter. I pointed out that successful IBIs were shorter if the grandmother were alive than if she were not.

In Blurton Jones (2006), I reported that, in contrast to grandmothers, grandfathers did not tend to live where their help to grandchildren would give them the greatest fitness benefits from their grandchildren. Here I could find no persuasive evidence that their survival had any effect on the survival of their grandchildren. In contrast, Strassmann and Garrard (2011) found a positive effect of maternal grandfathers even though they did not reside in the same compound as their grandchildren. I think this illustrates the importance of the "on the ground" details for understanding variation. Among these farmers, who live in patrilocal compounds, the maternal grandfather could be an important source of outside resources, transported from his compound to his grandchild's compound, and targeted at his grandchildren.

In their review "Who keeps children alive?", Sear and Mace (2008) find populations in which grandmothers do help, others in which they do not, and some in which paternal grandmothers hinder. Here we saw that Hadza paternal grandmothers as well as maternal grandmothers have a positive impact. Variability in effectiveness may arise from differences in the things grandmothers could do to help in different circumstances (see Fouts and Brookshire, 2009, for a useful discussion of variation in circumstances for helping). In some, childcare may be more practicable and more rewarding than foraging. In others, food processing may be more rewarding than foraging, and so on. The payoff from one activity may depend on the payoff to other potential helpers from this and other activities. The Hadza environment may be at an extreme of rich availability of wild plant foods, which may make foraging a particularly rewarding way to help. The Ache forest seems close to the other extreme, given the daily trekking life style, and the great effectiveness of hunting, there would seem to be little reward for women of any age spending much time pursuing plant foods. There may be little for Ache grandmothers to do but help carry the children and equipment during the day, help with the not very extensive food processing at night, and rather unsuccessfully intervene when child sacrifice became an issue. Nonetheless, Hill and Hurtado (1991, 1996) report positive effects of grandmothers on child survival. Both Hadza and !Kung mothers and grandmothers are confronted with quite extensive food processing, with baobab seeds and mongongo nuts, respectively, and among the !Kung and the western Hadza, the even more difficult-to-process and even more widely distributed marula nut (*Sclerocarya birrea*) (Peters, 1987, fig. 7). Nonetheless, Howell (2010, table 7.2) showed only negative effects of grandparent co-residence (mother's father and father's mother) on child BMI. Scelza (2009) reports on caretakers of Martu children aged under three years in the western desert of Australia. Although men hunt and women gather and hunt, some food is purchased and no data is offered on the amount that grandmothers forage. The focus of the study was direct care of small children. Grandmothers scored higher than fathers and siblings, anyone but mothers, in the more demanding forms of care such as feeding, bathing, soothing, or disciplining. Fouts and Brookshire (2009) report that in Aka Pygmy forest and village camps, grandmothers and great-aunts provided food to children as often as mothers did, and that weaned children were fed relatively more by great-aunts and grandmothers than by mothers. In their context, providing food appears to mean handing food to children, the source of the food is not given. Thus, although it appears that we have insufficient evidence to say whether grandmothers are effective helpers among hunter-gatherers as a whole, there are indications that they tend to behave in ways expected to benefit their grandchildren.

Lahdenpera *et al.* (2012) comment on the inconsistency of evidence for grandmother effects among hunter-gatherers, pointing to Howell's recent analysis of !Kung residence and nutritional status as assessed by BMI. We should keep two things in mind when thinking about Howell's analysis. First is Hill and Hurtado's point about helpers obscuring their own effects, which I addressed earlier and when looking at older women's co-residence (Blurton Jones, 2006). Howell does not report separately on grandmother survival and co-residence. Howell's sample apparently did not allow inclusion of effects of the age of child, mother, or grandparent, nor for tests of whether grandmothers were living where their help would be advantageous. Second, BMI may not be the best indicator of how well a young child is doing. As well as low body weight, low height can arise from under-nutrition, and compress the variation seen in BMI. BMI decreases during the crucial ages of one to five years, confounding improved survival chances with "worsening" BMI. Note that I saw the effects of the Hadza grandmother on weight, UAC, and triceps skinfolds, but not on BMI (SI 18.8).

Residence patterns may also shape cultural variation in grandmother activities. In the case of the very mobile Hadza (by which I mean that camps frequently change composition as well as location), the reproductive interest of older women seems to help shape residence (Blurton Jones, 2006). In other societies, the dominant interests of other classes like cattle-owning men, or the relative stability of fields and their ownership, may predominate and constrain the opportunities for older women to choose where to live. However, when reading about even the most patrilocal societies, one comes across indications that women try to be near their mothers, that their mother visits at the time of births, and the mother is where they go when they are divorced, or otherwise in trouble. For example, in Fouts and Brookshire (2009, p. 288) we read that "The Aka are loosely patrilocal and patrilineal. During the first two to seven years of marriage, families live matrilocally while husbands are performing bride-service, and often after bride-service couples chose to live matrilocally for various personal reasons." Residence classifications seem to skim over a lot of variation, flexibility, and complexity. Alvarez (2004) and Marlowe (2004) independently showed the complexity and Marlowe concluded "... contrary to the orthodox view, most foragers are not virilocal. Women frequently live with their kin, especially in the early years of marriage. Murdock's bilocal view is supported by the cross-cultural data and for just the reasons he cited: small, mobile, flexible groups that maintain ties with kin of both husband and wife are well suited to foraging for dispersed and seasonal foods" (Marlowe 2004, p. 283). This has since been overwhelmingly confirmed by Hill *et al.* (2011) in a survey of residence data from 32 recent hunting and foraging societies.

There are more issues of detail that could be looked at with a larger sample; for example, looking at better measures of grandmothers' opportunities and preferences about with whom to live. Another interesting opportunity may arise. Recent studies by Marlowe and his students may show us whether the "trough" in the Hadza age structure (described in Chapter 10) has generated a shortage of grandmothers, and whether this expected shortage was accompanied by changes in the patterns of help by the variety of potential helpers. I will present the data on men and siblings as potential helpers in Chapters 20 and 21, but first we must look at recent developments in grandmother research and theories about the evolution of post-reproductive life.

# 19 Grandmothers and competition between the generations

In this chapter, I report data on Hadza grandmothers and reproductive competition. In SI, I report Hadza data relevant to the paper by Fox *et al.* (2009), which is based on the pattern of inheritance of the X-chromosome (SI 19.1). I also discuss in SI the distinction between theories of "stopping early" and "living longer," models of the evolution of grandmothering, and offer simulations of evolution away from the current life history. The main business of this chapter arises from two important papers by Rufus Johnstone and Michael Cant on competition between women of different generations.

Following the descriptions of humans as cooperative breeders, Cant and Johnstone (2008) suggested that we look at a key feature of mammalian cooperative breeders, competition between generations, as a selection pressure for menopause. The core of Cant and Johnstone's argument is that if older women can acquire resources that are useful to younger kin, there must be competition over their allocation. In our papers (Hawkes *et al.*, 1997, 1998) and in Chapter 18, we suggested that Hadza grandmothers have their effect by providing food. Because food eaten by one child or grandchild is not available to be eaten by another, if the grandmother is still reproducing she should have less surplus food to give to grandchildren. The grandmother's effect on grandchild growth and survival should decrease as it is divided among more children and grandchildren. Despite the day-to-day collegiality of Hadza women, there is inevitable competition for the grandmother's food. Who would win the competition for her surplus, the grandmother or the younger generation?

In groups of cooperatively breeding mammals, such as the meerkat, mole rat, and some canids, there are a few females, or even only one, who breed and others who do not breed but remain in the group and contribute to the reproductive success (RS) of the breeders. Usually the breeders are older and the suppressed breeders are younger, and most are close kin. Cant and Johnstone proposed that in the human case, the older generation was suppressed. The key issue is the allocation of resources between the young kin and the older woman's continued childbearing and child-rearing. The outcome of the competition will depend on the fitness gains to each generation. Cant and Johnstone attend to allocation between the grandmother's own reproduction and that of her daughters and daughters-in-law.

Cant and Johnstone argued, in accord with majority opinion in earlier anthropological literature, that our ancestors were virilocal and most interaction between the generations of women was between women and their daughters-in-law. The daughter-in-law's children likely carry some of the paternal grandmother's

(PGM) genes and contribute to the grandmothers' fitness, but the mother-in-law's continued reproduction makes no contribution to the daughter-in-law's fitness. A game theory model confirmed that the daughter-in-law would therefore "win" a contest for resources allocated to reproduction. The mother-in-law would give up her own reproduction. Reproductive suppression should thus be found in the older females, especially if they have an alternative route to enhancing their fitness, by helping their descendants. The idea is attractive for its ingenuity.

Some initial doubts should be mentioned. Should we be trying to explain evolution of menopause (as a special endocrine mechanism, as "stopping early") or the elongated human lifespan ("living longer")? Whichever we favor (SI 19.2), Cant and Johnstone offer a new approach and a potential answer to questions about a trade-off between continued childbearing and greater provisioning of grandchildren. Their emphasis on virilocality should be questioned. Female dispersal is unusual for mammals but has long been suggested for chimpanzees and humans (Cant and Johnstone cite Ember, 1978, and Oota *et al.*, 2001). Cant and Johnstone also cite Alvarez (2004) and Marlowe (2004) as support for the predominance of female dispersion in human pre-history, although each of these authors argued clearly against this. Human residence is often quite pragmatic. Hill *et al.* (2011) show great variation and a tendency toward matrilocality in a sample of 32 ethnographically recorded hunter-gatherer populations. Wilkins and Marlowe (2006) showed that the genetic evidence for female dispersal was heavily dependent on post-agricultural events. Recent reports by Pusey and her co-workers (Murray *et al.*, 2007, 2009; Emery Thompson *et al.*, 2007; Langergraber *et al.*, 2013), for example, imply that chimpanzee dispersal and residence is much more complicated than simple female dispersal.We should think about other potential asymmetries that could have shaped the competition between generations.

## 19.1 Empirical tests for reproductive competition among Hadza women

We can try to test directly whether there are costs to reproduction of one generation caused by reproduction by the other generation. Strassmann (2011), Strassmann and Garrard (2011), Lahdenpera *et al.* (2012), and Mace and Alvergne (2012) did this with their data on farmers. Strassmann and Garrard limited their meta-analysis to patrilocal farmers, many of whom live in compounds containing a small number of related households (like those of some pastoralists). The studies looked at fertility of the two contestants and survival of their children during the brief period of "overlap," when either might bear a child or have small children to raise. Each shows evidence of competition between the PGM and her sons' wives. We can look for evidence of competition in the Hadza data in the same way.

The fieldworker among the Hadza will have two instant reactions to the idea that menopause represents breeding suppression. The first is only partly rhetorical. If anyone were suppressing a Hadza grandmother in any way, they would hear about it very loudly and clearly, and at some length. Tirades by older Hadza women are among the most memorable features of fieldwork among them. Everyone in camp can

**Photograph 19.1** A "midday cook up". After a few hours of digging tubers, women stop for lunch in the bush. Here the mother-in-law tends the fire roasting the tubers while the grandchild, on the daughter-in-law's lap, chews on a small piece of tuber. The foraging group of women sometimes break up into kin groups for their lunch stop. A distant group can be seen in the background of Photograph 18.2. © James F. O'Connell, 2015. Reproduced with permission.

hear the shouting, and even if they had not followed the details of the dispute, they would have identified the offending party as well as the offended. The second reaction notes that, having recovered from the verbal battery, either party can ignore the outburst, as they commonly do, or leave and join another camp anywhere they fancy. Leaving a Hadza camp entails little cost in removal effort, access to food or water, or relatedness to members of the new camp, and no obstacle to joining another one. The contrast with the closed groups of classical cooperative breeding mammals is extreme.

Because of the flexible residence, we should not consider Hadza grandmothers and daughters-in-law in isolation. Any daughter-in-law is likely to have access to a mother of her own. If she does not, then as reported in Blurton Jones (2006), she and her mother-in-law are more likely to live in the same camp (Photograph 19.1). In the last chapter I showed that PGM has a positive influence on her grandchildren's growth and survival. We can also look for competition between the generations more directly. To do this, I had to go beyond the data on interviewed women (few of whom were long past menopause) and extract the larger sample of potentially less reliable information on older women that is incorporated in the population register.

### 19.1.1 Period of overlap

I made a file that covers the lives of women from age 35, who were in the population register and seen in two or more censuses. Each record was a year in the life of a woman. The record included whether she gave birth or not during the year (based on birth year of her children in the population register), whether a son or daughter gave birth, her number of living children under age five years, and other details. This file allowed a look at the relationship between the older women's fertility and their children's reproduction. This new grandmother file could be linked to the annual fertility files of the younger, interviewed women in two ways. First, by adding the new data on the young woman's mother (MGM); second, by adding the new data on the young woman's current husband's mother (PGM). I could thus relate a woman's fertility to her mother's fertility and childcare burden, or to her mother-in-law's fertility and childcare burden. The new grandmother file was also linked to the child annual mortality hazard file in the same ways. With these, I could look at grandchildren's survival in relation to the fertility of each of their grandmothers, and the number of her own children under age five years whom the grandmother was caring for in each year.

In the sample, there were 1767 woman–years of data on 166 women aged over 35 and less than 50. Eighty-nine of them gave birth during this period, bearing 164 children. Fifty daughters gave birth during this period, to 157 children. Thirty-three sons had 79 births (tables at SI 19.4). There were just 17 years in which mother and daughter both gave birth, and four years when mother and daughter-in-law both gave birth. A useful measure ("anytocome") was whether the older woman had yet to bear her final recorded baby; she was recorded as "still fertile" in each year until the year in which her final recorded baby was born.

During the period of overlap, the number of under-five-year-olds by daughters passes two in 46 woman–years (2.6%). It passes one in 138 years (7.8%), and is greater than zero in 337 (19%). If we include the older woman's own under-five-year-olds, then in the average year there were 0.94 under-five-year-olds per grandmother, most of them having their own mother in addition to their grandmother. It may be significant that this period of overlap, the time when the older woman may be bearing and rearing her own small children while her grown children are beginning to bear the grandchildren, is a period during which there are few small children to be cared for. During the period of overlap, the average 45-year-old woman has about half a young grandchild by her daughters and a third by her sons.

Figure 19.1 shows the occurrence of births in the new grandmother file. The X-axis is years from the birth of the foundress grandmother herself. The Y-axis shows the probability of a birth to the foundress (the eventual grandmother), to her daughters, and to her sons. The first thing we can see is that there is a roughly 15-year period during which the chance of a Hadza woman and one of her daughters giving birth in the same year is greater than zero. The equivalent period for mother and daughters-in-law (her sons' wives) at just over 10 years, is shorter because sons first marry at an older age than daughters. The probability of the mother and daughter bearing a child in the same year is very low although we have seen such instances. The probabilities are lower for the mother and daughter-in-law.

Figure 19.1 also shows that the majority of births to the younger generation occur during the foundress's 50s and 60s. This may be the time when the grandmother's effort is most sternly tested. In the last chapter, Figure 18.3 showed that from age 50, about 70% of women have at least one living grandchild in the

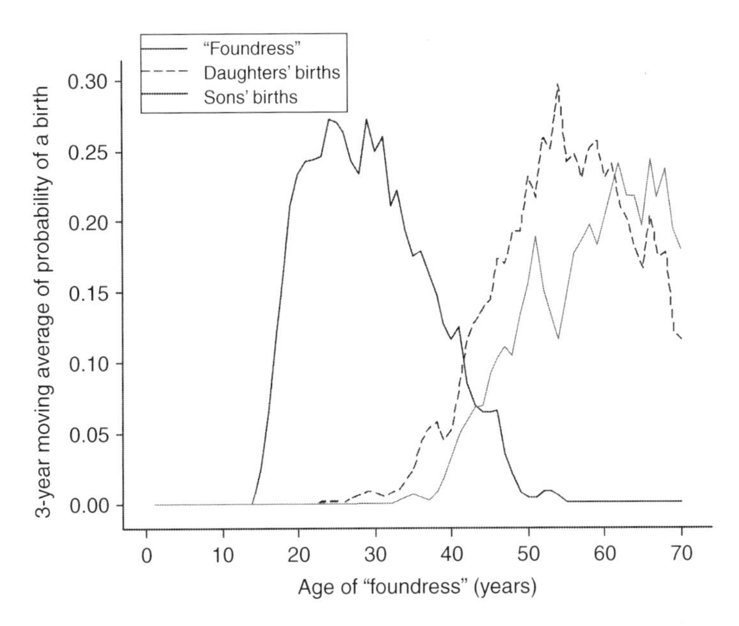


**Figure 19.1** Age at which women bear children and these children, in turn, bear grandchildren for the "foundress".

vulnerable under-five-year-old age group. The graph at SI 19.3 shows the number of children aged under five years. It peaks when the maternal grandmother (MGM) is between 55 and 65 years old, after the period in which the mother's and daughter's childbearing overlap. The average 60-year-old has more than one young grandchild by her daughters plus 0.8 by her sons. The 60-year-old has, in addition, a number of grandchildren aged more than five years. Thus, the number of under-five-year-old grandchildren that a woman has in any single year in this high mortality population is seldom large (Figure 18.3; SI 19.3). The number of older grandchildren is, of course, rather greater and their presence may be more important than I have allowed. Although they seldom die, they do not provide all their own food and their futures may be affected by grandmother help in ways I did not measure or analyze. They may be made to bear the brunt of any shortfall in food, being little affected by a shortfall that might put their youngest siblings at immediate risk.

Whether or not we choose to interpret these data as showing how efficiently the generations have become separated, they suggest that the period of overlap may not be a time during which the grandmother's work capacity is likely to be as severely stretched as it is later. If we expect competition to generate phenotypic adjustment of fertility, we must be assuming a significant short-term effect on each woman's energy budget and expecting this to affect her fertility, as a result of redistribution of the grandmother's resources. We may wonder whether the reduced foraging yield of the mothers with infants and the energetic demands of growing a fetus and of lactation, make the period of overlap more demanding than the later period of more numerous small grandchildren.

The data show that sample sizes during the period of overlap are small and it may be difficult to test for effects of competition convincingly. We should look liberally at the p-values of significance and pay attention to the confidence intervals for the odds ratios, which give a helpful impression of which way the results are trending. Additionally, because the samples are limited, I show here only the simplest analyses, attending to four variables: a child's death or survival, the occurrence of a birth, subsequent occurrence of a birth to the older woman, and the number of under-five-year-olds cared for ("burden5"). Other possible measures either give non-significant results, or tend to reflect the findings reported later.

### 19.1.2 Competition among the grandchildren for the grandmother's help?

The most basic use of the "census grandmother" file is to test whether there is any indication that the effect of grandmother help is "diluted" as it is spread across more grandchildren. We have proposed that grandmothers help by providing food. If this is so, their effectiveness should decrease as the food is divided among greater numbers of grandchildren. This would also be a basic requirement of any ideas about competition for the grandmother's help. Food kept for the grandmother's reproduction would not be available for the daughters' reproduction or grandchild nutrition. However, an effect of dilution of grandmother help has not been demonstrated with the analyses so far conducted, and clearly, it is a topic that requires more investigation than I can offer here.

I looked at deaths of children of the interviewed women, children who had a living MGM. The file included grandmothers of any age, thus covering their 50s and 60s when they have more grandchildren among whom to distribute their help. This file allows us to calculate the number of grandchildren of any age that the MGM has by daughters other than the subject child's mother. The result is a mean of 5.9 "cousins," ranging in age from zero to 20 years. The subject child is slightly more at risk the more "cousins" there are (b=0.065, p=0.006, OR=1.07, CI 1.02–1.12). We can also calculate the number of under-five-year-olds that these women have in each year of the subject child's life (mean 0.96, range 0 –4). This measure of the mother's sisters' "burden" of under-five-year-olds has a positive but far from significant relationship to the probability of child death. If we bias a new measure toward the vulnerable under-five-year-olds by adding the two measures, we also get a significant result (b=0.049, p=0.022, OR=1.05, CI 1.01–1.10). If there are more grandchildren, there is a higher chance of death of the subject child even when the MGM is alive. There is a cost to dilution of grandmother help.

### 19.1.3 Does fertility of one generation affect fertility of the other?

The older woman's fertility is declining during the period in which her daughters' or sons' fertility is increasing. If each data point is a year, a simple correlation would show a massive negative correlation between childbearing of one generation and the next. Thus, we have to concentrate on the small period of potential overlap (when the older woman is aged 35–50), and because fertility is so strongly related to age, control for the age of each woman. Regression models included the grandmother's age (linear because their fertility is declining in roughly linear fashion during this period), and age of younger women of childbearing age (cubic because that gives a good fit to the whole fertile period).

There were no negative effects of the mother's fertility on the daughter's fertility. Instead, the daughter had a higher probability of birth if her mother had yet to bear her final child (b=0.6170, p=0.042, OR=1.85, CI 1.02–3.36). A daughter-in-law had a non-significantly higher probability of a birth if her mother-in-law had more children aged under five years (b=0.4572, p=0.092, OR=1.58, CI 0.93–2.69).

When we look at the older woman's fertility, we find that she is more likely to have another birth after age 35 if a daughter has children aged less than five years (b=0.4057, p=0.009). She is also more likely to have another birth if she has already become a grandmother by her daughter (b=0.2658, p=0.079). There were no such results for a son. These results could arise if families differed in the length of their reproductive careers, the mother and daughter sharing the propensity for long or short careers. The results could also indicate some other synergy between the mother and daughter, as if their shared labor gave increasing returns to investment. In Chapter 20, we will see evidence for a positive effect of teenage children on their mother's late-career fertility.

### 19.1.4 Child survival and reproduction of the other generation

In this investigation, the age of the grandmother and the daughter is less important because child mortality has only weak relationships to the age of either. However, the age of the child must be controlled as there is a strong relationship to mortality that is best fit by age + $age^2$ + $age^3$.

A preliminary impression can be gained from the figures on the children born to the mother and the daughter in the same year. Thirteen different older women gave birth in the same year as a daughter at ages 36 to 53, four of them more than once. Four of the 17 grandchildren died. Five of the 17 children of the older woman died by the age of six years. At 9/34, 0.26 of those born dying by age six years, $l_6$=0.74 compared to $l_6$ estimated in Chapter 8 (SI 8.4 for sexes combined) at 0.63, hence, these death rates appear lower than those of the population as a whole.

**Survival of grandmother's children** There were no significant effects of daughter's or daughter-in-law's reproduction on the survival of the older woman's young children. The older women were older than 35 years, and I reported in the last chapter that women over 35 years have lower child mortality.

**Survival of the younger women's children** Children of young women were more at risk if their PGM has more small children of her own (b=0.7661, p=0.016, OR=2.15, CI 1.15–4.01), in the subset of the file in which PGM was alive. Thus, there may be some competition for PGM's productivity across a larger number children and grandchildren. Nonetheless, PGM appears to "win," and her daughter-in-law's children lose. If the file were restricted to records in which both grandmothers were alive, the result for PGM's burden was in the same direction but not significant (b=0.5836, p=0.130, OR=1.79, CI 0.84–3.82).

The MGM gives us the statistically most striking result, but not in the direction we would expect if there were competition between the generations. The more under-five-year-olds that the MGM has, the less likely her daughter's children are to die (b=–0.9487, p=0.006, OR=0.39, 0.20–0.77). This result, and the two previous results on fertility, not only suggest no competition but also a synergy between the mother and adult daughter. The surprising synergy between the mother and the daughter is discussed later. I now turn to data about another aspect of Cant and Johnstone's stimulating theory.

## 19.2 Do dispersal patterns and relatedness between group members shape the contest?

In most cooperative breeders, younger individuals delay or suppress breeding while helping a small number of older individuals breed. Costs of leaving the group are very high. By staying, the young individual has a better chance to become a breeder, or at least continue to help generate close kin. To Cant and Johnstone (2008), human menopause looked like an interesting reversal of this pattern; the older individuals stop breeding (stop competing) and help younger individuals breed. In a subsequent paper, they identified processes that might shape the age course of helping and competing.

Johnstone and Cant (2010) show that relatedness to group members can change with age, in directions that depend on dispersal and the mating system. They show that these age changes in relatedness can shape age changes in helping or competing. Two instances favor helping by the older individuals and breeding by the younger. One instance arises from female dispersal, and the other from low levels of dispersal with mating outside the group. They suggest the former as a match to ancestral human dispersal, and the latter as a match to the social organization of the other well-studied "menopausal" mammal, the killer whale (*Orcinus orca*). The two menopausal mammals can be accounted for by different dispersal and mating processes, which produce the same age-dependent balance of competition and helping.

Their argument is elegant (among the most elegant I can recall in a long career), clearly broad in scope, and draws attention to a neglected cost – the cost of competition with close kin. They can account for the absence of menopause in elephants, long-lived, with male dispersal; and probably in other male dispersal species, even some with demonstrated grandmother effects (such as the vervets described by Fairbanks and McGuire, 1986). That the care given by these grandmothers is not food and is likely to be non-depreciable, therefore less obviously competed for, may also be part of the reason these species continue to breed into extreme old age. We can try to match the Hadza kinship data to Johnstone and Cant's proposals. This seemed worthwhile despite two initial objections.

1. As previously mentioned, growing evidence contradicts the widely held belief that the ancestral human dispersal pattern was female dispersion, both on

ethnographic (Alvarez, 2004; Marlowe, 2004; Hill *et al.*, 2011) and genetic grounds (Wilkins and Marlowe, 2006). Female dispersal does seem abundant among herding and farming societies.

2. Among savanna hunter-gatherers, it is difficult to identify a social grouping equivalent to Johnstone and Cant's model "islands," a meerkat colony, or a whale pod, in which at least a recognizable core of individuals remain in the unit for much of their lifetime.

### 19.2.1 Hadza mobility and camp fluidity

I looked at camp compositions from census to census and was astonished at their variability. Contrary to my strong field impressions, there were few combinations of adult individuals seen together in more than one census. On average, over 70% of the people (over age 15) seen in camp with an adult woman, were not seen with her again in a later census. The number seen with her in several censuses was very small, and of course often comprise her husband, and her mother (or mother-in-law), or an adult daughter. Woodburn (1968b) remarked that an older woman who he had seen in most of his local censuses had been camped with 67 different people during the span of his visits. I found (SI 19.5) that among the 112 over-40-year-old Hadza woman in censuses, they had camped with an average of 62 different people over the space of a few years. As I implied earlier, costs of leaving a camp are very small. I was unable to persuade myself that a "group" such as the nuclear family, or the extended family, which has little spatial representation, would be a place to look for changes in relatedness with age.

### 19.2.2 Relatedness, age, and gender

Johnstone and Cant (2010, fig. 1) shows clear differences in the age course of relatedness of other individuals to breeding females. Of particular interest for the human case is the contrast between outcomes of female dispersal, and of male dispersal. When females disperse, female relatedness to other females in the group remains low at all ages. Male relatedness to females in the group increases with age, and at all ages is clearly higher than female relatedness to females. This is the scenario Cant and Johnstone (2008) link to evolution of menopause in humans. When males disperse, as in most mammals including most primates, female relatedness to females in the group is high and remains so for most of adult life, while male relatedness to females declines with age and is, at most ages, well below the female value.

I adapted the kinship program used in Chapter 6 to report the relatedness of each adult individual to each other adult individual in the same camp in each year of a census. I computed the mean relatedness between each category of person in their camp, or in the population (SI 19.6).

In every comparison of Hadza camp members, females were more related to females, including breeding-age females, than to males. Mean relatedness among camp members was higher than to members of other camps (Chapter 6).

Mean relatedness of females to female camp members over age 45 was higher, at 0.0658, than the relatedness of male camp members to the older females, at 0.0480 ($t=3.21$, $p=0.001$, $N=383$ females, 354 males). This female mean is 37% higher than the male mean. Some age changes in relatedness were statistically significant in regressions but all beta coefficients were trivially small, age accounting for at most 6% of the variance in relatedness. Furthermore, females under age 40 were more closely related to the females aged over 40 in their camp than to their female peers. This could suggest that young women are more likely to compete with each other than with older women.

The sex difference in relatedness was more compatible with the pattern derived by Johnstone and Cant from male dispersal than the pattern from female dispersal. In Chapter 6 we saw a very slight tendency for men, more than women, to change regions.

### 19.2.3 Another asymmetry between the generations – half-sisters

Can we offer any other evidence about Cant and Johnstone's argument that, in humans, the older generation could be expected to give way, other asymmetries that would favor the older woman conceding to the younger? Two are general to many larger long-lived mammals (Stearns, 1992): the low sensitivity of population growth to increases in fertility of older women (Chapter 12, and see Jones and Bliege Bird, 2014); the greater effect of age at the first birth than age at the last birth on a woman's RS (Chapter 13). A population, or sub-population of mutants, in which fertility of young women is increased by 1%, will grow much faster than a sub-population in which fertility of older women is increased by 1%. The Hadza population has been increasing rapidly, which tends to favor early reproduction over late reproduction, increasing the bias favoring reproduction by the younger generation. In Chapter 11, I argued that rapid increase may have been more frequent than stationary population among hunter-gatherers, with long-term stability only produced by randomly spaced crashes. This argument is no weaker than the argument for female dispersal.

A more interesting possibility, close to Cant and Johnstone's proposal, came from thinking about the effects of divorce and remarriage on the relatedness of Hadza women. A negative effect of sisters (discussed in the next chapter) appears to be entirely due to half-sisters. Full sisters made no difference when added to a model predicting a woman's number of living children from her age, births, and her mother's live children. The effect of including half-sisters in the model was negative and significant ($b=-0.5507$, $p=0.005$). Women with half-sisters accumulate living children more slowly than those with only full sisters.

Thus, among the Hadza, we can suggest another asymmetry: the rate of divorce and remarriage affects the relatedness between siblings. Women are bearing children for nearly 20 years but only 20% of Hadza marriages survive to 20 years (Chapter 15). A woman's family contains both full siblings and half-siblings. Although Hadza kinship terms make no distinction (Kirk Miller, personal communication, 2014), natural selection may. From the viewpoint of selection on the older woman, all her offspring are equally closely related to her and make equal contributions to her fitness (except for the timing issues discussed previously and in Jones and Bliege Bird 2014). However, from the point of view of an early child, the mother having her first birth at age 19 (close to the average for the Hadza, !Kung and Ache foragers), if her mother bears another child there is a high probability that it will be a half-sibling, not a full sibling. The inclusive fitness value of the mother's new baby to the maturing daughter is less than we might have expected and less than the daughter's relatedness to her own baby. In the same way as Cant and Johnstone argue for the asymmetry of PGM versus daughter-in-law contests, perhaps we can suggest that the frequency of remarriage produces an asymmetry of mother–daughter contests over resources to support their current reproduction. It also alters the symmetry of mother–son contests for the same reason. Son's mother's new children, as he begins to father his first child when he is about 25 years old, are unlikely to be his full siblings. There is a further complication for his mother (the PGM) because of the short duration of marriages. Her grandchildren may be borne by a variety of mothers. Thus, these grandchildren have a strong chance of being half-siblings to each other. This should not alter their inclusive fitness value to their PGM except in so far as it might accompany greater paternity uncertainty than in a more monogamous setting. In principle, I should be able to test whether there are negative effects between the mother's and the daughter's reproduction when the mother has remarried and her new children will be half-siblings to her newly matured daughter. However, subdividing samples is dangerous; we will never run out of questions, new data extraction would be required, and this book should not be further delayed.

## 19.3 Discussion of reproductive competition among Hadza women

We found some support for the view that because the grandmother's help takes the form of providing food, a divisible and depreciable resource, there is some dilution of the effectiveness of her help as the number of grandchildren increases. We should then expect to observe effects of reproductive competition between generations, yet we do not see a complete pattern of negative effects of each generation on the other.

While the relationship between child survival and PGM reproduction suggested competition, it appeared that the younger generation bore the cost. The relationship between the mother's and the daughter's reproduction appeared to be synergy more than competition. This observation conflicts with a starting assumption of grandmother hypotheses: that a post-childbearing woman, unencumbered by her own pregnancy or infants, can better acquire surplus resources that she directs toward younger kin. Grandmothers are thought able to help because they no longer direct resources to their own reproduction. We seem to have found, however, that even if the older woman bears another late baby, or still has one or two under-five-year-olds of her own to feed, there is no negative consequence for her daughter or her children. The strongest result was the unexpected positive association between survival of the daughter's children, and her mother's "burden" of living under-five-year-olds of her own. We can discuss several conceivable reasons for this surprising result.

### 19.3.1 Limited sample sizes, errors in age estimations

The sample sizes (shown in the tables in SI 19.4) are moderately large by the standards of some field studies but small when compared to studies of historical regional or national populations. The extended "population register sample" of older women ("census grandmothers") used here probably includes weaker information than was obtained for the interviewed women. For example, children of the oldest women may have been less likely to be recorded in the population register if they died as infants.

Ages are estimated (Chapter 4). The extended sample used in this chapter included many who were not intensively interviewed about information relevant to age estimation. These "census grandmothers'" ages had been estimated by how other people had reported them relative to their own age, and by the presence or absence in previous censuses and other criteria described in Chapter 4. Errors in estimated age of the mother or daughter could have exaggerated the period during which both women are raising children; for instance, if the census grandmothers were really younger than estimated while their daughters' ages were correctly estimated. In Chapter 7, the average age at a birth was calculated to be 28.75 for women with Hadza husbands. If the older woman was this age when she bore her children, and her children also bore children at this age, then the average older woman should see grandchildren born on average at twice the mean age, when she is 57.5 years old. Births to a son or daughter were predicted from grandmother age (a cubic relationship was the best fit) and the predicted values plotted against grandmother age. The peak is at 57 and 58 years old, exactly as expected. This gives some support for the accuracy of the age estimates of these "census" women.

### 19.3.2 A privileged cohort?

The women in their 40s were members of a scarce cohort, the dip in the age structure discussed in Chapter 10. Thus, they may have been a specially privileged cohort, with unusually few age mates (and unusually little competition or cooperation from them), and with an unusually high ratio of older helpers. They may also be survivors of a bottleneck – more robust than those who died. If the dip was caused by women leaving Hadza country via the Yaeda settlement in the 1960s, the women who stayed may have been more capable in the bush than those who had chosen to leave Hadza country, and therefore were more capable than average.

### 19.3.3 Fertile women have fertile daughters: shared "frailty"

Perhaps the result is because of some families benefiting from exceptional ability or exceptional shared circumstances.

1. Given women's modest adherence to the region of their birth, generations tend to inhabit the same more, or less, productive locations. I found no effects of the region on RS. Although abundance of game varies greatly between regions, the main plant foods are readily available in each (Chapter 2).

2. Mother and daughter are exposed to the same annual variation in rainfall or other facets of environmental wealth such as more fruit, or more game animals caught by men. The estimates of birth dates do not allow us to test for seasonal effects.
3. In the last chapter, multilevel regression showed that there were successful and unsuccessful grandmothers. In Chapter 7 we saw that there were successful and unsuccessful women. However, multilevel regression, "controlling" for grandmother identity, still showed the MGM "burden" to predict fewer deaths among her daughters' children ($b=-0.919$, $SE=0.311$, Wald statistic=8.73, $p<0.005$). In addition, when we looked at the younger generation's rate of producing live children, we found it is faster if their mothers had previously produced more live children, but slower if more of these were adult half-siblings (results given earlier). The data are compatible with the view that successful MGMs have successful daughters but that this does not explain away the "synergy" results reported previously.

### 19.3.4 Depreciable or non-depreciable care

In the last chapter, I showed that Hadza grandmothers enhance the growth and survival of their grandchildren and we have assumed that this was due to the food that they acquire. As foraging requires strength and energy, grandmothering might select for a longer vigorous adult lifespan. We do not see strong evidence of a disadvantage to the younger generation if the grandmother keeps enough food for her continued reproduction. Do grandmothers help in other ways, and would those select for an elongated vigorous lifespan?

Food is a simple example of depreciable care (Clutton-Brock, 1991); as each pot is divided among more individuals, each gets less. Other forms of care, like supervision and vigilance, are non-depreciable. I find it easier to think of the label "umbrella care" the analogy is that if the Hadza grandmother had an umbrella she could as easily shelter several children from the rain at no more cost than sheltering one. If grandmother care took this form, we would see little indication of competition for her resources, at least among those in the same camp. If grandmother care is non-depreciable, like the elephant matriarch which, we are told, may remember where water can be found in a drought, then her descendants would lose little from the grandmother's continued reproduction. In the next chapter, I discuss supervision as one way child helpers can be rewarding. While a 12-year-old girl takes all the toddlers to dig makalita, the mothers are spared the costs of dragging toddlers out to forage with them in addition to the sucklings they carry. It is likely that in some circumstances and some populations, grandmothers have their main effect through "baby-sitting." They may even be fed by the younger women in such cases, and perhaps indeed earn their keep by facilitating the young women's subsistence labor. Such help might select for longer life but not obviously for the strength and endurance displayed in foraging and food processing. Given the hours that we observe Hadza grandmothers to spend foraging and processing food (Hawkes *et al.*, 1997), it is hard to believe that they have their effects by baby-sitting and not by provisioning.

### 19.3.5 Late births only occur under favorable conditions: a full kit of helpers

Among 39 interviewed Hadza women aged 40 and over during the study period, 16 gave birth after the age of 40. Five gave birth more than once in their 40s, giving 22 births in all. For 20 of these births, the mother still had her mother or mother-in-law alive, while the remaining two had missing information on status of the newborn's grandmother. In addition, at 21 of the 22 births, the 40-year-old woman had at least one childless teenager. In the next chapter we will see that teenagers appear to have a strong positive effect on the fertility of their over-30-year-old mothers, just as Turke (1988) found on Ifaluk. Among the over-40s, women in longer-duration marriages had a higher probability of giving birth (b=0.062, p=0.015, OR=1.06, CI 1.01–1.12), a feature shared with those who gave birth in the same year as did their daughters. Long-lasting marriages imply more full siblings and fewer half-siblings in the family. Perhaps women only extend their childbearing into their 40s if they have a full complement of closely related helpers. Their birth rate in early quinquennia (20–24, 25–29 years of age) did not account for the probability of a birth in their 40s. Among the 77 women aged 40–55 who were measured, women who gave birth during this age were slightly (3 cm) but significantly taller and 1.5 kg heavier than other women of their age (SI 19.7). They showed no difference in upper arm circumference, triceps skinfold, body mass index, or rate of change in weight.

I had been inclined to view the few women who give birth in their 40s as the fortunate few who had resisted or avoided misfortunes of disease or the wear-and-tear of repeated childbearing. However, their full complement of helpers suggests we think about an influence of a favorable energy budget for these women. As we found no effect of grandmother or child helper presence on fertility of younger women (Chapters 18 and 20), this suggestion implies an increased sensitivity of the reproductive system of older women to their energy inputs and outputs. Thinking that an effect of energy balance on fertility might be the mechanism led me to look again at maternal depletion. As in Chapter 16, I could see no differences in the trajectories of weight or other measurements between the over-40s who bore a child and the over-40s who did not, although as reported earlier, those who gave birth were slightly taller and heavier.

On the time scale of evolution, older women may have suffered costs of competition with their younger kin, but in the contemporary Hadza situation, the costs of continued reproduction may be hidden by the apparently adaptive response of the elderly reproductive system. The elderly reproductive system allows a birth only when the costs will be minimal (whether from superior constitution or superior supply of helpers). The women who would suffer observable costs from another birth do not have another birth. Perhaps we are seeing a facultative response to reproductive competition. The observations should remind us of the difficulty of behavioral ecology without experimental manipulation. "Natural experiments" are sometimes not as "experimental" as we hoped. If the elderly Hadza reproductive system tends to perform in ways that minimize competition with younger kin, we should not expect direct evidence of competition such as increased risk to children of either generation.

Thus we could choose to regard the performance of the reproductive system as illustrating reproductive competition in action. Only by experimentally adding or removing children could we demonstrate effects of competition. This is an experiment we obviously will not do. Even the "natural experiment" comparing removal of a few children to boarding school in the 1990s is not entirely random. Parents resisted, and their final choices of who should go were far from random and the child's ability to help and the need for help were obvious criteria.

Thus, the absence of births without helpers could be viewed as supporting the reproductive competition theory. Generally, we view a theory that cannot be wrong as a bad theory. However, when we cannot manipulate the situation experimentally, we may sometimes be stuck with theories we cannot test. While some of us readily suggest that the behavior of agricultural peoples is less useful for exploring universal features of our species that probably evolved long ago (such as our life history), they may have their place. The "natural experiment" of the more constrained farmers' wives in their less flexible circumstances may be what we need in order to see the potential costs. As egalitarianism and sharing enabled IBI effects to be visible in the !Kung and Hadza and not obscured by wealth differentials, so sedentism may have allowed farmers to display the effects of dilution of the grandmother's help and the costliness of her own reproduction to her younger kin.

### 19.3.6 Mobility, flexibility, and the sharing ethic conceal competition

Despite possible weaknesses in the exact context, and the weak empirical support from the Hadza data, Cant and Johnstone's central logical point has to stand. Most studies and modeling efforts have failed to take into account the possible costs of an individual's reproduction to her kin. The difficulty of taking them into account arises, perhaps, from important differences between observed hunter-gatherers and true cooperative breeders.

Hadza cooperate in the sense that they share food and help each other but they cannot be classed as cooperative breeders *sensu Suricata*. In contrast to the classical examples, like meerkats, among the Hadza there are few costs for changing groups, and little evidence of competition between the generations. What would be the Hadza equivalent of a meerkat colony or whale pod? The patrilocal farmers' compounds discussed by Strassmann and Garrard (2011) are probably quite a good parallel to the closed "island populations" used as models by Johnstone and Cant (2010). A Hadza camp certainly is not. It is much too ephemeral. If one individual were imposing net detrimental costs on a second individual, the second would leave. Anthropologists have commonly described mobility as a means of resolving conflicts of interest. I cannot claim to have seen camp moves as conflict resolution in the way that Lee (1979, p. 397) has described among the !Kung and Chagnon and Bugos (1979) among the Yanomamo, but see no reason why it should not. Hadza are not immune to raising their voices and disputing noisily but such episodes usually seem to blow over rapidly. An accumulation of them, however, could indeed be a reason for moving. Mobility and changing camp composition may be one of the tools the Hadza (as well as other hunter-gatherers) have for lowering the costs of potential competition (also discussed in Blurton Jones, 2015). In addition, I have pointed out that Hadza grandmothers tend to be in a camp where they can best increase their fitness by help to grandchildren. This may not be possible when residence is less fluid than among the Hadza.

Hawkes *et al.* (1997) remark that all of the childbearing-aged women whose work schedules they recorded had a senior helper. Most were grandmothers, some MGM, others PGM. Others included a childless great-aunt or more remote kinswoman. It is likely that a young woman who needs the help of an older woman can find it from someone somewhere. There are also other potential helpers. While I have seen little evidence that husbands fill the domestic provisioner role that sociobiology has traditionally expected of them, Marlowe (2003) found that during 1995–1996 (his first field year), men whose wives had children under age three years brought more food to camp than did other men. Their wives brought home less than other categories of women, especially if they had a child aged under one year. He suggested that men were providing resources at a time when their wives and children were most vulnerable. I have speculated that men might also increase the frequency with which they bring small items of food to their house when the "trough" in the Hadza age structure results in a shortage of grandmothers. They might reduce the frequency with which they eat small prey in the bush, a venue of consumption reported by all observers from Woodburn (1968a) to Wood and Marlowe (2013).

All helpers may have some extra work capacity. A remarkable example is reported by Crittenden *et al.* (2013). Crittenden described the foraging output of a girl aged 10, whose father has long been physically handicapped by burn scars, and whose mother is regarded by researchers and Hadza alike as mentally slow. This girl brought home, on average, 1139 kcal/day but exceeded 10,000 kcal/day when figs were available, and shared the food with her younger siblings, parents, and occasionally her grandparents. Other girls of her age obtained about half her average amount. We do not know if this work is costing her impaired growth, or granting her the extra benefit of a reputation as an outstanding worker with well-fed siblings.

The sharing ethic, whatever theory we may hold about its origins (discussed in Blurton Jones, 2015), makes it likely that those suffering a shortfall will be able to make it up with donations from others for quite an extended period. This may be the largest factor masking the underlying competition between fitness-maximizing individuals.

### 19.3.7 Houses and differences between farmers and foragers

The published observations on grandmothers in agricultural societies are striking, and very surprising from my Hadza-biased viewpoint. Are hunter-gatherers different from the populations used in recent studies of grandmothers? The Hadza's high mortality and socially fluid lives might diffuse competition, and weaken the specificity of selection but we usually assume that over millennia of selection, very small differences will count. A bigger issue is houses. I mean this, and I am not alone in proposing important links between houses and sharing resources (Flannery, 1972, 2002). Houses and compounds influence, almost define, who knows what about whom, who shares what with whom, and who spends time with whom. A house is a place to bring your resources, to keep your personal goods, and a place to eat at least some of the food that arrives, and maybe all of it, without other people coming round and talking about why you did not give them enough. A warm-climate hunter-gatherer house is made of twigs and often is "see through" (Photograph 6.2), so is its open door. People are only indoors at night (or sometimes if they are sick). Everyone can see all the resources that come into camp (except a few small things carefully hidden that take a while to be noticed). Among hunter-gatherers, people do ask for food, and people do gossip about who did not give enough and who did not get enough.

The data on the weakness of reproductive competition among Hadza reinforce the points raised in Chapter 14 about the differences between the autonomy of Hadza women married to Hadza men, and the dependence of women married to farmers. The situation of savanna hunter-gatherer women differs radically from the situation of women in farmer societies. In the farming societies, by and large, men have taken control of access to most of the significant resources. Women may have no choice but to compete over the limited resources that come into their male-headed compound. Among the Hadza, or the !Kung, women's access to the staple foods is unrestricted. The limit is their own time and energy, and the proportion of the time and energy of other women (to whom resources are equally accessible) that they can enlist. In this situation women and their potential helpers can adjust their work to the level of potential competition. Competition may be ever present, or ever potentially present but there are many ways in which the forager lifestyle diffuses it and conceals it.

## 19.4 Models of the evolution of grandmothering: comments on the supplementary information

In SI 19.1, I report my attempt to replicate Fox *et al.* (2009, 2011) with Hadza data. When the sample of grandchildren is split by gender, results are not statistically significant but beta coefficients all indicate that grandmothers are beneficial both to daughters' children and to sons' children, even to their sons' sons. On the other hand, residence data, perhaps a good indication of investment, conforms to Fox's findings. If grandchildren live with the PGM, it is more often the granddaughter than the grandson.

In SI 19.2, I illustrate how plotting age in years shows the similarity in the fertile years of the human and chimpanzee. In contrast, if we plot reproduction by proportion of the lifespan, it looks as if humans cease reproduction exceptionally early, but longevity is left unaccounted for (see Cant and Johnstone, 2008, fig. 1). Comparing the fits of humans and killer whales (the other well-studied menopausal mammal) to those of Gosden and Telfer (1987), I suggest that killer whales may "stop early" while humans "live longer."

In SI 19.8, I show that daughters are producing grandchildren during the older woman's 50s and early 60s, declining thereafter. Because Hadza men marry later than women, and because some men start a new family in middle age, sons produce grandchildren later, at a rate that increases into the grandmother's early 70s, an age when we see surviving Hadza women becoming frail. I suggest that selection for longevity has acted via sons' children as well as daughter's children but that it has not acted to prolong life after the time when the daughters take over the grandmother's work themselves. There are very few great-grandmothers.

I discuss several of the models of evolution of grandmothering, their differences, and the shortcomings of their assumptions in SI 19.9. In my own simulations of the Hadza life history, I show the difficulty of evolving away from the current life history (SI 19.10). A graph (SI 19.10.1) shows how the effect of grandmother survival on population growth increases rapidly until women are in their 70s, but very little thereafter, suggesting (if I have correctly modeled selection via benefits to son's children) that in a Hadza-like population, selection for longevity would be weak beyond the age of 70. I also look at the trade-off between extended lifespan and higher early fertility on the assumption that resources must be allocated either to increasing early fertility, or to longevity. The aim was to show how large an increase in early fertility would outweigh the observed benefits of grandmother care. A 5% increase in fertility for every year of the childbearing period is worth a little more than extending vigorous, helping life from 40 to 45 years, and a little less than extending it to 50 years of age. The 5% increase in fertility entails a shortening of IBI, sufficient to make the population decrease. A trade-off between the pace of early childbearing and longer post-childbearing life may be a significant part of the puzzle of the evolution of grandmothering.

# 20 Children as helpers

In many societies, children take care of infants and supervise toddlers, while the mother works on crops, on food preparation, or supervises livestock (Weisner and Gallimore, 1977; Kramer, 2005a,b; Hrdy, 2009). In others, children are the main supervisors of the livestock, even at some cost to their growth (Sellen, 2000, p. 15). In a few, children do little more than play, but often in games related to subsistence and childcare (Briggs 1970; Draper and Cashdan, 1988). Sear and Mace (2008) report older children as effective helpers in five out of six "statistically adequate" studies. In much the same way as Hawkes *et al.* (1998) had suggested that the combination of the mother's and grandmother's productivity might allow the great fertility of humans compared to our nearest primate relatives, Kramer (2005a,b) and Kramer and Ellison (2010) point to the contributions of child helpers.

Child helpers originally attracted the attention of evolutionary-minded anthropologists because of an obvious analogy to avian helpers at the nest (Turke, 1988), and as a context in which to look at consequences of kin selection. Such analogies can be useful provocations but are seldom exact models. As well as paying attention to current ideas, each of us needs to think anew about likely costs and benefits in the setting in which we are working. The varied circumstances of different populations will shape different costs and benefits for the mother, for the child helper, and for the recipients of help.

My primary task in this chapter is to test whether Hadza children enhance the growth and survival of younger siblings, and the fertility of their mother. I set aside some obvious issues like help from cousins or uncles and aunts who are still children. Cousins may have their own young siblings, closer kin than our under-five-year-old subjects. The small and vulnerable children presumably gain from any help they can get, and Hrdy (2009, as well as Burkart *et al.*, 2009) and Hawkes (2014) draw attention to possible broad implications of selection on the ability of young children to enlist help from a variety of people. The mother also presumably gains from any help she can get, unless her fitness is reduced by her older children incurring serious costs from trying to help, as may be the case when child foraging is dangerous.

If we are to tackle our questions using data on variation between families, we must be concerned with intrinsic differences between the mothers. Does one observe more births to mothers whose first children survive because older children really are helpful, or because the most fertile and successful women have living teenagers, and continue childbearing into their later 30s? The mother whose first-born survives to be a helper will be different from the mother whose first-born does not. Testing for an effect of helpful siblings is much more difficult than testing for effects of grandmothers (who can be dead or alive due to factors independent of the mother or children), or fathers (who can be present or absent, although not always independent of qualities of their wives and children nor of the wish to stay together "for the good of the children"). Sibling helpers and helped are all products of the same mother, who has greater or lesser skills and opportunities, with more mouths to feed whenever she has more hands to feed them. Among the studies cited by Sear and Mace (2008), only their own Gambia study used multilevel statistics, which if set up the right way, can show and "control for" individual variation in fertility or hypothetical vigor.

Unless children are completely self-sufficient, the more older siblings there are who could help, the more there are for the mother to feed. Children of any age may always be a net economic drain, subsidized, as Kaplan *et al.* (2000) showed, by parents and older adults. If children, like the Hadza children, can acquire food that they consume for themselves, they make themselves "cheaper" to raise, but until self-sufficient, still require more resources from the mother than if they did not exist. Only if there are "economies of scale," ways in which older children's help can outweigh their cost, would we expect to see positive effects of child helpers. Perhaps only other forms of help, such as Kramer (2005b) discusses among her Mayan farmers, can allow additional children a positive effect on their mother's reproductive success. Help with non-depreciable care may be especially important.

## 20.1 What children can do to help

In the hunter-gatherer context, we are most likely to think about access to food. In many hunter-gatherer environments (arctic and desert), children are unable to obtain much food by their own activity, and everything they eat is provided by an adult. In others (Gould,1969; Bird and Bliege Bird, 2002, 2005; Tucker and Young, 2005), children collect tubers, shellfish, lizards. Draper argued that given the nature of !Kung subsistence, and the openness of everyone's lives (anyone can see whose child is heading for trouble), we should not expect effects of older siblings. The staple and very abundant food resources are located far from dry season water and there is little opportunity for children to forage successfully in the dry season (see also Blurton Jones *et al.*, 1994a). In the wet season, when people can camp near nut groves or *Grewia* stands, !Kung children do gather an as yet unmeasured amount of food. Hames and Draper (2004) showed no effects of early born girls on !Kung women's fertility or child survival; Hill and Hurtado (2009) found no effect of adult siblings, the only case in Sear and Mace's compilation with no sibling effect. In an elegant study of firewood collection in Malawian villagers and Tanzanian Maasai, Biran *et al.* (2004) showed that girls made a positive contribution to household labor by helping collect firewood, traveling some 2 km (Malawi) on trips lasting 241 mins, and 1 km in Simanjiro plain at 90 min/trip. Hadza girls also sometimes collect firewood but it is readily available on the edge of camp and little time is spent by girls or women in collecting firewood.

Our reports of successful food acquisition by Hadza children (Blurton Jones *et al.*, 1989, 1997; Hawkes *et al.*, 1995a) suggested that Hadza children might be capable of

**Photograph 20.1** By their own foraging, Hadza children make themselves "cheaper" to raise. However, do they produce enough to affect the growth or survival of their younger siblings? © James F. O'Connell, 2015. Reproduced with permission.

effective help as food providers (Photograph 20.1). My data on children's foraging mainly comprised return rates, how much a child collected per hour while targeting particular foods. These increase with age and strength, at different levels for different foods (Blurton Jones *et al.*, 1997, figs. 1, 2, 3). The minutes per day that children spent foraging also increased with age. It appeared to me that children around age 10 gathered about half their likely daily requirements. They still cannot get by without food contributions from their mothers or others. Children over age 12, however, seemed able to acquire as much or even more food than they required. This made me ask whether children of this age affect the survival of their younger siblings.

Crittenden (2009), Crittenden and Marlow (2008), and Crittenden *et al.* (2013) gathered similar data but also a larger sample of data on the amount of food brought into camp. As Hadza of all ages, and especially males, eat out of camp while foraging in the bush, Crittenden's data may be a much better indicator of which children are likely to be important for helping to feed the under-five-year-olds. The children who brought most into camp were in the age range of approximately 6–12 years, and they tended to share most with close kin. This provoked me into looking at data on full siblings separately from half-siblings, and on a younger age range of 6–12 years. Thus, I shall be reporting analyses on two sets of older siblings: (1) the 12–18-year-old "teenagers," (2) the 6–12-year-old siblings. The teenagers were any child, boy or girl, aged 12 and up to age 18, and who was not in school (boarding school). In the case of girls, I excluded the years before age 18 in which they had a baby of their own. Given the brevity of many Hadza marriages, it is not surprising that many of a

**Photograph 20.2** "All work and no play makes Jack a dull boy" goes the old saying. Hadza children are not obliged to work. They fill their days with many activities; here target practice, shooting at a log of soft wood in camp. © James F. O'Connell, 2015. Reproduced with permission.

woman's children are by different fathers, and I classified these as half-siblings. If the children are attributed to the same father, I classified them as full siblings.

Although much literature attends to girls as helpers, the Hadza make us look at boys too. Boys aged 6–12 forage for themselves as successfully as girls but tend to bring home less food (eating more of it out of camp). I have also seen teenage boys carry a toddler and tend him/her during the mother's departure on a foraging trip, or occupy themselves and entertain the few who are too small to carry a bow in target practice (Photograph 20.2). Toddlers were somehow never shot. The older boys seem to be very responsible about shooting, and I quite often saw boys wait, or direct a younger one to move somewhere safe. Boys also sometimes accompany women on their gathering trips (Photograph 20.3). These boys are referred to as guards. The qualification seems to be possession of a poisoned arrow, sometimes only one arrow! These boys range about, trying to shoot anything they see, but always arrive when the women sit down and light a fire to prepare their lunch in the bush.

Boys sometimes take younger siblings out of camp with them when they go about looking for sweat bee honey or shooting at birds. Girls and boys take small children with them to dig for makalita or collect baobab near camp. The vigilance of teenagers and older children may be useful as protection for small children, both from minor accidents (falls, burns, and scrapes, which can easily become infected), and to raise the alarm or help in hiding from visiting Datoga, who have been known to take children away to raise as herd boys. Girls frequently accompany their mothers when they go to collect water and sometimes collect water on their own; boys less often. Because mothers forage nearly every day, and grandmothers even more than mothers,

**Photograph 20.3** There are a variety of ways Hadza children could, and sometimes do, help; here heaving boulders while the mother pursues her underground puzzle in pursuit of tubers.


small children are left in camp without adult supervision for much of the time. There are usually one or two men at the men's place, working on arrows, talking, sleeping, who would respond to signs of severe peril but seem to provide no moment-to-moment supervision. It may be that the most important help that Hadza children give to their younger siblings is supervision during the several hours in which they are left in camp with neither mother nor grandmother.

Most of us who have looked at child helpers have assumed the benefits to the helper accrue through advantages to their close kin. There may be alternatives, including direct benefits to the helpers themselves. The child who leads a group of toddlers and younger children out to dig will be digging and eating at the same time. The costs are low. I noted (Blurton Jones *et al.*, 1997) that sometimes the teenagers suffered lower returns by accompanying adult women to dig tubers instead of foraging with the children. I suggested that by foraging, children might be gaining reputations as hard workers as much as they gain skills. They may, of course, simply find it best to comply with the mother's requests, although doing so while the mother is far out of sight for some hours seems a most un-Hadza-like abandonment of personal autonomy.

## 20.2 Availability of child helpers: are they there when they are needed?

Selection for helping will depend, according to commonly held ideas about helping, on whether there is anyone to help, how much a unit of help will increase their fitness, how closely related they are, how much help the helper can offer, and how costly it is to the helper (SI 20.1).

The under-five-year-olds are the most likely to die, and the survival of this age group has the biggest influence on population increase. It is likely that more care, either protection or food, will improve their chances. They are the most important targets of help for us to investigate. Population growth rate is also sensitive to the fertility of young women, so children's help to their mothers may be important if it gives them more siblings, and if it helps the mother keep more of the child's young siblings alive. Among the Hadza, about one-third of a child's siblings are half-siblings, because of frequent divorce and rapid remarriage.

There are some obvious limitations to the availability of child helpers. The harassed young mother of an infant and a three-year-old may have no older child helpers. The last-born child of an older mother will never have a new infant sibling to help with. The benefits to the mother, and to small children as recipients, accrue on a limited time schedule, and the benefits of being a helper may accrue only to a minority of children. From 1990 onward, a small number of Hadza children were taken away to boarding school. They usually stayed only a year and were not scored as available as a helper during their years in school (SI 20.2).

### 20.2.1 Who could the mother ask to help?

More child helpers will be available at some times in a woman's reproductive career than at other times. Figure 20.1 shows the number of under-five-year-olds, children

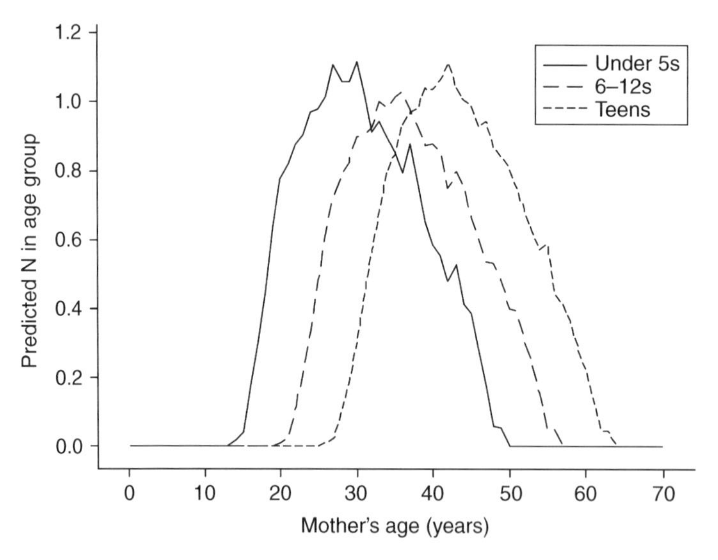


**Figure 20.1** Expected number of under-five-year-olds, children aged six to 12, and teenagers (aged 12–18) by mother's age, given observed average age-specific fertility and mortality. Many small children live with no teenage helper. Many teenagers live with no small siblings to help. Teenagers could be helpful to the mother only late in her childbearing career. The six- to 12-year-olds will be exposed to more opportunity to gain whatever benefits accrue from helping. N of living children of age y when the mother is aged x years was calculated as $ASF_{(x-y)} \times l_y$. Thus, N of living under-five-year-olds at the mother's age of 25 was ASF at (25-4) $\times$ $l_4$ plus ASF at (25-3) $\times$ $l_3$ plus ASF at (25-2) $\times$ $l_2$ plus ASF at (25-1) $\times$ $l_1$ plus ASF at 25 $\times$ 1. The number of living infants (aged zero to one year) is slightly overestimated.

aged 6–12, and teenagers (aged 12–18) that the average woman is expected to have at each point during her reproductive life.

In Figure 20.1, we can see that a woman aged <30 is unlikely to have any teenage helpers. About half of a woman's young children pass their early years before any siblings become teenagers (age 13). Many teenagers live during a time when the mother has ceased childbearing and their younger siblings have grown out of the early vulnerable years. Women in their 20s only have young children, and no teenagers who might help. Women from their mid-30s to mid-40s have a fairly even proportion of teenagers to youngsters. After this, all of their offspring will have become teenagers.

If younger children, anyone aged 6–12, for example, can be effective helpers, the potential advantage to the mother becomes more favorable. In Figure 20.1 we can see that many of a woman's under-five-year-olds will have a 6–12-year-old sibling available, although the youngest mothers will still be at a disadvantage. Clearly, it would pay the mother much earlier and for much more of her reproductive career if she can persuade her 6–12-year-olds to help, than if only her teenagers can help. The predicted and observed figures (taken from the women's annual fertility file) tell much the same story.

### 20.2.2 Who can the under-five-year-olds turn to for help?

The average under-five-year-old is more than twice as likely to have a sibling aged 6–12 than it is to have a teenage sibling. The young child had a sibling aged 6–12 in 47% of the years of the young child's record. It had a teenage sibling in only 21% of the record. About 30% of the older siblings are half-siblings (SI 20.3). Surprisingly, the probability that an under-five-year-old has either kind of sibling is lower than the probability that it has a living grandmother (88%, SI 20.4). High mortality shapes the number of child helpers as well as the number of grandmothers.

### 20.2.3 The child helper's view: is there anybody to help?

Figure 20.2 (data table at SI 20.5) shows the probability that a child has an under-five-year-old sibling that it could help. On average, in 76% of child–years of the 6–12-year-olds, there is a younger sibling under age five years. The average for the 12–18-year-olds is 64%. During the years seven to 10, the probability of having a young sibling aged <5 years is higher than 80%. For the older teenagers, the figure drops to about 50%. If an older sibling accrues benefits from helping siblings aged under five years (≦5), the opportunity for accruing these benefits is greater in middle childhood than later.

## 20.3 Child helpers and growth of younger siblings

Weight within the age group one to five years is linearly related to the child's age. The mother's weight (average of all the weighings of the mother as an adult,

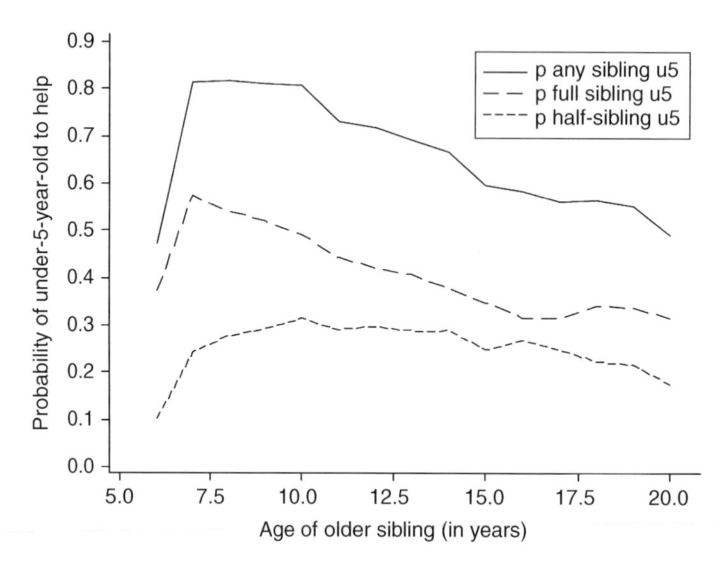


**Figure 20.2** Probability that a child helper of age x years has an under-five-year-old to help. The years seven to 10 present the greatest opportunity for a helpful child (Table of data in SI 20.5).

thus varying only between mothers) also makes a significant contribution to predicting the weight of the one- to five-year-olds ($b=0.0837$, $p<0.001$, $N=299$ child weighings); the father's weight does not. Thus, I controlled for these two measures in the regressions. Children of non-Hadza fathers tended to be bigger, hence I restricted the analyses to children of Hadza fathers.

Controlling for these variables, there were no consistent effects of older siblings on the growth of the under-five-year-olds in either single-level or multilevel regressions. Beta coefficients were positive but for most measures, far from statistically significant (SI 20.6). The data give little support for the idea that child helpers promote the growth of their younger siblings. This is a little surprising. Even though Hadza children do not acquire their entire daily nutritional requirements by their own efforts, one might suppose that the sacrifice of 300 g of baobab seed flour by a teenager, for example, would cost the teenager much less than it contributed to the diet of a toddler. Perhaps Hadza children are helpful in ways other than by being a forager.

## 20.4 Child helpers and child survival

I added the number of siblings of each sex, ages 6–12 or 12–18, available in each year of a child's life, to the child hazard file. I looked at survival of children aged zero to five years. The probability of child death changes with the mother's age at the birth of the child. It is high for mothers aged less than 20, stays level during the 20s and up to age 35, whereupon it declines. The women over 35 were less likely to lose their child than any other age group. In Chapter 18, we saw that the child survival effects of grandmothers were greatest on the younger women, and are negligible after age 35.

**Table 20.1** Child helpers and survival of the under-five-year-olds. Sample: women aged under 35 at birth of the under-five-year-old. Model for lines 1 to 4: child dead/alive = child age + child age$^2$ + mother age at birth + predictor. Model for line 5 includes preceding inter-birth interval, aimed to take account of the age gap between the younger and older sibling (I used an arbitrary score of five years for those who were first-borns). There were 185 deaths in 2318 child–years from 695 children of interviewed women. Similar results were obtained by multilevel regression

| Line | Predictor | b | p | OR | 95% CI |
|---|---|---|---|---|---|
| 1 | All 6–12-year-olds | -0.3817 | 0.010 | 0.68 | 0.51–0.91 |
| 2 | Any 6–12-year-olds | -0.6554 | 0.004 | 0.52 | 0.33–0.81 |
| 3 | Teenagers | 0.2359 | 0.284 | 1.27 | 0.82–1.95 |
| 4 | Any teenagers | 0.1535 | 0.625 | 1.17 | 0.63–2.16 |
| 5 | All 6–12-year-olds | -0.3823 | 0.011 | 0.68 | 0.51–0.91 |

When data on siblings were added to the model predicting child deaths, there was no significant effect in the whole sample. If the sample is restricted to women under age 35, there are some apparent effects of siblings aged 6–12 (Table 20.1). Children with one or more siblings aged between six and 12 were significantly more likely to survive.

In Chapters 17 and 18 we saw that inter-birth intervals and grandmothers had large and significant effects on child survival. When these are included in models to test the effect of child helpers, they dominate the picture. The effects of siblings aged 6–12 continue to be moderately strong but usually not significant, and with only modest effects on log-likelihoods (for example, number of sisters aged 6–12, b=-0.3509, p=0.160, OR=0.70 [0.43–1.15]); number of 6–12-year-olds, b=-0.3596, p=0.049, OR=0.70 (0.49–1.0). The data suggest that the older siblings have a beneficial effect (more siblings, lower probability of younger sibling death) but that it appears weak. When the sample is limited to the years when the mother's mother is dead, the effect of older siblings shows the same pattern but is not as strong as the paternal grandmother's effect. The data do not support the view that older siblings compensate for the loss of a grandmother.

Teenage siblings (aged 12–18) showed no effect among the young mothers (few of whom had a teenage child) or among the full sample of women, or among women aged 30 or more. We cannot claim that the child-rearing success of older mothers is a result of help from teenagers. In Chapter 19, however, I showed data that suggest the over-35-year-olds do not bear a child unless well equipped with helpers. Only those whose supply of helpers give them a high probability of keeping the new baby alive give birth at ages above 35.

## 20.5 Child helpers and mother's fertility

There appears to be a large positive effect of child helpers on women's fertility. I added the number of siblings in each age class in each year to the women's annual fertility hazard file. The first task was to determine whether the new baby would be the full sibling or a half-sibling of the older child. Siblings shared a mother but their relatedness to a new baby would also depend on the identity of their mother's current husband. I counted the number of older siblings whose father was the same as the woman's current husband. This means that the child would most likely be a full sibling of the next baby. They are also likely to be a full sibling of most of their mother's older children. Children whose father was a man to whom their mother is no longer married were also counted. They will probably not be full siblings of their mother's next child. They may be full siblings of some of their mother's younger children, or they may not be. However, we are looking for an effect on the mother's fertility and it may be most pertinent to know the chances of the child adding a full sibling to its collection of kin.

There was a positive association of presence and number of teenage children (aged 12–18) with greater fertility of their mother (probability of a birth in each year). Controlling for the woman's age (and age$^2$), the number of teenage children quite strongly predicts the probability of a birth. This is true both in the whole sample, and among women who ever had a birth (thus excluding the six women aged over 30 displaying "primary infertility"). The same result is obtained in the sub-sample of women aged over 30 (those old enough to have a teenage child if their early born children had survived) (SI 20.7). The graph in Figure 20.3 shows the rather striking result. The results for siblings aged 6–12 were similar, except that in

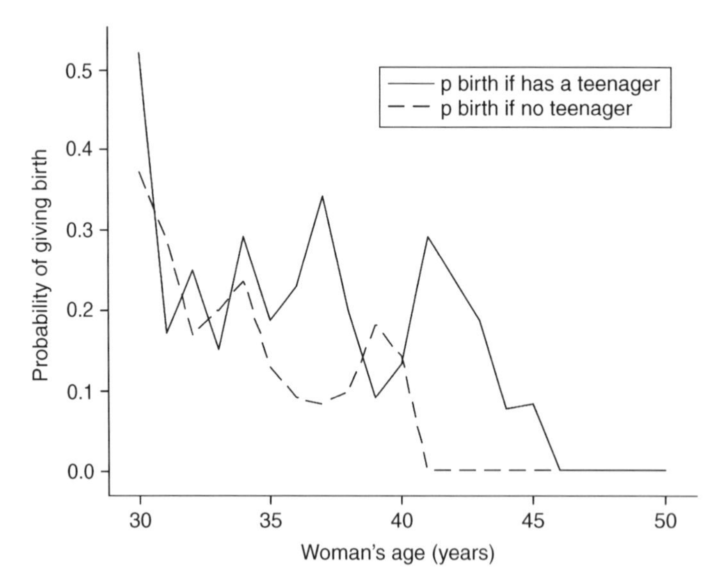


**Figure 20.3** Fertility of women aged 30 and upward. Comparing those who have at least one teenage child (solid line) and those who have none (dashed line). The six infertile women were excluded. The women's annual fertility file is examined to score the number of years with a birth or without a birth when there is at least one teenager, and the corresponding numbers when there is no teenager. The graph plots the proportion of years with a birth under each condition.

**Table 20.2** Results of multilevel logistic regression of probability of a birth, nested by mother. Siblings, especially teenage and full siblings of the new baby, are associated with a greater probability of a birth to a woman aged over 30. Model includes mother's age + age$^2$ + "brt25" (mother's birth rate at age 20–24.9). "Dad=hb" implies that the newborn is a full sibling of the helpers. Inclusion of brt25 accounts for much of the variation between mothers

| Line | Women ≧30 | b | SE | Wald | p |
|---|---|---|---|---|---|
| 1 | Teen siblings | 0.375 | 0.116 | 10.45 | <0.005 |
| 2 | Teen sibling Dad=hb | 0.482 | 0.122 | 15.61 | <0.001 |
| 3 | Siblings 6–12 | 0.184 | 0.098 | 3.53 | <0.100 |
| 4 | Siblings 6–12 Dad=hb | 0.206 | 0.094 | 4.80 | <0.05 |

the sample of all women-years, only the children of the current husband were significantly related to fertility (SI 20.7).

Before we accept these findings as evidence that child helpers actually increase their mother's fertility (or extend her reproductive career) we must explore alternatives. Strongest among these is that the presence of child helpers, especially teenagers, and the continuation of births in the woman's late 30s and 40s, are both results of an underlying greater fertility of the woman. I discussed this in the chapter on reproductive success (RS) (Chapter 13) and reported that birth rates early in the career predict birth rates later in the career, supporting the idea of individual variation in fertility. I tested one of the possibilities with logistic regression in Minitab. I added to the regression model each woman's birth rate during each quinquennium of her adult life before the age of 30. The apparent effect of teenagers was not greatly reduced by including these measures in the model. Adding other potential helpers to the model also has little effect on the contribution of the teenage measure.

Minitab gives us no way to control for the nested nature of these data, which is especially important when we suspect that we are really seeing an effect of differences between women and not an effect of current presence of teenage "helpers". Multilevel regression (Table 20.2), in which the annual hazard of fertility records are nested within women, should show us the effect of underlying differences between women on the contribution of teenagers. To my surprise, the random intercept model that allows for differences between women, shows an effect both of the number of 6–12-year-olds and teenagers after controlling for the mother's age (Table 20.2). Beta for teenagers was 0.375, SE 0.116, Wald test=10.45, $p<0.005$. When the sample was restricted to the 99 women aged over 30 (32 with no teenager, 67 with a teenager) the effect of teenagers remained significant (after controlling for the mother's age, age at the end of observation, previous birth rate ["brt25"], grandmother status, beta for teenagers was 0.361, SE=0.117, Wald test=9.52, $p<0.005>0.001$). In multilevel regression, the early birth rate measure greatly reduced the individual differences between mothers while leaving the effects of child helpers strongly significant.

I am left unable to dismiss the possibility that help from children somehow enables their mothers to bear and raise more children (further discussed later). These older mothers were surprisingly homogeneous as regards their fertility; the mother residuals given by Mlwin were surprisingly small. However, when I printed a list of the women who had no teenagers, a couple of impressions may be important. There are few such women (13 out of the 71 women over age 35; 32 out of the 99 women over age 30). The 13 aged over 35 seem to have had checkered careers. Four had been married to non-Hadza husbands. The over-30s with no teenager included the six apparently infertile women (excluded from Figure 20.3). One or two others had lived outside Hadza country. One had moved from region to region. They tended to have had shorter marriages. It was as if these women had had less control over their lives than the mothers of teenagers. One possibility is that these features add up to an increased opportunity to have acquired diseases of the reproductive tract that may have impaired their fertility later in life. By definition, those who bore no babies in their late 30s ended their careers earlier than the majority. It is also possible that these careers reflect the lack of close kin helpers whose effect on the older mother's energy budget I suggested in Chapter 19 was associated with bearing a child late in her career.

## 20.6 Mother's experience as a teenage helper

In Blurton Jones *et al.* (1997), I speculated on the reasons for Hadza children's hard work. Foremost, they get to eat when they want to, without waiting for the mother to come home at the end of her foraging excursion. Second, perhaps they are beginning to build reputations as hard or skillful workers. When the boys stop by, the girls dig harder. Third, perhaps the experience they gain at foraging, and at childcare, is useful in their adult lives. The last two were favored by the observation that on the days when girls went to dig tubers with the adult women, they foraged on "adult" foods but gained much lower calorie returns than on the days when they had stayed nearer camp with the younger children. They were taking the caloric loss perhaps to gain valuable experience of a part of adult life. But in general, Marlowe and my (2002) experiments on subsistence skills suggested these were learned rapidly, if somewhat dependent on developing sufficient strength, and did not support the view that the length of the human juvenile period had anything to do with learning to forage or hunt.

Perhaps a more useful aspect of children's time as a helper is their experience with smaller children. We can examine this because not all teenagers will experience babies and toddlers in the family. The teenager's birth order has an effect, and some mothers cease childbearing relatively early, depriving their early born children of small siblings with whom to gain experience as a "teenage helper."

The population register allowed me to assess whether each interviewed woman could have lived with small siblings during her teenage years. Did her mother bear children after this girl reached 12? This was scored 1 for "yes" and 0 for "no." To my surprise, this measure is a strong predictor of the survival of the woman's children.

Children whose mother had had the opportunity to be a "teenage helper," that is, she had been exposed to infant and toddler siblings after the age of 12, survived significantly better. Controlling for their age, the children of experienced mothers had a lower probability of dying (annual hazard file, b=–0.1486, p=0.008, OR=0.86 [0.77–0.96]). Because the score of mother experience does not vary and children are nested within mothers, I repeated the analysis with multilevel regression. The result remained significant. It appears that teenage girls acquire something useful to themselves from their "help" with younger siblings. Neither grandmother status nor the woman's age remove the effect of this experience.

## 20.7 Discussion: how can older siblings help? Effects on the older mothers?

There was a significant tendency for under-five-year-old children of young mothers (under age 35) to be more likely to survive if they had older siblings, especially siblings between the ages of six and 12. Multilevel regression showed that although individual mothers vary in their success at keeping children alive, this apparently does not explain away the effect of child helpers. The effect, like the grandmother effect, is mainly visible in the children of women younger than 35.

Hadza children acquire much of their own food, and share some of it with younger close kin. However, much of the time they still do not acquire enough for their own requirements, and their contribution to the feeding of younger siblings can only be quite limited. Their foraging makes them "cheaper" to raise than children who do not forage, but they are not without cost to the mother or other helpers. The food acquired by children could be useful to their younger siblings if, for example, 300 g of baobab seed paste gives a sufficiently greater effect on the growth or disease resistance of an under-five-year-old than its sacrifice costs the older child. Yet presence of siblings related inconsistently to growth measures. If siblings help, we cannot offer strong support to the idea that it is by helping their younger siblings grow and avoid the mortality risks of poor growth. That they seem to affect the young sibling's survival becomes surprising. What are they doing that helps?

We should consider the "labor saving" help that Kramer describes from her farmers, and look for help that could somehow outweigh the cost of feeding the helper. In farming societies, the child helpers enable the mother to work. Perhaps the older Hadza children's supervision of toddlers during the day does the same. Without it, the mothers might be inclined to take children of a greater age with them when they forage. Carrying them takes more of the mother's energy and can impair the mother's foraging (Marlowe, 2003, and Blurton Jones *et al.*, 1989, table 4, block 3 vs. block 4). Alternatively, grandmothers might stay home to tend the toddlers instead of foraging longer hours and more days than younger women. Supervising toddlers and the under-five-year-olds during the day may be effective and of the same cost whether there are few or many toddlers (up to a point – think of daycare staffing ratios!). In this case, when the child helper guards the toddlers, s/he uses non-depreciable "umbrella" care to facilitate the mother's depreciable "individual" care (Clutton-Brock, 1991, p. 8; Blurton Jones, 1993, p. 323). If Hadza children perform some form of "umbrella care," then as the mother's number of dependent children increases, the child helper's time "caretaking" becomes more valuable to the mother. If there are stronger positive effects of child helpers among some societies (farmers), it may be because there are more non-depreciable tasks to perform. Firewood collection may be an example, to judge from Biran *et al.* (2004, fig. 1), which shows decreasing per capita consumption with increasing household size. If the useful help given by older Hadza siblings is in "guarding" the toddlers, and this allows the mother and grandmother to forage more effectively and for longer hours, then the existence of one older sibling should be as effective as the existence of two, or even three. In response to this argument, I contrasted effects of having any 6–12-year-old with having none. This measure gives the largest and most significant associations with one- to five-year-olds' survival (Table 20.1, line 2).

If child helpers have their effect by "baby-sitting", allowing the mother to forage unencumbered by weanlings, and allowing grandmothers to forage rather than baby sit, we see a three-person "team." Hadza children's foraging then functions only weakly as "help," probably mainly just as a way to satisfy their own hunger. Nonetheless, their foraging near camp increases the rate at which food goes down the throats of a woman's children. Sometimes this outcome is maximized by another strategy. For example, among Hadza at Ilomo in fall 1988, a woman's children ate better if women and children took the long walk to the berry patches and stayed there long enough, than if each had stayed near camp and dug tubers (Hawkes *et al.*, 1995). In looking at the lesser amount of foraging by !Kung children (Blurton Jones *et al.*, 1994), we found that a concept of "team returns" (the combined calories per hour obtained by mother and child) helped make more sense of the observations. Among the !Kung in the dry season, the "team" of mother and children can eat better if older children stay home and process food (it can take many hours to crack one load of mongongo nuts).

"Team returns" perhaps point to one way of approaching aspects of the division of labor that Kramer (2005, and elsewhere) discusses. The term "division of labor" should be used only warily by the evolutionary minded. It implies a goal shared by a group, and there are few situations in which natural selection pulls all participants entirely in the same direction, although there are many situations in which the behavior that pays for one individual depends on the behavior that pays for other individuals. Our analyses of "team returns" relied on the mother and child both benefiting from the rate at which food became available to a child; but there may be many situations in which the mother's interests in her diversity of children do not line up with the interests of any single child. The study of children as helpers may be an example. Mating systems are undoubtedly another, more complex case. Supposed cases of "division of labor" may, as others have suggested, be examined better as possible instances of "mutualism," and examined for hidden "collective action" problems.

In the case of child helpers, the interests of the participants change as they age. The mother is probably most in need of helpers in the early years of raising her first two children, when she has fewest. An unmarried younger sister could be really helpful at this time. Children have most opportunity to influence their number of surviving siblings during their own middle childhood. The first or second child, if surviving, has the greatest chance to benefit from experience as a child helper. There is only a portion of the mother's child-rearing career when the situation is stable for any of the participants. Children, grandmothers, and fathers are not the only potential helpers of their youngest kin. In SI 20.8, the mother's and father's adult siblings are examined. Results are inconclusive but there is a tendency for the mother's brothers to have a positive effect relative to the mother's sisters, who we might regard as potential competitors.

Women with teenage children were more fertile than others, even after controlling for their age and individual fertility level, and earlier birth rates. This seems like strong evidence that teenage helpers promote women's fertility. Similarly striking is an observation about the births to the oldest women, those in their 40s, discussed in Chapter 19. All women who gave birth in their 40s had a living mother or mother-in-law at the time of the birth. They also had teenagers. In Chapter 19 I suggested that the otherwise unexplained success of the oldest women at keeping their children alive may be because they do not give birth to another child unless they are well-equipped with competent helpers. This would also help to account for the surprising finding reported in Chapter 19 that a woman and her daughter can give birth in the same year with no detriment to the infants. This "overlap" does not happen unless both are well furnished with helpers: mother, grandmother, and a teenager who was not the daughter giving birth (by definition, teenagers with their own baby were not included in the measure of teenage helpers). There may be interesting, even novel, implications of this observation for control of fertility in women when they are near the end of their childbearing career. Are the energy budgets of the older women more marginal? Are their reproductive systems more responsive to their energy budget? Even though multilevel regression controls for individual differences between women, my independent variables may have omitted some important features of women's lives. Again, the image of older women in control of their lives and younger women in difficulty comes to mind. Previously I reported listing the older women who had no teenage helpers and noted the variability in their lives. It was as if these women had been much less in control of their lives than the majority of older women. Yet again, these effects may not be unrelated to the helpers.

# 21 Husbands and fathers as helpers

> I never saw in this region such concerned mothers or such active family fathers as among the Wakindiga ...
>
> Obst, 1912, p. 25, p. 16 of translation by Gabrielle Kopahl

Do these active family fathers and husbands help women and children? By their hunting, Hadza men acquire a great deal of food, sometimes in dramatically large packages (Photograph 21.1). It all gets eaten (Photograph 21.2), usually with astonishing rapidity. Many people benefit from all this food (Photograph 21.3). Do husbands help the growth and survivorship of their children? Is that why men hunt? Is that why men hunt such large game? What about the small game that they take in modest amounts, and sometimes eat on their own out in the bush but sometimes bring back to camp (Woodburn, 1968a, p. 53; McDowell, 1981a; Marlowe, 2010; Wood and Marlowe, 2013, personal observation)? Marlowe has collected detailed observational data on direct childcare by Hadza men, and on men's foraging activity under various family circumstances. In this chapter, I will compare growth and survival of children with and without fathers, with step-fathers, with fathers often or rarely nominated as expert hunters, or with single mothers.

Fathers have attracted much attention in the evolutionary anthropology literature, usually being described as providers for wife and children. The fitness benefit men gain by using the food that they acquire to enhance the number and survival of their offspring has been held to be a major selective factor in the evolution of marriage, families, and other important features of our species. Provisioning has been suggested as functioning to gain access to a mate, to retain the mate, to increase her fertility, to increase the survival of her children. It is difficult, when several possible benefits are listed together, to test any one of them alone. Few tests were attempted until competing ideas came into the field. In their review, Sear and Mace (2008, pp. 5–8) remark "fathers frequently make no difference to child survival ... Even where fathers are important for child survival, it is not clear that the benefits they bring to children are the traditionally assumed benefits of provisioning and economic support ... indirect evidence that the importance of fathers lies at least partly in protecting children from other males comes from studies of the impact of the mother's divorce and remarriage." Among hunter-gatherers we see some quite strong evidence of an effect of Ache fathers on their children's survival (Hill and Hurtado, 1996), which was included in Sear and Mace's review, and an effect has also been claimed for the !Kung (Pennington and Harpending, 1988). These accompany reports of contrasts between step-fathers

**Photograph 21.1** Killing a large animal is a rare event for even the most expert hunter. A large animal yields a vast amount of meat. Many arrive at the kill site to carry meat home. © James F. O'Connell, 2015. Reproduced with permission.

and fathers, and of close involvement of fathers with their children, for instance among the Aka and Efe Pygmies (Hewlett, 1991b; Morelli and Tronick, 1992), and also among the Hadza (Marlowe, 1999, 2005).

In hunter-gatherer studies, the best-known alternative to the idea that men hunt as paternal investment has been proposed by Hawkes (1991), Hawkes *et al.* (1991), Bliege Bird *et al.* (2001), Bliege Bird and Bird (2008), and Smith and Bliege Bird (2000). It has become referred to as "show-off," or "costly signaling." I outlined its basis in Chapter 14. By pursuing large animals, which are shared widely, individual men become known as productive hunters and the source of extensive feasts. Because hunting is often difficult, energy consuming, and even dangerous, success may make a reliable signal of "quality," either phenotypic or heritable "good genes." The importance of the meat gained from their activity ensures that they are known. In this view, men's hunting need not bring advantages to their own children. It may give the men advantages in competing with other men. A part of the argument has been the tendency for men to pursue rare large game, at the apparent cost of neglecting small game, which data from the 1980s suggested might produce a more reliable daily income (Hawkes *et al.*, 1991).

**Photograph 21.2** Two boys carrying heavy loads of meat including intestines. Everything gets eaten, by many people, and very rapidly. © James F. O'Connell, 2015. Reproduced with permission.

**Photograph 21.3** Men, women, and children carry meat from a kill site, and we weighed the amount deposited at each house. We argue about the details of how hunting large animals promotes a man's reproductive success. We differ in emphasis on its effect on a man's wife and children, or on his relationships to other men and women. © James F. O'Connell, 2015. Reproduced with permission.

Alternative ideas in evolutionary anthropology about the origins of male care include protection from infanticide, and from predators, and that male care initially arose because it attracted and retained mates. For example, although step-fathers have been found to differ from fathers (Daly and Wilson, 1987, 1996; Marlowe, 1998, 1999, 2005), it has been suggested that the care that step-fathers give may best be categorized as "mating effort," maintaining access to the fertile mother of the step-children (Rohwer *et al.*, 1999; Lancaster and Kaplan, 2000). Recent literature (largely on forager–horticulturalists) usefully expands the perspectives on male care (Gurven and Hill, 2009; Gurven and von Rueden, 2010; von Rueden *et al.*, 2011; Winking and Gurven, 2011).

A wider array of alternatives can be found in the biological literature (Black, 1996; Gowaty, 1996; Reichard and Boesch, 2003; Kappeler and van Schaik, 2004; Kokko and Jennions, 2008; Alonzo, 2011), and are beginning to receive proper attention in anthropology (Borgerhoff Mulder, 2009). In an earlier paper (Blurton Jones *et al.*, 2000), our group added preliminary data on the Hadza and reports on the !Kung to an important comparison made by Hurtado and Hill (1992) between the Ache and Hiwi. Divorce rates did not associate with the "father effect," the cost of desertion to survival of children. They correlated better with a demographic variable "fertility units per male." Hurtado and Hill devised this measure as the number of women aged 15–40, multiplied by total fertility rate, and divided by the number of men aged 20–55. It reflects the number of conceptions for which men are competing and was intended as a measure of the potential benefits to a deserting male. We found that this was the best predictor of divorce rate in this tiny sample of four well-studied forager populations. The parenting/mating index (ratio of father effect to fertility units per male) did not predict divorce rate as it had for the Ache–Hiwi comparison. We suggested that male–male competition, especially when all men have lethal weapons at hand, might be a much more important influence on "monogamy" than had been acknowledged. Although, perhaps implicit in Hurtado and Hill's measure, we were unaware of the point Kokko and Jennions (2008) and others have made about costs of desertion. Under some conditions, the more males desert, the tougher the competition for new conceptions becomes. Desertion may be self-limiting. Kokko ultimately shows the importance of adult sex ratios for shaping the mating system, and suggests that adult sex ratios may sometimes be linked to mortality risks of adult life. Hadza men, who go out daily competing with lions for their prey, may be expected to suffer higher mortality than women (fig. SI 8.6). The implications of recent work, such as by Kokko, Gowaty, and others, will receive more attention when I try to round up the Hadza results in the next chapter. In this chapter, the emphasis is on assessing father effects in my larger final sample.

In the case of the Hadza, we see claims, from informants and ethnographers, that women need men to feed them (Bleek, 1930, notebooks), and may divorce a man who is an unsuccessful hunter or trader (Woodburn, 1968b). We hear from Hadza that the duty of men is to feed women, children, everybody. However, can we show effects of food acquired by men on mate retention, on women's health and fertility, and on children's growth or survival? For many, such distinctions are false. If mate retention is the main result, it might be thought to work only if there were some benefit that women obtained from men, from "being retained." There are other possibilities. For example, men may find it worthwhile not to interfere in a marriage if the husband is a successful hunter. Wrapped up in these issues is not only the reason for marriage but also the reasons for big game hunting. Humans differ from our closest relatives, and men differ from women, in the amount of hunting they do and in an apparent preference for hunting the largest animals. Why do they do this? The resources acquired clearly have some effect, but which parts of this effect most powerfully select for hunting, or for "being a couple," either now in the reproductive economic present, or previously during human pre-history?

We have seen several pieces of evidence about Hadza men's effect or lack of effect on their wives and children in the chapters on RS, marriage, and anthropometry, and I will summarize the evidence later. Here I want to report data on the presence of fathers and step-fathers and the growth and survival of their children and the fertility of their wives. If paternal investment is important, then we should see effects of the loss of the father on children's survival, and perhaps on their growth. Fathers should be more effective than step-fathers.

The general caution about helpers issued by Hill and Hurtado (1996) applies to fathers as well as other helpers. If helpers distribute their help to maximize their fitness, they may sometimes mask the effect of help. For instance, men might time their desertion to minimize its effect on the children's survival. Men might be welcomed quickly into a new marriage when their help as a step-father to small children is needed. If their departure removes effective help (potentially lowers the survival of their offspring), they must trade this off against the potential gains of competing for a new or additional mate. Changes in the effectiveness of their help, such as when children get older, should affect these trade-offs. As Sear and Mace (2008) remind us, if a father deserts, his contribution may be so rapidly replaced by other helpers, including a new husband, that the loss of his previous contribution leaves no trace. Although every subdivision of the sample or new independent variable added to a regression steals degrees of freedom, we can look at combinations of helpers.

## 21.1 What men can do to help their wives and children

Among the "filler questions" in the women's interviews was "whose job is it to: . . . ?" and another year, three separate questions, "what is the work of men/women/ children . . . ?" Answers to the first set were extremely stereotyped, so much so that I suspected some influence from settlement organizers or school teachers, rather than a direct reflection of the observable separateness of Hadza men's and women's lives. Virtually all women said that building the house, fetching water and firewood, and looking after children (including disciplining them) was the work of women and that men "hunt and rest." One or two (no more) women suggested that men could help with some of this work. In contrast to a woman who remarked that she was still waiting to see a man who got water, in one camp a woman said that men sometimes helped with this task because water was far away. Earlier during our visit I had seen men at this camp carrying water. Their pragmatism matches that of the men in Chavez *et al.*'s (1974) study in which rural Mexican fathers of children in the food supplemented group increased their interaction with their children partly in response to the child's greater activity.

As usual in anthropological interviews, the free form additions and comments by the subjects were the most interesting (to attempt to use rigorous questionnaires in an anthropological setting would undoubtedly waste some of the most creative opportunities our field can offer). Of their own accord, women confirmed anew much of what Woodburn has written about Hadza men and women. Husbands hunt and trade. If their hunting fails, they just keep trying. However, after consistent failure, a wife could look for someone new. A man's job includes carrying meat from a kill by another man. The food women contribute is "little food," meat is "real food." Only one women added that her husband would protect her if there were fierce animals around. A rare example heads Chapter 22. While grandmothers were described as digging, or by one woman as baby-sitting, grandfathers were described as sitting and growing old or getting baobab for the toddlers. Boys go after kanoa (sweat bee honey), girls help the mother. One woman said her husband could help look after the children, but if he did, she would need to get back early from gathering. Others said childcare was not men's work, unless it was to carry a sick child to hospital. Despite this, a young man who suffered from asthma and did not hunt was often to be seen carrying one of his younger children around camp. Marlowe (2003, p. 227) comments on this man and another who had broken his arm and stayed in camp with his children every day. The interviewed women may have exaggerated the gender roles but they are not out of the range of my day-to-day impressions. Women get food, water, and firewood every day. Men hunt, rest, and fiddle with their arrows.

Hadza men could do all the things that other helpers do for children, offer food, protection, instruction. After a day out with his wife, one man brought his haul of tubers to be weighed alongside those of his wife and the other women. His load dwarfed those of his wife and even the most vigorous grandmothers. During Marlowe's and my foraging experiments, men were able to excavate tubers faster than women (Blurton Jones and Marlowe, 2002). Men do sometimes bring home food, offer food directly to children, hold infants and toddlers, and allow children to watch while they make arrows, prepare skins, etc. Men do sometimes respond to screams or cries for help, greet strangers ahead of women or children, and travel about and return with trade goods. They do sometimes hold and interact with their small children but they do not spend much time on these pursuits (Marlowe, 1999, 2010, table 8.4, fig. 8.5). They could, but seldom do, perform other tasks that women and older children do, like collect water or firewood, or build a house. Any of these activities might increase the survival or growth of their children or their step-children, or the fertility of their wives. Nonetheless, as with all helpers, other avenues toward fitness may compete for their energy or time. While grandmothers have few other avenues to promote their own fitness, men have a striking alternative:

pursue more conceptions, either by acquiring a second wife, or by extra-marital affairs, and/or by switching from a less fertile or successful wife to a more fertile or successful mate. There is also a fitness gain to a man who can reduce the number of his wife's affairs. Opportunities for all of these activities may vary with circumstances. They will differ between locations, subsistence modes, mortality rates, and social pressures, between individuals, between seasons, and between populations.

## 21.2 Measures of father absence

When I write "fathers," I refer to the supposed genetic father of a child. Fathers were identified by the women. The interviewed women seemed quite ready to name men other than their current husband as fathers of some of their children. When I write "step-father," I refer to a man married to the child's mother but reported by her or others as not the genetic father of the child.

The father's status (alive=1, or dead=2, "dadstat") was extracted from the population register and filed along with the mother's status, and the grandmothers' status. Other measures were extracted from the mother's marriage history file. If the father is still married to the mother in any particular year of the child's life (such as when it was measured, died, or merely survived), the measure "dadin" scored 1 for that year. If he had divorced or died, "dadin" scored 0. Similar scores were derived for step-fathers ("stepin"=0/1), and for the mother alone (0=not alone, 1=alone) to show years in which she was not married. Because of the suspicion that marriages sometimes dissolved quickly after the death of a child, measures of marital status in the following year were also entered in each year of the record. They are labeled "dadinext," stepinext," and "momalonenext."

## 21.3 Availability of fathers

Despite the high divorce rate, 72% of children of Hadza husbands of interviewed women live with the father still married to the mother at age 5, 20% with a step-father, and 7.5% with a single mother. Forty-nine percent of children who live to age 15 have their father still living with their mother (Figure 21.1).

A much smaller number of children lose their fathers to death. Sixty-six of the fathers of the 695 children of interviewed women had died by the end of observation. The children's annual event file contains the information on years of a child's life in which its father is alive or dead. If their father is a Hadza, 8% of children have lost their father by age 5, 15% by age 10, 21% by age 15, and 24% by age 20 (data table in SI 21.1).

After a divorce, the man usually ends up living in a different camp from his ex-wife and children (Figure 21.1). For the under-five-year-olds, a divorced father was in the same camp as the children in 22% of the censuses that contained the locations of both the father and child. In the remaining 78%, he was in a different camp. For children aged five to nine years, their divorced father was still in camp in 20% of censuses, and for 10–14-year-olds, this figure was 16%. The children will get

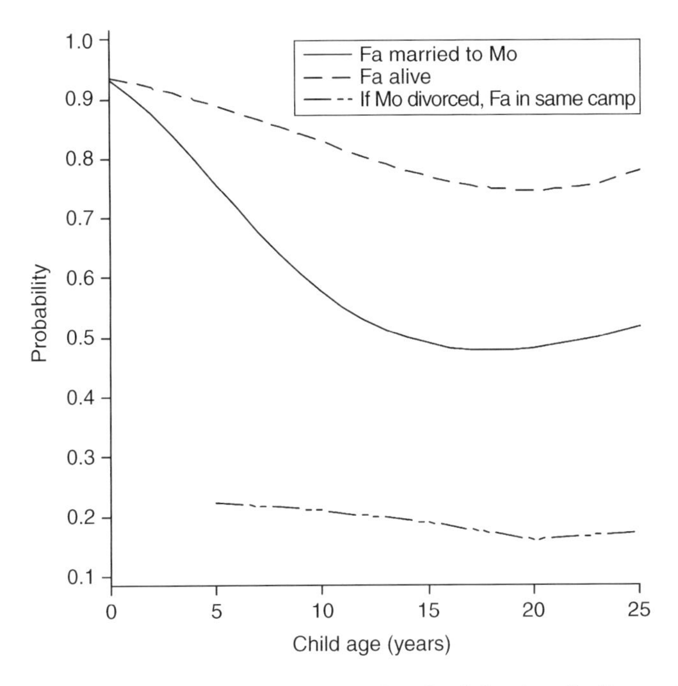


**Figure 21.1** Fit lines for probability that the father is still alive, and probability that the father is still married to the child's mother by age of child. Logistic regressions with child age + age$^2$ + age$^3$. Shown also, the probability estimated for each five-year age group, that if the parents are divorced, the father is still in the same camp as the mother and child (table SI 21.2).

the benefit of any large game their divorced father catches while he is in their camp, just like every other man's children, but they are unlikely to get whatever else had formerly accompanied his being married to their mother.

## 21.4 Fathers and child growth

The files that I used to look at effects of grandmothers and teenage siblings on growth also included the measures dadin, stepin, momalone. Among the children aged one to five years, no significant effects of the father's presence, or reputations upon growth were found that survived multilevel analysis, even when there was no maternal grandmother (SI 21.2 tables). The mother's current spouse's reputations had no relationship to child weight or other measures. This was true of unilevel regression analyses using Minitab, and multilevel analyses using Mlwin. Marlowe (2010, table 8.5) found that men with infants under age three years brought home more food, while their wives brought less. I found no relationship between weight and father's presence among the under-three-year-olds (controlled for nlog age to fit infant weight).

In contrast, among children in the age group 5–16, the presence of the father had a significant effect. The effect is stronger and more robust in the age range restricted to

**Table 21.1** Father presence and growth of older children, aged 5–12 (≧5 to <13). Measurements predicted by father presence "dadin" in three-level regression (father > child > observation). Model: measurement = age when measured + age$^2$ + mother's age at child's birth + N times in school + father presence or absence ("dadin"). Ns: 485 height measurements of 240 children of 122 fathers

| | b for father presence | SE | Wald | p | Ns |
|---|---|---|---|---|---|
| Height | 1.461 | 0.645 | 5.13 | <0.025 | 122/240/485 |
| Weight | 0.767 | 0.309 | 6.16 | <0.025 | 123/245/499 |
| UAC | 0.324 | 0.140 | 5.36 | <0.025 | 123/244/492 |
| Triceps | -0.046 | 0.216 | | n.s. | 108/215/431 |
| BMI | 0.076 | 0.135 | | n.s. | 122/240/485 |

5 to <13; that is, the period before girls (and some boys) begin the dramatic changes of teenage years. After appropriate control for age, (in Chapter 16 I reported that the best fit was the slightly upward curve given by age + age$^2$), mother's age at the child's birth, and school attendance (school attenders grow faster), the variable "dadin" (the biological father is currently married to the mother) has a significant positive relationship to the child's weight. With different numbers of measurements of each child, and different numbers of children of each woman and man, multilevel regression is essential. The "dadin" result for weight survives multilevel analysis (b=0.767, SE=0.309, Wald statistic=0 6.16, p=<0.025). In multilevel regression, dadin also had a significant positive relation to child height and upper arm circumference (UAC), but not to triceps skinfold nor body mass index (BMI) (Table 21.1; and SI 21.3).

Multilevel regressions can display the amount of variance due to variation between individual mothers or fathers (when the run is nested within fathers). Fathers are responsible for very little of the variation between weight measurements; mothers are responsible for rather more. After age at measurement has been added to the model, mother variance remains significant while father variance does not. This implies that there may be meaningful and interesting differences among mothers in their ability to raise larger children while an intrinsic property of fathers seems less relevant to child size. Nonetheless, the "dadin" finding remains quite robust, implying that the presence of the father is more important than his identity.

In this age group, if dadin=1, then the child's likely biological father has stayed with the mother until the age of weighing (more than five years), and these will be children of a marriage of near average or greater stability. Within this sample, children who were measured between the ages of five and 12.99, the length of the mother's current marriage had no significant effect when added to the regression models. The current marriage may not be the marriage to the child's father.

In this age group, the reputations of the father or step-father do not predict higher weight. The father's hunting reputation is negatively associated with child weight (b=−0.0348, p.04); for reputations of the mother's current spouse, the result is also negative but not significant. "Stepin" (the mother is married to a man who is not claimed as the biological father of the measured child) is negatively associated with child weight but with only marginal significance (SI 21.3).

## 21.5 Fathers and child survival

As in previous chapters, I used logistic regression to predict death or survival in each year of each of the 695 children's records. This allowed control for child age, mother's age at the birth of the child, and tests of additional independent variables such as father dead or alive, present or absent. Because I had the impression that men sometimes left in response to loss of a child, I added data fields that showed whether the father or step-father was present or absent in the year following the current year.

I compared the probability of a child's death in the years when the father was alive and still in the marriage (dadin=1) with the probability of death in the years when the father was no longer in the marriage (Table 21.2). It is important to control for the child's age as well as the mother's age. As a basic model, I entered age + $age^2$ + $age^3$ and mother's age at the birth of the child before the variables of interest. Logistic regression predicting child deaths shows a non-significant beneficial effect of the father's presence (lower probability of child death when the father was present) (Table 21.2). Multilevel logistic regression did not change the picture. Nor did excluding years when a step-father was present.

I wondered if limiting the sample to children whose maternal grandmother had died would reveal a stronger positive influence of the father's presence. The panel shown at SI 21.4 suggests that the absence of the mother's mother did not enhance the father effects.

An alternative approach used a Visual Basic program to build three Hadza child life tables, one using years of life when the father was present, another when the step-father was present, and the third for years when the mother was living with no husband. Resampling was used to test for differences. Survival was significantly lower with a step-father than with a probable biological father, but neither was significantly different from survival when the mother was alone (SI 21.5).

**Table 21.2** Probability of child death predicted by measures of father presence and absence. Sample comprises children of Hadza fathers. Because their subsistence is different, the children of the seven wage earners were excluded. Similar results are obtained for alternative samples. Logistic regressions controlling for age, $age^2$, $age^3$, mother's age at the birth of the child; all significant. Surprisingly, children of good hunters were more likely to die. 1089 child–years with father absent, 2780 with father present

| | b | p | OR | 95% CI | Deaths/child-years |
|---|---|---|---|---|---|
| Dadin | -0.2450 | 0.330 | 0.78 | 0.48–1.28 | 191/3869 |
| Stepin | 0.2355 | 0.399 | 1.27 | 0.73–2.19 | 166/3767 |
| Mom alone | 0.3223 | 0.468 | 1.38 | 0.58–3.3 | 191/3869 |
| Log hunt | 0.1266 | 0.009 | 1.13 | 1.03–1.25 | 191/3869 |
| Hunt >5 | 0.4276 | 0.009 | 1.53 | 1.11–2.11 | 191/3869 |

## 21.6 Father nominations and the "poor hunter" majority

In previous chapters we saw that Hadza men who become known as successful hunters of big game had greater RS. They remarried faster, and tended to get younger wives. Earlier we also saw no significant improvement in young children's nutritional status related to the father's nominations.

In Table 21.2 we can see that the fathers' nominations as an expert hunter predicted a greater probability of the death of their children, even though highly nominated men were more often still married to the child's mother. The children of expert hunters were more likely to die. Pro-rated hunter nominations were added to the logistic regression to predict child deaths. The beta coefficient for log-transformed hunting nominations was b=0.1266, p=0.009, OR=1.13 (1.03–1.25). If we use the categorization of men as having more than five nominations as an expert hunter, or fewer than five, we get the same significant effect of the father on child survival. Children of a good hunter are significantly more likely to die (Table 21.2 and Figure 21.2), even if their father was still married to their mother. If we use the far from normally distributed untransformed score, the result is b=0.0092, p=0.188, OR=1.01 (1.0–1.02). Whichever measure we attend to, these results give no support for the view that good hunters promote their children's survival. Results are the same if we restrict the sample to mothers aged under 35.

Contrary to the suggestion by Marlowe (2003, 2010, p. 150), which he based on the observation that men foraged more when their wife had an infant, in my data the detrimental effect of good hunters appeared to be strong on infants. Higher reputation as a hunter was associated with more deaths, log-hunt b=0.1793, p=0.005,

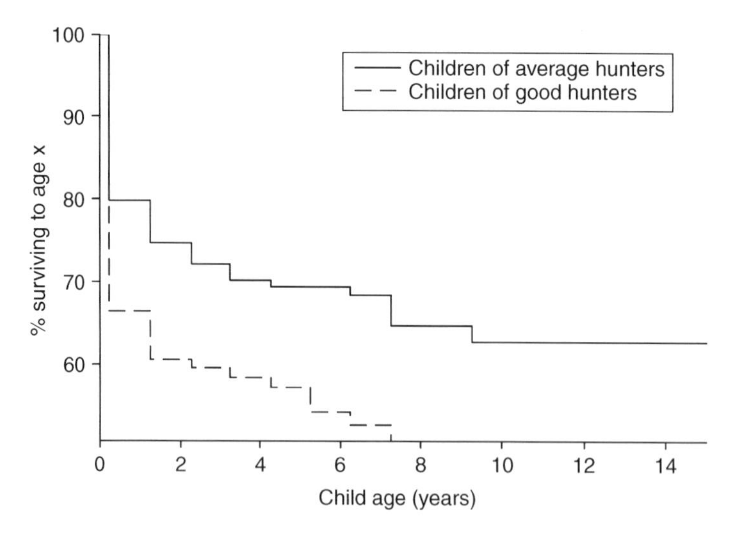


**Figure 21.2** Young children of expert hunters are at risk. Kaplan–Meier survival plot for child lifespan by father's hunting nominations. Sample restricted to children born between 1985 and 2000. 133 children of good hunters (pro-rated nominations ≧5) compared to 226 children of Hadza fathers with fewer nominations. Log-rank test 6.03, p=0.014, Wilcoxon=7.37, p=0.007. For the unrestricted sample log rank 5.63, p=0.018; Wilcoxon=5.16, p=0.023.

OR=1.2 (1.06–1.35). The effect of father presence (dadin) on infant death is not significant but perhaps suggestive (b=−0.4933, p=0.322, OR=0.61 (0.23–1.62), although we may note the very wide CI. Perhaps this finding can be reconciled with another of Marlowe's observations, that fathers of infants brought home more calories worth of food than men in other conditions, compensating for the lower food income of their wives. Perhaps the best hunters neglect this task of the new father, continuing their pursuit of the highest variance prey while other men increase their efforts to bring something home more often. Marlowe mentions that men with young children bring home honey more often than do others.

What if we remove the most nominated hunters from the sample? Relatively few men receive many nominations as a good hunter. What about the majority of men who receive few or no nominations, yet marry and have children? If we remove the 23 men nominated five or more times (the top 20% of men, removing 1081 child-years of data, leaving 4438 child-years remaining for the analysis), a beneficial relationship between the Hadza father's presence and children's deaths becomes evident (Figure 21.3). If, as usual, we control for child age, age$^2$, age$^3$, and for mother's age at the birth of the child, father presence favors child survival and becomes significant (dadin, b=−0.4655, p=0.049, OR=0.63 (0.40–1.0). Children of these men are less likely to die when their fathers are still in the marriage. If the sample is restricted to children under age six years (that is, the zero to five-year-olds), to make the analysis more closely comparable to that of grandmothers, these results are not changed in any meaningful way. (The non-significant results on growth of under-five-year-olds reported here were not changed by removing the good hunters.)

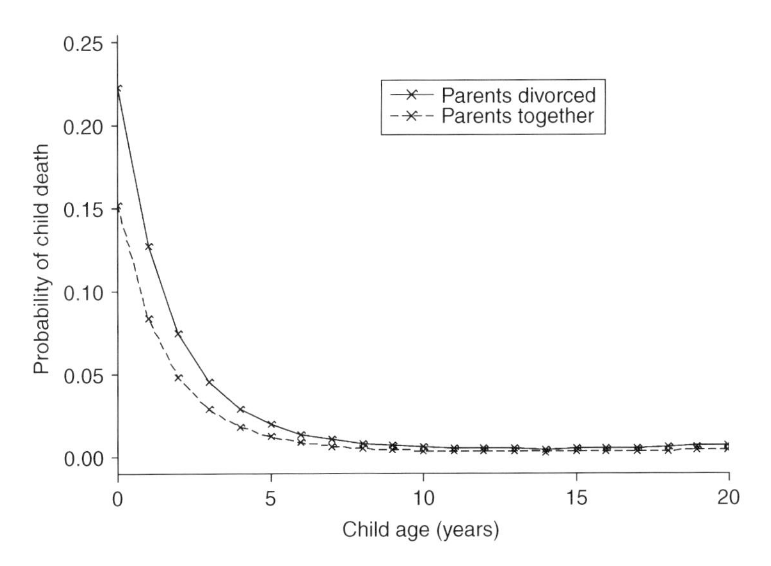


**Figure 21.3** Among the men nominated less than five times (the "poor hunter sample"), there is a lower probability of child death if the parents are together. Logistic regression model: child death = age + age$^2$ + age$^3$ + "dadin" (parents together), b=−0.4682, p=0.043, OR=0.63 (CI 0.40–0.99). 169 child deaths to 526 children.

These results support the idea that there may be a variety of male strategies within the Hadza population. The successful hunters are no benefit to their children's survival but achieve high RS. The "ordinary Joes" may benefit their children. This does not give them higher RS than the good hunters; it fails to compensate for their slower remarriage and wives more close to their own age. In the following discussion, we look at three more categories of men: "Swahilis (non-Hadzas)," "wage earners," and the few men with very high numbers of marriages. We still have to remember, however, that in all sub-samples, most child deaths occur when the father is still married to the mother, and sometimes child deaths precede father departures rather than follow them.

## 21.7 Do marriages break up in response to the death of a child?

Occasionally during the interviews, I gained the impression that a marriage had broken down because of the death of a child. There is a wide belief in the industrial nations that the death of a child threatens a marriage, and there is much literature on how to rescue the marriage. Empirical studies on the incidence of divorce following the death of a child are still relatively scarce. One excellent recent study shows clear effects of stillbirth on the dissolution of marriage or cohabitation in the United States (Gold *et al.*, 2010), and reports more than one other modern study showing an effect of child deaths on dissolution of marriages. "Divorce" following reproductive failure is frequently described in socially monogamous birds (Black, 1996). There is evidence that among some species of birds the re-pairings accomplish better RS, and that as pair duration increases, RS increases, and divorce contingent on reproductive failure becomes less common.

Desertion or divorce as a reaction to a child's death might surprise the theoretically inclined evolutionary anthropologist. From the viewpoint of merciless natural selection, one's first expectation would be that because the death of a child in a natural fertility population likely heralds a rapid return to ovulatory cycles, men would be inclined to stay with a woman to whom they already have socially recognized sexual access. Why would a man leave his wife just as she is about to become fertile again? There is a popular belief that people living in societies with high mortality do not grieve as we would. Nothing that I saw among the Hadza supports this notion. In the case of a child's death, not only is there the possible proximate threat to the marriage from incrimination and recrimination, but either partner may hold the loss of the child as a predictor of continued losses. The man might perceive failure to be a conscientious mother; the woman might perceive failure to be an assiduous provider or guardian. The accuracy of their implied prediction appears to be supported in Chapter 13. There we saw that the standard score of the proportion of children kept alive by a woman aged 20 or a man aged 25 was a good predictor of success at a subsequent and the final observation and of later RS.

The issue is important for interpreting the statistical "effect" of fathers. Because I cannot reliably divide time into durations of less than a year, any apparent

**Table 21.3** Predicting marriage breakup by child deaths. Logistic regressions in Minitab. Model is: Father left = child age + child age$^2$ + mother's age at birth + predictor. Sample was restricted to children of Hadza fathers but wage earners were included. 207 child deaths to 583 children. Smaller sample sizes as shown for the restricted sub-samples

| Sample | Predictor | b | p | OR | 95% CI | Departures/child-years observed/child deaths |
|---|---|---|---|---|---|---|
| Hadza dad | Child death | 0.5644 | 0.021 | 1.76 | 1.02–2.84 | 235/4673/207 |
| Child aged <6 Hadza dad | Child death | 0.5974 | 0.020 | 1.82 | 1.10–3.00 | 151/2270/187 |
| Child aged <6 Hadza dad and poor hunter | Child death | 0.6137 | 0.049 | 1.85 | 1.00–3.40 | 111/1671/118 |

tendency for child death to be more probable in the years in which the father is absent is difficult to interpret. It could arise either because absence of the father is dangerous to children, or because fathers tend to leave or be expelled in response to a child death. In no case can I determine which partner deserted, whether the father left or was expelled. Marlowe (2010, p. 179) reports that informants make strong but unsupportable statements about this. For example, men found out in an affair professed not to know why their wife threw them out.

Most child deaths occur while the child's father is still with the child's mother. We can look to see if there was a tendency for the father to be absent in the year following a child death. I extracted this information from the marital histories, and as mentioned earlier, added a variable to the child annual hazard files, "dadinext," denoting whether the father was still married to the mother (scores 1) or not (scores 0) in the year after each current year in the child file. Many of the fathers absent "next" year could also be absent "this" year. Thus, I derived a new measure "father left" by subtracting dadinext from dadin, and scoring 1 for those whose sum was 1, and 0 for all others. Knowing only that I knew little about the behavior of non-Hadza husbands (other than the unfavorable reports from Hadza women who had left them), I restricted the sample to children with Hadza fathers. The measure "father left" was predicted in logistic regressions by child age and (controlled for child age) by child death (Table 21.3). If a child died, then a present father was more likely to have left by the following year (Figure 21.4). Thus, there is a possibility that fathers sometimes leave in response to the death of a child. The father was more likely to leave if his wife's mother were dead (SI 21.6). It is also highly possible that both events, divorce and child death, followed separately from some dysfunction in the marriage. The dysfunction could include the man's failure to be an adequate provider or assistant caretaker. The topic cannot be exhaustively investigated with my data. Therefore, these results have to remain as a shadow over the evidence that Hadza fathers, even the poor hunters, make a difference to the survival of their children.

If either partner, consciously or unconsciously, takes a child death as a signal of a spouse's future proficiency, it would be a more effective response early in a marriage,

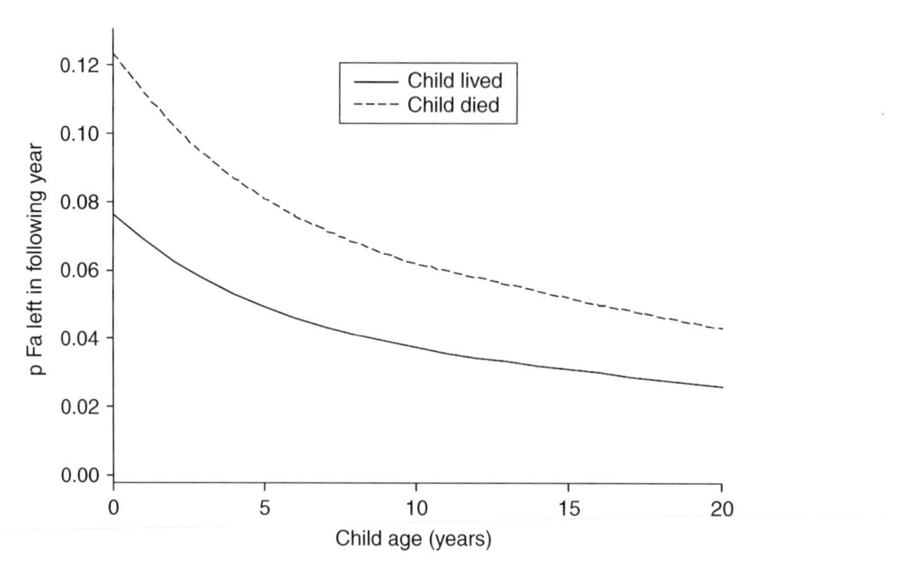


**Figure 21.4** Do marriages break up after a child's death? Probability that a Hadza father leaves the marriage (by child age) differs according to whether the child has died or not. Father left = child age + $age^2$ + $age^3$ + alive/died, b=0.5326, p=0.031, OR=1.70 (1.05–2.76).

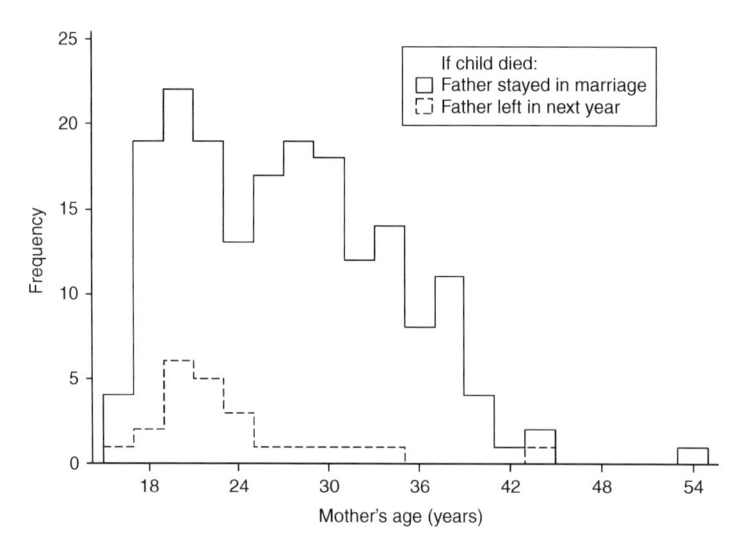


**Figure 21.5** Break-up of marriages after a child died by age of mother. A subset of the child annual file was made for only the years in which a child of a Hadza father died (207 cases). The distribution by mother's age is compared for the 23 cases, which were followed by the marriage breaking up and the 184 cases in which the marriage endured. Break-ups are more likely when the woman is young.

or with young parents who have no "track record" and a long segment of the reproductive career ahead of them. If the current partner is "below average," there is some chance that the next will be better. The bird literature reports greater probability of "divorce" in response to failure early in the "marriage," with a lower probability in a long-established "marriage" (Black, 1996). We might expect the "reproductive economics" of this to make even more sense in the much lower mortality, long-lived human population. Most of the cases of a Hadza marriage breaking up in the year after a child died concern marriages of women aged under 25 (Figure 21.5). Spearman's rho ($\rho$) suggests that this result is unlikely to have arisen by chance (Spearmans $\rho$=–0.15, N=207, t=2.42, df=205, p=<0.02 two-tailed).

## 21.8 Categories of men and alternative male strategies

Sixty-three out of the 695 children were children of just seven fathers who earned some kind of cash payment for work or an appointment; I call them "wage earners." They include three with appointments as community development officers, one employed on a fish farm, another employed on another European-owned farm, one a game officer, and my field assistant. A few other men had periods of employment as guards, trackers, and tour helpers, which usually led to protracted absence. They were not included in the wage earner category.

The seven Hadza men I classified as "wage earners" had more wives and more children, and may have kept more of their children alive than did other men. Being a wage earner goes with fewer child deaths (b=–0.4508, p=0.095, OR=0.64) after the usual controls for the child's and mother's ages. "Wage earners" and their wives tended to be more often found in the same place as previously seen. Their access to cash enabled them to buy food and other useful goods occasionally. Most hunted very little. They may thus have had access to the kind of resources that are more easily kept within their household.

One hundred and twelve of the children were attributed to a non-Hadza father, and while we always knew if he were still with the mother, we sometimes did not know whether a divorced non-Hadza husband had died. The father's tribe (non-Hadza or Hadza) had no effect on child survival (b=–0.1683, p=0.378, OR=0.85, CI 0.58–1.23). It had a very small effect on fertility (Chapter 7).

Some individual Hadza men stand out as notable exemplars of the "cad" strategy (Draper and Harpending, 1982), such as one man with six marriages and five "affairs." One outstanding, friendly, now elderly, Hadza face will come immediately to the mind of more experienced fieldworkers among the Hadza (Photograph 21.4). He did not appear in my quantitative analysis because, by the time our study began, he had "settled down" for many years. However, he had children by six different women (seven children alive, seven dead). His demeanor suggested nothing caddish; "charmers" would be a better label for several of these men. We know almost nothing about their psychology, and as usual, we should be on our guard against over-interpreting colloquial labels such as "cad" or "charmer." In the men's marriage summary file, I identified six men with an unusually high number of marriages for their age, reputations, and the length of observation (Chapter 15; SI 15.10). The list includes a son of the man we think of as epitomizing the strategy. He also comes across as far more charming than caddish. The standard scores of RS of these men are indistinguishable from those of the majority population. The number of marriages

**Photograph 21.4** A familiar and friendly face to many of the Hadza researchers. This man, here wearing a "meat jacket" on the way back from a kill site, had a record number of brief marriages but later settled down for a long-lasting marriage to a younger woman. We think of him as exemplifying one out of several possible male strategies discussed in the text. © James F. O'Connell, 2015. Reproduced with permission.

(number per year of the record) did not correlate with RS. This suggests that the "cad" strategy is unlikely to take over the population. The standard scores of the number of births of these men tended, not significantly, to be higher than the majority ($b=1.943$, $p=0.202$). Most interestingly, the standard scores for percentage survival of the children of these six men tend to be below the majority ($b=-3.957$, $p=0.021$). If we can believe a significant p-value for only six men, this is one of my few results that can be interpreted as compatible with the idea that men's continued presence positively affects their children's survival. The nomination scores of these men were not noteworthy. Marlowe (2010, p. 212) reports a similar observation about men with high numbers of marriages.

## 21.9 Step-fathers

Among Hadza, as among Americans and others, step-fathers appear to illustrate both dangers and benefits (Daly and Wilson, 1996). We have no case of infanticide by any man or woman among the Hadza, but SI 21.5 and Table 21.2 show that survival of Hadza children with a step-father is slightly but not significantly worse than without him. The growth of 6–12-year-old Hadza children appeared slightly worse when a step-father was in the family than when he was not. Most of the time, however, the more striking observation is that step-fathers are hard to distinguish from others. Men are not deterred from marrying a divorced woman by the number of small children she has (Chapter 15). Hadza report that step-fathers look after the children in the family regardless of paternity (Marlowe, 2010, p. 209), but admit to some differences in attitude if questioned further. Although Hadza men spent a very small percentage of the daylight hours near their children, differences between fathers and step-fathers are significant in Marlowe's data (1998, 1999, 2010, fig. 8.5). Step-fathers show significantly less "direct care" than biological fathers (controlling for the age of the child), and in Marlowe's data, they bring home less food, although both bring home a surprising amount of non-meat foods, at least an order of magnitude greater than seen during 1985–1986.

Step-fathers gain no biological fitness directly from their effects on their step-children. Whatever they gain is more likely from gaining and retaining access to their wife, giving themselves the opportunity for future offspring, and fulfilling the expectations of other Hadza with whatever benefits that may bring them. Because the number of small children a woman already has is a very good predictor of a birth, step-fathers do well to seek out women with small children. For the woman, the benefit is less clear. I have shown no benefit of the step-fathers to their wives. Presumably, continued childbearing is one outcome of the new marriage and we saw that the cost to existing children is small.

I cannot offer direct evidence to show how well Hadza step-fathers conform to the expectations of Rohwer *et al.* (1999) and Lancaster and Kaplan (2000). Their appearance as "active family fathers" makes them hard to distinguish from biological fathers, although both Marlowe and I have shown small negative effects of Hadza step-fathers. How much does male care function mainly to attract and keep a mate? Is the small excess of survival by children of biological fathers the true measure of the size of the benefits that accrue directly from paternal investment?

In introducing his data on foraging by step-fathers and fathers, Marlowe (1999, p. 403) suggests we should be alert for differences between these classes of men. For example, step-fathers may be less good hunters than fathers. His data suggest they treat their children differently, and that men forage less if they have more step-children. As his data show, many married women had both step-children and children of their current husband. A man who divorces and remarries is very likely to become an instant step-father. There are few differences between men who are at any moment a step-father, and men who are not. Especially, it is not possible to demonstrate that step-fathers are somehow "second class citizens." Men entering a marriage in which their wife has children under age seven are not significantly older than those marrying a woman who has no small children. In a man's first marriage, the median age of wives who already have children and those who do not is greater, but only by 1.8 years. Step-fathers have more nominations as a good hunter (mean=6.2 vs. 4.0, t=−1.83, p=0.068, df=191). Step-fathers' marriages last as long as other men's marriages, once we control for hunting nominations and wife's nominations (both of which are associated with longer marriages). Although there were a few youngish men, not having been married before, whose first marriage was to a much older woman (Chapter 15; SI 15.11), I cannot claim to have shown that step-fathers in general are unsuccessful competitors.

## 21.10 Discussion: did we find a "father effect"?

In this chapter, we have seen three pieces of data that can be taken to give weak support for paternal investment ideas.

1. Men can make a difference. Women's fertility and their children's survival do apparently vary with the husband's occupation: women married to non-Hadza men are very slightly more fertile; women married to Hadza men who earn a wage seem to have children who grow faster and are more likely to stay alive. These men are also much more likely to have two wives at one time than other Hadza men.
2. When I removed the expert hunters from the data, the father's presence seemed to enhance the survival of children. I previously discussed the idea that Hadza men can pursue a variety of strategies: expert hunter, philanderer, "regular Joe." The latter, seldom nominated as a good hunter, nevertheless hunt frequently, and describe themselves as hunters. Perhaps the less-nominated man may bring home a few baobab pods, and more than others do (but who does the costly processing?), or more honey, or spend time going after some birds, bringing these home more often than would a man widely recognized as an expert hunter. The data reported by Wood and Marlowe (2013) in their table S4 tends to support this suggestion, especially with respect to honey.
3. Among the 5–17-year-old children (and those aged 5–12), the father's presence in the marriage seemed to enhance the children's growth. This observation may have something to do with the duration of the marriage (which must have lasted at least five years if the child is over five years old and its father still present). The data are not very clear on this issue, the duration of marriages being confounded with the mothers' age. However, it leads me to wonder whether the Hadza categorization of adult women into tlakweko and paanakwiko, and even more importantly of men into elati or paanakwete has much more meaning than I have acknowledged. A paanakwete is a real adult, a responsible man. An elati is not, he is too young. Nevertheless, I find no simple relationship between, for example, child's weight and the age of the father. Older mothers keep more children alive and growing faster even without the grandmother's or teenager's help. Perhaps something about the stability of the marriage and the seniority of the husband contribute to, or merely reflect, the older (30–40-year-old) woman's competence. Scelza (2010) found that Martu men had an effect on their sons progress in their ceremonial careers, and on their RS. She suggests that we may have been missing important effects of fathers by concentrating on survival of young children. Perhaps one day Marlowe or his students will be able to look at the adolescent and adult careers of the sons of the men in my sample.

On balance, there is rather little evidence here that Hadza men enhance the growth and survival of their children and less that their hunting enhances the survival of their children or the fertility (SI 21.7) of their wives. By hunting large animals, men occasionally make available large amounts of food. A man's wife and children benefit from these but so do many other women, children, and men.

If we discard the paternal investment theory, we are left to wonder why Hadza women accept a husband, why they are angry if he leaves, what is attractive to them about a good hunter as opposed to a man who has friends or kin who are good hunters, for example. Could losing the advantages of "good genes" be worth burning down the house as, I heard it told, a Hadza woman did in response to her husband's frequent absences? Are they worth the cost to child mortality? Paternal investment seemed like a good answer to these questions. Women may prefer good hunters because of the reflected glory and influence, including that their kin will benefit from his successes. But so far I have been able to show no benefit to the wives of good hunters. Someone should test whether wives of expert hunters spend more time living with their closer kin (a quick look at the camp compositions in the series of censuses suggests that they may). A woman who can persuade her sister and her cousin to find a good hunter for a husband as well as hanging around hers, would be exposed to a much steadier supply of meat. The average time between catches of the average man in Hawkes and O' Connell's 1985–1986 observations (Hawkes *et al.*, 1991) was 29 days (this had increased to 107 days for the eastern Hadza by 2010, Wood and Marlowe, 2013, table 1, and S4). This is the rate that a woman married to an average hunter would experience. The wife of an expert hunter might eat meat a little more often. In either case, if another comparable hunter is added to the camp, the frequency with which she eats meat would be doubled, and she would wait half as long between meat meals on average. We should wonder if a superficial gender reversal of Levi-Strauss' "men farming women" to "women farming men" could be deepened into a productive understanding of how Hadza sex roles and reproductive strategies work. It may, at least, remind us of the often discussed social selection and biological market selection for generosity.

Perhaps it is just too difficult to see the effects of fathers, effects are there but we cannot see them; the sign of the regression coefficients for father absence suggest an effect but the confidence intervals overlap the zero effect. We cannot see them perhaps because each woman allocates resources received from men and allocates these resources between her own survival, producing more children, and ensuring the growth of the ones she already has. Her husband may be unable to control how much goes to the children, or whether it goes to his children more than to his step-children. Perhaps women work harder after divorce, which is why I looked at maternal depletion, where I found no indication of weight loss in single women in charge of small children. Perhaps we do not see the effects because, given that there are probably diminishing returns to every kind of care, Hadza grandmothers and child helpers can easily fulfill the tasks of helpers under all but rare circumstances. If only we could observe these rare circumstances often enough, we would see how effective fathers can be. All this may be true, but it neglects the fact that we do see effects of grandmothers and older children. If there are effects of fathers on child survival, they are small or very variable, and very small when compared to the effects of the other helpers. So how do Hadza men benefit from all that "hunting and resting," as the women describe it?

Perhaps the Hadza are just an exception. It would be strange, however, if the Hadza, living in an ecology as close to the embodiment of the often hypothetical environment of evolutionary adaptedness as we can observe directly (see Foley, 1996, for a full discussion of the EEA), were an exception to such a supposedly evolutionarily significant feature as paternal care and provisioning. Did our ancestors avoid the temptation of pursuing the large game that inhabit the terrain held to have contained the selective pressures that produced our species? Did they concentrate instead on the small game that might better ensure a daily supply for a woman and children? Perhaps they did, until projectile weapons became widely adopted; but then why adopt them? Perhaps, as large game finally disappear from Hadza country, or as the trough in the age structure produces a sudden shortage of grandmothers, we will see men change to a more domestic strategy. Smuggling in a small bird will become less bad form, bringing home some baobab less unmanly, and even trapping might become more acceptable as an adult activity. Perhaps, to judge from the data of Wood and Marlowe (2013), they already have. However, it would be a mistake to neglect the potential of the Bird–Smith–Hawkes costly signaling idea to provoke us into looking more closely into male activities that have left their mark so strikingly on the archaeological record, especially in recent millennia. Its relationship to the puzzle of public goods should also be pursued. Nor should we simply ignore alternative scenarios of the evolution of monogamy such as offered in Reichardt and Boesch (2003), or by Smuts (1992), Smuts and Smuts (1993), Gowaty (1996), and Hawkes (2004).

# 22 Variation among hunter-gatherers: evolutionary economics of monogamy, male competition, and the sharing ethic

... in utterly barbarous tribes the women have more power in choosing, rejecting, and tempting their lovers, or of afterwards changing their husbands, than might have been expected.

Darwin, 1871, 912

"Frankie! Bring the axe. This leopard wants to eat the baby." Hadza man to Marlowe, midnight, camped by the road somewhere in Hadzaland.

In this chapter, I attempt a summary and an introduction to some questions that arise from the Hadza research. I will also discuss our data on male and female reproductive strategies, an arena of much controversy. Darwin's surprise should usefully upset our prejudices, and the importance of its contrast with the barbarities of the Victorian English marriage system will become evident. Before we jump to conclusions about rare but vital "father effects," or become distracted by the image of a Hadza man without a bow by his side, note that as Frank Marlowe retold the events, the call came from the baby's very new step-father.

The Hadza data impinge on many issues in hunter-gatherer studies and biological anthropology. My range of issues may have been too wide. I must have missed many important references that could have simplified my task and clarified the results. My aim was to add more about the Hadza to the available data on hunter-gatherers. I cannot reanalyze all the important comparative accounts, such as by Walker *et al.* (2008), Kelly (2013), and others, but I will discuss Terashima's (1980) too long neglected suggestions about a forest–savanna dimension of warm climate hunter-gatherer variation. I will also suggest that human behavioral ecologists should not continue to neglect approaches to mating systems and to altruism that can widen our perspectives, and extend (with some associated risks) the aspects of behavior and society that can be tackled by an evolutionary ecology approach.

## 22.1 A summary of findings

My data support the general descriptions of many aspects of Hadza life by Woodburn (1968a,b, and subsequently) and Marlowe (2010). The richness of the Hadza environment (for hunting and gathering), the separateness of the sexes, their personal autonomy, and their mobility remain notable. The Hadza illustrate an egalitarian society in which food and other items are shared widely. Serial monogamy with rare polygamy is an adequate quick description of Hadza patterns of marriage. Many features of their marriages resemble those in large-scale societies (Chapter 15).

Evidence on Hadza ancient history and identity has accumulated dramatically (Mabulla, 2007; Tishkoff *et al.*, 2007). The possibility that, during climatic extremes, the rift and its volcanic highlands, generating large differences in altitude and therefore rainfall, has provided refugia among its summits or among springs at its foot was discussed in Chapter 2. Hadza continue to resist settlement and few have taken to agriculture (Chapter 3, and Blurton Jones, 2015). At the end of the twentieth century, the advent of tourism and its cash brought recognition to the Hadza but can be destructive, as when people stay where they can be found by tourists and traders alike, and the cash can then be converted to alcohol (Chapter 3). During my study, few Hadza left their homeland or married non-Hadza and about half of those who left came back, some bringing their children to raise as Hadza. Although the Hadza think of their country as divided into regions, anyone is free to live anywhere, and people move from one region to another quite frequently. Their mobility is extreme, both in frequency of camp moves and change in composition of camps, but within a preferred range of country.

Hadza mean fertility, at 6.1, remained the same between the 1950s and 2000, and is at the mean of several samples of non-industrial societies (Campbell and Wood, 1988; Bentley *et al.*, 1993a,b). Two features of age at the last birth were notable. Closed inter-birth intervals (IBIs) did not lengthen with age or parity. Despite low incidence of STDs, women simply ceased childbearing in their late 30s or early 40s with no "warnings" from elongated intervals. A striking correlate of the probability of a birth after age 35 was the supply of helpers, which seemed important even when individual "intrinsic" fertility was taken into account, and may imply a heightened role for variation in older women's energy budgets.

Mortality also appears to have changed little. A completely normal human demography and life history can be maintained by a hunter-gatherer lifestyle. The Hadza mortality data, with life expectancy at birth of 32.7 years, and at onset of childbearing another 42 years, and at age 45 a further 23 years, add to the collection of hunter-gatherer life tables that contrast sharply with archaeological estimates of hunter-gatherer mortality patterns. The contrast is probably related to method, not to a sudden modern but worldwide change in the life history of hunting and gathering peoples (SI 8.7). However, the contrast continues to attract attention, and it would certainly be helpful to be able to date the major evolutionary changes in human life history. As more genes influencing longevity and life history events become known, information on the antiquity of their variants may help resolve the debate.

Hunter-gatherer population dynamics remain an unsolved mystery. Severe fertility decline when energy budgets are stressed, and unpredictable high mortality (crashes) due to disease or warfare both remain possible solutions. We have only a few hints pointing toward a smoothly functioning density dependence. In the case of the Hadza and the Ache, we have strong evidence for rapid increase over a period of at least 50 years, and probably nearer to 100 years. In each case, there are good indications that these populations grew after cessation of the losses from capture or homicide at the hands of more powerful agricultural neighbors. The Hadza and Ache demonstrate the potential for hunting and gathering economies to fuel rapid increase. Population expansion may be less of a puzzle than some have supposed, and economic intensification less of a mystery and more feasible than others have argued.

For most of the twentieth century, Hadza seem to have been little influenced by their neighbors but they have had Cushitic-speaking farmer neighbors for some 4000 years, and the expanding Bantu population arrived nearby about 2000 years ago. The genetic studies imply a substantial, relatively recent, population bottleneck in the Hadza gene population. Yet the population of people who call themselves Hadza has been growing fast during the last 100 years. In Chapter 10, I suggested a resolution to this paradox but I collected little data on the fate of those who did not return from life "among the Swahilis." It may not be too late for others to fill the gap. There are still children of non-Hadza fathers who stayed outside the Hadza community but could be followed up. Is it true that either boys or girls have a more difficult life than their entirely non-Hadza neighbors, and rapidly leave fewer "Hadza" gene copies to be found by geneticists?

In Part II of the book, I tried to use differences between individuals to see how individual Hadza can maximize their reproductive success (RS). The demographic data support the common finding that population increase is most affected by survival of children under five years old, and by fertility of young women. A related individual helper who can promote either of these is likely to leave more descendants and gene copies than an individual who directs his/her help and resources elsewhere. In reporting the anthropometric data, collected for use as a proxy for fitness, I concentrated on a few more entertainingly disputable points. These include the special features of early childhood and the constancy of adult height and weight across five decades. Most aspects of Hadza growth fit the apparently universal patterns of human growth.

The chapters on helpers do not fit everyone's expectations. Children seem to be rather ineffective helpers in that the data do not clearly and consistently show they enhanced the growth or survival of their younger siblings. Hadza children, by their foraging, may make themselves cheaper to raise but it seems unlikely that they can generally be a net gain to their hunter-gatherer parents' reproduction (supporting Kaplan, 1994). Nonetheless, child helpers do seem to have a strong effect on their mother's late career fertility (also found by Turke, 1988).

Women past the childbearing years seemed to be a strong influence on child survival, especially when their daughters or daughters-in-law were young. A child's father's mother (PGM) was as effective as its mother's mother (MGM), and especially important if MGM had died. The Hadza data give mixed support to the idea that post-reproductive life evolved as a result of competition between the generations. Nonetheless, the quest for instances of competition should not be given up, although we should keep in mind that the "cooperative breeding" discussed as a characteristic of humans may be very different from cooperative breeding as described among "eusocial" mammals such as meerkats and mole rats. The extreme mobility of Hadza life is one crucial difference. In contrast to the classical mammal cases, among the Hadza the costs of leaving the current social group appear to be very small.

Among bygone debates about hunter-gatherers, the Kalahari debate was well addressed by Marlowe (2002) and discussed briefly in Chapter 3, and the influences of the modern world should always be considered in hunter-gatherer studies (Leacock, 1982, provided an excellent and temperate example). The modern world, by changing features of the environment, can provide interesting "natural experiments" that shed light on human adaptability and on core features of hunting and gathering societies (Codding and Kramer, 2015). The role of Hadza "encapsulation" among their more numerous neighbors was examined closely by Woodburn (1988). In Chapter 3, we saw that the "closeness" with which the Hadza are encapsulated is a feature of the late twentieth century. Any future studies of Hadza demography and behavior should keep account of the continuing encroachment upon them, include comparisons of the Hadza in more and less encroached parts of their land, and keep in mind factors such as alcohol with its already apparent effect on homicide rates. Even the consequences of any tendency toward sedentarization (to make it easy for tourists to find them) could be highly informative.

## 22.2 Variation among hunter-gatherers: a warm climate forest-desert dimension?

Hunter-gatherers vary in their behavior, their demography, and their habitats (Kelly, 2013). Evolutionary ecology should provoke us to account for the variation by attending to effects of local circumstances on the costs and benefits that result from behaving one way or another, and thus on the optimal, fitness-maximizing, course of behavior. The variation should teach us about how costs and benefits trade off to give different behavior in different circumstances.

I argued in Chapter 2 that Hadza represent the rich end of the African wooded savanna habitat with a rainfall of 500 mm/yr, numerous small water sources, and super-abundant plant foods. The !Kung and the G/wi teach us about a harder environment, while the Pygmies represent the high-rainfall lowland forests with more than 1200 mm/yr of rain. In between the rich Hadza wooded savanna and the forests of the Pygmies are the widespread Miombo woodlands (characterized by trees of the genera *Brachystegia*, *Isoberlinia*, and *Julbernardia*) with a rainfall of 700–1500 mm/yr but very poor soils, and from which we have only one account of living foragers. Terashima (1980) reported on the Bambote, who lived west of Lake Tanganyika, and who, while taller than the Pygmies, lived a relatively independent life hunting in Miombo forest (rainfall of 1000–1200 mm/yr) but traded with farmers for much of their plant food. This may immediately suggest a scarcity of plant foods in their habitat. In SI 22.1, I discuss my frustrated search for information on the plant and animal resources in Miombo woodland. The data do not suggest that Miombo, despite its higher rainfall, is a superior environment either for hunting or for plant foods. But the data do not exclude this habitat as completely inhospitable to foragers.

Terashima (1980), familiar with Japanese research on Pygmies, central Kalahari San, Hadza, and some Dorobo groups, makes interesting suggestions about variation between warm climate forest, savanna, and desert.

1. He describes variation in hunting methods that he ascribes partly to differences in opportunity to see animals and to track them. Forest animals tend to be smaller, more numerous, and more evenly distributed than the animals in the open environments, and this favors cooperative hunting by driving or using nets. Other influences on use of nets are discussed, for example, by Bailey and Aunger (1989b), Wilkie and Curran (1991), and Lupo and Schmitt (2002).
2. Terashima links mobility and flexibility of social structure to the same dimension. Where forest is more dense, groups occupy smaller areas and are more self-contained. Where habitat is open, groups wander farther, mingle, and change composition more extensively. Should we also expect people in higher-rainfall areas to have more warfare? Perhaps economic defensibility of territory (Dyson-Hudson and Smith, 1978) can be mapped onto Terashima's dimension. Terrain with numerous, evenly distributed, fast-replenishing small game may be more economically defendable than the habitat of wandering, scarce, large animals that sometimes congregate in herds, as a mobile patch.
3. Terashima suggests a link to residence rules, associating patrilocality with group hunting and exclusiveness, and associating flexible locality to solitary hunting and extensiveness. The data in Hill *et al.* (2011, table 1) seems to support a forest–savanna dimension from patrilocality among forest peoples to more flexible and matrilocal patterns among savanna and desert dwellers. Patrilocality may favor the "fraternal interest groups" argued by some to facilitate warfare. Data from the best known case, the Yanomamo, have recently been shown to favor an alternative process of formation of male alliances (MacFarlan *et al.*, 2014).
4. Terashima's picture includes variation in percentage of gathered plant food and hunted or trapped animal food. He cites Tanaka's observation that in the dry central Kalahari, people take 81.3% plant food and 18.7% meat. He expresses reservations about the percentage measure for the forest people (Bambote and Mbuti) because of their symbiosis with farmer neighbors, even though the Bambote dig tubers from *Dioscorea* and *Ipomoea*, especially when farmers have no surplus.

What leaps to mind is a comparison with the Ache (despite their extremely low population density; Hill and Hurtado, 1996, p. 49, 84). Their 83% animal foods and 17% plant food (the mirror image of the central Kalahari) comes under a rainfall of 1700–2000 mm/yr (Hill and Hurtado, 1996, p. 61) in forest that sounds very similar to that of the Bambote in its varying profile and density, and in its abundance of medium to small animals. Furthermore, much Ache hunting is conducted by groups of men flushing out hidden animals and clubbing or shooting them. Next comes to mind the contrast between the high rates of warfare and homicide among South American foragers (mostly forest peoples) and the lower rates among African foragers. Perhaps this contrast is not entirely the result of the more successful "pacification" by colonial powers in Africa (or by their predecessors in the height of the slave trade) but to the bias in the African forager sample toward desert and open savanna. Marlowe (2010, p. 167) notes the division of the Andamanese into several small tribes with a high sense of territoriality and perpetual warfare. This occurs in their forested and very high-rainfall habitat (roughly 3000 mm/yr).

An enterprising student with access to the several relevant databases that now exist could soon put Terashima's idea to the test. Issues like access to firearms, boats, and alcohol might make it difficult to incorporate some forager societies. Again, Australia might be an excellent second universe in which to test the idea, although only the extreme northeast had forest of any density, coincident with the 1500 mm/yr isohyet. Harris (1978) describes aboriginal populations in the Queensland rainforest from the writings of early visitors. Twelve tribes, speaking five different languages, occupied this 11,660 $km^2$ area. Estimates of population density went as high as 3.3/$km^2$ with an average of 2.14/$km^2$ estimated for 1800 and 5.34/$km^2$ for 1897. He reports that "Individuals belonging to the same tribe are usually on the best of terms but the different tribes are each other's mortal enemies." They held "fighting corroborees to settle disputes. At these corroborees the huge swords and shields were wielded in duels, but combatants were seldom killed." Unlike most Australian aborigines but like their African rainforest counterparts, they were of small stature. In contrast to other forest foragers, Harris estimates the animal resources to be quite limited, although he lists three terrestrial birds, 22 species of marsupial, and lizards and pythons as common prey. Plant foods were quite seasonal, but included nuts "on which the Aborigines to a large extent depended."

A wet/dry dimension may map less easily onto the demographic data. One might claim, by comparing the Ache to the Hadza to the !Kung to the G/wi, that fertility is higher in wetter, more forested environments. One might link this to the balance of animal to plant foods and the amount and nature of women's subsistence work (Blurton Jones, 1997b, in Hewlett, 1991, or the extent of the male contribution to diet, Marlowe, 2001). The data on Pygmy fertility, showing very great variation, only partially supports this suggestion. The number of parasites is higher in the forest and lower in deserts (Dunn, 1968, discussed in Chapter 2). The Pygmies suffer high adult mortality, and begin reproduction at an earlier age (Migliano *et al.*, 2007). The Ache live in forest but show lower mortality than the Hadza or the !Kung.

Whether a wet/dry dimension of hunter-gatherer biology and behavior does exist, these paragraphs serve to illustrate one aspect of the use of hunter-gatherer studies in thinking about the remote past. We know people are flexible in much of their behavior. Behavioral ecologists who study people have given us plenty of evidence that the behavior that results is often adaptive in the sense that it maximizes fitness under local conditions. Perhaps we can begin to identify some of the rules about the flexibility.

Marlowe (2010, ch. 10) identifies the Hadza as "the median foragers," but we should always ask for reasons for departures from the median. Dimensions such as the wet/dry dimension can suggest reasons for departures from the median. Understanding how behavior varies with rainfall may be helpful when we think about the extensive and frequent variation in past climates. We do not have to believe that we must identify a single environment of evolutionary adaptedness (EEA) for all characters in which just one survival and reproduction strategy evolved, and all others fell by the wayside or are not instructive when we build or test evolutionary scenarios. Most certainly we do not need to argue that "my hunter-gatherers are better than yours," more relevant to human origins, to the EEA, or to the expected built-in or universal psychology of the early Darwinian psychologists. It seems likely that when high rainfall extended across much of Africa, more of its inhabitants behaved more like the Ache, or the Pygmies and Bambote. When rainfall at Kalahari levels spread far and wide, more people were probably behaving more like the !Kung and G/wi, and some were seeking divergent refuges among high mountains, or balmy coastlines.

## 22.3 Men in competition; men in groups

Within the strongly egalitarian Hadza backdrop there is individual variation in fertility, child mortality, adult lifespan, and RS. Competition is not absent, but it is not on display.

The competitive context of men's lives is illustrated by the high variance in their RS reported in Chapter 13 (and discussed in SI 22.2). A few men raise three times the median number of children, while some men raise none. Any man who can reach the top end of the distribution has greatly enhanced his fitness. My data show that the competitive situation of men is important among Hadza, although in daily life its expression in obvious forms is very muted, even denied. Day-to-day relationships between men are peaceful and collegial. Men sit together working on arrows, sometimes working on each other's arrows, or smoking and talking for many hours on many days (SI 22.3). This takes a large portion of the 7.99 hours "manufacture and repair at home" reported by Hawkes *et al.*, (1997, table 1), and more in larger camps (with many men) than in smaller (Marlowe (1999, p. 63).

What would improve the chances of a man's success? Although everywhere, homicide is more abundant between males, but continuous open hostilities seem an unlikely route to success where each man is lethally armed, and in a society with "minimal politics" where undetected assassination is an everyday possibility (Woodburn, 1979). As Schacht *et al.* (2014) point out, a measure of male competition should not be confused with frequency of fighting, frequency of theft, frequency of desertion, or violence to women. There are many ways in which a man might "climb the tree." All depend upon how they affect the interests of women and other men.

The anthropological literature is rich with discussion of the ways men relate to each other despite, or because of, the force of selection for competition between them. With a few exceptions, human behavioral ecology has neglected this topic, despite very early beginnings by Tiger (1969), and by Bailey and Aunger (1989a) in one of the earliest quantitative observational studies of hunter-gatherer behavior. A recent paper by Rodseth (2012) leads the way toward linking behavioral ecology to some of the ideas in classical anthropology. The significance of Hadza men's relationships needs to be addressed in future research. The epeme feast, its secrecy, and separation from women and children is reminiscent of the literature on men's "secret societies" and the alliances among men, which Rodseth has brought to the attention of behavioral ecologists. Hadza are quite secretive about things "epeme," and fear other men's reactions to giving away the secrets. I did not push for an explanation. I was told that Marlowe has been told but on condition that he will not tell anyone else! Does the men's place not just act as an incidental mate guarding process (where are your rivals?), but with its lengthy interaction and its ceremonial aspects, support conventions about male–male competition? Does the solidarity of the men's place protect against the threat of the attractiveness of women to the alliances among men? Lars Smith once remarked that the epeme feast was "a feeble attempt at a men's resistance movement." He was only partly joking.

## 22.4 How do Hadza men gain superior reproductive success?

I have tried to look more closely than the traditional and over-generalized trade-off between mating effort and parenting effort, or between "cads and dads." First, I summarize results reported in previous chapters, and then begin to explore the extent to which the results may reflect different strategies, and ways in which they may support or contradict various theories of monogamy.

In contrast to women, men's number of births made a larger contribution to their RS, and there was no leveling of RS as the number of lifetime births increased (Chapter 13, Figure 13.3). Hadza men with greater RS spent a bigger proportion of their recorded adult life in a marriage, remarried more quickly after divorce, remarried younger wives, and had their first child at a younger age. The survival of their children made a smaller contribution to men's RS than it did for women. A few men had extra-marital affairs, and some started a second family in middle age (Chapters 13 and 15).

Remarrying a younger woman in middle age seemed an important part of some Hadza men's reproductive career. The peak of the fitted curve of age of men at remarriage was 39. In Chapter 7, I calculated that 41% of men's births occurred after the age of 40 and 14.5% after 50. Some 91 (53%) men over age 40 in the marriage summary file remarried. Among them, 25 married a woman aged under 30 (27.5%). These men increased their RS from 2.9 live children to 5.3 ($t=2.62$, $p=0.016$, $df=21$), almost doubling their RS. The rewards for those few who succeed in this competition with men of all ages are clearly very high. When I tried, in Chapter 12, to assess the contribution to population increase (sometimes used as an indicator of fitness), variation in the fertility of men aged 40–55 appeared to be rather important (SI 12.3).

Another measure that predicted high RS was hunting reputation. The few most-nominated hunters achieved their high RS by the means just listed and reported in Chapters 14 and 15. They appeared able to succeed in the competition. Women seem to prefer them. Was this because their wives and children benefited from the food that these men acquired? The question has been much debated.

Highly nominated Hadza men have more living children than other men, controlled for the man's age (high RS), but they achieve this by spending more of their adult life in a marriage to a woman of childbearing age (Chapter 15). They remarry more quickly than other men after divorce, and remarry younger women. They continue to bear children to a greater age, because they tend to have younger wives than other men of their age. They also appear to have more affairs (which gain them only 9% more births [Chapter 15]). These advantages all imply that the good hunters have more success in competition for women of fertile age. My data do not allow us to determine how much of this success is due to other men conceding to them, to kin encouraging the prospective brides, or to women's preferences. Nonetheless, I have been able to show no reproductive advantage to the wives or their children.

## 22.5 There must be a "meat effect," but is there a father effect? Is there a husband effect? Paternal care and paternal provisioning?

The dominant sociobiological idea about evolution of human mating systems has been that men work to provide for their families. Hunting and gathering represent a division of labor arising from bargaining over a shared goal of promoting the fitness of the father and mother. Gurven and Hill's (2009) study is the most comprehensive description and defense of this view. While they tended to emphasize the shared interest component of a bargaining model, others have found the conflicts of interest to generate more interesting hypotheses (Mock *et al.*, 1996; Houston *et al.*, 2005; Harrison *et al.*, 2009). The paternal provisioning view holds that men help raise their children and do so primarily by their subsistence work. Women seek good providers, and benefit from keeping them in a marriage, trading paternity for resources and childcare. The providers gain fitness by being granted higher paternity confidence, and by promoting the fertility of their wives and survival of their children. I found few signs of father effects, benefits that children receive from fathers, or "husband effects," benefits that women receive from husbands. I tried to test several expectations derived from the paternal investment view and few fared well. There was little indication of a father effect, and I conclude that a wider array of theories of marriage should be explored.

Hadza men acquire about 40% of food brought to camp and it all gets eaten very eagerly. Many people benefit from eating it. Thus, we may wonder whether the only benefit to the men who acquire the food is an increase in their children's fitness via the meat that they eat. They may gain by retaining their wife, who may adjust her workload or be able to recruit helpers accordingly. They may gain in relationships to other men, including their conceding to him in competition. We see little sign of man-to-man reciprocity; Hadza fit much better with the often described "generalized reciprocity" of hunters and gatherers, which is plain unreciprocated generosity (and more explanatory models are available than readers of human behavioral ecology may have come to expect).

A cut of 40% of the food intake would undoubtedly make a difference. However, it seems to be a difference that is so widely and evenly distributed that we can see no effect on individual women or their children due to the individual men responsible for the acquisition of this meat. Why would men acquire this meat if there is no selective advantage to doing so? The reputation data show that there is a selective advantage. Good hunters spend more of their adult life with a young fertile woman. But why? Is there no advantage to the women who accept them as husbands?

### Evidence about father effects

1. The percentage of adult life in a marriage accounts for a mere 0.4% of variance in a woman's child survival score (Figure 15.8b).
2. Paternal investment theory suggests that men would avoid parenting small children that are not their genetic offspring. Among the Hadza, small children proved no obstacle to their mother's remarriage; women with under-five-year-olds remarried faster than other women of their age (Chapter 15). They remarried men of equal status to their previous husband.
3. The father's departure from the marriage by divorce or death had no clear effect on the growth of the under-five-year-olds.
4. In contrast, there was a positive effect of the father's presence on growth of 5–13-year-olds, one of the few findings supporting the paternal investment theory. In other populations, evidence for men's effective patronage of their older sons has been reported (Scelza, 2010). It is difficult to see how Hadza fathers might have such opportunity.
5. The father's presence in the marriage had a non-significant association with the survival of children of any age (Table 21.2).
6. "Dadin" and "stepin" beta coefficients relate to child survival in the predicted direction but are far from significant in any model.
7. Step-fathers were only marginally worse for children than genetic fathers. In one analysis, I was able to show a very small difference between the survival of children with their father and with a step-father. Neither was significantly different from child survival when the mother was alone. While Marlowe (1999) showed statistically significant differences between the behavior to children and to step-children, men showed very little interaction with children. Marlowe commented that men may be step-fathers because they are poorer hunters, but I found no differences between step-fathers and fathers in age or reputation, probably a result of the high rate of divorce and remarriage.

### Evidence about an effect of husband on wife

1. In Chapter 15, I showed that in the Hadza population, the percentage of adult reproductive life (taken as ages 20–45 for women and 25–55 for men) in a marriage is more strongly related to men's RS than to women's. If marriage is to be analyzed as a bargain, men seem to get the best of it.
2. Affairs. Women married to two men gain nothing (Chapter 15). Women sharing a husband appear to be at no disadvantage.
3. There seemed to be no effect of the husband's presence in the marriage (dadin) on weight loss during the women's childbearing years (Chapter 16; SI 16.1). Weight of women of reproductive age decreases slightly with age but not with parity. None of the measures of the husband or marriage appear to affect the rate of loss.

4. Women's fertility. Data do not allow the comparison of women's fertility with or without a husband; any man linked to a birth is described by the woman as a husband. Analysis of the women's annual fertility hazard file suggested no increase of fertility within marriages when the husband was highly nominated. I found no significant negative relationship between husband reputation and IBIs. Highly nominated men tended to have wives with slightly and non-significantly longer IBIs.

Hadza men as a whole seem to make little difference to their wives and children. The data do not support the view of marriage as an exchange of resources for paternity. Men gain considerable paternity (number of births) from marriage, but on average, women gain next to no advantage for their children (Chapter 15; Figure 15.8b).

## 22.6 Alternative male strategies: different categories of men

Among other animals, an often-described outcome of competition is the evolution of alternative male reproductive strategies. It would be surprising if this were not matched by human ingenuity. In Chapter 15, I compared four categories of men: the expert hunters, just 20% of adult men; the "ordinary Joe" majority; a few "charmers" (a better label than "cads") who had many marriages but average RS; and six or so "wage earners" who had high RS, some from polygamy and perhaps from higher-than-average child survival.

Most surprising was the observation that survival of children of much-nominated hunters was worse than average. When I analyzed the 37 good hunters separately (measures described in Chapter 14), their presence was significantly detrimental to their children's survival. The remainder, the "ordinary Joes," all of whom would claim to be hunters and go out hunting every day but are not known for catching large animals, may have a positive effect on their children's survival (with p-values between 0.06 and 0.049, depending on the precise model or sub-sample analyzed). However, the sequence of events suggests that a number of their divorces were a reaction to the death of a child, not a cause of child deaths. It is also possible, and perhaps supported by the data in Wood and Marlowe (2014), that the less successful big game hunters tend to bring home more small animals, honey, and plant foods. Difference in focus on big game hunting (BGH) (to which the best hunters contribute most of the data) or the performance of the majority may account for differences in interpretations between Hawkes and colleagues, and Wood and Marlowe (2013).

Marlowe (2010, pp. 184–185) reports that women prefer a good hunter as a husband. I was surprised to find that several of the younger women appeared not to have clear ideas about who were good hunters. Apparently, Hadza women want to marry a good hunter but young women often do not know who the good hunters are. Instead, they may choose by qualities such as strength, patience, and intelligence. that may predict later success as a hunter. These may also be qualities of wider value in life, and perhaps reflect the heritable qualities that hunting success has been suggested to advertise. Sometimes Hadza women describe quite different factors that attracted them to their husband, "has nice words," "good to talk with," "his face." Among the few women I asked about husband qualities, bringing meat was often mentioned, but so was the claim that they would leave or throw him out if he "fought." This included: hit her, yelled at her for not bringing water, became a "black-hearted man," "lost his manners," which seemed to mean that he became inconsiderate. Their comments may have significance easily overlooked if we are too enthralled with the paternal investment theory.

## 22.7 If there is no father effect, and no husband effect, why is an errant husband worth burning down your house?

Despite their economic independence, Hadza women are angry if a husband cheats or leaves. Their reaction does sometimes include burning down the house, or demolishing it. At one camp where I stayed, a noisy argument broke out late in the evening. Children arrived at my fire, saying they were afraid. Eventually, everything quietened down but in the morning all that was left of one house was a circle of grass on the ground. The offending husband, who had previously come to me with the claim that people were saying he should be shot, giving no reason but complaining that it was one thing to grow old and die but another to be shot like an animal, had left camp with his new girl friend. The incident may remind us of Kohl-Larsen's 1958 (and others) claim that adultery could be punished by arrows, an outcome Kohl-Larsen seemed to feel was usually avoided.

Among Hadza women, I could find no economic or reproductive advantage associated with the husband's presence or with his nomination score. Hadza women's access to plant resources is unlimited, their consumption is limited only by their time and energy. Their access to meat seems little impaired by a husband's departure. They appear to get substantial shares of any large animal killed by a camp member. Hadza women may be as close as we will find in human societies to Gowaty's (1996) "unconstrained females," unconstrained in their access to resources vital to their survival. Hadza women seem to agree about their access to resources. Two of my "filler questions" in the reproductive history interviews were: "Would you have any problem for food if your husband went away for a while?" and "If you moved to a new area, would you have any problem getting food?" I used these filler questions only in one year because the answers were so uniform; the unfortunate result is a modest sample size. Fifteen out of 18 women replied that they would have no problem getting food if their husband went on safari. Fifteen out of a sample of 19 different women said they would have no problem in a new place. (Two women were asked both questions; the answers come from 37 different women). Several of these women, and three who I scored as saying there would be a problem, commented on the need to ask the local people. They said the local people could not refuse; one would ask primarily for advice. Husbands were occasionally present during interviews and two added the comment that they would ask about hunting, but for advice not permission.

Because of their independence, perhaps ironically, Hadza women's lifetime RS appears largely shaped by "classic female" features such as age at the first birth, fertility, and success at keeping children alive (Chapters 13 and 15). Optimal weaning decisions, or having a reproductive system that optimizes IBIs is also a contributor (Chapter 17), and their fitness is further promoted by subsequent success as a grandmother (Chapter 18).

We would think that a woman married to a good hunter eats meat more often than others. If good hunters tend to aggregate, there should be an even more evident effect. Although I have shown no benefit to women married to good hunters, I have not compared outcomes for women or children in camps with good hunters, or with only poor hunters, or in game scarce localities. Nor should we neglect potential advantages to a woman from her kin who are eager to join her husband's camp, or from her ability to influence which of her kin stay with her. These ideas could easily be tested by looking at data on co-residence.

Other possibilities come to mind, and were in early formulations of the "show-off" or signaling theory. Good hunters may acquire and guard wives more successfully because other men give way to them. The good hunters then continue to pursue their status goals at a cost to their children but without fear of losing their wives to other men. They could also be preferred by women merely because of their "good genes." Less good hunters may acquire wives, and keep them, because they bring home food more often but achieve fewer concessions by other men.

## 22.8 Alternative theories of monogamy and marriage

As Hadza father effects are at best weak, and there is plenty of reward for success in mating competition, why is monogamy the common condition? While some find father effects (like Winking and Gurven, 2011), other investigators have failed to find a father effect, yet primarily monogamous marriage exists in those societies as well. Among the Hadza, monogamy is serial but still monogamy: one man and one woman living together and sharing a house. Although affection is not openly displayed among Hadza, signs of attachment, intimacy, jealousy, and fury at cheating or desertion can be found in both men and women. The paternal investment theory of pair bonding seemed to account for the emotional aspects of marriage. Alternative theories have to face this challenge. As Gurven and Hill (2009) wrote with evident exasperation: "Long term pair-bonds cannot be explained by the signaling model alone because it implies that women should be indifferent to marrying good hunters and men should abandon their wives as their fecundity declines." While the first can be (too easily?) answered by the likely preference of unconstrained females for good genes, and some of the most reproductively successful Hadza men do abandon their wife late in her childbearing career, many do not. Alternative theories will not explain all the treasured features of marriage if we do not try, hopefully with testable rather than untestable ideas.

The classic sociobiological idea is one among several competing ideas about evolution of social and genetic monogamy (Reichard and Boesch, 2003). Recent papers on the evolution of monogamy discuss three of these: difficulty of guarding more than one dispersed female, protection against male infanticide, and male care. Others include monogamy as "failed polygamy" due to male–male competition, mate guarding consortships in which females gain by protection from coercion; and there are many more subdivisions and details within these theories that require tests not suggested by the male provisioning or division of labor theory.

Lukas and Clutton-Brock (2013) conclude that among mammals as a whole, the dispersed female theory appears to have been the primary precursor of monogamy. As in Brotherton and Komers' (2003) studies of the dik-dik, a shy, small deer-like neighbor of the Hadza everywhere in the bush, females are distributed too widely for males to be able to successfully guard more than one at a time against competing males. At first sight, this would seem to be a poor candidate for an origin of human monogamy, simply because so many primates live in groups, and so do most humans. However, Pusey and colleagues (Murray, 2007, 2009) describe chimpanzee females competing over preferred sites. Langergraber *et al.* (2013) show that associations between individual male and female chimpanzees can arise from their preferences for certain localities. They suggest this needs to be considered in connection with the early origins of human pair bonding. Given that Hadza move so often and are in camp with so many different people, one might also wonder if it is impossible for a man to "guard" more than one woman at a time. But in a Hadza-like demography, the average sibship size that includes a woman of reproductive age is about three (Chapter 9; SI 9.7). This implies that the average woman may have one sister and they may be expected to move around together (more so if they are full siblings, Chapters 19 and 20), and with their mother if she is still alive. The number of siblings would be fewer in a stationary population (neither increasing nor decreasing). The propositions of Langergraber *et al.* deserve continued attention. In my initial look at my data on "who lives with whom," it is striking that the husband and wife are together in more of my censuses than any other category of adult. A number of possibilities for research on who lives with whom come to mind.

Opie *et al.* (2013) argued that among primates, male infanticide has been the dominant precursor of monogamy. Male infanticide would impose a very high cost to male desertion. By staying with a mate, a male has the chance to prevent competing males from killing his offspring. Sear and Mace (2008), finding few cases in which men had been shown to affect child survival, suggested that infanticide prevention had a role in the origin of human monogamy. The one instance of a father effect that they found was the Ache, where there were incidents of sacrifice of children if the father died. However, these Ache sacrifices were sometimes of older girls "to help their father in the after life." They do not fit the primate (or lion) pattern of killing the smallest infants, which then results in females returning to fertility sooner than if they suckled their infant to normal weaning age. Hill and Hurtado (1996, p. 68) report that Ache men claimed that children were too expensive to support if their father were not contributing. We have no indication of any infanticide by men or women among the Hadza. Among other groups where infanticide was practiced (outside a context of warfare), it was usually the decision of the mother alone.

Although subsequently arguing the details of their differing claims about the primacy of female dispersion and male infanticide, both Lukas and Clutton-Brock (2013) and Opie *et al.* (2013) suggest that male care seems to have evolved secondary to monogamy. Alonzo (2011) showed that when females can chose their mates, male care can evolve. Gavrilets (2012) reaches a similar conclusion, in addition arguing that male care may appear first among lower ranking males, which may remind us of the less expert hunters discussed previously and in Chapter 21. These men were less likely to raise a subsequent second family than the expert hunters, but their presence, according to my statistical analyses, may have improved their children's chances of survival. Gowaty (1996), discussed later, implies that male care was more likely to be a mate choice criterion among females in poor condition or poor habitat. Others may wonder whether females could, if we go far enough back in time, exert any choice at all. Choice might be favored whenever females are the rarer sex, or when kin can assist the females themselves by tipping the balance of power between closely competing males (Raleigh and McGuire, 1989, provide a non-human primate analog, other instances are summarized by Smuts, 1992). Moreover, we might bear in mind the possibility that monogamy is merely an arrangement between males.

Hadza women probably derive some of their power to choose from their economic independence, and some from the support, or enthusiasm, of their kin. They may also gain power by the opportunity to threaten to "run away with a Swahili." Bailey (1988) suggested this may be an important feature of many observed modern hunter-gatherers. His quantitative example is the curtailed hunting time of Efe men and the general point that competing for mates may lead to less than optimal foraging. He also suggests that competition with villager men might lead to forager men specializing in wild resources that farmers and herders do not acquire. Some of the effects might be comparable to a strongly male-biased adult sex ratio. The opportunity for Hadza women to marry outsiders has increased only recently (Chapters 3 and 5), while Hadza have been recorded as killing large game since Obst's observations (Obst, 1912).

Smuts (1992) proposes that a protective role for an individual male may extend beyond resisting infanticide to include guarding against wider forms of male coercion. In other primates, male–female consortships are thought to protect females against dangerous and time-wasting interference and attempts at coercion by non-consorting males. Coercion by the husband himself can also be an issue, as Smuts points out, hence perhaps Hadza women's expressed preference for a peaceful husband. Smuts also reminds us that alliances of females can guard against male coercion, and no fieldworker among the Hadza would doubt the ability of Hadza women to display quite intimidating solidarity in support of kin. However, Hadza husbands and wives do not behave entirely like consorting primates. Most of the day they are apart, women foraging with other women, men fixing arrows and smoking at "the men's place" or out "walking about" (hunting, or visiting) on their own, or away all night at a hunting blind. Having a husband who might retaliate after being told of a transgression might or might not reduce the likelihood of injurious or time-wasting attempts at seduction, coercion, theft, or general hassling of women. It does not seem to prevent young, single, visiting Hadza men from helping themselves to women's hard-won foods. Somehow we have to understand and test the detailed evolutionary economics of the "respect" for other men's consortships that Smuts discusses, and many other components of her rather persuasive theory of the evolution of monogamy.

Monogamy as an outcome of male–male competition has been considered in other ways (Hawkes *et al.*, 1995b; Blurton Jones *et al.*, 2000). Male–male competition was missing from the early mate-desertion models. As Houston and McNamara (2002) and Kokko and Jennions (2008) have pointed out, the costs of desertion have to include the costs of competing for a new mate. As more males desert, the competition among the newly single males becomes more intense, if the newly deserted females are unavailable for conception. The implication that guarding a mate may be easier than acquiring one is tempting but should probably be examined carefully. It has also been argued that if females are scarce, males must comply with their choice criteria because desertion is again costly (there are more males to compete with), and staying is cheaper. This has usually been discussed in the context of preference for caring males. Gowaty (1996) points out that females may not always have this preference. The "unconstrained" female may choose on the basis of male genetic quality alone.

The simple and interesting model by Gowaty (1996) is primarily aimed at birds and some insects, although she does suggest we should consider it for humans as well. She distinguishes between constrained and unconstrained females. Unconstrained females have free access to all the resources they need for maintenance and survival, and we suppose, for reproduction and raising offspring. Constrained females do not. The unconstrained females should choose males by genetic quality, even if they all mate with the same male (as in a lekking system). Constrained females need improved access to resources, which makes them vulnerable to the efforts of lower quality males in what Gowaty calls "helpful coercion." Males can persuade the resource-constrained female to mate with them in exchange for useful resources (such as exclusive use of a productive territory), even at the cost to the female of mating with a lower "quality" male. Gowaty's fig. 2.2 (1996) suggests how female fitness might vary with habitat richness and female quality. A high-quality female in a rich habitat has the same fitness with or without a lasting male partner. In a poor habitat, or as a lower quality female, there may be a difference in fitness between those with a "helpful" male and those without. If we regard the Hadza habitat of 1985–2000 as rich (Chapter 2), then the presence of a male will make no difference, much as I seem to have found. In a less rich habitat, males may make more difference. In the drought year 1997, there were fewer Hadza divorces than in other years; perhaps not a mere coincidence. Among the !Kung, in a less rich environment, the divorce rate is lower, and a father effect has been claimed (on the limited basis that women married more than once had higher child mortality; Pennington and Harpending, 1988). Gowaty derives more predictions from this very simple model, including that variance in male RS will be higher in the rich habitat and lower in the poor habitat. In the poor habitat, females choose helpful males and more males gain by putting more effort into helpfully coercing females into accepting them. In the rich habitat and among high-quality females, males will be less able to helpfully coerce and more likely to have to suffer selection by their genetic quality, leaving more of them unchosen, and increasing the selection in favor of those who display their high quality. In Chapter 13 (Table 13.1), I reported the higher variance in Ache and Hadza men's RS and the lower variance among !Kung men. As Hadza habitat becomes impoverished at the hands of invading farmers and herders, if the impoverishment affects women's resources, we should expect to see women become more vulnerable to helpful coercion. We should be more likely to see father effects, and a lowering of variance in male RS.

Helpful coercion could also be discussed as "paternal provisioning" or "female extortion." Such entertaining terms are a treacherous blend of the usefully provocative and the wildly misleading. Bitter experience with "tolerated theft" (Hadza do not steal), and "show-off" (good hunters do not boast), teach quite firmly that the label can take on a life of its own, well removed from the original concept. "Pillars of society" might have given a better image for the expert hunters, until some Ibsen fan pointed out the seamy self-serving connotations of the label. My favorite was "the congressman theory of male strategies," but this has been outdated by legislation that makes it more difficult for your repeatedly elected and tolerated representative to bring valuable economic assets home to your district. The lesson is that we should try to spell out the implications of our ideas well enough that our readership is not too distracted by these homespun connotations.

Gowaty writes about another process that is important for anthropology, "resource sequestration." If a male can take control of resources, he can helpfully coerce females more easily. Gowaty gives territories and early migration to claim territories before females arrive, as avian examples. The Victorian British laws of marital property would be one of the most obvious and extreme human examples. Men had somehow taken control of resources. Women lost all economic independence upon marriage. Men legally controlled even the resources that women brought to the marriage. Under this utterly barbarous regime, divorce (rarely granted to women) left women destitute. A middle- and upper-class Victorian wife, therefore, was totally economically dependent on her husband. She had no alternative; leaving him was an idle threat at best. No wonder Darwin was surprised by his Fuegians. Many a wife in the rural developing world is in a not too dissimilar position from the Victorian wife. In many rural societies, the man owns the house and the fields and the livestock. Among herders, the men own the major portion of the livestock. However, two studies should make us wonder about this generalization. Among both the Herero (Pennington and Harpending, 1993) and the Datoga (Sellen *et al.*, 2000; Butovskaya, 2012), women own some of the livestock and apparently retain control of it. Datoga women are able to hold trials of men and fine them for transgressions. It would be interesting to know more about how women maintain their power in these cases, and in other cases, to know to what extent they are able to "work the system."

Hadza women are at an extreme. They can feed themselves by their own labor, and with some help from female kin can feed their children as well. In my data, I was unable to show more than a slender (and usually not statistically significant) handicap to a woman or to the survival and growth of her small children from male absence. Some young Hadza women marry non-Hadza men; the luxury presents, the house, and the food must seem attractive, but at least half of these women divorce and return to live among the Hadza (Chapters 5 and 14). In interviews, they made it clear that life as a farmer's wife was not to their liking. Being beaten was an affront; the lack of freedom onerous. They are lucky to be able to leave so easily. It is too hard to find them in the bush. I have seen it tried! I also came across a young non-Hadza woman who had the good fortune to know enough individual Hadza to enable her to flee into the bush and escape on the "underground" to relatives far from her husband's land.

There may be traces of "resource sequestration" even among the Hadza. Women are not meant to handle bows for example. Meat is spoken of as "real food," the plant food the women gather is not. The epeme meat is exclusively for men. Women are not supposed to witness or claim knowledge about the men's feast. Although it comprises all the pieces that Americans would not eat, the epeme meat incudes some of the fattiest parts of the animal. Recall Speth's (2010, and elsewhere) argument that fat is one of the most important nutrients obtained by hunting, and Sherry and Marlowe's (2007) observation that middle-aged Hadza men were able to accumulate some body fat.

Alternative theories of monogamy may awaken us to a wider array of possible origins, and possible variations, conflicts, and compromises in marriages. Thinking about a wider array of male strategies might give us similar assistance. Sadly but perhaps more realistically, the eventual picture of marriage and male strategies that emerges may not be quite as rosy as the picture originally painted by paternal investment and the division of labor.

## 22.9 "Show-off" and competitive altruism

Debate about hunter-gatherer reproductive strategies involves debate about the nature of food sharing. In this section, I suggest that here we have also been hobbled by attending to a greatly restricted sample of the available theories.

In Chapter 14, I introduced the "show-off" theory of big game hunting (BGH). Is BGH the best way for a man to feed his wife and children, or are there additional outcomes that reward the effort and dangers of hunting very large animals? In Chapter 21 and earlier, I reported that BGH's children are observed to suffer higher mortality than other men's children. In Hawkes *et al.* (1991), we suggested that children might do better with a small helping of meat almost daily than with gorging themselves at intervals of two weeks or more. Our data on hunting success and variance and Hawkes' experiment in which men were paid to pursue small game suggested that men could acquire small prey about one day in two if they tried. The sizes of these prey gave an average of 0.78 kg of meat/day. Large game was hit much less frequently, once in 29 hunter-days as a yearly average, more often in the dry season, less often in the wet months. Large prey were shared widely, which meant that a man's children ate from another man's kill more often than from his own, and each time they ate nearly as much as from their father's rare kill. It seemed unlikely that a man hunting big game would give an advantage to his children over those of other men. The details of the argument were elucidated in the report by Hawkes *et al.* (1991 and 2001), and have been set out again several times in response to counter-arguments. The issue is not directly about whether Hadza men show any paternal care, or fail to regard their children as important; after all, a man's children are his current fitness. It is more about whether BGH is a maximally effective childcare strategy, and whether it brings benefits to the hunter other than a good diet for his children. The abundant caricatures of the show-off ideas as "mating effort," as "cads," or as boastful womanizers, have distracted from important issues.

Hawkes proposed that men hunted large animals because of the attention it gained from the interested bystanders who ate most of the meat. She suggested several ways in which this might ultimately enhance a man's fitness. A successful hunter might be preferred as a neighbor and, in order to keep him in camp and his meat bonanzas coming, he might be deferred to. His wife and children might receive special treatment, encouraging him to stay and continue the good work. Hunting large, sometimes dangerous animals may signal his strength, ability, and cleverness in a way that cannot be faked and, therefore, may be a useful signal of his genetic quality to those who may prefer their children to have good genes. These same qualities may make him a potentially more costly opponent in disputes or competition for mates.

It would be difficult to fake the accomplishments of a good hunter. Hunting large and dangerous animals, in competition with their even more dangerous predators and would-be scavengers, is costly and this would make hunting success an honest signal of the required qualities. Lars Smith and I witnessed a short-lived attempt to fake success. In the gathering around our camp fire, a young man began to tell how he had killed a giraffe. He began to describe the kill. Others quickly saw through his story; where was the meat? Well, it hadn't actually died but he had hamstrung it. Where was that? Why don't we go out there and get it? That was the end of story. The young man was suitably embarrassed. Given the low frequency with which anyone kills a large animal, impressions of who is good at it and who is not will accumulate only slowly and erratically. Nonetheless, they do accumulate, perhaps largely by gossip. In Chapter 14, I showed that the accumulated reputations were consistent, and in the several small samples that allowed a test, they have correlated with observed large animal hunting success.

It was quickly pointed out that some of the proposed benefits to the good hunter entailed a secondary public goods problem (Smith, 1993). Why be the one to make concessions to the good hunter, his wife, or children to keep him in camp? Why not let someone else take the cost? Two of the benefits listed earlier would not suffer this problem; it is in the direct interests of each woman to choose a higher-quality mate, and in the interests of each man to choose his battles wisely.

My data show that good hunters succeed in competition with other men for younger, therefore more fertile, wives. Women show no reluctance to marry them. I was unable to show any benefit to the wives or their children, however. If others were making concessions to these women, they were of no measurable benefit to the women or their young children. Thus my data, as analyzed thus far, leave us with the still untested benefit of genetic quality to the women attracted to good hunters. Nor have I either confirmed or excluded the possibility that it is in the interests of other men to give way to good hunters in competition.

I have shown no benefits to women from being married to a good hunter. Perhaps being married to a good hunter gives women some of the social benefits that accrue to good hunters. It may display their attractiveness and some heritable quality that will increase the chance that their daughters are successful. They may be desired as good neighbors, perhaps able to influence the good hunter to stay around and continue the good work. I tried to test Hawkes' suggestion that good hunters get harder-working wives by using women's reputations. There was a small significant correlation between husband reputation and wife reputation. Women's reputations as hard workers mainly supported the observation that older women work the hardest. None of the conceivable benefits to wives of good hunters show in my record of women's RS.

This may not be the end of the story for the wider aspects of the show-off idea. There are unexplored aspects of the data, and recent theoretical arguments overcome the secondary public goods problem. Hadza women married to a good hunter presumably eat meat of large animals more often than do other women. We do not know, however, whether poor hunters also bring home a small animal or some honey more often. Wood and Marlowe (2013; table S1) seem to imply that they do bring back more honey. If women married to good hunters do eat more meat, it does not show in improved growth or survival of their children. Women married to a good hunter may be able to attract more of their close kin and friends to their camp and those kin may benefit. If they do, although grandmothers are very effective helpers (Chapter 18), they are not sufficiently attracted to the camp of a good hunter to generate an association between being a good hunter and his small children's growth or survival. There are other possibilities to do with camp composition and kin that may not show up in child growth or survival. It may be in the interests of each kin to persuade a young female relative to marry a good hunter, because they will more often be in camp with the good hunter. I have not tabulated child growth or survival in relation to co-residence with a good hunter who is not their father. My records of camp co-residence are widely spaced, being collected only at the time of each census, and may not be appropriate for this purpose.

## 22.10 Other theories of sharing: competitive altruism, partner choice, biological markets

Hadza say they share because their hearts are good. How are we to reconcile this with theoretical accounts of the economics and evolution of sharing, cooperating, helping? Little in this book has directly addressed this question, but no one who works with hunter-gatherers can escape it. The daily demands, lengthy charming explanations of why you owe just one more present, are painful and exhausting, the more so as every fieldworker becomes an advocate for "his people."

With her interest in "prestige" and her good neighbor suggestion, Hawkes anticipated the development of several important strands of work on evolution of cooperation. These include competitive altruism (Zahavi and Zahavi, 1997; Roberts, 1998, 2005; Sherrat and Roberts, 1998; Lotem *et al.*, 2002), "biological market" (Noe, 1994, 2001; MacFarlan *et al.*, 2012; Barclay, 2013), and partner choice with reputations (Milinski *et al.*, 2002; Semmann *et al.*, 2004, 2005; Fu *et al.*, 2008). These perspectives break the restrictions of the two-person prisoners' dilemma school of investigation of reciprocal altruism. If the "audience" members are free to choose who they take as a neighbor, and if living alone is costly (dangerous, for example, although we have known one youth who often lived alone, and one elderly couple was well known for living on their own, and Woodburn, 1968b, p. 105, lists camp sizes as ranging "from a single person to almost a hundred"), altruistic behavior can evolve. Third-party punishment is not needed; it is "built into" the choices of the actor and the audience and the costs of living alone. The process is more akin to sexual selection than to reciprocal altruism (Nesse, 2007). Signaling plays a key role in some of the models, which collectively seem to support exactly the interpretation that Hawkes (1993) and Hawkes and Bliege Bird (2002) gave for Hadza BGH and meat distribution. Because none of the references given here cite Hawkes, we may take them as independent support for the likelihood that natural selection can generate the scenario she outlined. They even make it clear that some of these systems can support the production of public goods by selfishly signaling individuals, a conclusion by Hawkes that some have found mysterious. These and related papers deserve much closer scrutiny than I have given them here, and may help us understand both the abundance of "generalized reciprocity" and the details of variation in other aspects of sharing that have been described in hunting and gathering societies.

Reciprocal altruism in its original narrow sense focuses on the contingency of help received from a specific individual on help given to that individual. The "competitive altruist" gains from widening his/her audience, including by displaying altruism to those who never reciprocate, just as we (and Woodburn, 1998) observed for Hadza hunters, and others before us have described among other hunter-gatherers (Sahlins, 1974, gave this the self-contradictory label "generalized reciprocity," which he contrasted with "balanced reciprocity"). Hunter-gatherer generosity to those in need, including those who never reciprocate, is no longer a theoretical problem that has to be squeezed into extensions of reciprocal altruism that attenuate the concept into meaninglessness. Reciprocal altruism probably exists, in particular contexts, such as when partners can be predictably located, and may be best used in its purist sense. Stretching it to include forager meat sharing may make it meaninglessly general. No one has specified how long a reciprocal altruist would be expected to wait for a reciprocation before deciding whether a "partner" is a worthwhile reciprocator, or deserving of punishment. If Hadza hunters are involved in reciprocal arrangements with individual partners (debated by Hawkes *et al.*, 2001; Gurven, 2004; Hawkes *et al.*, 2010), they have a long time to wait for their partner's next catch (a catch per 29 days in 1985–1986, a catch per 102 days among the eastern Hadza in 2005–2009; Wood and Marlowe, 2013). They have the added problem that their partner is quite likely living in a different camp or even a different region by the time s/he is in a position to reciprocate.

**Photograph 22.1** Mobility. Every camp move includes a decision about with whom to live. The man carries vulture feathers, valued for fletching some kinds of arrows. © James F. O'Connell, 2015. Reproduced with permission.

"Who lives in camp with whom" may be an important issue in Hadza life. The fluidity of camp composition and mobility of individuals implies that Hadza make continual choices about their neighbors (Photograph 22.1). This is a condition for a biological market and a far cry from the repeated prisoners dilemma. In a small, mobile population in which individuals are named, recognized, talked about, and remembered, the payoff for becoming known as a good neighbor may be substantial. While I have shown that reproductive benefits accrue to men who are recognized as good hunters, and that some women are known for being hard workers, I have largely neglected the reputations of women and the factors (additional to their husbands) that might make them "good neighbors." It may be important that we notice, when women sit down for their lunch in the bush, even when different kin groups of women have their own fires, a few roasted tubers are passed from one group to another. Perhaps women's varied and sometimes ambivalent responses to my questions about sharing birds of different sizes, reflect not just the crude economics of tolerated theft but the ethic of sharing and being known as a good neighbor, and its conflict with the temptation to hide or quickly consume a small item. As one of them said about keeping a bird for her children, "I can but it is bad" to refuse the others.

## 22.11 Prospects and hazards at an interface with social science insights

While it is easy to wave one's arms in favor of a biological market or partner choice account of hunter-gatherer sharing, we need to be able to explain why it is less evident in other societies and other species. There are more requirements for a good theory of sharing. We must account for some things being recognized as personal property (a bow, a digging stick) and seldom shared, others apparently recognized as private property but readily loaned (like a knife or pot), or given away like an article of clothing, and yet others apparently being shared reciprocally (Apicella *et al.*, 2012), and apparently some redistribution by games of chance (Woodburn, 1968a).

If fitness is influenced by with whom you share a camp, and with whom you share a camp depends on their reputation as a "good person," we may have a circumstance open to quite a variety of outcomes. What determines or limits what counts as being a good person? Could a population of individuals somehow "pick on" any random feature as the criterion for a good person? Is it ever likely to be someone who does you harm? According to some models, it is unlikely to be a feature that signals low quality. Those who have things to give away have done better than those who have none, and probably partly as a result of their "quality." In the Hadza case, and apparently the !Kung and the Martu, it is someone who works to give away food (Codding and Kramer, 2015). What about other public goods? Could it be someone who works out how to stop non-Hadzas destroying the land? Someone who deals usefully with outside authorities? If not, then why not? Hadza are conspicuously bad at such tasks. Economists have had plenty to say about the difficulty of solving public goods problems, and examples are in the public eye daily. Some suggest that the advertizing value of an act will be one influential feature. The individual who makes available a very large amount of meat is visible to many. The identity of the individual who figures out how to rescue the bucket from the well (a !Kung example) will not have a wide audience and his identity may become confused during the conversations about the achievement. How much of the "superstructure" of a society comprises measures, demonstrations, tests for who is a good person? Does the value of identifying good people shape some of the variation in which cultural traits are transmitted and which are not?

This section is long on "arm waving" and short on tight logic. We should try to determine what these proposals might imply: if everyone is trying to optimize their mix of neighbors, would all the good hunters live together? Would everyone be evenly distributed among good and less good individuals? Some mixture is hinted at by Gudo Mahiya's opinion that it was best to be a good hunter among few good hunters. If you were the only good hunter in camp, people came to expect too much of you. If you were among many good hunters, you were nobody special. Would the extreme mobility that we observe be another result, accidentally perpetuating the opportunity for "social selection"? Everyone is presumably constrained by ties of marriage and kinship. How much might these constrain the incentives for altruism? Are good neighbors subject to analogous constraints to good habitats? Why do we not see more evidence of direct competition and conflict over who lives in whose camp? Is there a useful analogy with models of foraging group size (Sibly, 1983; Smith, 1985, 1991) and joiners and leavers rules?

Social selection and signaling quality might result both in individuals who try to be generous, and bystanders who are eager to know who is generous and associate generosity with goodness. It may be that an "ethic," a trait and the common interest and approval of the trait, can easily arise from the biomarket processes. Does this mean that some of the traits we have neglected as inaccessible to an evolutionary ecology approach, abrogated to cultural transmission theory, are actually accessible to evolutionary ecology? An evolutionary ecology approach to "culture" may be able to do more than suggest that societal rules are just another arena in which old men can show off their erudition and manipulate the young.

It does not seem healthy that our field separates cultural transmission theory so sharply from natural selection and the adaptationist thinking of evolutionary ecology. From time to time, Boyd and Richerson (1985, and subsequently) have tried to remind us that cultural transmission theory rests on taking account of the fitness costs of acquiring information. However, most of the time we work as if these were quite separate fields. Some of what we currently think of as "randomly maladaptive, so probably culturally transmitted," and not to be explained by evolutionary ecology, may be nothing of the sort. Some of the rules, beliefs, and customs may be outcomes of partner choice, ways to compete over looking like a good person and with a discoverable ecological-economic root to them. Some may not be. This suggestion may offer a way to bridge the gap between evolutionary ecology and cultural transmission theory (and wider fields of thought about culture) and help account for some of the emotionality and power that surrounds culture and seems to be left out of cultural transmission theory. Evolutionary ecology may not be confined to studying society's infrastructure much longer but be creeping up on its "superstructure." We have too long ignored the lead offered by Smuts (1992 and elsewhere), but recently Bliege Bird *et al.* (2012) and Rodseth (2012) have set us good new leads. A close study of twenty-first century altruism models might give us more.

# References


Aime, C., Laval, G., Patin, E. *et al.* (2013). Human genetic data reveal contrasting demographic patterns between sedentary and nomadic populations that predate the emergence of farming. *Molecular Biology and Evolution* 30, 2629–2644.

Alberts, S. C., Buchan, J. C. and Altmann, J. (2006). Sexual selection in wild baboons: from mating opportunities to paternity success. *Animal Behaviour* 72, 1177–1196.

Alexander, R. D. (1979). *Darwinism and Human Affairs*. Seattle, WA: University of Washington Press.

Alonzo, S. H. (2011). Sexual selection favours male parental care, when females can choose. *Proceedings of the Royal Society of London B*. doi: 10.1098/rspb.2011.2237.

Altizer, S., Nunn, C. L., Thrall, P. H. *et al.* (2003). Social organization and parasite risk in mammals: integrating theory and empirical studies. *Annual Review of Ecology, Evolution, and Systematics* 34, 517–547.

Alvarez, H. P. (2004). Residence groups among hunter-gatherers: a view of the claims and evidence for patrilocal bands. In *Kinship and Behavior in Primates*, eds. B. Chapais and C. M. Berman. New York, NY: Oxford University Press, pp. 420–442.

Anderson, G. D. and Herlocker, D. J. (1973). Soil factors affecting the distribution of the vegetation types and their utilization by wild animals in Ngorongoro Crater, Tanzania. *Journal of Ecology* 61, 627–651.

Anderson, M. T. and Seifert, H. S. (2011). Opportunity and means: horizontal gene transfer from the human host to a bacterial pathogen. *mBio* 2, e00005–11.

Anderson, R. M. and May, R. M. (1991). *Infectious Diseases of Humans: Dynamics and Control*. Oxford: Oxford University Press.

Andrewartha, H. G. and Birch, L. C. (1982). *Selections from the Distribution and Abundance of Animals*. Chicago, IL: The University of Chicago Press.

Apicella, C. L., Marlowe, F. W., Fowler, J. H. *et al.* (2012). Social networks and cooperation in hunter-gatherers. *Nature* 481, 497–501.

Armon, P. J. (1979). Maternal deaths in the Kilimanjaro region of Tanzania. *Transactions of the Royal Society of Tropical Medicine and Hygiene* 73, 284–288.

Armstrong Schellenberg, J. R. M., Mrisho, M., Manzi, F. *et al.* (2008). Health and survival of young children in southern Tanzania. *BMC Public Health* 8, 194.

Arnold, S. J. and Wade, M. J. (1984a). On the measurement of natural and sexual selection: theory. *Evolution* 38, 709–719.

Arnold, S. J. and Wade, M. J. (1984b). On the measurement of natural and sexual selection: applications. *Evolution* 38, 720–734.

Bagnall, R. S. and Frier, B. W. (1994). *The Demography of Roman Egypt*. Cambridge: Cambridge University Press.

Bagshawe, F. J. (1924). The Peoples of the Happy Valley (East Africa). Part 1: The Aboriginal Races of Kondoa Irangi. *Journal of the African Society* 24, 25–33.

Bagshawe, F. J. (1925). The Peoples of the Happy Valley (East Africa). Part 2: The Kangeju. *Journal of the African Society* 24, 117–130.

Bailey, R. C. (1988). The significance of hypergyny for understanding subsistence behavior of contemporary hunters and gatherers. In *Diet and Subsistence: Current Archaeological Perspectives (Proceedings of the 19th Annual Chacmool Conference)*, eds. B. V. Kennedy and G. M. LeMoine. Calgary, AB: University of Calgary Archaeological Association, pp. 57–65.

Bailey, R. C. and Aunger, R. J. (1989a). Significance of the social relationships of Efe Pygmy men in the Ituri forest, Zaire. *American Journal of Physical Anthropology* 78, 495–507.

Bailey, R. C. and Aunger, R. (1989b). Net hunters vs. archers: variation in women's subsistence strategies in the Ituri forest. *Human Ecology* 17, 273–297.

Bailey, R. C. and Aunger, R. (1990). Humans as primates: the social relationships of Efe Pygmy men in comparative perspective. *International Journal of Primatology* 11, 127–146.

Bailey, R. C., Head, G., Jenike, M. *et al.* (1989). Hunting and gathering in tropical rain forest: is it possible? *American Anthropologist* 91, 59–82.

Bailey, R. C., Jenike, M. R., Ellison, P. T. *et al.* (1992). The ecology of birth seasonality among agriculturalists in central Africa. *Journal of Biosocial Science* 24, 393–412.

Bairagi, R., Chowdhury, M. K., Kim, Y. J. *et al.* (1985). Alternative anthropometric indicators of mortality. *American Journal of Clinical Nutrition* 42, 296–306.

Baker, S. J. K. (1974). A background to the study of drought in east Africa. *African Affairs* **73**, 170–177.

Bala, G. G. (1998). *Hadza Stories and Songs* (translated by Bonny Sands). Los Angeles, CA: Friends of the Hadzabe.

Barclay, G. W. (1958). *Techniques of Population Analysis*. New York, NY: John Wiley and Sons.

Barclay, P. (2013). Strategies for cooperation in biological markets, especially for humans. *Evolution and Human Behavior* **34**, 164–175.

Barclay, P. and Willer, R. (2007). Partner choice creates competitive altruism in humans. *Proceedings of the Royal Society of London B* **274**, 749–753.

Barnicot, N. A., Bennett, F. J., Woodburn, J. C. *et al.* (1972a). Blood pressure and serum cholesterol in the Hadza of Tanzania. *Human Biology* **44**, 87–116.

Barnicot, N. A., Mukherjee, D. P., Woodburn, J. C. *et al.* (1972b). Dermatoglyphics of the Hadza of Tanzania. *Human Biology* **44**, 621–648.

Barns, T. A. (1923). Ngorongoro, the giant crater; and the Gorilla, the giant ape. *Journal of the African Society* **22**, 179–188.

Bart, J. and Tornes, A. (1989). Importance of monogamous male birds in determining reproductive success. *Behavioral Ecology and Sociobiology* **24**, 109–116.

Baumann, O. (1894a). *Durch Massailand zur Nilquelle*. Berlin: Reimer (Reprinted Johnson Reprint Co., New York, NY, 1968).

Baumann, O. (1894b). Die kartographischen ergebnisse der Massai-Expedition des Deutschen Antisklaverei Comités. Dr. A. Petermanns Mitteilungen aus Justus Perthes' Geographischer Anstalt, Ergänzungsheft Nr. *111*.

Becker, N. S. A., Verdu, P., Hewlett, B. *et al.* (2010). Can life history trade-offs explain the evolution of short stature in human Pygmies? A response to Migliano *et al.* (2007). *Human Biology* **82**, 17–27.

Behrensmeyer, A. K. (2006). Climate changes and human evolution. *Science* **311**, 476–478.

Belovsky, G. E. (1988). An optimal foraging-based model of hunter-gatherer population dynamics. *Journal of Anthropological Archaeology* **7**, 329–372.

Belsey, M. A. (1976). The epidemiology of infertility: a review with particular reference to sub-Saharan Africa. *Bulletin of the World Health Organization* **54**, 319–341.

Belshaw, R., Pereira, V., Katzourakis, A. *et al.* (2004). Long-term reinfection of the human genome by endogenous retroviruses. *Proceedings of the National Academy of Sciences USA* **101**, 4894–4899.

Belsky, A. J. (1990). Tree/grass ratios in East African savannas: a comparison of existing models. *Journal of Biogeography* **17**, 483–489.

Bennett, F. J., Barnicot, N. A., Woodburn, J. C. *et al.* (1973). Studies on viral, bacterial, rickettsial, and treponemal diseases of the Hadza of Tanzania, and a note on injuries. *Human Biology* **45**, 243–272.

Bennett, F. J., Kagan, I. G., Barnicot, N. A. *et al.* (1970). Helminth and protozoal parasites of the Hadza of Tanzania. *Transactions of the Royal Society of Tropical Medicine and Hygiene* **64**, 857–880.

Bentley, G., Goldberg, T. and Jasienska, G. (1993a). The fertility of agricultural and non-agricultural traditional societies. *Population Studies* **47**, 269–281.

Bentley, G., Jasienska, G. and Goldberg T. (1993b). Is the fertility of agriculturalists higher than that of non-agriculturalists? *Current Anthropology* **34**, 778–785.

Bercovitch, F. B., Widdig, A. and Nurnberg, P. (2000). Maternal investment in rhesus macaques (*Macaca mulatta*): reproductive costs and consequences of raising sons. *Behavioral Ecology and Sociobiology* **48**, 1–11.

Berger, P. (1943). Uberlieferungen der Kindiga. *Africa (Berlin)* **2**, 92–122.

Betzig, L. (2012). Means, variances, and ranges in reproductive success: comparative evidence. *Evolution and Human Behavior* **33**, 309–317.

Biran, A., Abbot, J. and Mace, R. (2004). Families and firewood: a comparative analysis of the costs and benefits of children in firewood collection and use in two rural communities in sub-Saharan Africa. *Human Ecology* **32**, 1–25.

Bird, D. W. and Bliege Bird, R. (2002). Children on the reef: slow learning or strategic foraging? *Human Nature* **13**, 269–298.

Bird, D. W. and Bliege Bird, R. (2005). Martu children's hunting strategies in the Western Desert, Australia. In *Hunter-Gatherer Childhoods*, eds. B. S. Hewlett and M. E. Lamb. New Brunswick, NJ: Aldine Transaction, pp. 129–146.

Bird, D. W. and O'Connell, J. F. (2006). Behavioral ecology and archaeology. *Journal of Archaeological Research* **14**, 143–188.

Bird, R. (1999). Cooperation and conflict: the behavioral ecology of the sexual division of labor. *Evolutionary Anthropology* **8**, 65–75.

Birdsell, J. B. (1968). Some predictions for the Pleistocene based on equilibrium systems among recent hunter-gatherers. In *Man the Hunter*, eds. R. B. Lee and I. DeVore. Chicago, IL: Aldine, pp. 229–240.

Bittles, A. H. and Black, M. L. (2010). Consanguinity, human evolution, and complex diseases. *Proceedings of the National Academy of Sciences USA* **107** (Suppl. 1), 1779–1786.

Black, F. L., Hierholzer, W. J. and Pinheiro, F. (1974). Evidence for persistence of infectious agents in isolated human populations. *American Journal of Epidemiology* **100**, 230–250.

Black, J. M., ed. (1996). *Partnerships in Birds: the Study of Monogamy*. Oxford Ornithology Series. Oxford: Oxford University Press.

Blackburn, R. H. (1982). In the land of milk and honey: Okiek adaptations to their forests and neighbours. In *Politics and History in Band Societies*, eds. E. Leacock and R. Lee. Cambridge: Cambridge University Press, pp. 283–305.

Bleek, D. (1930). *Notebooks from Hadza Language Project, Lake Eyasi, Tanzania*. Cape Town: Capetown University Library.

Bleek, D. (1931a). The Hadzapi or Watindiga of Tanganyika territory. *Africa* **4**, 273–286.

Bleek, D. F. (1931b). Traces of former bushman occupation in Tanganyika Territory. *South African Journal of Science* **28**, 423–429.

Bliege Bird, R. and Bird, D. W. (2008). Why women hunt: risk and contemporary foraging in a western desert Aboriginal community. *Current Anthropology* **49**, 655–693.

Bliege Bird, R., Scelza, B., Bird, D. W. *et al.* (2012). The hierarchy of virtue: mutualism, altruism and signaling in Martu women's cooperative hunting. *Evolution and Human Behavior* **33**, 64–78.

Bliege Bird, R., Smith, E. A. and Bird, D. W. (2001). The hunting handicap: costly signaling in male foraging strategies. *Behavioral Ecology and Sociobiology* **50**, 9–19.

Blurton Jones, N. G. (1986). Bushman birth spacing: a test for optimal inter-birth intervals. *Ethology and Sociobiology* **7**, 91–105.

Blurton Jones, N. G. (1987). Bushman birth spacing: direct tests of some simple predictions. *Ethology and Sociobiology* **8**, 183–204.

Blurton Jones, N. G. (1989). The costs of children and the adaptive scheduling of births: towards a sociobiological perspective on demography. In *The Sociobiology of Sexual and Reproductive Strategies*, eds. A. Rasa, C. Vogel and E. Voland. London: Chapman and Hall, pp. 265–282.

Blurton Jones, N. G. (1993). The lives of hunter-gatherer children: effects of parental behavior and parental reproductive strategy. In *Juveniles: Comparative Socioecology*, eds. M. E. Pereira and L. A. Fairbanks. Oxford: Oxford University Press, pp. 309–326.

Blurton Jones, N. G. (1997). Too good to be true? Is there really a trade-off between number and care of offspring in human reproduction? In *Evolution and Human Behavior: a Critical Reader*, ed. L. Betzig. Oxford: Oxford University Press, pp. 83–86.

Blurton Jones, N. G. (2006). Contemporary hunter-gatherers and human life history evolution. In *The Evolution of Human Life History*, eds. K. Hawkes and R. R. Paine. Santa Fe, NM: SAR Press, pp. 231–266.

Blurton Jones, N. G. (2015). Why do so few Hadza farm? In *Twenty-first Century Hunters and Gatherers*. eds. K. Kramer and B. Codding. Santa Fe, NM: SAR Press.

Blurton Jones, N. G., Hawkes, K. and Draper, P. (1994). Foraging returns of !Kung adults and children: why didn't !Kung children forage? *Journal of Anthropological Research* **50**, 217–248.

Blurton Jones, N. G., Hawkes, K. and O'Connell, J. F. (1989). Modelling and measuring costs of children in two foraging societies. In *Comparative Socioecology*, eds. V. Standen and R. A. Foley. Oxford: Blackwell Scientific, pp. 367–390.

Blurton Jones, N. G., Hawkes, K. and O'Connell, J. F. (1996). The global process, and local ecology: how should we explain differences between the Hadza and the !Kung? In *Cultural Diversity in Twentieth Century Foragers*, ed. S. Kent. Cambridge: Cambridge University Press, pp. 159–187.

Blurton Jones, N. G., Hawkes, K. and O'Connell, J. F. (1997). Why do Hadza children forage? In *Genetic, Ethological and Evolutionary Perspectives on Human Development*, eds. N. L. Segal, G. E. Weisfeld and C. C. Weisfeld. Washington, DC: American Psychological Association, pp. 279–313.

Blurton Jones, N. G., Hawkes, K. and O'Connell, J. F. (2002). Antiquity of postreproductive life: are there modern impacts on hunter-gatherer postreproductive life spans? *American Journal of Human Biology* **14**, 184–205.

Blurton Jones, N. G., Hawkes, K. and O'Connell, J. F. (2005a). Hadza grandmothers as helpers: residence data. In *Grandmotherhood. The Evolutionary Significance of the Second Half of Female Life*, eds. E. Voland, A. Chasiotis and W. Schiefenhovel. New Brunswick, NJ: Rutgers University Press, pp. 160–176.

Blurton Jones, N. G., Hawkes, K. and O'Connell, J. F. (2005b). Older Hadza men and women as helpers. In *Hunter-Gatherer Childhoods*, eds. B. S. Hewlett and M. E. Lamb. New Brunswick, NJ: Aldine Transaction, pp. 214–236.

Blurton Jones, N. G. and Konner, M. J. (1976). !Kung knowledge of animal behavior. In *Kalahari Hunter-gatherers*, eds. R. B. Lee and I. DeVore. Cambridge, MA: Harvard University Press, pp. 325–348.

Blurton Jones, N. G. and Marlowe, F. W. (2002). Selection for delayed maturity: does it take 20 years to learn to hunt and gather? *Human Nature* 13, 199–238.

Blurton Jones, N. G., Marlowe, F. W., Hawkes, K. *et al.* (2000). Paternal investment and hunter-gatherer divorce rates. In *Adaptation and Human Behavior: An Anthropological Perspective*, eds. L. Cronk, W. Irons and N. Chagnon. New York, NY: Aldine de Gruyter, pp. 69–90.

Blurton Jones, N. G. and Sibly, R. M. (1978). Testing adaptiveness of culturally determined behaviour: Do bushman women maximise their reproductive success by spacing births widely and foraging seldom? In *Human Behaviour and Adaptation: Society for Study of Human Biology Symposium No 18*, eds. N. Blurton Jones and V. Reynolds. London: Taylor and Francis, pp. 135–158.

Blurton Jones, N. G., Smith, L. C., O'Connell, J. F. *et al.* (1992). Demography of the Hadza, an increasing and high density population of savanna foragers. *American Journal of Physical Anthropology* 89, 159–181.

Boehm, C. (1993). Egalitarian behavior and reverse dominance hierarchy. *Current Anthropology* 34, 227–254.

Bogin, B. (1988). The evolution of human childhood. *Bioscience* 40, 16–24.

Bogin, B. (1999). *Patterns of Human Growth*. Cambridge: Cambridge University Press.

Boone, J. L. (2002). Subsistence strategies and early human population history: an evolutionary ecological perspective. *World Archaeology* 34, 6–25.

Boone, J. L. and Kessler, K. (1999). More status or more children: social status, fertility reduction, and long-term fitness. *Evolution and Human Behavior* 20, 257–277.

Borgerhoff Mulder, M. (1988). The relevance of the polygyny threshold model to humans. In *Human Mating Patterns*, eds. C. G. N. Mascie-Taylor and A. J. Boyce. Cambridge: Cambridge University Press, pp. 84–102.

Borgerhoff Mulder, M. (1991). Human behavioral ecology. In *Behavioral Ecology: An Evolutionary Approach*, eds. J. R. Krebs and N. B. Davies. Oxford: Blackwell Scientific, pp. 69–98.

Borgerhoff Mulder, M. (1992). Demography of pastoralists: preliminary data on the Datoga of Tanzania. *Human Ecology* 20, 383–405.

Borgerhoff Mulder, M. (2009). Tradeoffs and sexual conflict over women's fertility preferences in Mpimbwe. *American Journal of Human Biology* 21, 478–487.

Borgerhoff Mulder, M., Sieff, D. and Merus, M. (1989). Disturbed ancestors: Datoga history in the Ngorongoro Crater. *Swara* 2(2), 32–35.

Borjeson, L. (2004). *A History under Siege. Intensive Agriculture in the Mbulu Highlands, Tanzania, 19th Century to the Present.* PhD thesis Stockholm: Department of Human Geography, Stockholm University.

Bowlby, J. (1969). *Attachment and Loss.* Vol. 1. *Attachment.* London: Hogarth Press.

Boyd, R. and Richerson, P. J. (1985). *Culture and the Evolutionary Process.* Chicago, IL: University of Chicago Press.

Briggs, J. L. (1970). *Never in Anger: Portrait of an Eskimo Family.* Cambridge, MA: Harvard University Press.

Brode, H. (1907). *Tippoo Tib: The Story of his Career in Central Africa.* London: Edward Arnold.

Brooke, C. (1967). Types of food shortages in Tanzania. *Geographical Review* 57, 333–357.

Brotherton, P. N. M. and Komers, P. E. (2003). Mate guarding and the evolution of social monogamy in mammals. In *Monogamy: Mating Strategies and Partnerships in Birds, Humans, and other Mammals*, eds. U. H. Reichard and C. Boesch. Cambridge: Cambridge University Press, pp. 42–58.

Brown, D. (1988). Components of lifetime reproductive success. In *Reproductive Success*, ed. T. H. Clutton-Brock. Chicago, IL: University of Chicago Press, pp. 439–453.

Brown, G. R., Laland, K. N. and Borgerhoff Mulder, M. (2009). Bateman's principles and human sex roles. *Trends in Ecology and Evolution* 24, 297–304.

Brown, L. and Cocheme, S. (1969). *A Study of Agro-climatology of the Highlands of East Africa.* Rome: Food and Agriculture Organization.

Buckle, L., Gallup, G. G. and Road, Z. (1996). Marriage as a reproductive contract: patterns of marriage, divorce, and remarriage. *Ethology and Sociobiology* 17, 363–378.

Bulmer, M. G. (1960). The twinning rate in Europe and Africa. *Annals of Human Genetics, London* 24, 121–125.

Burkart, J. M., Hrdy, S. B. and Van Schaik, C. P. (2009). Cooperative breeding and human cognitive evolution. *Evolutionary Anthropology* 18, 175–186.

Burstein, L. (1980). The analysis of multilevel data in educational research and evaluation. *Review of Research in Education* **8**, 158–233.

Burton, R. F. (1859). The lake regions of central Africa. *Journal of the Royal Geographical Society* **29**, 1–464.

Burtt, B. D. (1935). Observations on the genus Commiphora and its distribution in Tanganyika Territory. *Bulletin of Miscellaneous Information (Royal Botanic Gardens, Kew)* **1935**(3), 101–117.

Butovskaya, M. L. (2012). Wife battering and traditional methods of its control in contemporary Datoga pastoralists of Tanzania. *Journal of Aggression, Conflict and Peace Research* **4**, 28–44.

Butovskaya, M. L. (2013). Aggression and conflict resolution among the nomadic Hadza of Tanzania as compared with their pastoralist neighbors. In *War, Peace, and Human Nature: the Convergence of Evolutionary and Cultural Views*, ed. D. P. Fry. Oxford: Oxford University Press, pp. 278–296.

Caldwell, J. and Caldwell, B. (2003). Pretransitional population control and equilibrium. *Population Studies: A Journal of Demography* **57**, 199–215.

Caldwell, J. C. and Caldwell, P. (1983). The demographic evidence for the incidence and cause of abnormally low fertility in tropical Africa. *World Health Statistics Quarterly* **36**, 2–34.

Campbell, K. and Borner, M. (1995). Population trends and distribution of Serengeti herbivores: implications for management. In *Serengeti II Dynamics, Management, and Conservation of an Ecosystem*, eds. A. R. F. Sinclair and P. Arcese. Chicago, IL: The University of Chicago Press, pp. 117–145.

Campbell, K. L. and Wood, J. W. (1988). Fertility in traditional societies. In *Natural Human Fertility: Social and Biological Determinants*, eds. P. Diggory, M. Potts and S. Teper. London: MacMillan, pp. 39–69.

Campbell, M. C. and Tishkoff, S. A. (2008). African genetic diversity: implications for human demographic history, modern human origins, and complex disease mapping. *Annual Review of Genomics and Human Genetics* **9**, 403–433.

Cant, M. A. and Johnstone, R. A. (2008). Reproductive conflict and the separation of reproductive generations in humans. *Proceedings of the National Academy of Sciences USA* **105**, 5332–5336.

Carlaw, R. W. and Vaidya, K. (1983). Birth intervals and the survival of children to age five – some data from Nepal. *Journal of Tropical Pediatrics* **29**, 31–34.

Cashdan, E., Marlowe, F. W., Crittenden, A. *et al.* (2012). Sex differences in spatial cognition among Hadza foragers. *Evolution and Human Behavior* **33**, 274–284.

Caulfield, L. E., de Onis, M., Blossner, M. *et al.* (2004). Undernutrition as an underlying cause of child deaths associated with diarrhea, pneumonia, malaria, and measles. *American Journal of Clinical Nutrition* **80**, 193–198.

Cerling, T. E., Wynn, J. G., Andanje, S. A. *et al.* (2011). Woody cover and hominin environments in the past 6 million years. *Nature* **476**, 51–56.

Chagnon, N. A. (1968). *Yanomamo. The Fierce People.* New York, NY: Holt, Rinehart and Winston.

Chagnon, N. A. (1979). Is reproductive success equal in egalitarian societies? In *Evolutionary Biology and Human Behavior: An Anthropological Perspective*, eds. N. A. Chagon and W. Irons. North Scituate, MA: Duxbury Press, pp. 374–401.

Chagnon, N. A. and Bugos, P. E. (1979). Kin selection and conflict: an analysis of a Yanomamö ax fight. In *Evolutionary Biology and Human Social Behavior: An Anthropological Perspective*, eds. N. A. Chagnon and W. Irons. North Scituate, MA: Duxbury Press, pp. 213–238.

Chagnon, N. A. and Irons, W. (1979). *Evolutionary Biology and Human Social Behavior: an Anthropological Perspective.* North Scituate, MA: Duxbury Press.

Chang, C. (1982). Nomads without cattle: East African foragers in historical perspective. In *Politics and History in Band Societies*, eds. E. Leacock and R. B. Lee. Cambridge: Cambridge University Press, pp. 269–282.

Charnov, E. L. (1993). *Life History Invariants.* Oxford: Oxford University Press.

Chavez, A., Martinez, C. and Yaschine, T. (1974). The importance of nutrition and stimulation on child mental and social development. In *Early Malnutrition and Mental Development*, eds. J. Cravioto, L. Hambraeus and B. Vahlquist. Uppsala: Almquist and Wiksell, pp. 211–225.

Chrastil, E. R., Getz, W. M., Euler, H. A. *et al.* (2006). Paternity uncertainty overrides sex chromosome selection for preferential grandparenting. *Evolution and Human Behavior* **27**, 206–223.

Clutton-Brock, T. H., ed. (1988). *Reproductive Success.* Chicago, IL: University of Chicago Press.

Clutton-Brock, T. H. (1991). *The Evolution of Parental Care.* Princeton, NJ: Princeton University Press.

Clutton-Brock, T. H. and Isvaran, K. (2007). Sex differences in ageing in natural populations of vertebrates.

*Proceedings of the Royal Society of London B* **274**, 3097–3104.
Coale, A. and Demeny, P. (1983). *Regional Model Life Tables and Stable Populations*. New York, NY: Academic Press.
Coast, E. (2001). *Maasai Demography*. Unpublished PhD thesis. LSE Research Online. London: University of London.
Codding, B. and Kramer, K. L. (2015). *Why Forage? Hunters and Gatherers Living in the 21st Century*. Santa Fe, NM: School of Advanced Research.
Coe, M. J., Cumming, D. H. and Phillipson, J. (1976). Biomass and production of large African herbivores in relation to rainfall and primary production. *Oecologia (Berlin)* **22**, 341–354.
Cooper, B. (1949). The Kindiga. *Tanganyika Notes and Records* **27**, 8–15.
Cox, M. P., Morales, D. A., Woerner, A. E. *et al.* (2009). Autosomal resequence data reveal late stone age signals of population expansion in sub-Saharan African foraging and farming populations. *PLoS ONE* **4**, 6366.
Craig, M. H., Kleinschmidt, I., Nawn, J. B. *et al.* (2004). Exploring 30 years of malaria case data in KwaZulu-Natal, South Africa: part 1. The impact of climatic factors. *Tropical Medicine and International Health* **9**, 1247–1257.
Crittenden, A. N. (2009). *Allomaternal Care and Juvenile Foraging among the Hadza: Implications for the Evolution of Cooperative Breeding in Humans*. PhD thesis. San Diego, CA: University of California San Diego.
Crittenden, A. N. (2011). The importance of honey consumption in human evolution. *Food and Foodways* **19**, 257–273.
Crittenden, A. N., Conklin-Brittain, N. L., Zes, D. A. *et al.* (2013). Juvenile foraging among the Hadza: implications for human life history. *Evolution and Human Behavior* **34**, 299–304.
Crittenden, A. N. and Marlowe, F. W. (2008). Allomaternal care among the Hadza of Tanzania. *Human Nature* **19**, 249–262.
Cronk, L. (2004). *From Mukogodo to Maasai: Ethnicity and Cultural Change in Kenya*. Boulder, CO: Westview Press.
Crow, J. F. (1958). Some possibilities for measuring selection intensities in man. *Human Biology* **30**, 1–13.
Crowne, S. S., Gonsalves, K., Burrell, L. *et al.* (2011). Relationship between birth-spacing, child maltreatment, and child behavior and development outcomes among at-risk families. *Maternal and Child Health Journal*. doi: 10.1007/s10995-011-0909-3.
Daly, M. and Wilson, M. (1978). *Sex, Evolution, and Behavior*. North Scituate, MA: Duxbury Press.
Daly, M. and Wilson, M. (1987). Child abuse and other risks of not living with both parents. *Ethology and Sociobiology* **6**, 197–210.
Daly, M. and Wilson, M. (1988). *Homicide*. New York, NY: Aldine de Gruyter.
Daly, M. and Wilson, M. (1996). Violence against stepchildren. *Current Directions in Psychological Science* **5**(3), 77–81.
Daly, M. and Wilson, M. (2008). Is the "Cinderella Effect" controversial? A case study of evolution-minded research and critiques thereof. In *Foundation of Evolutionary Psychology*, eds. C. B. Crawford and D. Krebs. New York, NY: Psychology Press, pp. 383–400.
Danielson, E. R. (1961). A brief history of the Waniramba people up to the time of the German occupation. *Tanganyika Notes and Records* **56**, 67–78.
Darwin, C. (1871). *The Descent of Man and Selection in Relation to Sex*. London: John Murray.
Davies, N. B., Krebs, J. R. and West, S. A. (2012). *An Introduction to Behavioural Ecology*. Hoboken, NJ: Wiley-Blackwell.
Deevey, E. S. (1960). The human population. *Scientific American* **203**, 195–204.
Dempwolff, O. (1916–1917). Beitrage zur Kenntnis der Sprachen in Deutsch-Ostafrika. 12. Worter der Hatzasprache. *Zeitschrift fur Kolonialsprachen* **7**, 319–325.
De Vries, L. (1946). *German–English Dictionary for Students in Chemistry, Physics, Biology, Agriculture, and Related Sciences*. New York, NY: McGraw-Hill.
Dewey, K. G. and Cohen, R. J. (2007). Does birth spacing affect maternal or child nutritional status? A systematic literature review. *Maternal and Child Nutrition* **3**, 151–173.
Diallo, B. O., Joly, H. I., McKey, D. *et al.* (2007). Genetic diversity of *Tamarindus indica* populations: any clues on the origin from its current distribution? *African Journal of Biotechnology* **6**, 853–860.
Dominguez-Rodrigo, M., Diez-Martin, F., Mabulla, A. *et al.* (2007). The archaeology of the Middle Pleistocene deposits of Lake Eyasi. *Journal of African Archaeology* **5**, 47–78.
Dominguez-Rodrigo, M., Mabulla, A., Luque, L. *et al.* (2008). A new archaic *Homo sapiens* fossil from Lake Eyasi, Tanzania. *Journal of Human Evolution* **54**, 899–903.

Draper, P. and Cashdan, E. (1988). Technological change and child behavior among the !Kung. *Ethnology* **27**, 339–365.

Draper, P. and Harpending, H. (1982). Father absence and reproductive strategy: an evolutionary perspective. *Journal of Anthropological Research* **38**, 255–273.

Dunn, F. L. (1968). Epidemiological factors: health and disease in hunter-gatherers. In *Man the Hunter*, eds. R. B. Lee and I. DeVore. Chicago, IL: Aldine, pp. 221–228.

Durham, W. H. (1976). Resource competition and human aggression, part 1: a review of primitive war. *Quarterly Review of Biology* **52**, 385–415.

Dyson, T. (1977). *The Demography of the Hadza in Historical Perspective*. Edinburgh: African Historical Demography, Centre for African Studies, University of Edinburgh.

Dyson, T. (1991a). On the demography of south Asian famines. Part 1. *Population Studies* **45**, 5–25.

Dyson, T. (1991b). On the demography of south Asian famines. Part 2. *Population Studies* **45**, 279–297.

Dyson-Hudson, R. and Smith, E. A. (1978). Human territoriality: an ecological reassessment. *American Anthropologist* **80**, 21–41.

Early, J. D. and Headland, T. N. (1998). *Population Dynamics of a Philippine Rain Forest People: The San Ildefonso Agta*. Gainsesville, FL: University Press of Florida.

Elkan, P. W., Parnell, R. and Smith, J. L. D. (2009). A die-off of large ungulates following a *Stomoxys* biting fly out-break in lowland forest, northern Republic of Congo. *African Journal of Ecology* **47**, 528–536.

Ellison, P. T. (2001). *On Fertile Ground*. Cambridge, MA: Harvard University Press.

Ember, C. R. (1978). Myths about hunter-gatherers. *Ethnology* **17**, 439–448.

Emery Thompson, M., Jones, J. H., Pusey, A. E. *et al.* (2007). Aging and fertility in wild chimpanzees: implications for the evolution of menopause. *Current Biology* **17**, 2150–2156.

England, P. and McClintock, E. A. (2009). The gendered double standard of aging in US marriage markets. *Population and Development Review* **35**, 797–816.

Erb, J. D. and Boyce, M. S. (1999). Distribution of population declines in large mammals. *Conservation Biology* **13**, 199–201.

Eveleth, P. B. and Tanner, J. M. (1976). *Worldwide Variation in Human Growth*. Cambridge: Cambridge University Press.

Fairbanks, L. A. (1996). Individual differences in maternal style: causes and consequences for mothers and offspring. *Advances in the Study of Behavior* **25**, 579–611.

Fairbanks, L. A. and McGuire, M. T. (1986). Age, reproductive value, and dominance-related behavior in Vervet monkey females: cross generational influences on social relationships and reproduction. *Animal Behavior* **34**, 1710–1721.

Farler, J. P. (1882). Native routes in East Africa from Pangani to the Masai country and the Victoria Nyanza. *Proceedings of the Royal Geographical Society and Monthly Record of Geography* **4**, 730–742.

Ferguson, R. B. and Whitehead, N. L. (1992). *War in the Tribal Zone*. Santa Fe, NM: SAR Press.

Finch, C. E. and Stanford, C. B. (2004). Meat-adaptive genes and the evolution of slower aging in humans. *Quarterly Review of Biology* **79**(1), 3–50.

Fisher, H. (1989). Evolution of human serial pairbonding. *American Journal of Physical Anthropology* **78**, 331–354.

Fix, A. G. (1989). Semai Senoi mortality: two-census method. *American Journal of Human Biology* **1**, 471–477.

Flannery, K. V. (1972). The origins of the village as a settlement type in Mesoamerica and the near east: a comparative study. In *Man, Settlement and Urbanism*, eds. P. J. Ucko, R. Tringham and G. W. Dimbleby. London: Duckworth, pp. 23–53.

Flannery, K. V. (2002). The origins of the village revisited: from nuclear to extended households. *American Antiquity* **67**, 417–433.

Flannery, K. V. and Marcus, J. (2003). The origin of war: new 14C dates from ancient Mexico. *Proceedings of the National Academy of Sciences USA* **100**, 11 801–11 805.

Fleer, H. E. (1981). Teleconnections of rainfall anomalies in the tropics and subtropics. In *Monsoon Dynamics*, eds. J. Lighthill and R. P. Pearce. Cambridge: Cambridge University Press, pp. 5–18.

Foley, R. A. (1982). A reconsideration of the role of predation on large mammals in tropical hunter-gatherer adaptation. *Man, New Series* **17**, 383–402.

Foley, R. A. (1996). The adaptive legacy of human evolution: a search for the environment of evolutionary adaptedness. *Evolutionary Anthropology* **4**, 194–203.

Ford, K., Huffman, S. L., Chowdhury, A. K. M. A. *et al.* (1989). Birth-interval dynamics in rural Bangladesh and maternal weight. *Demography* **26**, 425–437.

Fosbrooke, H. A. (1950). A proto-historic burial, Naberera Masai district, Tanganyika Territory. *South African Archaeological Bulletin* **5**, 105–107.

Fosbrooke, H. A. (1956). A stone age tribe in Tanganyika. *South African Archaeological Bulletin* **11**, 3–8.

Fosbrooke, H. A. (1972). *Ngorongoro – The Eighth Wonder*. London: Andre Deutsch.

Foster, A., Ebinger, C., Mbede, E. *et al.* (1997). Tectonic development of the northern Tanzanian sector of the East African rift system. *Journal of the Geological Society* **154**, 689–700.

Fouts, H. N. and Brookshire, R. A. (2009). Who feeds children? A child's-eye-view of caregiver feeding patterns among the Aka foragers in Congo. *Social Science and Medicine* **69**, 285–292.

Fox, M., Johow, J. and Knapp, L. A. (2011). The selfish grandma gene: the roles of the X-chromosome and paternity uncertainty in the evolution of grandmothering behavior and longevity. *International Journal of Evolutionary Biology* Article ID 165919.

Fox, M., Sear, R., Beise, J. *et al.* (2009). Grandma plays favorites: X-chromosome relatedness and sex-specific childhood mortality. *Proceedings of the Royal Society of London B*. doi: 10.1098/rspb.2009.1660.

Frank, O. (1983). Infertility in sub-Saharan Africa: estimates and implications. *Population and Development Review* **9**, 137–144.

Fretwell, S. D. (1972). *Populations in a Seasonal Environment*. Princeton, NJ: Princeton University Press.

Fry, D. P. and Soderberg, P. (2013). Lethal aggression in mobile forager bands and implications for the origins of war. *Science* **341**, 270–273.

Fu, F., Hauert, C., Nowak, M. A. *et al.* (2008). Reputation-based partner choice promotes cooperation in social networks. *Physical Review E* **78**, 026117.

Gage, T. (1998). The comparative demography of primates: with some comments on the evolution of life histories. *Annual Review of Anthropology* **27**, 197–221.

Gage, T. B., Dyke, B. and MacCluer, J. W. (1986). Estimating mortality level for small populations: an evaluation of a pair of two-census methods. *Population Studies* **40**, 263–273.

Gage, T. B., Dyke, B. and Riviere, P. G. (1984). The population dynamics and fertility of the Trio of Surinam: an application of a two-census method. *Human Biology* **56**, 691–701.

Galdikas, B. M. F. and Wood, J. W. (1990). Birth spacing in humans and apes. *American Journal of Physical Anthropology* **83**, 185–191.

Galvin, K. A., Boone, R. B., Smith, N. M. *et al.* (2001). Impacts of climate variability on East African pastoralists: linking social science and remote sensing. *Climate Research* **19**, 161–172.

Gangestad, S. W. and Simpson, J. A. (1990). Toward an evolutionary history of female sociosexual variation. *Journal of Personality* **58**, 69–96.

Gangestad, S. W. and Simpson, J. A. (2000). The evolution of human mating: trade-offs and strategic pluralism. *Behavioral and Brain Sciences* **23**, 675–687.

Garcia-Moro, C., Hernandez, M. and Lalueza, C. (1997). Estimation of the optimum density of the Selk'nam from Tierra del Fuego: inference about human dynamics in extreme environments. *American Journal of Human Biology* **9**, 699–708.

Garenne, M. (2002). Sex ratios at birth in African populations: a review of survey data. *Human Biology* **74**, 889–900.

Garenne, M. L., Leclerc, P. M. and Matthews, A. P. (2011). Parameterisation of the transition to first marriage with the Picrate model. *South African Journal of Demography* **12**, 109–124.

Gavrilets, S. (2012). Human origins and the transition from promiscuity to pair-bonding. *Proceedings of the National Academy of Sciences USA* **109**, 9923–9928.

Gebreselassie, T. and Mishra, V. (2011). Spousal agreement on preferred waiting time to next birth in sub-Saharan Africa. *Journal of Biosocial Science* **43**, 385–400.

Giblin, J. (1990). Trypanosomiasis control in African history: an evaded issue? *The Journal of African History* **31**, 59–80.

Gillman, T. (1936). A population map of Tanganyika Territory. *Geographical Review* **26**, 353–375.

Gillson, L. (2006). A 'large infrequent disturbance' in an East African savanna. *African Journal of Ecology* **44**, 458–467.

Githeko, A. K. and Ndegewa, W. (2001). Predicting malaria epidemics in the Kenyan highlands using climate data: a tool for decision-makers. *Global Change and Human Health* **2**, 54–63.

Glewwe, P. and Miguel, E. A. (2008). The impact of child health and nutrition on education in less developed countries. In *Handbook of Development Economics*, eds. T. P. Schultz and J. Strauss. New York, NY: Elsevier, pp. 3561–3606.

Godfray, H. C. and Rees, M. (2002). Population growth rates: issues and an application. *Philosophical Transactions of the Royal Society of London B* **357**, 1307–1319.

Gold, K. J., Sen, A. and Hayward, R. A. (2010). Marriage and cohabitation outcomes after pregnancy loss. *Pediatrics* **125**, 1202–1207.

Goldstein, H. (1984). The methodology of school comparisons. *Oxford Review of Education* **10**, 69–74.

Goldstein, H. (1986). Multilevel mixed linear model analysis using iterative generalized least squares. *Biometrika* **73**, 43–56.

Goldstein, H. (1987). *Multilevel Models in Educational and Social Research*. New York, NY: Oxford University Press.

Gosden, R. G. and Telfer, E. (1987). Numbers of follicles and oocytes in mammalian ovaries and their allometric relationships. *Journal of Zoology (London)* **211**, 169–175.

Götzen, G. A. (1895). *Durch Afrika von Ost nach West*. Berlin: Reimer.

Gould, R. A. (1969). *Yiwara: Foragers of the Australian Desert*. New York, NY: Scribner.

Gowaty, P. A. (1996). Battle of the sexes and origins of monogamy. In *Partnerships in Birds*, ed. J. M. Black. Oxford: Oxford University Press, pp. 21–52.

Grafen, A. (1988). On the uses of data on lifetime reproductive success. In *Reproductive Success*, ed. T. H. Clutton-Brock. Chicago, IL: The University of Chicago Press, pp. 454–471.

Graves, B. M. (2010). Ritualized combat as an indicator of intrasexual selection effects on male life history evolution. *American Journal of Human Biology* **22**, 45–49.

Gurven, M. (2004). To give and to give not: the behavioral ecology of human food transfers. *Behavioral and Brain Sciences* **27**, 543–583.

Gurven, M. and Hill, K. (2009). Why do men hunt? A reevaluation of "Man the Hunter" and the sexual division of labor. *Current Anthropology* **50**, 51–74.

Gurven, M. and Kaplan, H. (2007). Longevity among hunter-gatherers: a cross-cultural examination. *Population and Development Review* **33**, 321–365.

Gurven, M. and von Rueden, C. (2010). Hunting, social status and biological fitness. *Biodemography and Social Biology* **53**, 81–99.

Gurven, M. and Walker, R. S. (2006). Energetic demand of multiple dependents and the evolution of slow human growth. *Proceedings of the Royal Society of London B* **273**, 835–841.

Gutierrez, M. C., Brisse, S., Brosch, R. *et al.* (2005). Ancient origin and gene mosaicism of the progenitor of *Mycobacterium tuberculosis*. *PLoS Pathogens* **1**, 55–61.

Guyatt, H. L. and Snow, R. W. (2004). Impact of malaria during pregnancy on low birth weight in sub-Saharan Africa. *Clinical Microbiology Reviews* **17**, 760–769.

Haig, D. (2014). Troubled sleep. Night waking, breast-feeding and parent–offspring conflict. *Evolution, Medicine, and Public Health* **2014**, 32–39.

Hames, R. and Draper, P. (2004). Women's work, child care, and helpers-at-the-nest in a hunter-gatherer society. *Human Nature* **15**, 319–341.

Hamilton, W. D. (1966). The moulding of senescence by natural selection. *Journal of Theoretical Biology* **12**, 12–45.

Hammel, E. A. and Howell, N. (1987). Research in population and culture: an evolutionary framework. *Current Anthropology* **28**, 141–160.

Hammer, M. L. A. and Foley, R. A. (1996). Longevity and life history in hominid evolution. *Human Evolution* **11**, 61–66.

Harder, T. C., Kenter, M., Appel, M. J. G. *et al.* (1995). Phylogenetic evidence of canine distemper virus in Serengeti's lions. *Vaccine* **13**, 521–523.

Harpending, H. C. (1994). Infertility and forager demography. *American Journal of Physical Anthropology* **93**, 385–390.

Harpending, H. C., Draper, P. and Pennington, R. (1990). Cultural evolution, parental care and mortality. In *Disease in Populations in Transition: Anthropological and Epidemiological Perspectives*, eds. A. C. Swedlund and G. J. Armelagos. New York, NY: Bergin and Garvey, pp. 251–265.

Harpending, H. C., Sherry, S. T., Rogers, A. R. *et al.* (1993). The genetic structure of ancient human populations. *Current Anthropology* **34**, 483–496.

Harper, K. N., Ocampo, P. S., Steiner, B. M. *et al.* (2008). On the origin of the Treponematoses: a phylogenetic approach. *PLoS Neglected Tropical Diseases* **2**, 1–13.

Harris, D. R. (1978). Adaptation to a tropical rain-forest environment: aboriginal subsistence in northeastern Queensland. In *Human Behaviour and Adaptation*, eds. N. G. Blurton Jones and V. Reynolds. London: Taylor and Francis, pp. 113–134.

Harrison, E., Barta, Z., Cuthill, I. *et al.* (2009). How is sexual conflict over parental care resolved? A meta-analysis. *Journal of Evolutionary Biology* **22**, 1800–1812.

Hassan, F. A. (1973). On mechanisms of population growth during the Neolithic. *Current Anthropology* **14**, 535–542.

Hassan, F. A. (1975). Determinants of the size, density, and growth rates of hunting-gathering populations.

In *Population, Ecology, and Social Evolution*, ed. S. Polgar. The Hague: Mouton, pp. 27–52.
Hassan, F. A. (1978). Demographic archaeology. *Advances in Archaeological Method and Theory* 1, 49–103.
Hassan, F. A. (1981). *Demographic Archaeology*. New York, NY: Academic Press.
Hawkes, K. (1991). Showing-off: tests of an hypothesis about men's foraging goals. *Ethology and Sociobiology* 12, 29–54.
Hawkes, K. (1993). Why hunter-gatherers work: an ancient version of the problem of public goods. *Current Anthropology* 34, 341–361.
Hawkes, K. (2003). Grandmothers and the evolution of human longevity. *American Journal of Human Biology* 15, 380–400.
Hawkes, K. (2004). Mating, parenting, and the evolution of human pair bonds. In *Kinship and Behavior in Primates*, eds. B. Chapais and C. M. Berman. Oxford: Oxford University Press, pp. 443–473.
Hawkes, K. (2006). Slow life histories and human evolution. In *The Evolution of Human Life History*. eds. K. Hawkes and R. R. Paine. Santa Fe, NM: SAR Press, pp. 45–93.
Hawkes, K. (2014). Primate sociality to human cooperation, why us not them? *Human Nature* 25, 28–48.
Hawkes, K. and Bliege Bird, R. (2002). Showing off, handicap signaling, and the evolution of men's work. *Evolutionary Anthropology* 11, 58–67.
Hawkes, K. and Blurton Jones, N. G. (2005). Human age structures, paleodemography, and the grandmother hypothesis. In *Grandmotherhood*, eds. E. Voland, A. Chasiotis and W. Schiefenhovel. New Brunswick, NJ: Rutgers University Press, pp. 118–140.
Hawkes, K., Hill, K. and O'Connell, J. F. (1982). Why hunters gather: optimal foraging and the Ache of eastern Paraguay. *American Ethnologist* 9, 379–398.
Hawkes, K., Kim, P. S., Kennedy, B. *et al.* (2011). A reappraisal of grandmothering and natural selection. *Proceedings of the Royal Society of London B* 278, 1936–1938.
Hawkes, K., O'Connell, J. F. and Blurton Jones, N. G. (1989). Hardworking Hadza grandmothers. In *Comparative Socioecology*, eds. V. Standen and R. A. Foley. Oxford: Blackwell, pp. 341–366.
Hawkes, K., O'Connell, J. F. and Blurton Jones, N. G. (1991). Hunting income patterns among the Hadza: big game, common goods, foraging goals and the evolution of the human diet. *Philosophical Transactions of the Royal Society of London B* 334, 243–251.
Hawkes, K., O'Connell, J. F. and Blurton Jones, N. G. (1995a). Hadza children's foraging: juvenile dependency, social arrangements, and mobility among hunter-gatherers. *Current Anthropology* 36, 688–700.
Hawkes, K., O'Connell, J. F. and Blurton Jones, N. G. (1997). Hadza women's time allocation, offspring provisioning, and the evolution of long post-menopausal lifespans. *Current Anthropology* 38, 551–577.
Hawkes, K., O'Connell, J. F. and Blurton Jones, N. G. (2001). Hadza meat sharing. *Evolution and Human Behavior* 22, 113–142.
Hawkes, K., O'Connell, J. F. and Blurton Jones, N. G. (2014). More lessons from the Hadza about men's work. *Human Nature*. doi: 10.1007/s12110-014-9212-5.
Hawkes, K., O'Connell, J. F., Blurton Jones, N. G. *et al.* (1998). Grandmothering, menopause, and the evolution of human life histories. *Proceedings of the National Academy of Sciences USA* 95, 1336–1339.
Hawkes, K., O'Connell, J. F. and Coxworth, J. E. (2010). Family provisioning is not the only reason men hunt. *Current Anthropology* 51, 259–264.
Hawkes, K., Rogers, A. R. and Charnov, E. L. (1995b). The male's dilemma: increased offspring production is more paternity to steal. *Evolutionary Ecology* 9, 662–677.
Hawkes, K. and Smith, K. R. (2010). Do women stop early? Similarities in fertility decline between humans and chimpanzees. *Annals of the New York Academy of Sciences* 1204, 1796–1805.
Headland, T. N. (1987). The wild yam question: how well could independent hunter-gatherers live in a tropical rain forest ecosystem? *Human Ecology* 15, 463–491.
Headland, T. N. and Reid, L. A. (1989). Hunter-gatherers and their neighbors from prehistory to the present. *Current Anthropology* 30, 43–66.
Henn, B. A., Gignoux, C. R., Jobin, M. *et al.* (2011). Hunter-gatherer genomic diversity suggests a southern African origin for modern humans. *Proceedings of the National Academy of Sciences USA* 108, 5154–5162.
Henrich, J., Boyd, R. and Richerson, P. J. (2012). The puzzle of monogamous marriage. *Philosophical Transactions of the Royal Society of London B* 367, 657–669.
Hewlett, B. S. (1991a). Demography and childcare in preindustrial societies. *Journal of Anthropological Research* 47, 1–37.
Hewlett, B. S. (1991b). *Intimate Fathers: the Nature and Context of Aka Pygmy Paternal Infant Care*. Ann Arbor, MI: University of Michigan Press.

Hiernaux, J. and Hartono, D. B. (1980). Physical measurements of the adult Hadza of Tanzania. *Annals of Human Biology* **7**, 339–346.

Higley, J. D., Mehlmann, P. T., Poland, R. E. *et al.* (1996). CSF testosterone and 5-HIAA correlate with different types of aggressive behaviors. *Biological Psychiatry* **40**, 1067–1082.

Hill, K. (1982). Hunting and human evolution. *Journal of Human Evolution* **11**, 521–544.

Hill, K. and Hurtado, A. M. (1991). The evolution of reproductive senescence and menopause in human females. *Human Nature* **2**, 313–350.

Hill, K. and Hurtado, A. M. (1996). *Ache Life History: the Ecology and Demography of a Foraging People.* New York, NY: Aldine de Gruyter.

Hill, K. and Hurtado, A. M. (2009). Cooperative breeding in South American hunter-gatherers. *Proceedings of the Royal Society of London B* **276**, 3863–3870.

Hill, K., Hurtado, A. M. and Walker, R. S. (2007). High adult mortality among Hiwi hunter-gatherers: implications for human evolution. *Journal of Human Evolution* **52**, 443–454.

Hill, K. and Kintigh, K. (2009). Can anthropologists distinguish good and poor hunters? Implications for hunting hypotheses, sharing conventions, and cultural transmission. *Current Anthropology* **50**, 369–377.

Hill, K., Walker, R. S., Bozicevic, M. *et al.* (2011). Coresidence patterns in hunter-gatherer societies show unique human social structure. *Science* **331**, 1286–1289.

Hinde, A. (1998). *Demographic Methods.* London: Arnold.

Hinde, K. (2009). Richer milk for sons but more milk for daughters: sex-biased investment during lactation varies with maternal life history in Rhesus monkeys. *American Journal of Human Biology* **21**, 512–519.

Hirst, M. A. (1972). Tribal mixture and migration in Tanzania: an evaluation and analysis of census tribal data. *Canadian Geographer* **16**, 230–248.

Hladik, A., Bahuchet, S., Ducatillion, C. *et al.* (1984). The tuberous plants of the central African rain forest. *Revue d'Ecologie (la Terre et la Vie)* **39**, 249–290.

Hobcraft, J., McDonald, J. W. and Rutstein, S. O. (1985). Demographic determinants of infant and early child mortality: a comparative analysis. *Population Studies* **39**, 363–385.

Homewood, K., Kristjanson, P. and Chevenix-Trench, P., eds. (2009). *Staying Maasai? Livelihoods, Conservation and Development in East African Rangelands.* Studies in Human Ecology and Adaptation. New York, NY: Springer.

Houston, A. I. and McNamara, J. M. (2002). A self-consistent approach to paternity and parental effort. *Philosophical Transactions of the Royal Society of London B* **357**, 351–362.

Houston, A. I., Szekely, T. and McNamara, J. M. (2005). Conflict between parents over care. *Trends in Ecology and Evolution* **20**, 33–38.

Howell, N. (1976). Toward a uniformitarian theory of human paleo-demography. *Journal of Human Evolution* **5**, 25–40.

Howell, N. (1979). *Demography of the Dobe Area !Kung.* New York, NY: Academic Press.

Howell, N. (1982). Village composition implied by a paleodemographic life table: the Libben Site. *American Journal of Physical Anthropology* **59**, 263–269.

Howell, N. (2000). *Demography of the Dobe !Kung.* New York, NY: Aldine de Gruyter.

Howell, N. (2010). *Life Histories of the Dobe !Kung.* Berkeley, CA: University of California Press.

Hrdy, S. B. (2009). *Mothers and Others.* Cambridge, MA: Harvard University Press.

Hubbell, S. P. and Johnson, L. K. (1987). Environmental variance in lifetime mating success, mate choice, and sexual selection. *American Naturalist* **130**, 91–112.

Huntingford, S. W. B. (1953). The southern Nilo-Hamites. *Ethnographic survey of Africa: East Central Africa* **8**, 132–135.

Hurtado, A. M. and Hill, K. (1992). Paternal effect on offspring survivorship among Ache and Hiwi hunter-gatherers: implications for modelling pair-bond stability. In *Father–Child Relations: Cultural and Biosocial Contexts*, ed. B. Hewlett. New York, NY: Aldine de Gruyter, pp. 31–55.

Hussein, K., Sumberg, J. and Seddon, D. (1999). Increasing violent conflict between herders and farmers in Africa: claims and evidence. *Development Policy Review* **17**, 397–418.

Ichikawa, M. (1980). The utilization of wild food plants by the Suiei Dorobo in Northern Kenya. *Journal of the Anthropological Society Nippon* **88**, 25–48.

Iliffe, J. (1979). *A Modern History of Tanganyika.* Cambridge: Cambridge University Press.

Iliffe, J. (1995). *Africans. The History of a Continent.* Cambridge, Cambridge University Press.

Jackson, C. H. N. (1945). Comparative studies of the habitat requirements of tsetse fly species. *Journal of Animal Ecology* **14**, 46–51.

Jaeger, F. (1911). *Das Hochland der Riesenkrater und die umliegenden Hochländer Deutsch-Ostafrikas*. Berlin: Mittler.

James, W. H. (1984). The sex ratio of black births. *Annals of Human Biology* **11**, 39–44.

Janson, C. H. and van Schaik, C. P. (1993). Ecological risk aversion in juvenile primates: slow and steady wins the race. In *Juvenile Primates*, eds. M. E. Pereira and L. A. Fairbanks. New York, NY: Oxford University Press, pp. 57–74.

Jelliffe, D. B., Woodburn, J. C., Bennett, F. J. *et al.* (1962). The children of the Hadza hunters. *Tropical Paediatrics* **60**, 907–913.

Jenike, M. R. (1988). *Seasonal Hunger among Tropical Africans: The Lese Case*. PhD thesis. Los Angeles, CA: University of California Los Angeles.

Jenike, M. R. (1995). Variation in body fat and muscle mass: responses to seasonal hunger among tropical horticulturists, Zaire. *Ecology of Food and Nutrition* **34**, 227–249.

Jenike, M. R. (1996). Activity reduction as an adaptive response to seasonal hunger. *American Journal of Human Biology* **8**, 517–534.

Johnson, F. (1923). Notes on Kiniramba. *Bantu Studies* **2**, 167–192.

Johnstone, R. and Cant, M. A. (2010). The evolution of menopause in cetaceans and humans: the role of demography. *Proceedings of the Royal Society of London B* **277**, 3765–3771.

Jones, G., Steketee, R. W., Black, R. E. *et al.* (2003). How many child deaths can we prevent this year? *The Lancet* **362**, 65–71.

Jones, J. H. (2009). The force of selection on the human life cycle. *Evolution and Human Behavior* **30**, 305–314.

Jones, J. H. and Bliege Bird, R. (2014). The marginal valuation of fertility. *Evolution and Human Behavior* **35**, 65–71.

Jones, J. H., Wilson, M. L., Murray, C. *et al.* (2010). Phenotypic quality influences fertility in Gombe chimpanzees. *Journal of Animal Ecology* **79**, 1262–1269.

Jones, K. P., Walker, L. C., Anderson, D. *et al.* (2007). Depletion of ovarian follicles with age in chimpanzees: similarities to humans. *Biology of Reproduction* **77**, 247–251.

Judge, D. S. and Carey, J. R. (2000). Postreproductive life predicted by primate patterns. *Journal of Gerontology* **55A**, B201–B209.

Kaare, B. and Woodburn, J. (1999). Hadza. In *The Cambridge Encyclopedia of Hunters and Gatherers*, eds. R. B. Lee and R. Daly. Cambridge: Cambridge University Press, pp. 200–204.

Kamatou, G. P. P., Vermaak, I. and Viljoen, A. M. (2011). An updated review of *Adansonia digitata*: a commercially important African tree. *South African Journal of Botany* **77**, 908–919.

Kaplan, H. (1994). Evolutionary, and wealth flows theories of fertility. Empirical tests and new models. *Population and Development Review* **20**, 753–791.

Kaplan, H. (1997). The evolution of the human life course. In *Between Zeus and the Salmon: the Biodemography of Longevity*, eds. K. W. Wachter and C. E. Finch. Washington, DC: National Academy of Sciences Press, pp. 175–211.

Kaplan, H., Hill, K., Lancaster, J. *et al.* (2000). A theory of human life history evolution: diet, intelligence, and longevity. *Evolutionary Anthropology* **9**, 149–186.

Kaplan, S. N. (1992). *Top Executive Rewards and Firm Performance: A Comparison of Japan and the US*. Cambridge, MA: National Bureau of Economic Research.

Kappeler, P. and van Schaik, C., eds. (2004). *Sexual Selection in Primates*. Cambridge: Cambridge University Press.

Kashaigili, J. J. (2010). *Assessment of Groundwater Availability and its Current and Potential Use and Impacts in Tanzania*. Report prepared for the International Water Management Institute (IWMI). Morogoro: Sokoine University of Agriculture.

Keckler, C. N. W. (1997). Catastrophic mortality in simulations of forager age-at-death: where did all the human go? In *Integrating Archaeological Demography: Multidisciplinary Approaches to Prehistoric Populations*, ed. R. Paine. Carbondale, IL: Southern Illinois University Press, pp. 205–227.

Kelly, R. C. (2000). *Warless Societies and the Origin of War*. Ann Arbor, MI: University of Michigan Press.

Kelly, R. C. (2005). The evolution of lethal intergroup violence. *Proceedings of the National Academy of Sciences USA* **102**, 15 294–19 298.

Kelly, R. L. (1995). *The Foraging Spectrum*. Washington, DC: Smithsonian Institution Press.

Kelly, R. L. (2013). *The Lifeways of Hunter-Gatherers*. Cambridge: Cambridge University Press.

Kim, P. S., Coxworth, J. E. and Hawkes, K. (2012). Increased longevity evolves from grandmothering. *Proceedings of the Royal Society of London B* **279**, 4880–4884.

Kinoshita, F. (1998). Mortality crises in the Tokugawa period – a view from Shumon Aratame-Cho in northeastern Japan. *Japan Review* **10**, 53–71.

Kittler, R., Kayser, M. and Stoneking, M. (2003). Molecular evolution of *Pediculus humanus* and the origin of clothing. *Current Biology* 13, 1414–1417.

Kjekshus, H. (1977). *Ecology Control and Economic Development in East African History: The Case of Tanganyika 1850–1950*. London: Heineman.

Knauft, B. M. (1991). Violence and sociality in human evolution. *Current Anthropology* 32, 391–428.

Knell, R. J. and Webberley, K. M. (2004). Sexually transmitted diseases of insects: distribution, evolution, ecology and host behaviour. *Biological Reviews* 79, 557–581.

Knight, A., Underhill, P. A., Mortensen H. M. *et al.* (2003). African Y chromosomes and mtDNA divergence provides insight into the history of click languages. *Current Biology* 13, 464–473.

Kohl-Larsen, L. (1958). *Wildbeuter in Ost-Afrika: Die Tindiga, ein Jager- und Sammlervolk*. Berlin: Dietrich Reimer.

Kokko, H. and Jennions, M. D. (2008). Parental investment, sexual selection and sex ratios. *Journal of Evolutionary Biology* 21, 919–948.

Kokko, H., Klug, H. and Jennions, M. D. (2012). Unifying cornerstones of sexual selection: operational sex ratio, Bateman gradient and the scope for competitive investment. *Ecology Letters* 15, 1340–1351.

Konner, M. J. (1972). Aspects of the developmental ethology of a foraging people. In *Ethological Studies of Child Behavior*, ed. N. G. Blurton Jones. Cambridge: Cambridge University Press, pp. 285–304.

Konner, M. J. (2010). *The Evolution of Childhood*. Cambridge, MA: Harvard University Press.

Konner, M. J. and Worthman, C. (1980). Nursing frequency, gonadal function, and birth spacing among !Kung hunter-gatherers. *Science* 207, 788–791.

Kozlowski, J. and Wiegert, R. G. (1987). Optimal age and size at maturity in annuals and perennials with determinate growth. *Evolutionary Ecology* 1, 231–244.

Kramer, K. L. (2005a). Children's help and the pace of reproduction: cooperative breeding in humans. *Evolutionary Anthropology* 14, 224–237.

Kramer, K. L. (2005b). *Maya Children: Helpers at the Farm*. Cambridge, MA: Harvard University Press.

Kramer, K. L. (2008). Early sexual maturity among Pume foragers of Venezuela: fitness implications of teen motherhood. *American Journal of Physical Anthropology* 136, 338–350.

Kramer, K. L. and Ellison, P. T. (2010). Pooled energy budgets: resituating human energy allocation trade-offs. *Evolutionary Anthropology* 19, 136–147.

Kramer, K. L. and Greaves, R. D. (2007). Changing patterns of infant mortality and maternal fertility among Pume foragers and horticulturalists. *American Anthropologist* 109, 713–726.

Krapf, J. L. (1860). *Travels and Researches and Missionary Labours during an Eighteen Years Residence in East Africa*. Abingdon: Cass and Co.

Kratz, C. A. (1994). *Affecting Performance. Meaning, Movement, and Experience in Okiek Women's Initiation*. Washington, DC: Smithsonian Institution Press.

Krebs, J. R. and Davies, N. B., eds. (1978). *Behavioural Ecology: An Evolutionary Approach*. Oxford: Blackwell Science.

Kreider, R. M. and Fields, J. M. (2002). *Number, Timing, and Duration of Marriages and Divorces: 1996. Current Population Reports*. Washington, DC: US Census Bureau, P70–80, 1–20.

Kruger, D. J. and Nesse, R. M. (2006). An evolutionary life-history framework for understanding sex differences in human mortality rates. *Human Nature* 17, 74–97.

Kusimba, C. M. (2004). Archaeology of slavery in East Africa. *African Archaeological Review* 21, 59–88.

Lack, D. (1954). *The Natural Regulation of Animal Numbers*. Oxford: Oxford University Press.

Lack, D. (1966). *Population Studies with Birds*. Oxford: Clarendon Press.

Laden, G. and Wrangham, R. (2005). The rise of the hominids as an adaptive shift in fallback foods: plant underground storage organs (USOs) and the Australopith origins. *Journal of Human Evolution* 49, 482–498.

Lahdenpera, M., Gillespie, D. O. S., Lummaa, V. *et al.* (2012). Severe intergenerational reproductive conflict and the evolution of menopause. *Ecology Letters* 15, 1283–1290.

Lahdenpera, M., Lummaa, V., Hella, S. *et al.* (2004). Fitness benefits of prolonged post-reproductive lifespan in women. *Nature* 428, 178–181.

Lakatos, I., ed. (1970). *Falsification and the Methodology of Scientific Research Programs*. Criticism and the Methodology of Knowledge: Proceedings of the International Colloquium in the Philosophy of Science, London, 1965. Cambridge: Cambridge University Press.

Lakatos, I. and Zahar, E. (1975). Why did Copernicus' research program supersede Ptolemy's? In

*The Copernican Achievement*, ed. R. S. Westman. Berkeley, CA: University of California Press, pp. 354–383.

Lamberti, L. M., Fischer Walker, C. L., Noiman, A. *et al.* (2011). Breastfeeding and the risk for diarrhea morbidity and mortality. *BioMed Central Public Health* 11(Suppl. 3), S15.

Lancaster, J. B. and Kaplan, H. S. (2000). Parenting other men's children: costs, benefits, and consequences. In *Adaptation and Human Behavior*, eds. L. Cronk, N. Chagnon and W. Irons. New York, NY: Aldine, pp. 179–201.

Lane, C. (1996). *Pastures Lost: Barabaig Economy, Resource Tenure and the Alienation of their Land in Tanzania.* Nairobi: International Institute for Environment and Development, Initiatives Press.

Langergraber, K. E., Mitani, J. C., Watts, D. P. *et al.* (2013). Male-female socio-spatial relationships and reproduction in wild chimpanzees. *Behavioral Ecology and Sociobiology* 67, 861–873.

Larsen, U. (2000). Primary and secondary infertility in sub-Saharan Africa. *International Journal of Epidemiology* 29, 285–291.

Larsen, U. (2003). Infertility in Central Africa. *Tropical Medicine and International Health* 8, 354–367.

Larsen, U. and Hollos, M. (2003). Women's empowerment and fertility decline among the Pare of Kilimanjaro region, Northern Tanzania. *Social Science and Medicine* 57, 1099–1115.

Laslett, P. (1995). Necessary knowledge: age and aging in the societies of the past. In *Aging in the Past: Demography, Society, and Old Age*, eds. D. I. Kertzer and P. Laslett. Berkeley, CA: University of California Press, pp. 3–77.

Lawi, Y. Q. (1999). Where physical and ideological landscapes meet: landscape use and ecological knowledge in Iraqw, northern Tanzania, 1920s–1950s. *International Journal of African Historical Studies* 32, 281–310.

Le Gall, B., Nonnotte, P., Rolet, J. *et al.* (2008). Rift propagation at craton margin. Distribution of faulting and volcanism in the north Tanzanian divergence (East Africa) during Neogene times. *Tectonphysics* 448, 1–19.

Leacock, E. (1982). Relations of production in band society. In *Politics and History in Band Societies*, eds. E. Leacock and R. Lee. Cambridge: Cambridge University Press, pp. 159–170.

Lee, P. C., Majluf, P., Gordon, I. J. (1991). Growth, weaning and maternal investment from a comparative perspective. *Journal of Zoology, London* 225, 99–114.

Lee, R. B. (1969). Eating Christmas in the Kalahari. *Natural History (December)*, 14–22, 60–63.

Lee, R. B. (1972). Population growth and the beginnings of sedentary life among the !Kung bushmen. In *Population Growth: Anthropological Implications*, ed. B. Spooner. Cambridge, MA: MIT Press, pp. 328–342.

Lee, R. B. (1979). *The !Kung San.* Cambridge: Cambridge University Press.

Lee, R. B. and DeVore, I. (1968). *Man the Hunter.* Chicago, IL: Aldine.

Lee, R. B. and DeVore, I., eds. (1976). *Kalahari Hunter-gatherers.* Cambridge, MA: Harvard University Press.

Lee, R. B. and Guenther, M. (1993). Problems in Kalahari historical ethnography and the tolerance of error. *History in Africa* 20, 185–235.

Lee, R. D. (1987). Population dynamics of humans and other animals. *Demography* 24, 443–465.

Lemnge, M. M., Kamugisha, M. L., Njunwa, K. J. *et al.* (2001). Exploratory study of malaria situation in Hanang and Babati districts after reported malaria epidemic: 1. Health facility based information on malaria morbidity and mortality. *Tanzania Health Research Bulletin* 3(2), 18–24.

Liker, A., Freckleton, R. P. and Szekeley, T. (2014). Divorce and infidelity are associated with skewed adult sex ratios in birds. *Current Biology* 24, 880–884.

Linder, H. P., Lovett, J., Mutke, J. M. *et al.* (2005). A numerical re-evaluation of the sub-Saharan phytochoria of mainland Africa. *Biologist Skrifter* 55, 229–252.

Lindsay, S. W., Bodker, R., Malima, R. *et al.* (2000). Effect of 1997–98 El Niño on highland malaria in Tanzania. *The Lancet* 355, 989–990.

Livingstone, F. B. (1958). The distribution of the sickle cell gene in Liberia. *American Journal of Human Genetics* 10, 33–41.

Lockhard, J. and Adams, R. M. (1981). Human serial polygyny: demographic, reproductive, marital, and divorce data. *Ethology and Sociobiology* 2, 177–186.

Loibooki, M., Hofer, H., Campbell, K. L. I. *et al.* (2002). Bushmeat hunting by communities adjacent to the Serengeti National Park, Tanzania: the importance of livestock ownership and alternative sources of protein and income. *Environmental Conservation* 29, 391–398.

Longbottom, D. and Coulter, L. J. (2003). Animal chlamydioses and zoonotic implications. *Journal of Comparative Pathology* **128**, 217–244.

Lorenz, K. (1941). Vergleichende bewegungstudien an Anatinen. *Journal of Ornithology* **89**, 194–294.

Lotem, A., Fishman, M. A. and Stone, L. (2002). From reciprocity to unconditional altruism through signalling benefits. *Proceedings of the Royal Society of London B* **270**, 199–205.

Lovejoy, C. O. (1981). The origins of man. *Science* **211**, 341–350.

Lukas, D. and Clutton-Brock, T. H. (2013). The evolution of social monogamy in mammals. *Science* **341**, 526–530.

Lupo, K. D. and Schmitt, D. N. (2002). Upper palaeolithic net-hunting, small prey exploitation, and women's work effort: a view from the ethnographic and ethnoarchaeological record of the Congo basin. *Journal of Archaeological Method and Theory* **9**, 147–179.

Lutz, W. and Qiang, R. (2002). Determinants of human population growth. *Philosophical Transactions of the Royal Society of London B* **357**, 1197–1210.

Lycett, J. E., Henzi, P. S. and Barrett, L. (1998). Maternal investment in mountain baboons and the hypothesis of reduced care. *Behavioral Ecology and Sociobiology* **42**, 49–56.

Mabulla, A. (2003). Archaeological implications of Hadzabe forager land use in the Eyasi basin, northern Tanzania. In *East African Archaeology: Foragers, Potters, Smiths and Traders*, eds. C. M. Kusimba and S. Kusimba. Philadeplhia, PA: University of Pennsylvania Museum of Archaeology and Anthropology, pp. 33–58.

Mabulla, A. Z. P. (2007). Hunting and foraging in the Eyasi Basin, Northern Tanzania: past, present and future prospects. *African Archaeology Review* **24**, 15–33.

MacDonald, K. (1995). Evolution, the 5 factor model, and levels of personality. *Journal of Personality* **63**, 525–567.

Mace, R. and Alvergne, A. (2012). Female reproductive competition within families in rural Gambia. *Proceedings of the Royal Society of London B* **279**, 2219–2227.

MacFarlan, S. J., Remiker, M. and Quinlan, R. (2012). Competitive altruism explains labor exchange variation in a Dominican community. *Current Anthropology* **53**, 118–124.

MacFarlan, S. J., Walker, R. S., Flinn, M. V. and Chagnon, N. A. (2014). Lethal coalitionary aggression and long-term alliance formation among Yanomamo men. *Proceedings of the National Academy of Sciences USA* **111**(47), 16 662–16 669.

Makinson, C. (1985). The health consequences of teenage fertility. *Family Planning Perspectives* **17**, 132–139.

Manson, J. H. and Wrangham, R. W. (1991). Intergroup aggression in chimpanzees and humans. *Current Anthropology* **32**, 369–390.

Marchi, E., Kanapin, A., Byott, M. *et al.* (2013). Neanderthal and Denisovan retroviruses in modern humans. *Current Biology* **23**, R994–R995.

Margulis, S. W., Altmann, J. and Ober, C. (1993). Sex-biased lactational duration in a human population and its reproductive costs. *Behavioral Ecology and Sociobiology* **32**, 41–45.

Marlowe, F. W. (1998). *Paternal Investment and Mating Effort: Paternal Care Among Hadza Foragers of Tanzania*. PhD thesis. Los Angeles, CA: University of California Los Angeles.

Marlowe, F. W. (1999). Male care and mating effort among Hadza foragers. *Behavioral Ecology and Sociobiology* **46**, 57–64.

Marlowe, F. W. (2000). The patriarch hypothesis. *Human Nature* **11**, 27–42.

Marlowe, F.W. (2001). Male contribution to diet and female reproductive success among foragers. *Current Anthropology* **42**, 755–760.

Marlowe, F. W. (2002). Why the Hadza are still hunter-gatherers. In *Ethnicity, Hunter-Gatherers, and the "Other": Association or Assimilation in Africa*, ed. S. Kent. Washington, DC: Smithsonian Institution Press, pp. 247–275.

Marlowe, F. W. (2003). A critical period for provisioning by Hadza men: implications for pair bonding. *Evolution and Human Behavior* **24**, 217–229.

Marlowe, F. W. (2004). Marital residence among foragers. *Current Anthropology* **45**, 277–284.

Marlowe, F. W. (2005). Who tends Hadza children? In *Hunter-Gatherer Childhoods*, eds. B. S. Hewlett and M. E. Lamb. New Brunswick, NJ: Aldine Transaction, pp. 177–190.

Marlowe, F. W. (2010). *The Hadza Hunter-Gatherers of Tanzania*. Berkeley, CA: University of California Press.

Marlowe, F. W. and Berbesque, J. C. (2009). Tubers as fallback foods and their impact on Hadza hunter-gatherers. *American Journal of Physical Anthropology* **140**, 751–758.

Marlowe, F. W. and Berbesque, J. C. (2012). The human operational sex ratio: effects of concealed ovulation, and menopause on mate competition. *Journal of Human Evolution* **63**, 834–842.

Marshall, F. (2001). Agriculture and use of wild and weedy greens by the Piik ap Oom Okiek of Kenya. *Economic Botany* 55, 32–46.

Martin, R. M. (2001). Commentary: does breastfeeding for longer cause children to be shorter? *International Journal of Epidemiology* 30, 481–484.

Martorell, R., Horta, B. L., Adair, L. S. *et al.* (2010). Weight gain in the first two years of life is an important predictor of schooling outcomes in pooled analyses from five birth cohorts from low- and middle-income countries. *Journal of Nutrition* 140, 348–354.

Mayaud, P., Grosskurth, H., Changalucha, J. *et al.* (1995). Risk assessment and other screening options for gonorrhoea and chlamydial infections in women attending rural antenatal climics. *Bulletin of the World Health Organization* 73, 621–630.

Mayer, J. and Meese, E. (2005). Human endogenous retroviruses in the primate lineage and their influence on host genomes. *Cytogenetic and Genome Research* 110, 448–456.

McDowell, W. (1981a). *A Brief History of the Mangola Hadza.* Dar es Salaam: The Rift Valley Project, Ministry of Information and Culture Research Division, Republic of Tanzania.

McDowell, W. (1981b). *Hadza Traditional Economy and its Prospects for Development.* Dar es Salaam: The Rift Valley Project, Ministry of Information and Culture Research Division, Republic of Tanzania.

McGuire, M. T., Raleigh, M. J. and Pollack, D. B. (1994). Personality features in Vervet monkeys: the effects of sex, age, social status, and group composition. *American Journal of Primatology* 33, 1–13.

McNeilly, A. S. (2001). Neuroendocrine changes and fertility in breast-feeding women. *Progress in Brain Research* 133, 207–214.

McNeilly, A. S. and McNeilly, J. R. (1979). Effects of lactation on fertility. *British Medical Bulletin* 35, 151–154.

Meegan, M., Morley, D. C. and Brown, R. (1994). Child weighing by the unschooled: a report of a controlled study of growth monitoring over 12 months of Maasai children using direct recording scales. *Transactions of the Royal Society of Tropical Medicine and Hygiene* 88, 635–637.

Mehlman, M. (1988). *Later Quaternary Archaeological Sequences in Northern Tanzania.* PhD thesis. Champaign-Urbana, IL: University of Illinois.

Meindertsma, J. D. and Kessler, J. J. (1997). *Towards Better Use of Environmental Resources: A Planning Document of Mbulu and Karatu Districts, Tanzania.* Mbulu: Netherlands Economic Institute.

Mendez, M. A. and Adair, L. S. (1999). Severity and timing of stunting in the first two years of life affect performance on cognitive tests in late childhood. *Journal of Nutrition* 129, 1555–1562.

Migliano, A. B., Vinicius, l. and Lahr, M. M. (2007). Life history trade-offs explain the evolution of human pygmies. *Proceedings of the National Academy of Sciences USA* 104, 20 216–20 219.

Milinski, M., Semmann, D. and Krambeck, H.-J. (2002). Reputation helps solve the 'tragedy of the commons'. *Nature* 415, 424–426.

Mills, C. (2011). When do professional golfers reach their peak? *www.think-golf.info/archives/59/when do professional golfers reach their peak?*

Mitani, J. C., Gros-Luis, J. and Richards, A. F. (1996). Sexual dimorphism, the operational sex ratio, and the intensity of male competition in polygynous primates. *American Naturalist* 147, 966–980.

Mock, D. W., Schwagmeyer, P. L. and Parker, G. A. (1996). The model family. In *Partnerships in Birds. The Study of Monogamy*, ed. J. M. Black. Oxford: Oxford University Press, pp. 53–69.

Moisi, J. C., Gatakaa, H., Noor, A. M. *et al.* (2010). Geographic access to care is not a determinant of child mortality in a rural Kenyan setting with high health facility density. *BMC Public Health* 10, 142.

Molleson, T., Cox, M., Waldron, H. H. *et al.* (1993). *The Spitalfields Project.* Vol. 2. *The Anthropology.* York: Council for British Archaeology.

Monahan, C. M. (1998). The Hadza carcass transport debate revisited and its archaeological implications. *Journal of Archaeological Science* 25, 405–424.

Monden, C. W. S. and Smits, J. (2009). Maternal height and child mortality in 42 developing countries. *American Journal of Human Biology* 21, 305–311.

Morell, V. (1994). Mystery ailment strikes Serengeti lions. *Science* 264, 1404.

Morelli, G. A. and Tronick, E. Z. (1992). Efe fathers: one among many? A comparison of forager children's involvement with fathers and other males. *Social Development* 1, 36–54.

Morley, D. (1963). A medical service for children under five years of age in West Africa. *Transactions of the Royal Society of Tropical Medicine and Hygiene* 57, 82–87.

Mturi, A. J. (1997). The determinants of birth intervals among non-contracepting Tanzanian women. *African Population Studies* 12, ep97011.

Mturi, A. J. and Curtis, S. L. (1995). The determinants of infant and child mortality in Tanzania. *Health Policy and Planning* 10, 384–394.

Mturi, A. J. and Hinde, P. R. A. (1994). Fertility decline in Tanzania. *Journal of Biosocial Science* **26**, 529–538.

Muller, M. N. and Wrangham, R. W., eds. (2009). *Sexual Coercion in Primates and Humans*. Cambridge, MA: Harvard University Press.

Murray, C. M., Lonsdorf, E. V., Eberly, L. E. *et al.* (2009). Reproductive energetics in free-living female chimpanzees (*Pan troglodytes schweinfurthii*). *Behavioral Ecology* **20**, 1211–1216.

Murray, C. M., Mane, S. and Pusey, A. E. (2007). Dominance rank influences female space use in wild chimpanzees, *Pan troglodytes*: toward an ideal despotic distribution. *Animal Behaviour* **74**, 1795–1804.

Mwangi, E. K., Stevenson, P., Gettinby, G. *et al.* (1998). Susceptibility to trypanosomosis of three *Bos indicus* cattle breeds in areas of differing tsetse fly challenge. *Veterinary Parasitology* **79**, 1–17.

Ndagala, D. K. (1991). The unmaking of the Datoga: decreasing resources and increasing conflict in rural Tanzania. *Nomadic Peoples* **28**, 71–82.

Nesse, R. M. (2007). Runaway social selection for displays of partner value and altruism. *Biological Theory* **2**, 143–155.

Nettle, D. (2006). The evolution of personality variation in humans and other animals. *American Psychologist* **61**, 622–631.

Niklas, K. J. (2001). Size matters. *Trends in Ecology and Evolution* **16**, 468.

Noe, R. and Hammerstein, P. (1994). Biological markets: supply and demand determine the effect of partner choice in cooperation. *Behavioral Ecology and Sociobiology* **35**, 1–11.

Noe, R., van Hooff, J. A. R. A. M. and Hammerstein, P. (2001). *Economics in Nature: Social Dilemmas, Mate Choice and Biological Markets*. Cambridge: Cambridge University Press.

Norton, M. (2005). New evidence on birth spacing: promising findings for improving newborn, infant, child, and maternal health. *International Journal of Gynecology and Obstetrics* **89**, 51–56.

Norton-Griffiths, M. (1978). *Counting Animals*. Nairobi: African Wildlife Leadership Foundation.

Norton-Griffiths, M., Herlocker, D. and Pennycuik, L. (1975). The patterns of rainfall in the Serengeti ecosystem. *East African Wildlife Journal* **13**, 347–374.

Nunes, A., Nogueira, P. J., Borrego, M. J. *et al.* (2008). *Chlamydia trachomatis* diversity viewed as a tissue-specific coevolutionary arms race. *Genome Biology* **9** (R153), 1–13.

Nunn, C. L., Gittleman, J. L. and Antonovics, J. (2000). Promiscuity and the primate immune system. *Science* **290**, 1168–1170.

Nurse, G. T., Tanaka, N., MacNab, G. *et al.* (1973). Non-venereal syphilis and Australian antigen among the G/wi and G//ana San of the Central Kalahari Game Reserve, Botswana. *Central African Journal of Medicine* **19**, 207–213.

O'Connell, J. F. and Hawkes, K. (1981). Alyawara plant use and optimal foraging theory. In *Hunter-Gatherer Foraging Strategies*, eds. B. Winterhalder and E. A. Smith. Chicago, IL: The University of Chicago Press, pp. 99–125.

O'Connell, J. F., Hawkes, K. and Blurton Jones, N. G. (1988a). Hadza scavenging: implications for Plio-pleistocene hominid subsistence. *Current Anthropology* **29**, 356–363.

O'Connell, J. F., Hawkes, K. and Blurton Jones, N. G. (1988b). Hadza hunting, butchering and bone transport and their archaeological implications. *Journal of Anthropological Research* **44**, 113–161.

O'Connell, J. F., Hawkes, K. and Blurton Jones, N. G. (1990). Re-analysis of large animal body part transport among the Hadza. *Journal of Archaeological Science* **17**, 301–316.

O'Connell, J. F., Hawkes, K. and Blurton Jones, N. G. (1991). Distribution of refuse-producing activities at Hadza residential base camps: implications for analyses of archaeological site structure. In *The Interpretation of Archaeological Spatial Patterning*, eds. E. M. Kroll and T. D. Price. New York, NY: Plenum, pp. 61–75.

O'Connell, J. F., Hawkes, K. and Blurton Jones, N. G. (1992). Patterns in the distribution, site structure and assemblage composition of Hadza kill-butchering sites. *Journal of Archaeological Science* **19**, 319–345.

O'Connell, J. F., Hawkes, K. and Blurton Jones, N. G. (1999). Grandmothering and the evolution of *Homo erectus*. *Journal of Human Evolution* **36**, 461–485.

Obst, E. (1912). Von Mkalama ins Land der Wakindiga. *Mitteilungen der Geographischen Geselleschaft in Hamburg* **26**, 3–45.

Obst, E. (1915). Das abflusslose rumpfschollenland im nordöstlichen Deutsch-Ostafrika. *Mitteilungen der Geographischen Gesellschaft in Hamburg* **29**, 7–268.

Oeppen, J. and Vaupel, J. W. (2002). Broken limits to life expectancy. *Science* **296**, 1029–1031.

Oli, M. K. and Dobson, F. S. (2003). The relative importance of life-history variables to population growth rate in mammals: Cole's prediction revisited. *American Naturalist* **161**, 422–440.

Olsen, B. E., Hinderaker, S. G., Lie, R. T. *et al.* (2002). Maternal mortality in northern rural Tanzania: assessing the completeness of various information sources. *Acta Obstetricia et Gynecologica Scandinavica* **81**, 301–307.

Oota, H., Pakendorf, B., Weiss, G., *et al.* (2005). Recent origin and cultural reversion of a hunter-gatherer group. *PLoS Biology* **3**, 536–542.

Oota, H., Settheetham-Ishida, W., Tiwawech, D. *et al.* (2001). Human mtDNA and Y chromosome variation is correlated with matrilocal versus patrilocal residence. *Nature Genetics* **29**, 20–21.

Opie, C., Atkinson, Q. D., Dunbar, R. I. M. *et al.* (2013). Male infanticide leads to social monogamy in primates. *Proceedings of the National Academy of Sciences USA* **110**, 13 328–13 332.

Orubuloye, I. O. and Caldwell, J. C. (1975). The impact of public health services on mortality: a study of mortality differentials in a rural area of Nigeria. *Population Studies* **29**, 259–272.

Otterbein, K. F. (2004). *How War Began*. College Station, TX: Texas A&M University Press.

Packer, C., Herbst, L., Pusey, A. E. *et al.* (1988). Reproductive success of lions. In *Reproductive Success*, ed. T. H. Clutton-Brock. Chicago, IL: University of Chicago Press, pp. 363–383.

Paine, R. R., ed. (1997). *Integrating Archaeological Demography: Multidisciplinary Approaches to Prehistoric Population*. Occasional Paper. Carbondale, IL, Center for Archaeological Investigations.

Paine, R. R. (2000). If a population crashes in prehistory, and there is no paleodemographer there to hear it, does it make a sound? *American Journal of Physical Anthropology* **112**, 181–190.

Paine, R. R. and Boldsen, J. L. (2006). Paleodemographic data and why understanding holocene demography is essential to understanding human life history evolution in the Pleistocene. In *The Evolution of Human Life History*, eds. K. Hawkes and R. R. Paine. Santa Fe, NM: School of American Research Press.

Patrut, A., von Reden, K. F., Lowy, D. A. *et al.* (2007). Radiocarbon dating of a very large African baobab. *Tree Physiology* **27**, 1569–1574.

Patz, J. A., Campbell-Lendrum, D., Holloway, T. *et al.* (2005). Impact of regional climate change on human health. *Nature* **438**, 310–317.

Pennington, R. (2001). Hunter-gatherer demography. In *Hunter-Gatherers. Biosocial Society Symposium series 13*, eds. C. Panter-Brick, R. Layton and P. Rowley-Conwy. Cambridge: Cambridge University Press, pp. 170–204.

Pennington, R. and Harpending, H. C. (1988). Fitness and fertility among Kalahari !Kung. *American Journal of Physical Anthropology* **77**, 303–319.

Pennington, R. and Harpending, H. C. (1993). *The Structure of an African Pastoralist Community: Demography, History, and Ecology of the Ngamiland Herero*. Oxford: Clarendon Press.

Pennycuik, L. and Norton-Griffiths, M. (1976). Fluctuations in the rainfall of the Serengeti ecosystem, Tanzania. *Journal of Biogeography* **3**, 125–140.

Peters, C. R. (1987). Nut-like oil seeds: food for monkeys, chimpanzees, humans, and probably ape-men. *American Journal of Physical Anthropology* **73**, 333–363.

Pfeiffer, S. (2012). Conditions for evolution of small adult body size in southern Africa. *Current Anthropology* **53**, S383–S394.

Phillips, J. E. (1983). African smoking and pipes. *The Journal of African History* **24**, 303–319.

Pollet, T. V. and Nettle, D. (2008). Taller women do better in a stressed environment: height and reproductive success in rural Guatemalan women. *American Journal of Human Biology* **20**(3), 264–269.

Popkin, B. M., Adair, L., Akin, J. S. *et al.* (1990). Breastfeeding and diarrheal morbidity. *Pediatrics* **86**, 874–882.

Porter, C. C. and Marlowe, F. W. (2006). How marginal are forager habitats? *Journal of Archaeological Science* **38**, 59–68.

Potter, J. E. (1977). Problems in using birth-history analysis to estimate trends in fertility. *Population Studies* **31**, 335–364.

Potter, J. E. (1988). Birth spacing and child survival: a cautionary note regarding the evidence from the WFS. *Population Studies* **42**, 443–450.

Potts, R. (1998). Variability selection in hominid evolution. *Evolutionary Anthropology* **7**, 81–96.

Prentice, A. M., Whitehead, R. G., Watkinson, M. *et al.* (1983). Prenatal dietary supplementation of African women and birth-weight. *The Lancet* (5 March 1983), 489–492.

Preston, S. H. and Haines, M. R. (1991). *Fatal Years: Child Mortality in Late Nineteenth-Century America*. Princeton, NJ: Princeton University Press.

Prins, H. H. T. and Douglas-Hamilton, I. (1990). Stability in a multi-species assemblage of large herbivores in East Africa. *Oecologia* **83**, 392–400.

Prins, H. H. T. and Loth, P. E. (1988). Rainfall patterns as background to plant phenology in northern Tanzania. *Journal of Biogeography* **15**, 451–463.

Prins, H. H. T. and van der Jeugd, H. P. (1993). Herbivore population crashes and woodland structure in East Africa. *Journal of Ecology* **81**, 305–314.

Prins, H. H. T. and Weyerhaeuser, F. J. (1987). Epidemics in populations of wild ruminants: anthrax and impala, rinderpest and buffalo in Lake Manyara National Park, Tanzania. *Oikos* **49**, 28–38.

Psouni, E., Janke, A. and Gerwicz, M. (2012). Impact of carnivory on human development and evolution revealed by a new unifying model of weaning in mammals. *PLoS ONE* **7**, e32452.

Quinlan, R. J. (2007). Human parental effort and environmental risk. *Proceedings of the Royal Society of London B* **274**, 121–125.

Raleigh, M. J. and McGuire, M. T. (1989). Female influences on male dominance acquisition in captive vervet monkeys, *Cercopithecus aethiops sabaeus*. *Animal Behaviour* **38**, 59–67.

Rasbash, J., Steele, E., Browne, W. *et al.* (2003). *A User's Guide to MlWin Version 2.0*. London: UK Center for Multilevel Modelling, Institute of Education, University of London.

Reche, O. (1914). Die ethnographische Sammlung. Das abflusslose rumpfschollenland im nordöstlichen Deutsch-Ostafrika. *Mitteilungen der Geographischen Gesellschaft in Hamburg* **24**, 251–266.

Reed, D. L., Currier, R. W., Walton, S. F. *et al.* (2011). The evolution of infectious agents in relation to sex in animals and humans: brief discussions of some individual organisms. *Annals of the New York Academy of Sciences* **1230**, 74–107.

Reichard, U. H. and Boesch, C., eds. (2003). *Monogamy: Mating Strategies and Partnerships in Birds, Humans and other Mammals*. Cambridge: Cambridge University Press.

Rice, W. R., Gavrilets, S. and Friberg, U. (2010). The evolution of sex-specific grandparental harm. *Proceedings of the Royal Society of London B* **277**, 2727–2735.

Rich, S. M., Leendertz, F. H., Xu, G. *et al.* (2009). The origin of malignant malaria. *Proceedings of the National Academy of Sciences USA* **106**, 14 902–14 907.

Robbins, L. H., Campbell, A. C., Brook, G. A. *et al.* (2012). The antiquity of the bow and arrow in the Kalahari desert: bone points from White Paintings rock shelter, Botswana. *Journal of African Archaeology* **10**, 7–20.

Roberts, G. (1998). Competitive altruism: from reciprocity to the handicap principle. *Proceedings of the Royal Society of London B* **265**, 427–431.

Roberts, G. (2005). Cooperation through interdependence. *Animal Behaviour* **70**, 901–908.

Rodseth, L. (2012). From bachelor threat to fraternal security: male associations and modular organization in human societies. *International Journal of Primatology* **33**, 1194–1214.

Roff, D. A. (1992). *The Evolution of Life Histories: Theory and Analysis*. New York, NY: Chapman and Hall.

Rogers, A. R. (1992). Resources and population dynamics. In *Evolutionary Ecology and Human Behavior*, eds. E. A. Smith and B. Winterhalder. New York, NY: Aldine de Gruyter, pp. 375–402.

Rohwer, S., Herron, J. C. and Daly, M. (1999). Stepparental behavior as mating effort in birds and other animals. *Evolution and Human Behavior* **20**, 367–390.

Ronsmans, C. (1996). Birth spacing and child survival in rural Senegal. *International Journal of Epidemiology* **25**, 989–996.

Rose, S. (1999). Evolutionary psychology: biology impoverished. *Interdisciplinary Science Reviews* **24**, 175–178.

Rutstein, S. O. (2005). Effects of preceding birth intervals on neonatal, infant and under-five years mortality and nutritional status in developing countries: evidence from the demographic and health surveys. *International Journal of Gynecology and Obstetrics* **89**(Suppl. 1), S7–S24.

Ryner, M., Holmgren, K. and Taylor, D. (2008). A record of vegetation dynamics and lake level changes from Lake Emakat, northern Tanzania, during the last c. 1200 years. *Journal of Paleolimnology* **40**, 583–601.

Sahlins, M. (1974). *Stone Age Economics*. London: Tavistock Publications.

Sands, B. (1995). *Evaluating Claims of Distant Linguistic Relationships: The Case of Khoisan*. PhD thesis. Los Angeles, CA: University of California Los Angeles.

Sands, B. (1998). The linguistic relationship between Hadza and Khoisan. In *Language, Identity, and Conceptualization among the Khoisan* (Quellen zur Khoisan-Forschung/Research in Khoisan Studies 15), ed. M. Schladt. Cologne: Rudiger Köppe, pp. 266–283.

Sands, B., Maddieson, I. and Ladefoged, P. (1993). The phonetic structures of Hadza. *UCLA Working Papers in Phonetics* **84**, 67–88.

Sattenspiel, L., Mobarry, A. and Herring, D. A. (2000). Modeling the influence of settlement structure on the spread of influenza among communities. *American Journal of Human Biology* **12**, 736–748.

Scelza, B. A. (2009). The grandmaternal niche: critical caretaking among Martu aborigines. *American Journal of Human Biology* **21**, 448–454.

Scelza, B. A. (2010). Fathers' presence speeds the social and reproductive careers of sons. *Current Anthropology* **51**, 295–303.

Schacht, R., Rauch, K. L. and Borgerhoff Mulder, M. (2014). Too many men: the violence problem? *Trends in Ecology and Evolution* **29**, 214–222.

Scheinfeldt, L. B., Sol, S. and Tishkoff, S. A. (2010). Working toward a synthesis of archaeological, linguistic, and genetic data for inferring African population history. *Proceedings of the National Academy of Sciences USA* **107**(Suppl. 2), 8931–8938.

Schnee, H. (1920). *Deutsches Kolonial-Lexikon.* Leipzig: Quelle and Meyer.

Schnorr, S. L., Candela, M., Rampelli, S. *et al.* (2014). Gut microbiome of the Hadza hunter-gatherers. *Nature Communications* **5**, 3654.

Schoeninger, M. J., Bunn, H. T., Murray, S. S. and Marlett, J. A. (2001). Composition of tubers used by Hadza foragers of Tanzania. *Journal of Food Composition and Analysis* **14**, 15–25.

Schrire, C. (1980). An inquiry into the evolutionary status and apparent identity of San hunter-gatherers. *Human Ecology* **8**, 9–32.

Schultz, J. (1971). *Agrarlandschaftliche Veranderungen in Tanzania (Mbulu/Hanang Districts).* Hamburg: Weltform.

Schwandt, M. L., Lindell, S. G., Sjoberg, R. L. *et al.* (2010). Gene-environment interactions and response to social intrusion in male and female Rhesus macaques. *Biological Psychiatry* **67**, 323–330.

Scrimshaw, N. S. (2003). Historical concepts of interactions, synergism and antagonism between nutrition and infection. *Journal of Nutrition* **133**, 316S–321S.

Scudder, T. (1962). *The Ecology of the Gwembe Tonga.* Manchester: Manchester University Press.

Sear, R. and Mace, R. (2008). Who keeps children alive? A review of the effects of kin on child survival. *Evolution and Human Behavior* **29**, 1–18.

Sear, R., Mace, R. and McGregor, I. A. (2003a). The effects of kin on female fertility in rural Gambia. *Evolution and Human Behavior* **24**, 25–42.

Sear, R., Mace, R. and McGregor, I. A. (2003b). A life-history analysis of fertility rates in rural Gambia: evidence for trade-offs or phenotypic correlations? In *The Biodemography of Human Reproduction and Fertility*, eds. J. Rodgers and H.-P. Kohler. Boston, MA: Kluwer, pp. 135–160.

Sellen, D. W. (2000). Age, sex and anthropometric status of children in an African pastoral community. *Annals of Human Biology* **27**, 345–365.

Sellen, D. W. (2006). Lactation, complementary feeding, and human life history. In *The Evolution of Human Life History*, eds. K. Hawkes and R. R. Paine. Santa Fe, NM: SAR Press, pp. 155–196.

Sellen, D. W., Borgerhoff Mulder, M. and Sieff, D. F. (2000). Fertility, offspring quality, and wealth in Datoga pastoralists. In *Adaptation and Human Behavior*, eds. L. Cronk, N. Chagnon and W. Irons. New York, NY: Aldine de Gruyter, pp. 91–114.

Sellen, D. W. and Smay, D. B. (2001). Relationship between subsistence and age at weaning in "pre-industrial societies". *Human Nature* **12**, 47–87.

Semmann, D., Krambeck, H. and Milinski, M. (2004). Strategic investment in reputation. *Behavioral Ecology and Sociobiology* **56**, 248–252.

Semmann, D., Krambeck, H. and Milinski, M. (2005). Reputation is valuable within and outside one's own social group. *Behavioral Ecology and Sociobiology* **57**, 611–616.

Senior, H. S. (1938). Sukuma salt caravans to Lake Eyasi. *Tanganyika Notes and Records* **6**, 87–90.

Setel, P. W., Whiting, D. R., Hemed, Y. *et al.* (2006). Validity of verbal autopsy procedures for determining cause of death in Tanzania. *Tropical Medicine and International Health* **11**, 681–696.

Sheldon, B. G. (1993). Sexually transmitted disease in birds: occurrence and evolutionary significance. *Philosophical Transactions of the Royal Society of London B* **339**, 491–497.

Sherratt, T. N. and Roberts, G. (1998). The evolution of generosity and choosiness in cooperative exchanges. *Journal of Theoretical Biology* **193**, 167–177.

Sherry, D. S. and Marlowe, F. W. (2007). Anthropometric data indicate nutritional homogeneity in Hadza foragers of Tanzania. *American Journal of Human Biology* **19**, 107–118.

Shuster, S. M. and Wade, M. J. (2003). *Mating Systems and Strategies.* Princeton, NJ: Princeton University Press.

Sibly, R. M. (1983). Optimal group size is unstable. *Animal Behaviour* **31**, 947–948.

Sibly, R. M., Barker, D., Denham, M. C. *et al.* (2005). On the regulation of populations of mammals, birds, fish, and insects. *Science* **309**, 607–610.

Sibly, R. M., Hone, J. and Clutton-Brock, T. H. (2002). Introduction to population growth rate: determining factors and role in population regulation. *Philosophical Transactions of the Royal Society of London B* **357**, 1149–1151.

Silk, J. B. (1990). Sources of variation in inter-birth intervals among captive bonnett macaques (*Macaca radiata*). *American Journal of Physical Anthropology* **82**, 213–230.

Simondon, K. B., Costes, R., Delaunay, V. *et al.* (2001). Children's height, health and appetite influence mothers' weaning decisions in rural Senegal. *International Journal of Epidemiology* 30, 476–481.

Simpson, M. J. A., Simpson, A. E., Hooley, J. *et al.* (1981). Infant-related influences on birth intervals in rhesus monkeys. *Nature* 290, 49–51.

Sinclair, A. R. E. (1979). Dynamics of the Serengeti ecosystem. In *Serengeti. Dynamics of an Ecosystem*, eds. A. R. E. Sinclair and M. Norton-Griffiths. Chicago, IL: University of Chicago Press, pp. 1–30.

Sinclair, A. R. E. and Arcese, P., eds. (1995). *Serengeti II. Dynamics, Management, and Conservation of an Ecosystem*. Chicago, IL: University of Chicago Press.

Skolnick, M. H. and Cannings, C. (1972). Natural regulation of numbers in primitive human populations. *Nature* 239, 287–288.

Smith, B. H. (1992). Life history and the evolution of human maturation. *Evolutionary Anthropology* 1, 134–142.

Smith, B. H., Crummett, T. L. and Brandt, K. L. (1994). Ages of eruption of primate teeth: a compendium for aging individuals and comparing life histories. *Yearbook of Physical Anthropology* 37, 177–231.

Smith, C. C. and Fretwell, S. D. (1974). The optimal balance between size and number of offspring. *American Naturalist* 108, 499–506.

Smith, E. A. (1985). Inuit foraging groups: some simple models incorporating conflicts of interest, relatedness, and central sharing. *Ethology and Sociobiology* 6, 27–47.

Smith, E. A. (1991). *Inujjuamiut Foraging Strategies*. New York, NY: Aldine de Gruyter.

Smith, E. A. (1993). Comment on "Why hunters work" by Kristen Hawkes. *Current Anthropology* 34, 356.

Smith, E. A. and Bliege Bird, R. (2000). Turtle hunting and tombstone opening: public generosity as costly signaling. *Evolution and Human Behavior* 21, 245–261.

Smith, E. A., Borgerhoff Mulder, M. and Hill, K. (2001). Controversies in the evolutionary social sciences: a guide for the perplexed. *Trends in Ecology and Evolution* 16, 128–135.

Smith, E. A. and Winterhalder, B., eds. (1992). *Evolutionary Ecology and Human Behavior*. New York, NY: Aldine de Gruyter.

Smith, T. M., Machanda, Z., Bernard, A. B. *et al.* (2013). First molar eruption, weaning, and life history in living wild chimpanzees. *Proceedings of the National Academy of Sciences USA* 110, 2787–2791.

Smuts, B. B. (1992). Male aggression against women: an evolutionary perspective. *Human Nature* 3, 1–44.

Smuts, B. B. and Smuts, R. T. (1993). Male aggression and sexual coercion of females in nonhuman primates and other mammals: evidence and theoretical implications. *Advances in the Study of Behavior* 22, 1–63.

Snyder, K. A. (2005). *The Iraqw of Tanzania*. Cambridge, MA: Westview Press.

Speth, J. D. (2010). *The Paleoanthropology and Archaeology of Big-Game Hunting*. New York, NY: Springer.

Spinage, C. A. (1973). A review of ivory exploitation and elephant population trends in Africa. *East African Wildlife Journal* 11, 281–289.

Spinage, C. A. (2012). *African Ecology*. Berlin: Springer.

Stager, J. C., Ryves, D. B., Chase, B. M. *et al.* (2011). Catastrophic drought in the Afro-Asian monsoon region during Heinrich event 1. *Science* 331, 1299–1302.

Stearns, S. C. (1992). *The Evolution of Life Histories*. Oxford: Oxford University Press.

Stearns, S. C. and Koella, J. (1986). The evolution of phenotypic plasticity in life history traits: predictions for norms of reaction for age- and size-at-maturity. *Evolution* 40, 893–913.

Stevens, A., Morissette, J., Woodburn, J. *et al.* (1977). The inbreeding coefficients of the Hadza. *Annals of Human Biology* 4, 219–223.

Stewart, K. J. (1988). Suckling and lactational anoestrus in wild gorillas (*Gorilla gorilla*). *Journal of Reproduction and Fertility* 83, 627–634.

Stewart, K. M. (2009). Effects of secondary tuber harvest on populations of devil's claw (*Harpagophytum procumbens*) in the Kalahari savannas of South Africa. *African Journal of Ecology* 48, 146–154.

Stock, J. T. and Migliano, A. B. (2009). Stature, mortality, and life history among indigenous populations of the Andaman Islands, 1871–1986. *Current Anthropology* 50, 713–725.

Stone, A. C., Wilbur, A. K., Buikstra, J. E. *et al.* (2009). Tuberculosis and leprosy in perspective. *Yearbook of Physical Anthropology* 52, 66–94.

Stoner, C., Caro, T., Mduma, S. *et al.* (2006). Changes in large herbivore populations across large areas of Tanzania. *African Journal of Ecology* 45, 202–215.

Stover, J. and Ross, J. (2010). How increased contraceptive use has reduced maternal mortality. *Maternal and Child Health Journal* 14, 687–695.

Strassmann, B. I. (2011). Cooperation and competition in a cliff-dwelling people. *Proceedings of the National Academy of Sciences USA* **108**, 10894–10901.

Strassmann, B. I. and Garrard, W. M. (2011). Alternatives to the grandmother hypothesis. *Human Nature* **22**, 201–222.

Strassmann, B. I. and Gillespie, B. (2002). Life history theory, fertility and reproductive success in humans. *Proceedings of the Royal Society of London B* **269**, 553–562.

Strassmann, B. I. and Gillespie, B. (2003). How to measure reproductive success? *American Journal of Human Biology* **15**, 361–369.

Subramanian, S. V., Ackerson, L. K., Smith, G. D. *et al.* (2009). Association of maternal height with child mortality, anthropometric failure, and anemia in India. *Journal of the American Medical Association* **301**, 1691–1701.

Sugiyama, M. S. (2014). Fitness costs of warfare for women. *Human Nature* **25**, 476–495.

Suttles, W. (1968). Coping with abundance: subsistence on the Northwest coast. In *Man the Hunter*, eds. R. B. Lee and I. DeVore. Chicago, IL: Aldine, pp. 56–68.

Sutton, J. E. G. (1986). The irrigation and manuring of the Engaruka field system. *Azania* **21**, 27–51.

Sutton, J. E. G. (1989). Towards a history of cultivating the fields. *Azania* **24**, 98–112.

Sutton, J. E. G. (1990). *A Thousand Years of East Africa*. Nairobi: British Institute in East Africa, London: Thames and Hudson.

Sutton, J. E. G. (1993). Becoming Maasailand. In *Being Maasai: Ethnicity and Identity in East Africa*, eds. T. Spear and R. Waller. London: James Currey, pp. 38–60.

Szekeley, T., Weissing, F. J. and Komdeur, J. (2014). Adult sex ratio variations: implications for breeding system evolution. *Journal of Evolutionary Biology* **27**, 1500–1512.

Takahata, Y., Koyama, N., Huffman, M. A., Norikoshi, K. and Suzuki, H. (1995). Are daughters more costly to produce for Japanese macaque mothers? Sex of the offspring and subsequent inter-birth intervals. *Primates* **36**, 571–574.

Tanaka, J. (1980). *The San Hunter-Gatherers of the Kalahari*. Tokyo: University of Tokyo Press.

Tanner, J. M. (1962). *Growth at Adolescence*. Oxford: Blackwell Scientific Publications.

Tanner, J. M. (1964). *The Physique of the Olympic Athlete*. London: Allen and Unwin.

Terashima, H. (1980). Hunting life of the Bambote: an anthropological study of hunter-gatherers in a wooded savanna. *Senri Ethnological Studies* **6**, 223–268.

Testart, A. (1982). The significance of food storage among hunter-gatherers: residence patterns, population densities, and social inequalitites. *Current Anthropology* **23**, 523–537.

Thapa, S., Short, R. V. and Potts, M. (1988). Breast feeding, birth spacing and their effects on child survival. *Nature* **335**, 679–682.

Thornton, R. (1987). *American Indian Holocaust and Survival: A Population History since 1492*. Norman, OK: University of Oklahoma Press.

Tiger, L. (1969). *Men in Groups*. New York, NY: Random House.

Tills, D., Kopeć, A., Warlow, A. *et al.* (1982). Blood group, protein, and red cell enzyme polymorphisms of the Hadza of Tanzania. *Human Genetics* **61**, 52–59.

Tinbergen, N. (1963). On aims and methods of ethology. *Zietschrift fur Tierpsychologie* **20**, 410–433.

Tishkoff, S. A., Gonder, M. K., Henn, B. M. *et al.* (2007). History of click-speaking populations of Africa inferred from mtDNA and Y chromosome genetic variation. *Molecular Biology and Evolution* **24**, 2180–2195.

Tishkoff, S. A., Reed, F. A., Friedlaender, F. R. *et al.* (2009). The genetic structure and history of Africans and African Americans. *Science* **324**, 1035–1044.

Tomikawa, M. (1970). The distribution and the migrations of the Datoga tribe: the sociological distinctions of the Datoga society in Mangola area. *Kyoto University African Studies* **5**, 1–46.

Tomita, K. (1966). The sources of food for the Hadzapi tribe: the life of a hunting tribe in East Africa. *Kyoto University African Studies* **1**, 157–173.

Trivers, R. L. (1972). Parental investment and sexual selection. In *Sexual Selection and the Descent of Man*, ed. B. Campbell. Chicago, IL: Aldine, pp. 139–179.

Trivers, R. L. (1974). Parent–offspring conflict. *American Zoologist* **14**, 249–264.

Tucker, B. and Young, A. G. (2005). Growing up Mikea: Children's time allocation and tuber foraging in southwestern Madagascar. In *Hunter-Gatherer Childhoods*, eds. B. S. Hewlett and M. E. Lamb. New Brunswick, NJ: Aldine Transactions, pp. 147–174.

Turchin, P. (2009). Long-term population cycles in human societies. *Annals of the New York Academy of Sciences* **1162**, 1–17.

Turke, P. W. (1988). Helpers at the nest: childcare networks on Ifaluk. In *Human Reproductive Behavior: A Darwinian Perspective*, eds. L. Betzig, M. Borgerhoof

Mulder and P. Turke. Cambridge: Cambridge University Press, pp. 173–188.

Twisk, J. W. R. (2006). *Applied Multilevel Analysis.* Cambridge: Cambridge University Press.

Valeggia, C. and Ellison, P. T. (2009). Interactions between metabolic and reproductive functions in the resumption of postpartum fecundity. *American Journal of Human Biology* **21**, 559–566.

Van Ginneken, J. K. (1974). Prolonged breastfeeding as a birth spacing method. *Studies in Family Planning* **5**, 201–206.

Van Schaik, C. P. and Kappeler, P. M. (2003). The evolution of social monogamy in primates. In *Monogamy: Mating Strategies and Partnerships in Birds, Humans and other Mammals*, eds. U. H. Reichard and C. Boesch. Cambridge: Cambridge University Press, pp. 59–80.

Vayda, A. P. (1969). Expansion and warfare among swidden agriculturalists. In *Environment and Cultural Behavior*, ed. A. P. Vayda. Garden City, NY: Natural History Press, American Museum of Natural History, pp. 202–216.

Vincent, A. S. (1985a). *Wild Tubers as a Harvestable Resource in the East African Savannas: Ecological and Ethnographic Studies. Anthropology.* PhD thesis Berkeley, CA: University of California Berkeley.

Vincent, A. S. (1985b). Plant foods in savanna environments: a preliminary report of tubers eaten by the Hadza of northern Tanzania. *World Archaeology* **17**, 131–148.

Vitzthum, V. J. (2009). The ecology and evolutionary endocrinology of reproduction in the human female. *Yearbook of Physical Anthropology* **52**, 95–136.

Voland, E., Chasiotis, A. and Schiefenhovel, W., eds. (2005). *Grandmotherhood: The Evolutionary Significance of the Second Half of Female Life.* New Brunswick, NJ: Rutgers University Press.

Von Rueden, C., Gurven, M. and Kaplan, H. (2011). Why do men seek status? Fitness payoffs to dominance and prestige. *Proceedings of the Royal Society of London B* **278**, 2223–2232.

Von Rueden, C., Gurven, M., Kaplan, H. and Stieglitz, J. (2014). Leadership in an egalitarian society. *Human Nature* **25**, 538–566.

Wade, M. J. and Arnold, S. J. (1980). The intensity of sexual selection in relation to male sexual behavior, female choice, and sperm precedence. *Animal Behavior* **28**, 446–461.

Wade, M. J. and Shuster, S. M. (2005). Don't throw Bateman out with the bathwater! *Integrative and Comparative Biology* **45**, 945–951.

Wakefield, T. (1870). Routes of native caravans from the coast to the interior of Eastern Africa, chiefly from information given by Saidi bin Ahedi, a native of a district near Gazi, in Udigo, a little north of Zanzibar. *Journal of the Royal Geographic Society of London* **40**, 303–339.

Walker, P. L., Johnson, J. R. and Lambert, P. M. (1988). Age and sex biases in the preservation of human skeletal remains. *American Journal of Physical Anthropology* **76**, 183–188.

Walker, R. S., Gurven, M., Burger, O. *et al.* (2008). The trade-off between number and size of offspring in humans and other primates. *Proceedings of the Royal Society of London B* **275**, 827–833.

Walker, R. S., Gurven, M., Hill, K. *et al.* (2006). Growth rates and life histories in twenty-two small-scale societies. *American Journal of Human Biology* **18**, 295–311.

Walker, R. S. and Hamilton, M. J. (2008). Life-history consequences of density dependence and the evolution of human body size. *Current Anthropology* **49**, 115–122.

Walker, R. S., Hill, K., Kaplan, H. *et al.* (2002). Age dependency of strength, skill, and hunting ability among the Ache of Paraguay. *Journal of Human Evolution* **42**, 639–657.

Walters, S. L. (2008). *Fertility, Mortality and Marriage in Northwest Tanzania, 1920–1970: A Demographic Study using Parish Registers.* PhD thesis. Cambridge: University of Cambridge.

Ward, R. E. (1999). *Messengers of Love.* Kearney, NE: Morris Publishing.

Washburn, S. L. and DeVore, I. (1961). Social behavior of baboons and early man. In *Social Life of Early Man*, ed. S. L. Washburn. Chicago, IL: Aldine, pp. 91–105.

Weisner, T. and Gallimore, R. (1977). My brother's keeper: child and sibling caretaking. *Current Anthropology* **18**, 169–190.

Weiss, K. M. (1973). Demographic models for anthropology. (Society for American Archaeology, Memoir 27). *American Antiquity* **38**, 1–88.

Weiss, K. M. (1981). Evolutionary perspectives on human aging. In *Other Ways of Growing Old*, eds. P. T. Amoss and S. Harrell. Stanford, CA: Stanford University Press, pp. 25–58.

Wells, S. (2006). *Deep Ancestry: Inside the Genographic Project.* Washington, DC: National Geographic.

Werther, C. W. (1894). *Zum Victoria Nyanza.* Berlin: Paetel.

Werther, C. W. (1898). *Die mittleren Hochländer des nördlichen Deutsch-Ost-Afrika. Wissenschaftliche*

*Ergebnisse der Irangi-Expedition 1896–1897*. Berlin: Paetel.

Westerberg, L.-O., Holmgren, K., Borjeson, L. *et al.* (2010). The development of the ancient irrigation system at Engaruka, northern Tanzania: physical and societal factors. *Geographical Journal* **175**, 304–318.

Wheelwright, N. T. (1986). A seven-year study of individual variation in fruit production in tropical bird-dispersed tree species in the family Lauraceae. In *Frugivores and Seed Dispersal*, eds. A. Estrada and T. H. Fleming. Dordrecht: W. Junk, pp. 19–35.

Wiessner, P. (2014). The rift between science and humanism: what's data got to do with it? Wenner Gren conference: *Towards Integrating Anthropology: Niche Construction*, Cultural Institutions and History, Sintra, Portugal.

Wilkie, D. S. and Curran, B. (1991). Why do Mbuti hunters use nets? Ungulate hunting efficiency of archers and net-hunters in the Ituri rain forest. *American Anthropologist* **93**, 680–689.

Wilkie, D. S., Morelli, G., Rotberg, F. *et al.* (1999). Wetter isn't better: global warming and food security in the Congo Basin. *Global Environmental Change* **9**, 323–328.

Wilkins, J. F. and Marlowe, F. W. (2006). Sex-biased migration in humans: what should we expect from genetic data? *BioEssays* **28**, 290–300.

Wilmsen, E. (1989). *Land Filled with Flies. A Political Economy of the Kalahari*. Chicago, IL: University of Chicago Press.

Wilson, C. (1984). Natural fertility in pre-industrial England, 1600–1799. *Population Studies* **38**, 225–240.

Wilson, E. O. (1975). *Sociobiology: The New Synthesis*. Cambridge, MA: Harvard University Press.

Winking, J. and Gurven, M. (2011). The total cost of father desertion. *American Journal of Human Biology* **23**, 755–763.

Winter, E. H. and Molyneaux, L. (1963). Population patterns and problems among the Iraqw. *Ethnology* **2**, 490–505.

Winterhalder, B. (1993). Work, resources and population in foraging societies. *Man, New Series* **28**, 321–340.

Winterhalder, B., Baillargeon, W., Cappelletto, F. *et al.* (1988). The population ecology of hunter-gatherers and their prey. *Journal of Anthropological Archaeology* **7**, 289–328.

Winterhalder, B. and Smith, E. A., eds. (1981). *Hunter-Gatherer Foraging Strategies*. Prehistoric Archeology and Ecology Series. Chicago, IL: University of Chicago Press.

Wolfe, N. D., Dunavan, C. P. and Diamond, J. (2007). Origins of major human infectious diseases. *Nature* **447**, 279–283.

Wolfe, N. D., Switzer, W. M., Carr, J. K. *et al.* (2004). Naturally acquired simian retrovirus infections in central African hunters. *The Lancet* **363**, 932–937.

Wood, B. M. and Marlowe, F. W. (2013). Household and kin provisioning by Hadza men. *Human Nature* **24**, 280–317.

Wood, B. M. and Marlowe, F. W. (2014). Toward a reality-based understanding of Hadza men's work. *Human Nature* **25**, 620–630.

Wood, J. and Smouse, P. (1982). A method for analyzing density-dependent vital rates with an application to the Gainj of Papua New Guinea. *American Journal of Physical Anthropology* **58**, 403–411.

Wood, J. W. (1994). *Dynamics of Human Reproduction*. New York, NY: Aldine de Gruyter.

Wood, J. W., Lai, D., Johnson, P. L. *et al.* (1985). Lactation and birth spacing in highland New Guinea. *Journal of Biosocial Science Suppl* **9**, 159–173.

Woodburn, J. C. (1962). The future of the Tindiga: a short account of the present position and possibilities for the future of a hunting tribe in Tanzana. *Tanganyika Notes and Records* **58–59**, 268–273.

Woodburn, J. C. (1964). *The Social Organisation of the Hadza of North Tanganyika*. PhD thesis. Cambridge: University of Cambridge.

Woodburn, J. C. (1968a). An introduction to Hadza ecology. In *Man the Hunter*, eds. R. B. Lee and I. DeVore. Chicago, IL: Aldine, pp. 49–55.

Woodburn, J. C. (1968b). Stability and flexibility in Hadza residential groupings. In *Man the Hunter*, eds. R. B. Lee and I. DeVore. Chicago, IL: Aldine, pp. 103–110.

Woodburn, J. C. (1970). *Hunters and Gatherers. The Material Culture of the Nomadic Hadza*. London: The British Museum.

Woodburn, J. C. (1972). Ecology, nomadic movement and the composition of the local group among hunters and gatherers: an East African example and its implications. In *Man, Settlement and Urbanism*, eds. P. J. Ucko, R. Tringham and G. W. Dimbleby. London: Duckworth, pp. 193–206.

Woodburn, J. C. (1979). Minimal politics: the political organization of the Hadza of north Tanzania. In *Politics and Leadership: A Comparative Perspective*, eds. W. Shack and P. Cohen. Oxford: Clarendon Press, pp. 244–266.

Woodburn, J. C. (1982). Egalitarian societies. *Man, New Series* **17**, 431–451.

Woodburn, J. C. (1988). African hunter-gatherer social organisation: is it best understood as a product of encapsulation? In *Hunters and Gatherers*, Vol. 1. *History, Evolution and Social Change*, eds. T. Ingold, D. Riches and J. C. Woodburn. Oxford: Berg, pp. 31–64.

Woodburn, J. C. (1998). 'Sharing is not a form of exchange': an analysis of property-sharing in immediate-return hunter-gatherer societies. In *Property Relations: Renewing Anthropological Tradition*, ed. C. M. Hann. Cambridge: Cambridge University Press, pp. 48–63.

Wrangham, R. W. (2009). *Catching Fire. How Cooking made us Human*. New York, NY: Basic Books.

Wrangham, R. W., Wilson, M. L. and Muller, M. N. (2006). Comparative rates of violence in chimpanzees and humans. *Primates* **47**, 14–26.

Wrigley, E. A. (1966). Family limitation in pre-industrial England. *Economic History Review* **19**, 82–109.

Wynne-Edwards, V. C. (1962). *Animal Dispersion in Relation to Social Behaviour*. Edinburgh: Oliver and Boyd.

Yahya-Malima, K. I., Olsen, B. E., Matee, M. I. *et al.* (2006). The silent HIV epidemic among pregnant women within rural northern Tanzania. *BMC Public Health* **6**, 109.

Yanda, P. Z. and Madulu, N. F. (2005). Water resource management and biodiversity conservation in the Eastern Rift Valley Lakes, Northern Tanzania. *Physics and Chemistry of the Earth* **30**, 717–725.

Yerusalmy, J. (1945). On the interval between successive births and its effect on survival of infant 1. An indirect method of study. *Human Biology* **17**, 65–106.

Young, T. P. (1994). Natural die-offs of large mammals: implications for conservation. *Conservation Biology* **8**, 410–418.

Yuan, I.-C. (1931). Life tables for a southern Chinese family from 1365 to 1849. *Human Biology* **3**, 157–179.

Zahavi, A. and Zahavi, A. (1997). *The Handicap Principle*. Oxford: Oxford University Press.

Zhao, Z. (1997). Long-term mortality patterns in Chinese history: evidence from a recorded clan population. *Population Studies* **51**, 117–127.

Zheng, H.-X., Yan, S., Qin, Z.-D. *et al.* (2012). MtDNA analysis of global populations support that major population expansions began before Neolithic time. *Scientific Reports* **2**, 745.

Zhou, G. N., Minakawa, A. K., Githeko, A. K. *et al.* (2004). Association between climate variability and malaria epidemics in the East African highlands. *Proceedings of the National Academy of Sciences USA* **101**, 2375–2380.

Zhu, Q. and Bingham, G. P. (2011). Human readiness to throw: the size-weight illusion is not an illusion when picking the best objects to throw. *Evolution and Human Behavior* **32**, 288–293.

# Index

[Note: page numbers in bold refer to tables, figures or plates cited.]